Methods of Experimental Physics

VOLUME 1

CLASSICAL METHODS

METHODS OF EXPERIMENTAL PHYSICS:

L. Marton, *Editor-in-Chief*

Claire Marton, *Assistant Editor*

1. Classical Methods, 1959
   Edited by Immanuel Estermann
2. Electronic Methods
   Edited by E. Bleuler *and* R. O. Haxby
3. Molecular Physics
   Edited by Dudley Williams
4. Atomic and Electron Physics
   Edited by Vernon W. Hughes *and* Howard L. Schultz
5. Nuclear Physics
   Edited by Luke C. L. Yuan *and* Chien-Shiung Wu
6. Solid State Physics (*in two parts*), 1959
   Edited by K. Lark-Horovitz *and* Vivian A. Johnson

Volume 1

Classical Methods

Edited by

IMMANUEL ESTERMANN

Office of Naval Research, Washington, D.C.

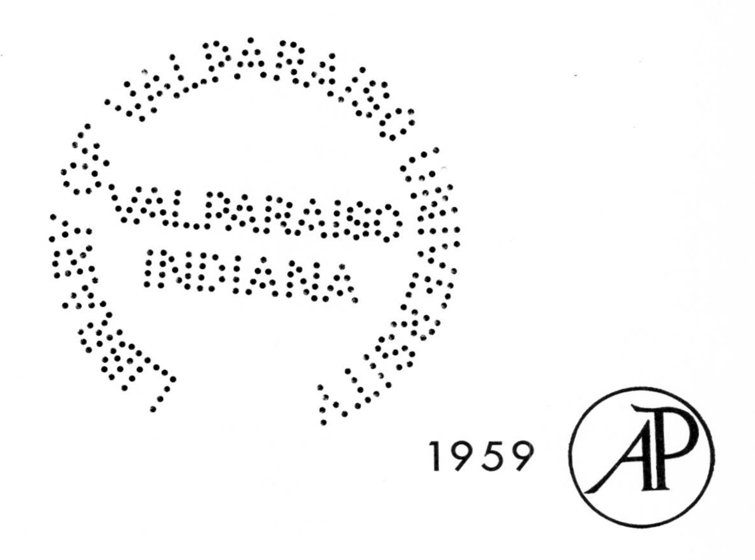

1959

ACADEMIC PRESS · New York and London

111 Fifth Avenue
New York 3, N. Y.

United Kingdom edition published by
ACADEMIC PRESS INC. (London) Ltd.
40 Pall Mall, London, S. W. 1

Library of Congress Catalog Card Number: 59-7686

PRINTED IN THE UNITED STATES OF AMERICA

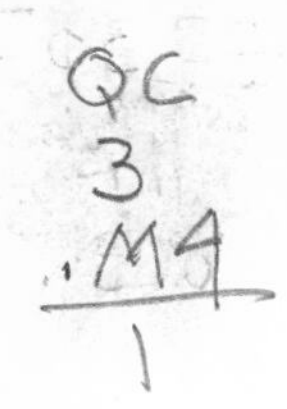

CONTRIBUTORS

Anthony Arrott, *Physics Department, Scientific Laboratory, Ford Motor Company, Dearborn, Michigan*

D. A. Bromley, *Physics Division, Atomic Energy of Canada Ltd., Chalk River, Ontario, Canada*

E. Richard Cohen, *Atomics International, A Division of North American Aviation, Inc., Canoga Park, California*

Immanuel Estermann, *Office of Naval Research, Washington, D.C.*

Michael Ference, Jr., *Director, Scientific Laboratory, Ford Motor Company, Dearborn, Michigan*

Simeon A. Friedberg, *Department of Physics, Carnegie Institute of Technology, Pittsburgh, Pennsylvania*

L. J. Giacoletto, *Electrical Department, Scientific Laboratory, Ford Motor Company, Dearborn, Michigan*

J. E. Goldman, *Physics and Chemistry Departments, Scientific Laboratory, Ford Motor Company, Dearborn, Michigan*

John R. Holmes, *Physics Department, University of Southern California, Los Angeles, California*

F. C. Hurlbut, *Division of Aeronautical Sciences, University of California, Berkeley, California*

Paul M. Marcus, *Department of Physics, Carnegie Institute of Technology, Pittsburgh, Pennsylvania*

Sidney Reed, *Office of Naval Research, Washington, D.C.*

Vincent Salmon, *Stanford Research Institute, Menlo Park, California*

G. L. Weissler, *Physics Department, University of Southern California, Los Angeles, California*

Sigmund Weissmann, *College of Engineering, Rutgers University, New Brunswick, New Jersey*

FOREWORD

The *Methods of Experimental Physics* is the outgrowth of a discussion, several years ago, of the need in the American literature of physics for a handbook type volume giving the experimental physicist a convenient guide in his work. In the course of this discussion the conclusion was reached that, with the present development of physics, the material would have to be subdivided into several volumes, each volume covering a sufficiently large field without too much overlapping. The scope and basic philosophies of this publishing venture were outlined in the original editorial policy statement:

"The experimental physicist—the person for whom this book should be written—is normally a specialist, working in a relatively narrow domain. He is presumed to know his own specialty very well. In the course of his work, however, he is often confronted with the necessity of using methods borrowed from neighboring fields, with which he is less familiar. In such cases he may have to make a literature search or ask advice from a specialist in the neighboring field. The existence of a concise presentation of the most important methods used in experimental physics would considerably simplify the task.

"The methods used may be purely experimental, but very often they are partly theoretical or computational. They may be qualitative, or may require the ultimate in accuracy. In an unfamiliar field, the experimenter will need a guide to the ways and means best adapted to his own investigation.

"Thus the book should be a concise, well illustrated presentation of the most important methods, or general principles, needed by the experimenter, complete with basic references for further reading. Indication of limitations of both applicability and accuracy is an important part of the presentation. Information about the interpretation of experiments, about the evaluation of errors, and about the validity of approximations should also be given. The book should not be merely a description of laboratory techniques, nor should it be a catalog of instruments.

"These volumes should be written so as to be of value to all research workers who use physical methods. Finally, the volumes should be organized in such a way that they will provide essential tools for graduate students in physics."

Having outlined the policy and scope of this work, the next task was to find editors who could bring the series to fruitful completion. We were successful in securing the collaboration of outstanding scientists to serve as editors for the volumes, and it is a pleasure to be able to list their names on the title page of this first volume of the series.

It remains to be judged by the reader how far we have achieved the goal set in our editorial policy. The volume editors and I sincerely hope that we have achieved it to a reasonable extent. Together with them I would like to invite the opinions of all our colleagues to let us know what omissions there are and what errors there may be, so that in future editions we may correct them. Uppermost in our minds has been the desire to provide the research worker and the graduate student with a good and useful tool to aid in their research, because we firmly believe that the conception of emphasizing the *method,* with a concurrent neglect of gadgetry, is an untried idea in the American literature of physics.

Finally, it is my pleasant duty to thank publicly all those who helped in the realization of these goals. The officials and staff of Academic Press Inc. were tireless in ironing out difficulties and providing a stimulating collaboration. The greatest part of the work on this first volume was on the shoulders of Dr. Immanuel Estermann; no amount of thanks can express my appreciation for the devotion and knowledge which he brought to the task. Editorial details on the home front were handled skillfully by Mrs. Claire Marton. Last, and most important, I extend my best thanks to the authors, whose contributions have created this volume.

L. Marton

Washington, D.C.
February, 1959

PREFACE TO VOLUME I

The rational subdivision of the science of physics into components is becoming increasingly difficult. In addition to the "classical" fields, such as mechanics, heat, sound, light, electricity, and magnetism, there have developed in the last decades a number of functional specialties, such as nuclear and solid state physics, which cut across the classical subdivisions and have added their own experimental methods for the pursuit of their objectives. In the determination of the material to be included in this volume, the editor was guided by the following considerations: The content should be of interest to physicists working in all the specialties, and beyond that, to researchers in other scientific fields for which physical methods are becoming more and more important. On the other hand, methods and techniques developed primarily for the various functional branches of physics have been reserved for the other volumes in the series.

It is obvious that the volume title "Classical Methods" is not to be interpreted as an antonym to modern methods. The term refers to those methods which are related to the general subdivisions of classical physics, but the editor and the authors have endeavored to include the most recent developments in experimental technology and theoretical justification, leaving out those techniques which have only historical interest. It was not intended to produce a "cook book" which gives a detailed description of favorite recipes but rather a "guide book" which points out the advantages, capabilities and limitations of the various methods and thus enables the user to select those which appear to be appropriate for his particular problem. Numerous references leading to more detailed descriptions of the different techniques are expected to provide additional information where needed. Methods and techniques customarily taught in elementary physics courses are either passed over or treated very briefly.

Completing a measurement is only one part of a research objective; the correct interpretation of the result is at least of equal importance. We have, therefore, included relatively detailed discussions of the theoretical significance of the various concepts and their relationships to the quantities that are actually measured. Thus a relatively large portion of this volume contains theoretical material which the authors felt to be necessary for the precise clarification of the concepts and for the proper interpretation of the measurements.

While the arrangement of the various parts and chapters follows a uniform style, no attempt has been made to suppress the individuality of the authors. Each part or chapter represents the approach of a different person to a common problem. Some authors are putting more emphasis on the experimental, others more on the theoretical side of their subject. It is the editor's hope that this divergence of attitude, which extends even to the preference of different systems of units by different authors, will make the book more readable and enjoyable.

The volume editor wishes to express his gratitude to the general editor and to the authors for their excellent cooperation and to the publisher and his staff for their patient assistance in coping with the many problems connected with the preparation of this volume.

I. ESTERMANN

February, 1959

CONTENTS

6. Heat and Thermodynamics

7. Optics

8. Electricity

9. Magnetism

1. EVALUATION OF MEASUREMENT*

1.1. General Rules

In a concise expression of the results of the measurement of a physical quantity, three pieces of information should be given: a number, a numerical statement of reliability, and an appropriate set of units.† The number is generally an estimate expressed in a finite set of digits (the exceptions are numbers which are exact by arbitrary definition, or mathematical constants such as the base of natural logarithms) reflecting the limited accuracy of physical measurement. The statement of reliability is usually written as plus or minus one, or at *most* two digits in units of the last digit of the number, together with a sufficient explanation to allow interpretation. In particular, one should state how many measurements were employed in the determination of the number and of its reliability. As will be seen below, this is of great value in the critical comparison of the results of different experiments, and in their combination with results of previous work. The number of digits that can be read is indicated by the smallest scale division, the *least count* of an instrument. Usually, one additional digit can be estimated between scale divisions. No more—and no less—digits should be recorded than can be read reproducibly.

To remove ambiguities, a standard form may be used for the recording of data: the decimal point is put just after the first nonzero digit, and the number is multiplied by the appropriate power of ten. Every digit is then understood to be significant. The magnitude of error is automatically indicated by the number of significant digits. The final result will generally have one more significant digit than the individual readings. This procedure implies that one should not round off readings. Any round-off increases the error. In the course of computation, round-off may be inevitable. A brief discussion of errors so introduced, with further references, is given in Chapter 1.6.

For the estimation of the best value of the desired quantity and of the significance of the result, statistical techniques are used. The terms "best" and "significant" should be understood in a technical sense: i.e., "best" and "significant" according to some statistical criterion. The criteria applied depend on assumptions which may or may not be true: attention should be paid to their validity. In the following, only a prescription of the techniques can be given. For this reason, a word of warning is in order:

† Units are discussed in Part 2 of this volume.

* Part 1 is by **Sidney Reed.**

these techniques, properly used, can improve the understanding of the results and the judgment of their worth—but they are not a substitute for thought.[1] It should be emphasized that work in certain fields, e.g., cosmic ray or high energy physics, requires more complete attention to statistical techniques in the planning and interpretation of experiments than can be discussed here.[2] The assistance of a statistician in such work will often be indispensable.

1.2. Errors

1.2.1. Systematic Errors, Accuracy

Statements about reliability of a measurement require assessment of the accuracy and of the precision of the work. Accuracy implies precision and, beyond that, freedom from what long usage has termed systematic errors. Examples best indicate the nature of systematic errors: (1) The torsion constant of a quartz fiber depends on the weight suspended by it (Bearden). This was overlooked in the course of the work aimed at the determination of the charge of the electron by Millikan. Consequently the value of air viscosity, which was used in this determination and which in turn was obtained using a quartz fiber to provide torque, probably included a systematic error because the suspended mass used in the experiment differed from that with which the torsion constant was determined.[3] (2) Miller performed a long series of experiments at Mount Wilson to detect a dependence of light speed on the direction of the earth's motion. A small dependence seemed to be present. A recent analysis of his data indicated a probable correlation of his results with heating of the building due to sunlight.[4] Here the systematic error was periodic with a long period. (3) A systematic misinterpretation which is not a gross mistake may be termed a systematic error. Michelson apparently omitted to correct some of his work on the velocity of light for the difference between group and phase velocity.[3]

In general, systematic errors are definite functions of experimental

[1] Further discussion and references are given by E. B. Wilson, Jr., "An Introduction to Scientific Research." McGraw-Hill, New York, 1952. Chapters on errors of physical measurement are given also by H. Cramer, "Elements of the Theory of Probability and its Applications." Wiley, New York, 1955; and in N. Arley and K. R. Buch, "Probability and Statistics." Wiley, New York, 1950.

[2] See, for example, L. Jánossy, "Cosmic Rays." Oxford Univ. Press, London and New York, 1953.

[3] R. T. Birge, *Nuovo cimento* **6**, Suppl. No. 1, 44 (1957).

[4] R. S. Shankland, S. W. McCuskey, F. C. Leone, and G. Kuerti, *Revs. Modern Phys.* **27**, 167 (1955). Miller was aware of this possibility, but could not pin it down.

method, instruments, or environmental conditions. If detected they can usually be corrected. Sometimes a single correction will be adequate for the entire work and can be applied at the end. The caliber of investigators cited in the above examples should be a warning that constant, or slowly varying systematic errors are hard to detect. The crucial test is the comparison of measurements of the same quantity obtained from different experiments, using different principles.

1.2.2. Accidental Errors, Precision

Precision implies close reproducibility of the results of successive individual measurements. It is assumed that, in general, there is a variation from measurement to measurement. This scatter of data is usually considered due to accidental errors; it is imagined that the experiment is aimed at a constant quantity, superimposed on which there is a random sum of small effects, independent of each other and of the quantity itself, and which are responsible for the variation of the results. Absence of variation is not necessarily an indication of precision; it may be due simply to an excessively large least count of the instrument used (see Section 1.4.1). Indefinite refinement of scale, which is possible in imagination, is limited by the fluctuation phenomena of microphysics. The ultimate degree of attainable precision is set by the probabilistic features of these fluctuations.[5]

1.3. Statistical Methods

To analyze accidental errors, the actual data are imagined to be a random selection, one for each measurement, of values from a large reference distribution which could be generated by infinite repetition of the experiment. In statistical terms, this is a finite sample from a "parent distribution" (p.d.). For reasons of mathematical convenience, it is usual to assume that the p.d. can be approximated satisfactorily by an analytic function (p.d.f.) having two or three parameters. Our finite data sample permits, at most, to estimate the p.d.f. parameters representing the true value and the precision of the measurement.

In most cases, a reasonable, explicit assumption of a definite form of the p.d.f. is desirable. Which form should be taken depends on a preliminary assessment of the probabilistic features of the experiment. If the errors are accidental in the sense described in Section 1.2.2 above, a normal (see

[5] R. B. Barnes and S. Silverman, *Revs. Modern Phys.* **6,** 162 (1934). A recent summary is given by C. W. McCrombie, *Repts. Progr. in Phys.* **16,** 266 (1953).

Section 1.3.1) distribution function (n.d.f.) is appropriate. If the experiment is directly concerned with probabilistic phenomena, e.g., counting experiments in nuclear physics, the Poisson or some other discrete probability distribution function may be chosen. In this case the measurements are generally indirect[6] (see Chapter 1.5).

To obtain an estimate of the best value one does not need to assume any particular p.d.f.; e.g., an estimation using least squares can be made.[7] A sharp quantitative statement of the statistical significance of a difference between two "best" estimates of the same quantity cannot be made, however, without assuming a definite form for the p.d.f.

1.3.1. Mean Value and Variance

The fraction of readings $dN(x)/N$ drawn from the p.d.f. $f(x)$ lying in the range between x and $x + dx$ is

$$dN(x)/N = f(x)\,dx. \tag{1.3.1}$$

The function $f(x)$ is normalized: $\int f(x)\,dx = 1$. The average of any function $g(x)$, denoted by $\langle g(x)\rangle$, is defined by $\langle g(x)\rangle = \int g(x)f(x)\,dx$. The range of integration may, for mathematical convenience, extend in both directions to infinity. Of special importance are the average of x called the *mean*

$$\langle x\rangle = \int xf(x)\,dx \tag{1.3.2}$$

and the average of $(x - \langle x\rangle)^2$, called the *dispersion* or *variance* of x

$$\sigma^2(x) = \int (x - \langle x\rangle)^2 f(x)\,dx. \tag{1.3.3}$$

The square root of the variance $\sigma(x)$ is called *standard deviation* or sometimes *standard error*. It is a measure of the spread of the data and thus of the precision. An important example of a p.d.f., often assumed to apply to accidental errors, is the Gaussian or *normal distribution* (n.d.f.):

$$f_1(x) = [\sqrt{2\pi\sigma^2(x)}]^{-1}\exp[-(x - \langle x\rangle)^2/2\sigma^2(x)] \tag{1.3.4}$$

characterized by two parameters, the mean $\langle x\rangle$ and the variance $\sigma^2(x)$. N measurements x_i allow the formation of the *sample mean*

$$\bar{x} = N^{-1}\sum_{i=1}^{N} x_i \tag{1.3.5}$$

and the *sample variance*

$$s^2(x) = (N - 1)^{-1}\sum_{i=1}^{N} (x_i - \bar{x})^2. \tag{1.3.6}$$

[6] M. Annis, W. Cheston, and H. Primakoff, *Revs. Modern Phys.* **25**, 818 (1953).

[7] E. R. Cohen, *Revs. Modern Phys.* **25**, 709 (1953).

If our sample is regarded as one of many equally reliable, independent samples of N measurements each, the means of such samples will fluctuate. The variance of the means, $\sigma^2(\bar{x})$ is related to the variance of individual measurements by

$$\sigma^2(\bar{x}) = \sigma^2(x) \cdot N^{-1}.$$

The mean has the property of being the value of a parameter a which minimizes $\Sigma_{i=1}^{N}(x_i - a)^2$. On the grounds of consistency, one expects that in some sense $\bar{x}$ converges to $<x>$ as $N \rightarrow \infty$.* For computation, it is useful to subtract a constant A of the order of size of x_i, so that

$$\bar{x} - A = N^{-1} \sum_{i=1}^{N} (x_i - A) \tag{1.3.7}$$

and

$$s^2(x) = (N - 1)^{-1} \left[\sum_{i=1}^{N} (x_i - A)^2 - N(\bar{x} - A)^2 \right]. \tag{1.3.8}$$

1.3.2. Statistical Control of Measurements

The use of any p.d. implies that the data may be regarded as drawn at random from it. There are statistical tests for this implication,[1] but in the case of data scatter because of accidental errors, a rough "control chart" can assist in detecting systematic departures which are functions of time. Such a chart may be made by plotting, on the abscissa, the order (in time) of the reading, and on the ordinate, the reading itself. If there is previous information on the scatter of the data using the same instrument under similar conditions, so that $\sigma(x)$ is known, one can, at least tentatively, draw lines on the chart at $\bar{x} \pm 3\sigma$ which should, if the data are in control, bracket practically all the points. It is quite valuable to have such a chart associated with a precise instrument.

If no previous information is available, one should take a number of points, draw lines at $\bar{x} \pm 3s$ and continue for a few more readings in order to see whether the additional data fall between these lines. If it appears that randomness is a fair assumption, one can use the function "chi-square"[8] to test the fit of an assumed p.d.f. Chi-square, or χ^2, is defined as

$$\chi^2 = \text{sum of} \left\{ \frac{(\text{observed value} - \text{value expected from p.d.f.})^2}{\text{value expected from p.d.f.}} \right\}$$

* This is so in the technical sense of convergence in probability; see, e.g., Cramer, reference 1.

[8] Examples of the use of χ^2 are given by Wilson, reference 1, p. 200; by Arley and Buch, reference 1, p. 209. A critical discussion is given by W. G. Cochran, *Ann. Math. Statistics* **23,** 315 (1952).

and is tabulated as function of the number of degrees of freedom. Here the number of degrees of freedom equals the number of terms in the sum minus one plus the number of p.d.f. parameters which must be estimated from the data; in the case of a n.d.f., the number of degrees of freedom is the number of terms minus three. It is generally necessary to group the observed data and the corresponding values from the p.d.f. into cells.[8] For moderate numbers of readings, say 20 or so, χ^2 will only show substantial discrepancies between the data and the proposed p.d.f. From the table of χ^2 one can find the probability that a value of χ^2 at least as large as that computed could have arisen by chance. If the computed value has a low probability, this is a signal to look for systematic errors.[9]

1.4. Direct Measurements

It is useful to distinguish between direct measurements, such as can be made of length, time, or electrical current; and indirect measurements, in which the quantity in question can be calculated from measurement of other quantities. In the latter case the law of connection between the quantities measured and sought may also be in question. In such a case, one has first to decide whether the proposed relation holds for any values of the quantities (establishment of the law of connection), and if so, to make as good an estimate as possible of the quantity desired.[6] In the case of direct measurements only the latter problem needs to be solved. This simple situation will be discussed first.[10] There are several cases, depending on what information is available at the start.

1.4.1. Errors of Direct Measurements

If one has information at the start of the experiment regarding the variance of readings of the measuring instrument under similar conditions, the following procedure can be employed: One can draw up a control chart, using the previous $\sigma(x)$ together with the mean $\bar{x}$ of a short pre-

[9] A brief table (I) of χ^2 is given in the Appendix to this Part. More extensive tables are readily found, e.g. in recent editions of the "Handbook of Chemistry and Physics," 39th ed. Chemical Rubber, Cleveland, Ohio, 1957–1958. Most tables are abridged versions of those due to R. A. Fisher and F. Yates "Statistical Tables for Biological, Agricultural and Medical Research." Oliver and Boyd, Edinburgh and London, 1953.

[10] A valuable, readable discussion is given by W. E. Deming and R. T. Birge, *Revs. Modern Phys.* **6**, 119 (1934).

liminary run. If subsequent readings appear to be in statistical control, i.e. if the points fall between the lines at $\bar{x} \pm 3\sigma$, one can terminate the process at a definite number of readings which depends on the precision desired. One can then say that the most likely value of $<x>$ is given by the mean $\bar{x}$, and that the reliability of this estimate is such that the probability is one-half that the interval between $\bar{x} - 0.67\sigma/\sqrt{N}$ and $\bar{x} + 0.67\sigma/\sqrt{N}$ contains $<x>$. The precision increases with N in the sense that the interval having a definite probability of containing $<x>$ narrows proportional to $N^{-\frac{1}{2}}$.* In this case, the interval length is sharply defined for fixed N and probability. But this is only so if $\sigma(x)$ is known *a priori*, or if the situation discussed in one of the next two cases prevails:

(a) If there is a long string of readings (several hundred) at hand, all taken, as far as can be judged, under statistical control, one can from this data alone determine $s(x)$ as a good approximation to $\sigma(x)$. One can then proceed as described in the preceding paragraph; the influence of previous data will be small.

(b) In most practical situations, one has probably only small numbers of readings. The case just mentioned can sometimes be approached by combining a reasonable number of readings with previously obtained data. To do this, one regards these small sets of data as drawn at random from a set of n.d.f.'s having possibly different means, but the same σ. It is here assumed that the previous work is summarized into m sets and that the numbers of measurements in each set, $n_1, \ldots n_m$, and the corresponding sample variances $s_1, \ldots s_m$ are available. An estimate of the value of σ^2 is then:

$$\sigma^2(x) \approx \left\{ \sum_{i=1}^{m} [(n_i - 1)s_i^2(x)] \right\} \Big/ (N - m) \qquad (1.4.1)$$

and its dispersion can be estimated as $\sigma(s) \approx \sigma(x)/\sqrt{2(N - m)}$. In such cases the data of an additional short run should be combined with the preceding data in estimating the new over-all variance $\sigma^2(x)$ by simply adding the appropriate values of $(n_{m+1} - 1)s^2_{m+1}$. The standard deviation so obtained applies now to *any* of the samples, and as the number of these increases, provided the hypothesis of the data being drawn from a common population with a given value of σ is well founded, the value of σ will become more and more reliable.

In the cases just described, it is important to know whether or not it is reasonable to use all or only some of the different sets of data. A way of

* An interval of this type is called a *confidence interval.* It should be distinguished from a *tolerance interval* which will contain a definite fraction of the population, e.g., a single observation. (See Arley and Buch, reference 1, p. 168.)

assessing this has been given by Bartlett and an example is discussed by Youden.[11] The prescription is as follows: One forms the expression B for m sets, as stated above:

$$B = \left[(N - m)\ln\left(\frac{\Sigma(n_i - 1)s_i^2}{N - m}\right) - \sum (n_i - 1)\ln s_i \right] \times \left[1 + \frac{\sum\left(\frac{1}{n_i - 1} - \frac{1}{N - m}\right)}{3(m - 1)} \right]^{-1} . \tag{1.4.2}$$

The expression B has a chi-square distribution for $m - 1$ degrees of freedom if all sets of data have p.d.f.'s with the same σ. If the computed B has, according to a table of χ^2, a low probability, all the m sets of data selected are not likely to have the same σ.

Frequently the only information available at the start is that provided by the data itself. If the data seem to be in statistical control, one can make statements about the probability of bracketing the p.d.f. mean which differ from those possible when $\sigma(x)$ is known. The levels of probability now depend on the number of data points in the sample since the intervals bracketing $<x>$ can now vary in length from sample to sample of the same size. The type of statement that can be made for this case is that the best estimate of $<x>$ is $\bar{x}$, and that the probability is $1 - P$ that the interval between

$$\bar{x} - \frac{t(P,f)s(x)}{\sqrt{N}} \text{ and } \bar{x} + \frac{t(P,f)s(x)}{\sqrt{N}}$$

will include, on the average of many samples of size N, the p.d.f. mean $<x>$. The function $t(P,f)$ is known in the statistical literature as Student's or Fisher's t and is tabulated giving probability levels P as a function of f the number of degrees of freedom. In this case f is $N - 1$.[12]

This statement indicates that the precision increases, i.e., that the interval narrows as N increases. It might be inferred that with repetition of the experiment the precision of the mean would improve indefinitely. That would be so under ideal circumstances but in actual experiments the probability of occurrence of systematic errors increases with the length of the investigation.

Another limitation of the precision is due to the least count of the instruments used. All the statements above depend on the approximation

[11] W. J. Youden, "Statistical Methods for Chemists." Wiley, New York, 1951.

[12] A brief table of t (Table II) is given in the Appendix to this Part.

of a continuous parent distribution function. This approximation improves with the ratio of σ to the least count w. ($w \leq \frac{1}{4}\sigma$ is usually considered reasonable.)[13]

If w is large, and the readings come out constant, statistical methods cannot be applied. If there are random errors in the data, one can improve the precision by repetition, but a very large number of measurements are required if $\sigma \approx w$.*

In order to be fairly sure the accidental errors are in a state of statistical control, a certain number of measurements is necessary (how many depends on the investigation, but the number must be fairly large). Once such a state has been reached, further measurements can be added to refine the precision of the mean, but on the hypothesis that each added measurement can be joined with the preceding ones. For this reason a continuously running control chart is very useful.

1.4.2. Rejection of Data

A closely related problem is the rejection of data. Often some criterion for doing so is given, based on analysis of the data alone. This is not recommended. A measured point should not be excluded on statistical grounds alone. If a control chart as suggested in the previous paragraph is used, and if a point or set of points seem out of control, the experimental conditions should be carefully re-examined, and if no assignable causes can be found, the point should be remeasured after settings have been remade, etc. This presumes that some analysis is made in the course of the experiment itself, which is the procedure to be strongly recommended. Unless assignable causes can be found, the suspected data should be retained. In any case a clear statement of the situation is imperative, and a procedure once adopted should be consistently followed.[14]

1.4.3. Weighted Measurements, Internal and External Consistency

A set of n observations is usually assigned a statistical weight W inversely proportional to the variance of their mean, $W \propto [\sigma(\bar{x})]^{-2}$, the

* If the least count is large, attention must be paid to providing a random character for the data. For example, an improvement in the precision of a measurement of length can be obtained by putting the ends of an object at random between the least count marks of a scale rather than always setting one end to coincide with one of the markers.

[13] See chapter by C. Eisenhart, *in* "Techniques of Statistical Analysis" (C. Eisenhart, M. W. Hastay, and W. A. Wallis, eds.) McGraw-Hill, New York, 1947; see also, J. Cameron, *in* "Fundamental Formulas of Physics" (D. Menzel, ed.) Prentice-Hall, New York, 1955.

[14] A discussion is given by N. Arley, *Kgl. Danske Videnskab. Selskab., Mat.-fys. Medd.* **18,** No. 3 (1949).

factor of proportionality corresponding to the variance of a set of measurements having (an arbitrary) unit weight. This assignment is consistent with the fundamental basis of weighting, namely that the arithmetic mean of n simple measurements has n times the weight of a single measurement.[7] In this case, the factor of proportionality is the variance $\sigma^2(x)$ and

$$W = \frac{\sigma^2(x)}{\sigma^2(\bar{x})} = n.$$

If there are m sets of measurements of a physical quantity, all having the same precision, i.e., the same σ, with, respectively, $n_1, n_2, \ldots n_m$ observations in the m sets and with corresponding arithmetic means

$$\bar{x}_i = \sum_{j=1}^{n_j} \frac{x_{ij}}{n_j}$$

the over-all weighted mean, or grand mean, is

$$\bar{\bar{x}} = \Sigma W_i\bar{x}_i/\Sigma W_i = \Sigma n_i\bar{x}_i/\Sigma n_i = \sum_j \sum_i x_{ij}/\Sigma n_i. \tag{1.4.3}$$

If the variances of the several sets of measurements differ, we have

$$\bar{\bar{x}} = \frac{\Sigma n_i \sigma_i^{-2} \bar{x}_i}{\Sigma \sigma_i^{-2}}. \tag{1.4.4}$$

The sum of weighted squared deviations from a given parameter ξ, $\Sigma_i W_i(\bar{x}_i - \xi)^2$, has a minimum value for $\xi = \bar{\bar{x}}$. For purposes of computation, it is convenient to use the identity

$$\Sigma W_i(\bar{x}_i - \bar{\bar{x}})^2 = \Sigma W_i \bar{x}_i^2 - \bar{\bar{x}}^2 \Sigma W_i. \tag{1.4.5}$$

As in the case of the mean of single measurements, the precision of $\bar{\bar{x}}$ can be estimated as follows: the probability is $1 - P$ that the interval between $\bar{\bar{x}} - t(P,f)s(\bar{\bar{x}})/\sqrt{N-1}$ and $\bar{\bar{x}} + t(P,f)s(\bar{\bar{x}})/\sqrt{N-1}$ contains the $\langle x \rangle$ for the entire set of $\Sigma n_i = N$ measurements, where $s^2(\bar{\bar{x}}) = \Sigma W_i(\bar{x}_i - \bar{\bar{x}})^2/\Sigma W_i$ and $t(P,f)$ is the t function with $f = N - 1$.

The assumption that the means of the different sets of measurements can be regarded as chosen at random from an n.d.f. of variance σ^2 can be checked by computing $\Sigma W_i(\bar{x}_i - \bar{\bar{x}})^2/\Sigma W_i$ and comparing with values in a table of chi-square with $m - 1$ degrees of freedom. If the computed value corresponds to a low probability, there may be a systematic error in some of the sets of measurements.

If σ^2 is not known *a priori*, we can use the ratio F of the estimate of σ^2

from the m sets to that of all the individual measurements taken together —see Eq. (1.4.1)

$$F(P, N - m, m - 1) = \frac{\Sigma W_i(\bar{x}_i - \bar{\bar{x}})^2}{\left(\sum \frac{(n_i - 1)s_i^2}{N - m}\right)(m - 1)} \tag{1.4.6}$$

(under the assumption that all x are drawn from a n.d.f.). This function is tabulated in statistical tables as the F distribution with $m - 1$ degrees of freedom in the numerator and $N - m$ degrees of freedom in the denominator. Tables of F give (low) probability levels that values of F would be found by chance at least as large as those tabulated. The probability for Table III in the Appendix is 5%. If the computed F from Eq. (1.4.6) exceeds that found in the table for the appropriate numbers of degrees of freedom in numerator and denominator, this is an indication that systematic errors may be present in some of the data.*

The procedures just described, in which a kind of consistency test is performed, are of considerable importance. It involves comparison of two estimates of the spreads, one formed by taking the more or less natural grouping of the data and estimating the variance of the distribution of weighted means of the groups (numerator); the other by estimating the over-all variance of the combined groups (denominator). R. T. Birge, who used this ratio systematically in his early work on the fundamental constants,[15] called it the ratio of external to internal consistency. The analysis of variance[1] involves a systematic use of the F ratio with subgroups of data, in order to detect and isolate significant differences. In a careful investigation, this technique can aid in the isolation and identification of systematic errors.† [16]

1.4.4. Significance of Results

Of particular importance is the comparison of the best estimates from two experiments using different physical principles to measure the same quantity. Under the assumption that both are under statistical control, i.e., their data may be regarded as drawn, at random, from normal p.d.f.'s and that these p.d.f.'s have different means but the same variance σ^2, one

* One must use available information carefully when using a table of F. Youden gives some advice on this (see reference 11, pp. 20, 21, 29).

† An analysis of this type is given in references 4 and 16.

[15] R. T. Birge, *Phys. Rev.* **40,** 207 (1932).

[16] E. R. Cohen, J. W. M. DuMond, T. W. Layton, and J. S. Rollett, *Revs. Modern Phys.* **27,** 363 (1955); E. R. Cohen and J. W. M. DuMond, *in* "Handbuch der Physik—Encyclopedia of Physics" (S. Flügge, ed.), Vol. 35. Springer, Berlin, 1957.

can proceed as follows: if n_1 and n_2 are the numbers of measurements, $\bar{x}_1$ and $\bar{x}_2$ their means, and s_1^2 and s_2^2 their variances, the expression

$$t = t(P, n_1 + n_2 - 2)$$
$$= (\bar{x}_1 - \bar{x}_2) \left[\frac{(n_1 - 1)s_1^2 + (n_2 - 1)s_2^2}{n_1 + n_2 - 2} \right]^{-\frac{1}{2}} \times \left[\frac{n_1 n_2}{n_1 + n_2} \right]^{\frac{1}{2}} \quad (1.4.7)$$

should have Student's t distribution with $n_1 + n_2 - 2$ degrees of freedom. If the value of t from Eq. (1.4.7) has a low level of probability for the appropriate number of degrees of freedom, the probable existence of a systematic, as opposed to an accidental or chance difference between $\bar{x}_1$ and $\bar{x}_2$ is indicated.

In some cases, the results from these experiments will have unequal precision, so that they cannot be assumed to be samples drawn from populations having one and the same variance. This situation is discussed by Fisher (Behrens' test).[9]

1.5. Indirect Measurement

Errors of indirect measurement occur when the desired quantities are obtained from those measured with the aid of equations. These equations may be the expression of an established physical relation, or may have to be established by means of the data itself.

1.5.1. Propagation of Errors

A simple case of indirect errors is the propagation or compounding of absolute errors. The desired quantity z is related to the directly measured different quantities $x^{(1)}, x^{(2)}, \ldots x^{(m)}$ by a known function,

$$z = f(x^{(1)} \cdots x^{m}).$$

Using Taylor's theorem to expand around the true values z^0, $x^{0(i)}$, one obtains

$$z - z^0 = \sum_j \left\{ \frac{\partial f}{\partial x^{(j)}} (x^{(j)} - x^{0(j)}) \right\}$$
$$+ \sum_{ij} \frac{1}{2} \left\{ \frac{\partial^2 f}{\partial x^{(i)} \, \partial x^{(j)}} (x^{(j)} - x^{0(j)})(x^{(i)} - x^{0(i)}) + \cdots \right\}. \quad (1.5.1)$$

Squaring, averaging over all d.f.'s for the $x^{(i)}$, and retaining only the lowest order terms, one has

$$\begin{aligned}\sigma^2(z) \approx <(z - z^0)^2> &= \sum \frac{\partial f}{\partial x^{(i)}} \frac{\partial f}{\partial x^{(j)}} <(x^{(i)} - x^{0(i)})(x^{(j)} - x^{0(j)})> \\ &= \sum \left(\frac{\partial f}{\partial x^{(i)}}\right)^2 \sigma_i{}^2 + \sum \frac{\partial f}{\partial x^{(i)}} \frac{\partial f}{\partial x^{(j)}} \rho_{ij}\sigma_i\sigma_j \qquad (1.5.2)\end{aligned}$$

where the correlation coefficient ρ_{ij} is defined as*

$$\rho_{ij} = \frac{<(x^{(i)} - x^{0(i)})(x^{(i)} - x^{0(j)})>}{[\sigma^2(x^{(i)})\sigma^2(x^{(j)})]^{\frac{1}{2}}}.$$

The derivatives are evaluated at $x^{0(i)}$ or at the nearest approximation thereto. The approximation resulting from neglecting the higher order terms may be poor, for example if any of the higher derivatives, e.g., $\partial^k f/\partial x^{(i)k}$ multiplied by the kth moment $<(x^{(i)} - x^{0(i)})^k>$ is comparable with the lowest order terms. If the errors in the $x^{(i)}$ are independent, the equation has the form usually given

$$\sigma^2(z) = <(z - z^0)^2> = \sum \left(\frac{\partial f}{\partial x^{(j)}}\right)^2 \sigma^2(x^{(j)}). \qquad (1.5.3)$$

One can assign a measure of error σ to z on the basis of preassigned error values in the $x^{(j)}$ using the above formulas. The interpretation of the statistical significance of the error in z, however, requires some knowledge of the distribution of the errors.

There are several practical conclusions to be drawn from this formula. One is that in an error analysis of this type, the $x^{(i)}$ should be carefully examined for independence. If the $x^{(i)}$ are obtained from a least squares fit of all the data jointly, for example, or if in the computation of the $x^{(i)}$ before insertion into the formula some common error-contributing components are used, e.g., one of the physical constants, the correlation coefficient ρ_{ij} will in general not be zero.[16] The accurate determination of the physical constants themselves requires a knowledge of the correlation coefficients in addition to that of the standard errors.

Another practical consequence refers to the design of the experiment. The best distribution of errors among the quantities $x^{(i)}$ is that in which all the errors are equal. As a consequence, effort should not be spent on further refining the measurements expected to be most accurate, but on those expected to be least accurate in order to approach the optimum condition.

* An alternative procedure to using the correlation coefficients is to make a linear transformation so that Eq. (1.5.1) becomes a linear combination of independent quantities.

A third consequence occurs in the transformation of weights that must be made when transforming a variable in order to bring an equation into a more tractable form for the determination of its parameters by least squares fitting. For example $y = a^{bx}$ is often transformed into a straight line by setting $y' = \ln y = \ln a + bx$ for the determination of a and b by least squares procedures. (See Chapter 1.4.) If the error in x is assumed to be negligible and if that in y is constant, each value of y' does not have equal weight but must be weighted according to

$$W(y') = (\sigma^2(y'))^{-1} \approx \left[\left(\frac{1}{y}\right)^2 \sigma^2(y)\right]^{-1} \tag{1.5.4}$$

or, more generally

$$y' = f(y); \qquad W(y') = \left[\left(\frac{\partial f}{\partial y}\right)^2 \sigma^2(y)\right]^{-1}. \tag{1.5.5}$$

It may be of importance, in connection with the above discussion, to decide whether the correlation coefficient is significantly different from zero. It is assumed that there are equal numbers of measurements, n, of each quantity, and that the joint distribution of, say, $x^{(i)}$, $x^{(j)}$ is a bivariate normal distribution.[19] With the expressions

$$t = \left|\frac{r\sqrt{n-2}}{\sqrt{1-r^2}}\right|; \qquad r = \frac{\Sigma_x(x^{(i)} - \bar{x}^{(i)})(x^{(j)} - \bar{x}^{(j)})}{s^2(x^{(i)})s^2(x^{(j)})} \tag{1.5.6}$$

values of t can be computed and compared with tabulated values for $n - 2$ degrees of freedom. If the computed t is large and thus corresponds to a low probability, this indicates the correlation is probably significant and that it should be computed.

A good description of the approach to the general problem of indirect measurements is given by Cohen and DuMond.[16]

1.5.2. Least Squares

We are now discussing the following situation: A set of data y is generated by changing values of a variable x. In the general case, both x and y are subject to errors. We assume a certain functional relation between y and x, and attempt to estimate values of its parameters and to estimate how well it fits the data. A rather clear case is that in which there is some physical reason for believing that the functional relation has a definite form.

It is important to have some idea of the over-all behavior of the data, in order not to waste time trying to fit the wrong type of function. The data should be plotted first. It may be that to start arbitrarily with a definite kind of function, e.g., a polynomial, is really a mistake and the data would be much better fitted by a sum of exponentials, for example. In order to interpret a data plot, some acquaintance with the behavior of simple functions is helpful.[17]

There are two general classes of least squares approximation functions, one represented by the polynomial functions, and the other by periodic functions. A special case of polynomial function, the linear or straight line relation, is important enough to deserve separate discussion.

The usual treatment, and the one given below, is that wherein the errors in y are assumed to be much greater than those in x, so that x may be taken as exact. Even so, the errors in y may vary with x, and this may be known *a priori* or not. If the errors in both x and y are comparable, only a partial analysis is available.[18]

If there is assumed to be a linear relation between an exact, but unknown, Y and x, and if all the measurements are given equal weight, this relation could be written

$$Y = \alpha + \beta x$$

whereas the approximation to this relation is

$$\hat{y} = a + bx.$$

With given data y_i, a and b are determined by the least squares' condition that the difference $\Sigma_i(y_i - \hat{y}(x_i))^2$ be minimized with respect to a and b. This leads to

$$\left.\begin{aligned} a &= \frac{\bar{y}(\Sigma x_i^2) - \bar{x}(\Sigma x_i y_i)}{\Sigma(x_i - \bar{x})^2} = \bar{y} - b\bar{x} \\ b &= \frac{(\Sigma x_i y_i) - n\bar{x}\bar{y}}{\Sigma(x_i - \bar{x})^2} = \frac{\Sigma(x_i - \bar{x})(y_i - \bar{y})}{\Sigma(x_i - \bar{x})^2}. \end{aligned}\right\} \tag{1.5.7}$$

For equispaced x data, the above formulas have the simpler form

$$b = \frac{6}{n(n^2 - 1)} [(y_n - y_1)(n - 1) + (y_{n-1} - y_2)(n - 3) + \cdots] \tag{1.5.8}$$

which can be written in the easily memorizable form, suggested by Birge[20]

$$b = \frac{(y_n - y_1)(n - 1) + (y_{n-1} - y_2)(n - 3) + (y_{n-2} - y_3)(n - 5) + \cdots}{(n - 1)^2 + (n - 3)^2 + (n - 5)^2 + \cdots}. \tag{1.5.9}$$

The intercept is

$$a = \bar{y} - \frac{n + 1}{2} b. \tag{1.5.10}$$

[17] Graphs of simple functions are given by A. G. Worthing and J. Geffner, "Treatment of Experimental Data." Wiley, New York, 1943.

[18] A. Wald, *Ann. Math. Statistics* **11,** 284 (1940); J. Neyman and E. M. Scott, *ibid.* **22,** 352 (1941); **23,** 135 (1953).

The differences $\hat{y} - y$ are assumed to be taken from a n.d.f., of which an approximation to the variance is given by

$$s^2(y) = \frac{\left(\sum x_i^2 - n\bar{x}^2\right)\left(\sum y_i^2 - n\bar{y}^2\right) - \left(\sum x_i \sum y_i - \frac{\Sigma x_i \Sigma y_i}{n}\right)^2}{(n-2)(\Sigma x_i^2 - n\bar{x}^2)}$$

$$= \frac{1}{n-2}\left[\sum (y_i - \bar{y})^2 - \frac{[\Sigma(x_i - \bar{x})(y_i - \bar{y})]^2}{\Sigma(x_i - \bar{x})^2}\right]. \tag{1.5.11}$$

If the coefficients a and b are assumed to be normally distributed, they have variances approximated by

$$\left.\begin{aligned} \sigma^2(b) &\approx s^2(b) = s^2(y)/(\Sigma x_i^2 - n\bar{x}^2) \\ \sigma^2(a) &\approx s^2(a) = s^2(y)\Sigma x_i^2/(n\Sigma x_i^2 - n^2\bar{x}^2). \end{aligned}\right\} \tag{1.5.12}$$

However, a and b are not statistically independent. For this and other reasons it is often advantageous to work with the relation

$$y' = a' + b'(x - \bar{x})$$

instead of that above. Following an analogous minimization procedure one has for n data values x_i

$$\left.\begin{aligned} a' &= \bar{y} \qquad\qquad s^2(a') = s^2(y')/n \\ b' &= \frac{\Sigma y_i(x_i - \bar{x})}{\Sigma(x_i - \bar{x})^2} \qquad s^2(b') = s^2(y')/(\Sigma(x_i - \bar{x})^2) \\ (n-2)s^2(y') &= \Sigma(x_i - y')^2 \\ &= (\Sigma(x_i - \bar{x})^2)^{-1}[(\Sigma(x_i - \bar{x})^2)(\Sigma y_i^2 - n\bar{y}^2) - (\Sigma y_i(x_i - \bar{x}))^2] \end{aligned}\right\} \tag{1.5.13}$$

but now the coefficients a' and b' are statistically independent. The y intercept is given for the first relation by a and for the second by $a' - b'\bar{x}$. The $x - \bar{x}$ intercept, if extrapolation to it is significant, i.e., has physical meaning, is given by $-a'/b'$. Under the normal n.d.f. assumption, etc., the variance of a'/b', $\sigma^2(a'/b')$ is approximated by

$$\sigma^2(a'/b') \approx (a'/b')^2\left(\frac{\sigma^2(a')}{a'^2} + \frac{\sigma^2(b')}{b'^2}\right). \tag{1.5.14}$$

If it is known *a priori* that the linear relation is the special one for which $a = 0$, these formulas are modified to

$$\begin{aligned} a &= 0 \qquad\qquad s^2(a) = 0 \\ b &= \Sigma x_i y_i/\Sigma x_i^2 \qquad s^2(b) = s^2/\Sigma x_i^2 \\ s^2(y) &= \frac{1}{n-1}[\Sigma x_i^2 \Sigma y_i^2 - (\Sigma x_i y_i)^2]/\Sigma x_i^2. \end{aligned} \tag{1.5.15}$$

If each x_i has a weight W_i the formulas above can be modified straightforwardly.*

If the y intercept is found to be near the origin, there may be a question whether the intercept differs significantly, in the statistical sense, from zero. A test of this is to compute $t = a/s(a)$ and compare with tabulated values of t with $n - 2$ degrees of freedom. If this value (of t) has a low probability, this indicates the probable existence of a nonzero intercept.

If one wishes to assess the agreement of a value b with a preassigned value of the slope β, assuming it is certain that the line goes through the origin, one can form $t = (\beta - b)/s(b)$ and use a table of Student's t as usual with $n - 2$ degrees of freedom. To decide, for lines of the form $y = a^{(1)} + b^{(1)}x$ and $y' = a^{(2)} + b^{(2)}x$, the question whether these lines have a significantly different slope, the variance of $b^{(1)} - b^{(2)}$ is estimated by

$$\sigma^2(b^{(1)} - b^{(2)}) \approx \left(\frac{s^{(1)2} + s^{(2)2}}{n_1 + n_2 - 4}\right)\left(\frac{1}{\Sigma x_i^{(1)2} - \bar{x}^2\Sigma x_i^{(1)}} + \frac{1}{\Sigma x_i^{(2)2} - \bar{x}^2\Sigma x_i^{(2)}}\right) \tag{1.5.16}$$

and one can compute t by dividing $b^{(1)} - b^{(2)}$ by the square root of this expression. This value of t can be compared with those in the table for $n_1 + n_2 - 4$ degrees of freedom. For lines known to pass through the origin, the variance of $b^{(1)} - b^{(2)}$ is similarly estimated by

$$\sigma^2(b^{(1)} - b^{(2)}) \approx \left(\frac{s^{(1)2} + s^{(2)2}}{n_1 + n_2 - 2}\right)\left(\frac{1}{\Sigma x_i^{(1)2}} + \frac{1}{\Sigma x_i^{(2)2}}\right). \tag{1.5.17}$$

If a simple control chart shows a tendency towards a trend, one can test whether the data can be fitted to a linear function with a slope $\neq 0$ by forming

$$t = \Sigma(x_i - \bar{x})(y_i - \bar{y})/s(y) \tag{1.5.18}$$

where $s(y)$ is taken from Eq. (1.5.11) and using a table of Student's t with $N - 2$ degrees of freedom. If the computed t corresponds to a low probability, a trend is likely to exist.

If, assuming negligible error in x, a polynomial of the nth order $f(x)$ is considered to be appropriate to fit the data y, e.g.

$$y = f(x) = \sum_{k=1}^{n} a_k x^k \tag{1.5.19}$$

the purpose of the least squares procedure is to determine the polynomial

* See Cameron.[13]

coefficients a_k so as to minimize

$$Q = \sum_{i=1}^{N} \left(y - \sum_{k=1}^{n} a_k x_i^k\right)^2.$$

(It is assumed that the number of data points N far exceeds n.) This gives the sequence of linear equations

$$\sum_{j=1}^{N} \left(y_j - \sum_{k=1}^{n} a_k x_j^k\right) x_j^l = 0; \qquad l = 1, 2, \ldots n. \tag{1.5.20}$$

The minimal value of Q is then

$$Q_{\min}^{(n)} = \sum_{j=1}^{N} y_j \left(y_j - \sum_{k=1}^{n} a_k x_j^k\right).$$

An estimate of the goodness of fit is then given by computing $Q_{\min}^{(n)}/(N - n)$. If the deviations from the assumed function are due to random fluctuations alone, $Q_{\min}^{(n)}$ should behave as χ^2 with $N - n$ degrees of freedom. How large an exponent n one should carry, or to put it in another way, how many coefficients are statistically significant, can be estimated by forming the F-ratio

$$F = (Q_{\min}^{(n)} - Q_{\min}^{(n+1)})/Q_{\min}^{(n+1)}(N - n - 1). \tag{1.5.21}$$

If this exceeds the tabulated value of F for one degree of freedom in the numerator and $N - n - 1$ degrees of freedom in the denominator, corresponding to a low probability level, it is an indication that there is a significant improvement in the fit by going to power $n + 1$. The variance of the coefficients a, assuming these are normally distributed, can be estimated from formulas given in statistical texts.[19] The computational work in the estimation of statistical significance and the variance of a coefficient a is unwieldy since one has to recompute everything when going from one power of the polynomial to the next.[19]

At least in this respect the use of orthogonal polynomials has some advantages. For data taken at equal spacings of x, combinatorial, or finite-difference methods can be used. An extensive discussion, including numerical examples, can be found in the article of Birge and Weinberg.[20]

In general, the coefficients of the orthogonal polynomials P_k are chosen to minimize

$$Q = \sum_{j} \left(y_j - \sum_{k} b_k P_k(x_j)\right)^2.$$

[19] M. G. Kendall, "The Advanced Theory of Statistics," Vol. 2. Griffin, London, 1948.

[20] R. T. Birge, *Revs. Modern Phys.* **19**, 298 (1947); with appendix by J. M. Weinberg.

This leads to

$$b_k = \sum_j y_j P_k(x_j) \Big/ \sum_j (P_k(x_j))^2$$

and using the definition of orthogonality, i.e., that $\Sigma_j P_k(x_j) P_l(x_j) = \delta_{kl}$, we can write the minimal value of Q, $Q_{\min}$ as $\Sigma_j y_j^2 - \Sigma_k (b_k \, \Sigma_j (P_k(x_j))^2)$. The statistical significance of a coefficient b_{n+1} can be tested by forming the F-ratio

$$F = \frac{Q_{\min}^{(n)} - Q_{\min}^{(n+1)}}{Q_{\min}^{(n+1)}} \left(\frac{1}{N - n - 1} \right) \tag{1.5.22}$$

and using a table of F values, with 1 and $N - n - 1$ degrees of freedom in numerator and denominator, respectively. If the computed F has a larger value than that in the table, one should include at least b_{n+1} and possibly more terms. An estimate of the variance of one of the coefficients b, assuming these are normally distributed, is given by

$$\sigma^2(b_k) \approx \frac{Q_{\min,k}}{\Sigma_x P_k^2(x)} \frac{n}{n - k - 1}. \tag{1.5.23}$$

By means of a transformation of the independent variable it may be possible to use the formulas for polynomial fitting to handle a wider class of situations. For example by putting $x = e^z$, $z = \ln x$ one can fit a function of the type $1 + a_1 e^z + a_2 e^{2z} + a_n e^{nz}$ with the polynomial type formulas. If it seems appropriate to try to fit a function of the type

$$y = C_1 e^{\alpha_1 x} + C_2 e^{\alpha_2 x} \qquad \text{with} \qquad x_1 \neq \alpha_2$$

so that y depends on the variables x in a nonlinear fashion, one has to be careful that small errors in α do not lead to misleading results.[21] Periodic analysis is discussed extensively in standard texts.[21]

1.6. Errors of Computation

Recent progress in automatic computation has given much incentive to the discussion of numerical analysis and errors of computation, particularly the errors of solving large sets of linear equations. Introductory discussions with extensive further references are given by Hildebrand.[21] Errors of linear equation solving are treated by Dwyer.[22] The publication

[21] F. B. Hildebrand, "Introduction to Numerical Analysis." McGraw-Hill, New York, 1956. A useful index is given in the appendix of this text, as well as an extensive bibliography.

[22] P. S. Dwyer, "Linear Computations." Wiley, New York, 1951.

"Mathematical Tables and Other Aids to Computation" (MTAC)[23] contains useful material, including lists of recent tables.[24]

Apart from gross errors, or mistakes, the errors of computation can be classed into: (1) the propagation of errors initially present; (2) the generation of errors in detailed steps of the computation, now commonly called round-off errors; and (3) the type of error due to cutting off, at a finite number of steps, a computation involving an infinite limiting process or asymptotic series, called truncation error. Although these types of error combine in complicated ways, fortunately the over-all error of the computation can be estimated by a simple addition. In extensive computations, the distribution of errors has to be obtained from statistical considerations in order to make a practical over-all estimate.[21]

There are some accepted rules for good computational design. The procedure should allow the computer to find and get rid of mistakes, to estimate bounds for other errors, and to check the final result. There should be a minimum number of individual operations, including look-ups, resetting of scales, and arithmetic operations. Approximate operations, such as division and root extraction, should be eliminated or at least put off as far as possible towards the end of the calculation. An assessment should be made of the number of digits required at each step to maintain the desired over-all accuracy.

Regarding calculational aids of moderate accuracy, a few general statements can be made. For multiplication and division, tables of logarithms to 4, 5, 7, and 10 decimal places are available. A calculating machine which multiplies and divides is usually more convenient. Tables of powers of x, sin x, cos x, exp x and other special functions are also available. A comprehensive index to 1946 has been published,[24] and more recent lists can be found in MTAC.[23] The ordinary ten inch slide rule has a relative accuracy of about 0.3% per operation of multiplication or division, and can only handle numbers having three significant figures. The relative accuracy, but not the convenience, of a slide rule improves roughly proportional to the length of scale. Automatic desk calculating machines are often so designed that a fixed number of decimal places can be carried throughout. In such a case the number having the largest number of decimal places should fix the position of the machine. In calculation with numbers having a definite number of significant figures, one additional digit should be carried.

It is not recommended that numbers be systematically rounded off in

[23] "Mathematical Tables and Other Aids to Computation" (MTAC). National Research Council, Washington, D.C.; published as a quarterly since 1943.

[24] A. J. Fletcher *et al.*, "Index of Mathematical Tables," McGraw-Hill, New York, 1946.

order to eliminate doubtful digits during the course of a calculation. Doing so generally increases the error.

There are conventional rules of rounding off followed in many tables and in circumstances where rounding off is inevitable and under the control of the computer. These are given here so that an estimate may be made of the errors thereby introduced. To round off a number, in the conventional sense, to n digits in the decimal system, the nth digit is increased by 1 if the $n + 1$st digit is greater than 5, and left unaltered if the $n + 1$st digit is less than 5. If the $n + 1$st digit is equal to 5, the nth is left unaltered if even. If odd, it is increased by 1. The rounded-off number is said to have n significant digits or figures, and the error of the rounded-off number is not larger than one half unit in the nth digit. If two numbers are rounded off to the same set of n significant figures, their difference does not exceed one unit in the nth place from the true value x rounded off to n significant figures. In multiplication, raising to powers, or division, the relative errors and the number of significant figures are important, but not the decimal place. In addition or subtraction the absolute values of errors and the location of the decimal point do matter.

The subtraction of nearly equal numbers, causing a loss in significant figures, should be avoided, or made explicit if possible. Thus if xy and ux are nearly equal, but y and u are not, it is better to multiply x by $(y - u)$ than to form $xy - ux$.

Numerical analysis of the arithmetic operations* gives the following results: If $x\dagger$ represents a number x rounded off to n significant figures, the error in the nth significant digit of x^m with m = any real number, is less than $10^n[(1 + 5 \cdot 10^{-n})^m - 1]$. If $y\dagger$ is also rounded off to n significant figures, the product $x\dagger y\dagger$ differs from the true value xy by less than 4 in units of the nth digit. When handling an (algebraic) sum of numbers having the same number of significant figures, but different decimal positions, a good rule is to round off the ones with more decimal positions to one place beyond that of those with the least decimal positions.

If an auxiliary table of entry differences is given to facilitate interpolation, the difference between rounded off values (such as those usually given in tables) has less "round-off" error than the difference between entries before these are rounded off.

* See Eisenhart[13] for a brief discussion of errors of interpolation; see also reference 21.

APPENDIX

TABLE I. The χ^2-Distribution*

Probability $P \rightarrow$		0.95	0.1	0.05	0.01	0.001
Degrees of freedom ↓	f					
	1	0.00393	2.706	3.841	6.635	10.827
	2	0.103	4.605	5.991	9.210	13.815
	3	0.352	6.251	7.815	11.341	16.268
	4	0.711	7.779	9.488	13.277	18.465
	5	1.145	9.236	11.070	15.086	20.517
	6	1.635	10.645	12.592	16.812	22.457
	7	2.167	12.017	14.067	18.475	24.322
	8	2.733	13.362	15.507	20.090	26.125
	9	3.325	14.684	16.919	21.666	27.877
	10	3.940	15.987	18.307	23.209	29.588
	11	4.575	17.275	19.675	24.725	31.264
	12	5.226	18.549	21.026	26.217	32.909
	13	5.892	19.812	22.362	27.688	34.528
	14	6.571	21.046	23.685	29.141	36.123
	15	7.261	22.307	24.996	30.578	37.697
	16	7.962	23.542	26.296	32.000	39.252
	17	8.672	24.769	27.587	33.409	40.790
	18	9.390	25.989	28.869	34.805	42.312
	19	10.117	27.204	30.144	36.191	43.820
	20	10.851	28.412	31.410	37.566	45.315
	21	11.591	29.615	32.671	38.932	46.797
	22	12.338	30.813	33.924	40.289	48.268
	23	13.091	32.007	35.172	41.638	49.728
	24	13.848	33.196	36.415	42.980	51.179
	25	14.611	34.382	37.652	44.314	52.620
	26	15.379	35.563	38.885	45.642	54.052
	27	16.151	36.741	40.113	46.963	55.476
	28	16.928	37.916	41.337	48.278	56.893
	29	17.708	39.087	42.557	49.588	58.302
	30	18.493	40.256	43.773	50.892	59.703

* This table is abbreviated from Table IV of Fisher and Yates, "Statistical Tables for Biological, Agricultural and Medical Research." Oliver and Boyd, Edinburgh and London, 1953, by kind permission of the authors and publishers.

TABLE II. Student's t-Function*

Probability $P \rightarrow$		0.50	0.10	0.05	0.01
Degrees	f				
of	1	1.000	6.34	12.71	63.66
freedom f	2	0.816	2.92	4.30	9.92
↓	3	0.765	2.35	3.18	5.84
	4	0.741	2.13	2.78	4.60
	5	0.727	2.02	2.57	4.03
	6	0.718	1.94	2.45	3.71
	7	0.711	1.90	2.36	3.50
	8	0.706	1.86	2.31	3.36
	9	0.703	1.83	2.26	3.25
	10	0.700	1.81	2.23	3.17
	12	0.695	1.78	2.18	3.06
	14	0.692	1.76	2.14	2.98
	16	0.690	1.75	2.12	2.92
	18	0.688	1.73	2.10	2.88
	20	0.687	1.72	2.09	2.84
	25	0.684	1.71	2.06	2.79
	30	0.683	1.70	2.04	2.75
	40	0.681	1.68	2.02	2.71
	50	0.679	1.68	2.01	2.68
	70	0.678	1.67	2.00	2.65
	100	0.677	1.66	1.98	2.63
	∞	0.674	1.64	1.96	2.58

* This table is abbreviated from Table V of Fisher and Yates, "Statistical Tables for Biological, Agricultural and Medical Research." Oliver and Boyd, Edinburgh and London, 1953, by kind permission of the authors and publishers.

TABLE III. The F-Distribution for Probability $P = 0.05$*

Degrees of freedom in denominator ↓ f_2 / Degrees of freedom in numerator → f_1	1	2	3	4	5	6	8	12	24	∞
1	161.4	199.5	215.7	224.6	230.2	234.0	238.9	243.9	249.0	254.3
2	18.51	19.00	19.16	19.25	19.30	19.33	19.37	19.41	19.45	19.50
3	10.13	9.55	9.28	9.12	9.01	8.94	8.84	8.74	8.64	8.53
4	7.71	6.94	6.59	6.39	6.26	6.16	6.04	5.91	5.77	5.63
5	6.61	5.79	5.41	5.19	5.05	4.95	4.82	4.68	4.53	4.36
6	5.99	5.14	4.76	4.53	4.39	4.28	4.15	4.00	3.84	3.67
7	5.59	4.74	4.35	4.12	3.97	3.87	3.73	3.57	3.41	3.23
8	5.32	4.46	4.07	3.84	3.69	3.58	3.44	3.28	3.12	2.93
9	5.12	4.26	3.86	3.63	3.48	3.37	3.23	3.07	2.90	2.71
10	4.96	4.10	3.71	3.48	3.33	3.22	3.07	2.91	2.74	2.54
11	4.84	3.98	3.59	3.36	3.20	3.09	2.95	2.79	2.61	2.40
12	4.75	3.88	3.49	3.26	3.11	3.00	2.85	2.69	2.50	2.30
13	4.67	3.80	3.41	3.18	3.02	2.92	2.77	2.60	2.42	2.21
14	4.60	3.74	3.34	3.11	2.96	2.85	2.70	2.53	2.35	2.13
15	4.54	3.68	3.29	3.06	2.90	2.79	2.64	2.48	2.29	2.07
16	4.49	3.63	3.24	3.01	2.85	2.74	2.59	2.42	2.24	2.01
17	4.45	3.59	3.20	2.96	2.81	2.70	2.55	2.38	2.19	1.96
18	4.41	3.55	3.16	2.93	2.77	2.66	2.51	2.34	2.15	1.92
19	4.38	3.52	3.13	2.90	2.74	2.63	2.48	2.31	2.11	1.88
20	4.35	3.49	3.10	2.87	2.71	2.60	2.45	2.28	2.08	1.84
21	4.32	3.47	3.07	2.84	2.68	2.57	2.42	2.25	2.05	1.81
22	4.30	3.44	3.05	2.82	2.66	2.55	2.40	2.23	2.03	1.78
23	4.28	3.42	3.03	2.80	2.64	2.53	2.38	2.20	2.00	1.76
24	4.26	3.40	3.01	2.78	2.62	2.51	2.36	2.18	1.98	1.73
25	4.24	3.38	2.99	2.76	2.60	2.49	2.34	2.16	1.96	1.71
26	4.22	3.37	2.98	2.74	2.59	2.47	2.32	2.15	1.95	1.69
27	4.21	3.35	2.96	2.73	2.57	2.46	2.30	2.13	1.93	1.67
28	4.20	3.34	2.95	2.71	2.56	2.44	2.29	2.12	1.91	1.65
29	4.18	3.33	2.93	2.70	2.54	2.43	2.28	2.10	1.90	1.64
30	4.17	3.32	2.92	2.69	2.53	2.42	2.27	2.09	1.89	1.62
40	4.08	3.23	2.84	2.61	2.45	2.34	2.18	2.00	1.79	1.51
60	4.00	3.15	2.76	2.52	2.37	2.25	2.10	1.92	1.70	1.39
120	3.92	3.07	2.68	2.45	2.29	2.17	2.02	1.83	1.61	1.25
∞	3.84	2.99	2.60	2.37	2.21	2.09	1.94	1.75	1.52	1.00

* This table is abbreviated from Table V of Fisher and Yates, "Statistical Tables for Biological, Agricultural and Medical Research." Oliver and Boyd, Edinburgh and London, 1953, by kind permission of the authors and publishers.

2. FUNDAMENTAL UNITS AND CONSTANTS*

2.1. Systems of Units

2.1.1. Definition of Units

Measurement is basic to the development of a quantitative description of physics. Experimental physics deals primarily with the determination of numerical values to be assigned to physical quantities or physical phenomena. These numerical values are obtained in two basically different ways. On the one hand we have simple counting in which a sequence of events is set into one-to-one correspondence with the natural numbers; on the other, we have a comparison of two similar physical magnitudes. The value of a physical quantity (as distinct from an enumeration of events) is determined by using a standard magnitude and finding the ratio of the desired magnitude to the standard. This standard is the *unit* by which the desired quantity is measured. The ratio of these two magnitudes is usually noncommensurate.

2.1.1.1. Derived Units. It is possible to define a unit individually and independently for each kind of physical quantity but this leads to an inconvenient system. As an example of this, the area of a rectangle is usually defined as equal to the product of its base and its height. However, this definition is correct only if the unit of area is related to the unit of length in such a way that a square the length of whose side is one unit will have an area of one unit. The unit of area is therefore *derived* from the unit of length. Examples of such units are sq ft (ft^2) or sq cm (cm^2). Here the exponent notation clearly indicates the dependent character of the unit.

We can, however, measure area in acres or in circular mils; in this case the formula for area is

$$A = kab \qquad \text{or} \qquad A = Cr^2$$

where a and b are the base and height of the rectangle and A is the area. The proportionality constant k depends then on the units being used. If A is the area of a circle of radius r, the constant C is equal to π only if area and radius are measured in proper units. If the radius is measured in mils and the area in circular mils, we have $C = 4$. A similar example of more direct interest to physics is the definition of units of force. The force on a body is proportional to the product of its mass and its acceleration. By an appropriate choice of this proportionality constant the unit of force may be derived. Two choices are common; the unit force is defined as that

* Part 2 is by **E. Richard Cohen.**

acting when a unit mass has a unit acceleration, or alternately the unit force is defined as that force which acts to give a unit mass the acceleration of free fall in the earth's gravitational field. (The latter definition is useful in many engineering applications but for precise measurement it has the disadvantage that the unit of force is dependent upon position on the earth's surface.) It is also possible to define the unit of force independently of Newton's laws, in which case the constant of proportionality (rather than the magnitude of the unit) must be determined by experiment.

2.1.1.2. Fundamental Units. Although physical laws allow one to define one unit in terms of others, there must be a fundamental set of units which cannot be connected by physical law. In this sense these units are independent. The number of independent units depend on the state of our knowledge of physics; thus, prior to the development of the mechanical theory of heat and the statistical mechanical description of gases, temperature was a physical quantity unrelated to length, mass, and time, and hence the degree (on whatever scale) was an independent unit. It is now possible, although not always convenient, to express temperature in terms of the random kinetic energy per molecule. Thus the measurement of temperature in degrees rather than in terms of ergs/molecule is a convenience similar to that of measuring area in acres. The experimentally determined proportionality constant relating temperature and energy is Boltzmann's constant.

2.1.2. CGS and Other Systems of Units

The adoption of length, mass, and time as fundamental quantities is not a unique one and hence the definition of units in terms of the units of length, mass, and time is also arbitrary; it is, however, convenient.

The units generally used in experimental physics are based on the metric system. Various modifications have been proposed and used for different purposes. The usual system of physics is the centimeter, gram, second (cgs) system or its modification, the meter, kilogram, second (mks) system. The English (fps) system, which is based on the foot, pound, and second, respectively, as the units of length, mass, and time, is used primarily in mechanical engineering but has only a limited use in experimental physics.

2.2. Definition of the Fundamental Units

2.2.1. Length

2.2.1.1. Metric Units. The fundamental unit of length in the metric system is the meter. In its original conception it was intended to represent 1 ten-millionth of the earth's quadrant on the meridian through Paris. It was also originally intended to define the unit of mass (kilogram) as the mass of 1 cubic decimeter of pure water. Both of these definitions have, however, been abandoned in favor of standards which are more easily measured and preserved.

The meter is defined as the distance between two engraved lines on a platinum–iridium bar maintained at the International Bureau of Weights and Measures at Sèvres, France when the bar is supported in a prescribed manner at the temperature of melting ice, and at standard atmospheric pressure (760 mm of Hg). A supplementary definition of the meter in terms of the wavelength of the red cadmium line was adopted in 1927 by the Seventh General International Conference on Weights and Measures. This definition is[1]

$$1 \text{ wavelength} = 6438.4696 \times 10^{-10} \text{ meter.} \tag{2.2.1}$$

The wavelength is measured in air with careful specifications of the temperature, pressure, and humidity as well as specifications of the operation of the light source.

There has been some tendency to use this wavelength to define the exact value of the angstrom unit and leave the exact relation between the meter and the angstrom as a constant to be determined by measurement. The present view, however, is to define the angstrom as exactly 10^{-10} meters and to abandon the meter bar at Sèvres as the primary standard. There are, however, disadvantages to the use of the cadmium red line as the defining standard, and the green line of Hg^{198} or a line of the Kr^{84} spectrum have been suggested as alternates. At the present writing no firm decision has been reached in regard to which spectral line should be used as the basic standard for the angstrom.[2]

Until such time as a new standard is determined we may use Eq. (2.2.1) as the definition of the meter.

In the fields of X-ray crystallography the X-unit is more useful than the angstrom because it is defined in terms of crystal spacings and affords

[1] J. R. Benoit, C. Fabry, and A. Perot, *Trav. et mém. bur. intern. poids et mésures* **15**, 131 (1913).

[2] E. C. Crittenden, Report on the Tenth General Conference on Weights and Measures, *Science* **120**, 1007 (1954).

a scale on which relative lengths can be accurately determined. The X-unit was intended to be 10^{-3} Å but because of the early inaccurate value of Avogadro's number, Siegbahn's original measurement[3] of the effective grating space of calcite for first order reflection is to be considered instead as the definition of the X-unit.

$$d = 3029.040 \text{ X-units at } 18°\text{C (first order).} \qquad (2.2.2)$$

In 1947 a committee composed of Sir Lawrence Bragg, M. Siegbahn B. E. Warren, and H. Lipson recommended[4] for general adoption the relationship 1 X-unit = $(1.00202 \pm 0.00003) \times 10^{-11}$ cm. It has, however, been pointed out subsequently[5] that this value was heavily influenced by data that are subject to a serious systematic error which could be as large as several parts per hundred thousand, and the best estimate available, as of January 1957, is the relation

$$1 \text{ X-unit} = 1.00204 \pm 0.00002 \times 10^{-11} \text{ cm.} \qquad (2.2.3)$$

2.2.1.2. English Units.* In England the inch and the foot are still maintained as units of length, but in the United States there is no "English" standard of length. Although the United States nominally uses English units, these units are defined by law in terms of metric standards, and the fundamental standard of length is the U.S. prototype meter bar. The inch is then defined as

$$1 \text{ in.} \equiv \frac{1}{39.37} \text{ meter} = 2.54000508 \text{ cm.} \qquad (2.2.4)$$

The British inch is slightly smaller (2.8 ppm)

$$1 \text{ British in.} = 2.5399980 \text{ cm.} \qquad (2.2.5)$$

2.2.2. Mass

The original philosophy in establishing the metric system was to have all units logically related. Thus, the unit of mass was to be defined as the mass of a cubic decimeter of pure water at maximum density. However, this definition has since been replaced by the platinum–iridium standard, the international prototype kilogram. It is then a matter of experimental determination to establish the maximum density of water. This was done by Henning and Jaeger (corrected by Guillaume)[6]

$$\rho_{\max} = 0.999972 \pm 0.000004 \text{ gm cm}^{-3}. \qquad (2.2.6)$$

* See page 52 concerning the definition of the inch.

[3] M. Siegbahn, "Spektroscopie der Röntgenstrahlen." Springer, Berlin, 1931.

[4] Sir Lawrence Bragg, *Acta Cryst.* **1,** 46 (1948); *J. Sci. Instr.* **24,** 27 (1947).

[5] J. W. M. DuMond and E. R. Cohen, *Phys. Rev.* **103,** 1583 (1956).

[6] F. Henning and W. Jaeger, *in* "Handbuch der Physik" (H. Geiger and K. Scheel, eds.), Vol. 2, p. 491. Springer, Berlin, 1933.

There are, as a result, two independent units of volume in the metric system. The meter (centimeter or millimeter) may be used to define a volume in the usual way; but alternately, the liter is defined as the volume of 1 kg of water at its maximum density and hence one has the relationship:

$$1 \text{ liter} = 1000.028 \pm 0.004 \text{ cm}^3. \tag{2.2.6a}$$

In the past chemists have made a distinction between the symbol cm^3, and the symbol cc. Fortunately, the designation ml (milliliters) has almost completely replaced the ambiguous cc.

The U.S. pound is defined in terms of the prototype kilogram maintained at the Bureau of Standards:

$$1 \text{ lb} = 453.5924277 \text{ gm}. \tag{2.2.7}$$

2.2.3. Time

The second is the fundamental unit of time in both the English and metric systems. The mean solar day, which is the average interval between successive meridian passages of the sun, is divided into 86,400 sec. The Tenth General Conference on Weights and Measures meeting in Paris in 1954 tentatively adopted[2] the second "as the fraction, 1/31556925.975 of the tropical year 1900.0." This definition of the second must of course be implemented by techniques which can measure time to an accuracy (of the order of one part in 10^{10}) required to make this definition significant. "Atomic clocks" can actually achieve, and perhaps even surpass, this accuracy. Such time standards are capable of much greater constancy and reproducibility than is the rotation of the earth, which is known to have variations of the order of several parts in a billion in comparison to sidereal time standards. The fact that the present definition of the second is in terms of the tropical year at the epoch 1900.0 is a reflection of the problem of the variability of the year as well as of the existence of more accurate time standards against which this variation can be measured. Without such standards it would be meaningless to refer the year to a specific epoch. Atomic clocks, such as the cesium atom clock, the ammonia molecule clock, or the Maser in which accuracies of 1 part in 10^{12} may ultimately be attained, will certainly replace the rotation of the earth as the time standard eventually. In addition to increased accuracy, there is the attractive feature that a determination can be made in a few minutes. This is in comparison with the astronomical determinations which require years for the collection and analysis of data and then leave only a time standard in retrospect.

2.2.4. Force

The units of force are closely related to the definition of force and to Newton's second law of motion, which states that the rate of change of momentum of a particle is directly proportional to the net force acting on the particle. Thus, the acceleration of a standard mass is a measure of the force. The unit of force in the metric system is either the dyne, the force necessary to give a mass of 1 gm an acceleration of 1 cm sec^{-2}, or the newton, the force required to give a mass of 1 kg an acceleration of 1 meter sec^{-1}. The newton is therefore 10^5 dynes.

In the British system the corresponding unit is the poundal, a force which imparts an acceleration of 1 ft sec^{-2} to a mass of 1 lb. In engineering, but rarely in physics, there is some use for a gravitational system of units in which force is the fundamental standard and mass is a derived unit. The unit of force is the pound weight, which is the force exerted by gravity on the U.S. standard pound. The unit of mass is then definable in terms of this force. It is that mass which will be given an acceleration of 1 ft sec^{-2} by a force of 1 lb wt. Obviously, this unit of mass (which is called a "slug" for want of a better name) is equal to g_0 lb. The constant g_0 is numerically equal to the acceleration of gravity in ft sec^{-2}. If g_0 is understood to be the local acceleration, then the unit of force and the unit of mass vary from point to point, a situation which has obvious disadvantages. If on the other hand, the constant g_0 is specified as the acceleration of gravity at Washington, the gravitational units differ only by an arbitrary constant from the absolute units; the former then have no logical advantage over the latter and have, in fact, the disadvantage that the arbitrary constant involved is not known with the accuracy with which the standards themselves are maintained. The gravitational system is, therefore, one of inherently lower accuracy than the absolute system, and its only advantage appears to be a dubious pedagogical one.

TABLE IA. Units and Conversion Factors: Length*

1 X-unit	= $(1.00204 \pm 0.00002) \times 10^{-11}$ cm
1 angstrom	= 10^{-8} cm
1 mil	= 0.00254 cm
1 cm	= 0.032808333 ft
1 in.	= 2.54000508 cm
1 ft	= 30.480061 cm
1 yd	= 36 in. = 3 ft = 91.440183 cm = 0.91440183 meter
1 meter	= 100 cm = 39.37 in. = 3.2808333 ft = 1.0936111 yd
1 km	= 1000 meters = 0.62136995 mi
1 mi	= 5,280 ft = 1.60934722 km

* See page 52 concerning the definition of the inch.

TABLE IB. Units and Conversion Factors: Mass

1 gm	= 0.035273957 oz = 0.00220462234 lb
1 oz	= 28.3495265 gm = 0.0625 lb
1 lb	= 453.5924277 gm = 16 oz
1 kg	= 2.20462234 lb
1 slug	= 32.1729 lb = 14.5934 kg
1 ton	= 2000 lb = 907.184855 kg
1 long ton	= 2240 lb = 1016.046937 kg

TABLE IC. Units and Conversion Factors: Time

1 min	= 60 sec
1 hr	= 60 min = 3600 sec
1 day	= 24 hr = 1440 min = 86,400 sec
1 sidereal day	= 86164.09054 sec = $23^{hr}\ 56^{min}\ 4.09054^{sec}$
1 tropical year	= $365^{day}\ 5^{hr}\ 48^{min}\ 45.975^{sec}$

TABLE ID. Units and Conversion Factors: Force

1 dyne	= $0.72329985 \times 10^{-4}$ poundal = 10^{-5} newton
	= 2.24809×10^{-6} lb

TABLE IE. Units and Conversion Factors: Energy

1 joule	= 10^7 ergs = 23.730262 ft poundal
	= 0.73756 ft lb = 10197.2 gm cm
1 ft poundal	= 4.2140285×10^5 ergs = 0.042140285 joule
1 ft lb	= 13825.52485 gm cm
1 cal (15°)	= $(4.1854 \pm 0.0004) \times 10^7$ ergs = 4.1854 ± 0.0004 joule
1 cal (thermochemical)	= 4.1840 joule (definition)
	= 41.29287 cm^3 atmos = 0.04129171 liter atmos
1 Btu	= 251.98 cal = 1054.8 joule
	= 25030.7 ft poundals
	= 777.98 ft lb = 10.4047 liter atmos = 0.3676 ft^3 atmos
	= 3.9292×10^{-4} hp hr
1 cal (mean)	= 0.0039685 Btu = 0.001459 ft^3 atmos
	= 99.334 ft poundals = 3.0874 ft lb
	= 0.42685 kg meter = 4.18596 joule

2.2.5. Temperature

The concept of heat and cold is a subjective physiological and psychological one which achieves objective definition from the observed change in physical properties with temperature. The history of the development of the physical concepts and of the methods of measurement of temperature can be found in the standard textbooks on thermodynamics.

A temperature scale requires for its definition two fixed points. In the older Fahrenheit and Celsius* scales these are the freezing point and boiling point of pure water at standard atmospheric pressure. In the Fahrenheit scale the fixed points are designated as 32° and 212°, respectively; in the Celsius scale, 0° and 100°, and in the Reaumur scale (little used), 0° and 80°.

The definition of temperature as "what you read on a thermometer" is more than an epigram, since the international temperature scale is defined simply by specifying the thermometer and the methods to be used for its calibration. For more complete details the reader is referred to Chapter 6.1 or to the National Bureau of Standards *Journal of Research.*[7] The 1948 international scale is defined in such a way as to be as nearly as possible equivalent to Kelvin's absolute thermodynamic temperature scale. This latter scale is based on the behavior of an ideal perfect gas, and hence defines temperature directly in terms of the energy content of such a gas. It is the only temperature scale of fundamental physical significance since it is independent of the properties of any particular substance.

The absolute temperature scale was originally defined to be the Celsius scale translated to read zero at absolute zero, but the Tenth General Conference on Weights and Measures (1954) revised the definition to use the zero point as one of the fixed points. The other fixed point is the temperature of the triple point of water "by assigning to it the temperature 273.16° Kelvin exactly."[8] The triple point is chosen rather than the ice point because, in this way, the temperature is uniquely determined without the necessity of specifying a pressure or saturation of water with air as in the international scale. The value assigned to the triple point was chosen in order to obtain the best possible agreement between the thermodynamic and the international scales.

The ice point is $0.0100 \pm 0.0002°C$ below the triple point so that the thermodynamic ice point is $273.1500 \pm 0.0002°K$. The temperature of absolute zero is $-273.16 \pm 0.01°C$ so that the new Kelvin degree is slightly larger (36 ppm) than the centigrade degree.

$$1°C = 0.999964 \pm 0.000036°K. \tag{2.2.8}$$

The boiling point of water, which is not now fixed by the definition of the

* Formerly also called centigrade scale. The Celsius scale is no longer defined in terms of the boiling point of water and is therefore not strictly "centigrade."

[7] *J. Research Natl. Bur. Standards* **42,** 209 (1949); see also: J. A. Hall, *in* "Temperature: Its Measurement and Control in Science and Industry" (H. C. Wolfe, ed.), Vol. 2, p. 115. Reinhold, New York, 1955.

[8] E. C. Crittenden, Report on the Tenth General Conference on Weights and Measures, *Science* **120,** 1007 (1954).

thermodynamic scale, is 373.1464 ± 0.0036°K. Although no recommendation has been made, it is quite probable that the Celsius scale will be abandoned in the strict sense and will continue only as a secondary scale defined in terms of the absolute Kelvin scale. The freezing point and boiling point of water are

$$\begin{array}{llll} \text{F.P.} & 0.0000 \pm 0.0002°\text{C} & = & 273.1500 \pm 0.0002°\text{K} \\ \text{B.P.} & 99.9964 \pm 0.0036°\text{C} & = & 373.1464 \pm 0.0036°\text{K}. \end{array} \qquad (2.2.9)$$

2.2.6. Electrical Units

Absolute electrical units may be defined in terms of the laws of force between charges or between current elements or, alternately, arbitrary standards can be set up to establish these units. In this latter case, of course, the conversion factors relating electrical and mechanical quantities must be determined experimentally.

In 1908, the International Conference on Electrical Units and Standards established absolute electromagnetic units to be consistent with the cgs system of mechanical units. The unit current is defined as that current which, flowing in an arc of unit radius and unit length will produce a force of 1 dyne on a unit magnetic pole at its center. The unit magnetic pole is in turn defined as such that two magnetic poles of unit strength at unit distance exert a force of 1 dyne on each other. The absolute volt and absolute ohm are then to be defined in terms of the power dissipated in a resistor by the passage of a unit current.

At the time, experimental techniques were incapable of providing adequate standards with these definitions and the conference, therefore, also recommended, "as a system of units representing [the electromagnetic cgs unit] and sufficiently near to them to be adopted for the purpose of electrical measurements," the adoption of the international ampere, ohm, and volt. The international ampere was defined as the current which would deposit 0.001118 gm of silver per sec on the cathode of a specified silver voltameter operated under specified conditions. The international ohm was defined as the resistance of a column of mercury having specified dimensions and at a specified temperature. The international volt is then defined from Ohm's law. This also leads to the definition of the international watt as the energy dissipated per second by a current of 1 amp flowing through a resistance of 1 ohm. It is not to be expected that the international watt would be exactly equal to 10^7 ergs/sec and conversion factors were, therefore, required to relate the international values to their absolute counterparts. In addition, comparisons were not made in practice with the international standards but with prototypes maintained at the Standards Laboratories of the principal participating nations.

In October 1946, the techniques of establishing absolute units having reached a sufficiently high degree of precision, action was taken by the International Committee to abolish the international units and to certify all electrical standards directly in absolute units. The following ratios were established as those governing the change:

$$1 \text{ mean international ohm} = 1.00049 \text{ absolute ohm}$$
$$1 \text{ mean international volt} = 1.00034 \text{ absolute volt.}$$

Thus, since January 1, 1948 (the effective date of the change), the U.S. Bureau of Standards (and all other world laboratories as well) has certified all standard cells and resistances in absolute units. Because of small drifts which had occurred with time, the U.S. prototype international units did not agree with the mean international units quoted above. These differences were established by intercomparisons at the International Bureau at Sèvres, and the conversion factors between absolute units and the U.S. international units, as certified by the U.S. Bureau of Standards prior to January 1, 1948, are given in Table II.

TABLE II. Ratios of U.S. NBS International Electrical Units to Absolute Electromagnetic Units

1 U.S. international ohm	=	1.000495 absolute ohm
1 U.S. international volt	=	1.000330 absolute volt
1 U.S. international ampere	=	0.999835 absolute ampere
1 U.S. international coulomb	=	0.999835 absolute coulomb
1 U.S. international henry	=	1.000495 absolute henry
1 U.S. international farad	=	0.999505 absolute farad
1 U.S. international watt	=	1.000165 absolute watt

Since January 1, 1948, the U.S. National Bureau of Standards has been certifying all electrical quantities directly in terms of absolute units. Further details about electrical and magnetic units are to be found in Chapter 8.1 of this volume.

2.3. Fundamental Constants

2.3.1. Gravitational Constant

The force with which any two bodies in the universe attract each other is, according to Newton's law of gravitation, proportional to the square of the distance between them. The constant of proportionality is a universal constant. This constant, G, and the velocity of light, c, are the most fundamental constants of the universe. The general theory of relativity attempts to explain the constant G in terms of the structure of the universe itself, but such considerations lie beyond the scope of the present survey.

Early attempts to measure the constant of gravitation made use of the change in the direction of the force of gravity as a result of gravitational attraction of a mountain, or the change in magnitude of the force of gravity at the top and bottom of a deep mine shaft. The first method was used by N. Maskelyne, Astronomer Royal of England, in 1774. He obtained the result $G = 7.4 \times 10^{-8}$ dynes cm^2 gm^{-2}. In 1854, Sir George Airy, in his time also Astronomer Royal, measured the change in the period of a pendulum at the top and bottom of a mine shaft, and obtained the value $G = 5.7 \times 10^{-8}$ dynes cm^2 gm^{-2}.

The most accurate measurements of G, however, are direct laboratory determinations of the force of attraction between two bodies of accurately known mass and geometry. The static methods measure the change in deflection of a beam or torsion balance when a heavy known mass is brought close to the mass being weighed. More accurate still are the dynamic methods in which the change in period of a horizontal torsion pendulum is measured when known masses are placed in such a position as to provide an additional restoring force, or in a position where the gravitational force opposes the torsional restoring force. The most accurate of these determinations are those of Heyl[9] and of Heyl and Chrzanowski[10] at the U.S. National Bureau of Standards in which they obtain

$$G = 6.670 \pm 0.008 \times 10^{-8} \text{ dyne cm}^2 \text{ gm}^{-2}. \tag{2.3.1}$$

2.3.1.1. The Acceleration of Gravity. The gravitational attraction of the earth for all objects on its surface is such a fundamental observational fact that its true significance was overlooked for centuries. The acceleration with which a body falls at the earth's surface is a composite of two forces: the gravitational attraction of the earth's mass and the centrifugal

[9] P. R. Heyl, *J. Research Natl. Bur. Standards* **5**, 1243 (1930).

[10] P. R. Heyl and P. Chrzanowski, *J. Research Natl. Bur. Standards* **29**, 1 (1942).

force associated with the rotation of the earth. The gravitational component is itself not constant over the surface of the earth because of the departure of the figure of the earth's surface from a perfect sphere. The international gravity formula[11] gives the combined result of these two effects

$$g = 978.0495[1 + 0.005289 \sin^2 \theta - 0.0000073 \sin^2 2\theta] \text{cm sec}^{-2} \quad (2.3.2a)$$

where g is the "acceleration of gravity" at sea level at latitude θ.

Heiskanen[12] gives a somewhat different expression for the international gravity formula:

$$g = 978.0490[1 + 0.0052884 \sin^2 \theta - 0.0000059 \sin^2 2\theta] \text{cm sec}^{-2}. \quad (2.3.2b)$$

The acceleration also decreases with altitude above sea level at the rate

$$\frac{dg}{dh} = -0.3086 \times 10^{-5} \text{ sec}^{-2} \quad (2.3.3)$$

or 0.3085 cm sec^{-2} for each kilometer increase in altitude above sea level.

The international gravity formula does not give the actual value of g to be measured at points on the earth's surface but the value of g associated with an idealized geoid which best fits the average shape of the earth. Heiskanen[12] also quotes a modified formula which is intended to represent a triaxial ellipsoidal earth with a longitudinal variation in g. In general, the "gravity anomaly" must be determined at each point on the surface of the earth. This is the excess of the true gravity (reduced to sea level) over the value given by the international formula.

2.3.2. Velocity of Light

Attempts to measure the velocity of light go back to Galileo. The first experimental evidence for a finite velocity of light was the work of Ole Römer on the periodic variation in the apparent period of revolution of the first satellite of Jupiter. In 1849, H. L. Fizeau made the first reasonably accurate measurement of the velocity of light using a rotating toothed wheel, and found a value of the order of 300,000 km sec^{-1}.

The developments in electronics during World War II greatly stimulated work on the measurement of the velocity of light, c. In attempting to use radar for navigational purposes Aslakson[13] showed that the velocity

[11] Smithsonian Physical Tables, 9th rev. ed. (W. E. Forsythe, ed.), p. 714. Smithsonian Institution, Washington, D.C., 1954.

[12] D. E. Gray, ed., "American Institute of Physics Handbook," p. **2**–98. McGraw-Hill, New York, 1957.

[13] C. J. Aslakson, *Nature* **164,** 711 (1949); *Trans. Am. Geophys. Union* **30,** 475 (1949).

of light must be 10 to 20 km/sec higher than the previously accepted value. Since then, E. Bergstrand[14] developed the Kerr cell modulation method into a precise geodetic instrument (the geodimeter), and has obtained a value

$$c = 299793.1 \pm 0.3 \text{ km sec}^{-1}. \tag{2.3.4a}$$

MacKenzie,[15] using the geodimeter over a geodetic base line in Scotland, obtained the value

$$c = 299792.4 \pm 0.5 \text{ km sec}^{-1}. \tag{2.3.4b}$$

In addition, Froome[16] at the National Physical Laboratory, Teddington, England, measured the velocity of light at microwave frequencies (24,005 Mc sec^{-1}) and obtained the value

$$c = 299793.0 \pm 0.3 \text{ km sec}^{-1}. \tag{2.3.4c}$$

The same author later published the results of a new determination of the velocity of light using an improved version of his microwave interferometer.[16a] This new instrument operates at a frequency of 72,000 Mc sec^{-1} (a wavelength of approximately 4 mm). In the course of this new work an error was discovered in the calibration of the one-meter length-standard used with the prototype interferometer and this has caused the published value obtained with that instrument to be high by 0.2 km sec^{-1}.

The final value obtained with the new instrument is

$$c = 299792.50 \pm 0.10 \text{ km sec}^{-1} \tag{2.3.4d}$$

in good agreement with the corrected preliminary value (299792.75 $\pm$ 0.30).

2.3.3. Normal Molar Volume; Gas Constant

The normal molar volume of an ideal gas is the volume occupied by one gram molecule at 0°C under a pressure of 1 atmos. For an ideal gas, this volume is independent of the molecular weight of the gas.

The ideal gas law is

$$pv = R_0 T \tag{2.3.5}$$

where v is the volume per mole of gas, p is the pressure exerted by the gas, R_0 is the gas constant, and T the absolute temperature.

[14] E. Bergstrand, *Nature* **163,** 338 (1949); **165,** 405 (1950); *Arkiv Fysik* **2,** 119 (1950).

[15] I. C. C. MacKenzie, Ordnance Survey Professional Papers, No. 19. Her Majesty's Stationery Office, London, England, 1954.

[16] K. D. Froome, *Proc. Roy. Soc.* **A213,** 123 (1952); **A223,** 195 (1954).

[16a] K. D. Froome, *Nature* **181,** 258 (1958).

The Tenth General Conference on Weights and Measures in October, 1954, established normal atmospheric pressure to be, by definition[17]

$$1 \text{ atmos} = 1{,}013{,}250 \text{ dyne cm}^{-2}. \tag{2.3.6}$$

This replaces the earlier definition (1927) of the normal atmosphere as[18] "the pressure due to a column of mercury 760 mm high, having a mass of 13.5951 gm cm^{-3} subject to a gravitational acceleration of 980.665 cm sec^{-2} and equal to 1013250 dynes cm^{-2}." Although this earlier definition appears to be an operational one it is not, since it specifies exactly the result of all the necessary measurements.* Hence the present simpler definition is logically the more preferable.

Any real gas satisfies Eq. (2.3.5) only at zero pressures, and the determination of the molar volume must, therefore, be made by measuring the behavior of a gas (usually oxygen) over a wide range of pressures and extrapolating the data to zero pressure. The results of several precise measurements have been analyzed by Birge;[19] converted to the physical scale of atomic weights the result is

$$V_0 = 22420.7 \pm 0.9 \text{ cm}^3 \text{ atmos mole}^{-1}. \tag{2.3.7}$$

The gas constant follows immediately from this since $R_0 = V_0/T_0$ and $T_0 = 273.1500 \pm 0.0002°\text{K}$

$$R_0 = (8.31696 \pm 0.00034) \times 10^7 \text{ erg mole}^{-1} \text{ deg}^{-1}. \tag{2.3.8}$$

2.3.4. Atomic Masses

Atomic masses as measured by the physicist are defined as the mass of the neutral isotopic atom relative to the mass of the neutral atom of O^{16}, defined to have the exact value 16.000000. Excellent tables of atomic masses have been prepared by Wapstra[20] and Huizenga[21] and others.

The chemical scale of atomic weights, in which the normal isotopic mixture of oxygen isotopes is assigned the weight 16.000, is equivocal because of the variability in isotopic composition depending on the source of the oxygen. The significance of the measurements by Swartout and Dole[22] on the density of water prepared from atmospheric oxygen (which showed that it was heavier than water distilled from fresh water lakes

* As a purely arithmetical exercise the product of the three factors is *not* equal to the stated answer.

[17] E. C. Crittenden, *Science* **120,** 1007 (1954).

[18] G. K. Burgess, *J. Research Natl. Bur. Standards* **1,** 635 (1928).

[19] R. T. Birge, *Repts. Progr. in Phys.* **8,** 90 (1942).

[20] A. H. Wapstra, *Physica* **21,** 367 (1955); **21,** 385 (1955).

[21] J. R. Huizenga, *Physica* **21,** 410 (1955).

[22] S. A. Swartout and M. Dole, *J. Am. Chem. Soc.* **61,** 2025 (1939).

by 6.6 ppm), appears to have been not fully recognized at the time. This corresponds to an increase of 60 ppm in the abundance of O^{18} or a change of the O^{18}/O^{16} ratio from a "standard" 510 to 495. These measurements are also slightly confused by a possible change in deuterium content, although to ascribe even 1 ppm of the density change to a variation of deuterium content would involve a decrease of the H/D ratio from 6900 to 6500.

The measured abundance ratio O^{16}/O^{18} in oxygen varies[23] from approximately 495 to 515, which corresponds to a variation in the ratio of chemical to physical scales of atomic weight from 1.000278 to 1.000268. The former figure corresponds to oxygen obtained from the atmosphere or from limestone rocks, whereas the latter figure is found for oxygen from iron ores or water. There thus appears to be a differing isotopic pattern in organic and inorganic oxygen.

The most satisfactory solution appears to be the abandonment of the chemical scale as one which can be defined operationally and the substitution of the physical scale of atomic weights as the only basic scale. The chemical scale would then be defined as a derived system by *defining* the conversion factor between the two scales as 1.000275. This value has in fact been used since 1939 by the International Commission to convert to the chemical scale measurements made on the physical scale.

2.3.5. Avogadro's Constant

The mass of 1 atomic mass unit is $1/N$, the reciprocal of Avogadro's constant. We thus define Avogadro's constant as the number of fictitious molecules of unit molecular weight required to yield a portion of matter whose mass is equal to the unit mass. Thus N is a dimensional quantity with dimensions, $[M^{-1}]$. This point has often been overlooked or improperly applied.

Direct experimental determinations of Avogadro's constant are based primarily on the role which this constant plays as the link between macroscopic and microscopic physics. Faraday's constant of electrochemistry, F, is the net electric charge contained in 1 gm equivalent of univalent ions; it is immediately evident that this is simply Ne, where N is Avogadro's constant and e is the charge on the electron. Similarly, the macroscopic density of a crystal is given simply by M/Nv where M is the molecular weight and v is the volume per molecule.

The first accurate determination of the value of Avogadro's constant came with Millikan's measurement (in the famous "oil-drop" experiment) of the charge on the electron. Unfortunately, the error of 0.6% in

[23] A. O. C. Nier, *Phys. Rev.* **77**, 792 (1950).

Millikan's value because of an unsuspected error in the viscosity of air (see Section 1.1.1) has been perpetuated in the Siegbahn scale of X-ray wavelengths. The chain of causality here ought to be briefly reviewed.

Since $F = Ne$, Avogadro's constant follows directly, as soon as e is determined, as $N = F/e$. If we consider now a simple cubic crystal such as rock salt, with a spacing between lattice points equal to d, the volume per molecule, NaCl, is just $2d^3$. The lattice spacing, d, is then given by

$$d = (M/2N\rho)^{\frac{1}{3}} = (Me/2\,\rho F)^{\frac{1}{3}}. \qquad (2.3.9)$$

Thus an error of 0.6% in e implied an error of 0.2% in all wavelengths which are computed from the comparison with crystal spacings. (Actually the Siegbahn scale as defined in Section 2.2.1 is based on calcite rather than rock salt. Because of the rhombohedral structure of calcite Eq. (2.3.9) would be modified to include a geometrical shape factor which depends on the dihedral angle between the cleavage faces of the calcite crystal, but the dependence on the measured value of the electronic charge is unaltered.) It is best at present, however, to consider the Siegbahn scale as a completely arbitrary one and to determine by experiment the value of the conversion factor Λ (see Section 2.2.1.1). In this way all of the careful and very precise measurements of X-ray wavelengths (about 3000 are tabulated in the tables of Cauchois and Hulubei[24]) measured by the method of crystal diffraction can be maintained as a consistent set of data in which the ratio of two lengths are accurately known to within a few parts in 10^5 or better.

Since, however, the conversion factor between X-ray wavelengths measured in X-units and the same lengths expressed in milliangstroms is uncertain, it is necessary to represent this fact in quoting the results of measurements. Thus, a measurement of Avogadro's constant in terms of the macroscopic density of a crystal and the grating space of the same crystal measured by X-ray diffraction determines not Avogadro's constant directly, but rather the combination $N\Lambda^3$ where Λ is the conversion factor between X-units and milliangstroms. The best direct measurement of this quantity is[25]

$$N\Lambda^3 = 6061.79 \pm 0.23 \times 10^{20}\ \text{mole}^{-1}\ \text{(direct measurement)}. \qquad (2.3.10a)$$

The best value, based on both direct and indirect evidence is

$$N\Lambda^3 = 6061.62 \pm 0.22 \times 10^{20}\ \text{mole}^{-1}\ \text{(best value)} \qquad (2.3.10b)$$

[24] Y. Cauchois and H. Hulubei, "Longueurs d'Onde des Émissions X et des Discontinuités d'Absorbtion," Tables de Constantes Sélectionées, Vol. 1. Masson, Paris, 1947.

[25] E. R. Cohen and J. W. M. DuMond, *in* "Handbuch der Physik—Encyclopedia of Physics" (S. Flügge, ed.), Vol. 35, p. 1. Springer, Berlin, 1957.

and Avogadro's constant is given (as the result of a least-squares fitting of all pertinent available data)

$$N = 6024.9 \pm 0.2 \times 10^{20} \text{ mole}^{-1}. \quad (2.3.11)$$

2.3.5.1. Boltzmann's Constant. The universal gas constant, R_0, represents the energy per degree per mole; when this is expressed as energy per molecule one obtains Boltzmann's constant

$$k = R_0/N = 13804.4 \pm 0.7 \times 10^{-20} \text{ erg deg}^{-1}. \quad (2.3.12)$$

2.3.6. Faraday Constant

The faraday is that quantity of electric charge associated with 1 mole of univalent ions. In electrochemistry it is defined as the quantity of electricity whose flow is associated in electrolysis with one equivalent weight of atoms or molecules reacting electrolytically at one of the electrodes. The equivalent weight, W, is equal to the molecular weight divided by the valence state of the molecule $W = M/v$. Thus, if a charge q is associated in electrolysis with a mass m of a pure substance

$$m/q = E = W/F. \quad (2.3.13)$$

Measurements of the electrochemical equivalent, E, for silver carried out at the U.S. National Bureau of Standards yielded the value $F = 9651.29 \pm 0.19$ emu mole^{-1} when converted to the physical scale of atomic weights.

However, a study of the work on the silver voltameter[26] shows clearly that the emphasis at the time was not so much one of accuracy as it was reproducibility. The major purpose for the work was the establishment of a reproducible standard for the measurement of electric charge in terms of the mass of silver deposited on an electrode in a standard voltameter. There are several possible sources of error in the silver voltameter if one asks for accuracy rather than merely precision and reproducibility. The measured gain in weight of the cathode may not be due entirely to deposited silver; there may be a more or less fixed fraction of other atoms deposited with the silver from the electrolyte (it is perhaps remarkable that the *loss* of silver from the *anode* was not measured). Some of the deposited silver may have redissolved or otherwise sloughed off before weighing. Because of isotopic effects, the mean atomic weight of the deposited silver may not be the same as the atomic weight of "natural" silver.

Because of the difficulties associated with the silver voltameter,

[26] E. B. Rosa and G. W. Vinal, *Bull. Natl. Bur. Standards* **13,** 479 (1916–1917).

S. J. Bates[27] suggested the use of an iodine system. Although it was not recognized at the time (1912), iodine has the distinct advantage of being monoisotopic and hence there can be no change in molecular weight in electrolysis. The iodine reaction is also "purer" than the silver reaction, and the amounts of iodine involved in the reactions at each electrode can be measured. The value obtained by Bates and Vinal[28] as corrected by Vinal, and with a later correction[29] for changes in the value of the atomic mass of iodine, is $F = 9652.15 \pm 0.13$ emu mole^{-1}.

The direct measurements are, therefore, in disagreement since they differ by almost four times the standard error of the difference. The indirect determination, in fact, strongly indicates that the silver voltameter value is indeed in error; the best value,[17] based on all pertinent data, is

$$F = 9652.15 \pm 0.11 \text{ emu mole}^{-1}. \tag{2.3.14}$$

2.3.7. The Rydberg Constant and the Fine Structure Constant

The first great success of the Bohr theory of the atom was the explanation of the formula for the Balmer and Lyman series for hydrogen and the derivation of a theoretical formula for the Rydberg constant. Actually the Rydberg constant, which refers to an infinitely heavy nucleus, cannot be directly measured; it is deduced from measurements of the Balmer and Lyman series for hydrogen, deuterium, and helium. Its value, as recomputed by Cohen,[30] is

$$R_\infty = \frac{2\pi^2 me^4}{h^3 c} = 109737.309 \pm 0.012 \text{ cm}^{-1}. \tag{2.3.15}$$

Another constant intimately involved in the description of atomic spectra is the fine structure constant of Sommerfeld, $\alpha = 2\pi e^2/hc$. We shall not discuss here the possibility of developing a theory of the universe in which the value of α may be determined *a priori*, such as has been discussed by Eddington,[31] but shall consider its value to be determined only by experiment. The most accurate measurement of the fine structure constant comes from the work of Willis Lamb, Jr. and his collaborators[32]

[27] S. J. Bates, Ph.D. Thesis, University of Illinois, Urbana, Illinois, 1912. (unpublished).

[28] S. J. Bates and G. W. Vinal, *J. A. Chem. Soc.* **36,** 916 (1914).

[29] G. W. Vinal (1949) and J. A. Hipple (1951), private communications.

[30] E. R. Cohen, *Phys. Rev.* **88,** 353 (1952).

[31] A. S. Eddington, "Fundamental Theory," Cambridge Univ. Press, London and New York, 1946.

[32] W. E. Lamb, Jr. and R. C. Retherford, *Phys. Rev.* **79,** 549 (1950); **81,** 222 (1951); W. E. Lamb, Jr., *ibid.* **85,** 259 (1952); W. E. Lamb, Jr. and R. C. Retherford, *ibid.* **86,** 1014 (1952); S. Triebwasser, E. S. Dayhoff, and W. E. Lamb, Jr., *ibid.* **89,** 98 (1953); E. S. Dayhoff, S. Triebwasser, and W. E. Lamb, Jr., *ibid.* **89,** 106 (1953).

on the energy levels of hydrogen and deuterium. Dayhoff, Triebwasser, and Lamb measured the separation in frequency units between the $2^2P_{\frac{1}{2}}$ and $2^2P_{\frac{3}{2}}$ states of deuterium and obtained

$$\Delta E = (10971.59 \pm 0.10) \times 10^6 \text{ Mc/sec.} \tag{2.3.16}$$

This cannot be directly equated to $1/16\ \alpha^2 R_\infty c$ because of three important corrections. This simple formula is only a first approximation to the Dirac equation values of the energy levels; in the first place, the Rydberg constant R_∞ must be replaced by the Rydberg constant for deuterium R_{D} which differs from R_∞ by the ratio of the deuteron mass to the mass of the neutral deuterium atom, and secondly, an additional correction of relative amount $5/8\ \alpha^2$ must be included as the next order term in the Dirac theory. The third correction, on the other hand, is not implicit in the Dirac theory of the hydrogen atom but represents a basic alteration to it. This is the anomalous magnetic moment of the electron and is properly identified with the Lamb–Retherford shifts in the energy of the hydrogen energy levels. The correct theoretical expression is given by

$$\Delta E = \frac{1}{16} \alpha^2 R_{\text{D}} c \left(1 + \frac{5}{8} \alpha^2\right)\left(2 \frac{\mu_e}{\mu_0} - 1\right) \tag{2.3.17}$$

where $\mu_e/\mu_0 = 1 + (\alpha/2\pi) - 2.973(\alpha^2/\pi^2) + \cdots$ is the ratio of the electron magnetic moment to the Dirac moment. The inferred value of the fine structure constant is, therefore,[17]

$$1/\alpha = hc/2\pi e^2 = 137.0373 \pm 0.0006. \tag{2.3.18}$$

2.4. Determination of Best Values

The numerical values of the physical constants and conversion factors listed in the previous section are interrelated. It is almost axiomatic that all measurements of physical constants will depend on the values of other constants and conversion factors as well. These same constants and conversion factors will affect the measured values of other physical constants and, in general, the total number of relationships which can be measured involving various physical constants is much greater than the number of physical constants to be determined. Thus, there exists a set of data which, in fact, overdetermine the values of the physical constants. One is thus faced with a situation which is essentially a set of N equations in q unknowns, with N greater than q. If these equations are not arithmetically consistent no true solution of the set is possible. It is known, however, that the numerical values measured in an experiment are subject to errors. If we assume, therefore, that one certain set involving q of the N equations is correct, we may determine the errors associated with the other $N - q$ equations. This is the usual procedure by which the overdetermined system can be avoided. However, if there is no clear choice as to which equations are correct and which are in error, it is necessary to ascribe errors to all of the equations and to use a procedure such as least-squares to determine the "best" values (see Section 2.4.2).

2.4.1. Related Precision Measurements

In addition to the physical constants described in the preceding sections there are several measurements of quantities which represent relationships between constants and which, therefore, have a direct bearing on the numerical values of these constants. Several of these relationships will be briefly mentioned.

2.4.1.1. Gyromagnetic Ratio of the Proton. Atomic particles exhibit the properties of angular momentum (spin) and magnetic dipole moment. Therefore, when a proton is placed in a magnetic field there is (in the classical picture) a torque exerted on the proton dipoles which tend to align them with the field. Because of the spin, they respond to the torque by precessing around the direction of the field with a frequency proportional to the external field. This constant of proportionality has been measured at the U.S. National Bureau of Standards with the result[33]

$$\gamma_p = \omega_p/B = 26752.3 \pm 0.6 \text{ rad sec}^{-1} \text{ gauss}^{-1}. \tag{2.4.1}$$

2.4.1.2. Magnetic Moment of the Proton. A particle with a charge to mass ratio, e/m, in a magnetic field of constant intensity, B, executes circular orbits with an angular frequency, ω, given by $\omega = Be/(mc)$. If now the frequency of the gyromagnetic resonance is used to measure the magnetic field in which protons have a cyclotron resonance frequency, $\omega_c = Be/(m_p c)$, we find that the ratio of these two frequencies is just the magnetic moment of the proton expressed in nuclear magnetons,

$$\mu = \omega_p/\omega_c = \gamma_p m_p c/e. \tag{2.4.2}$$

The magnetic moment measured here must still be corrected for the diamagnetic effect of the electrons in the molecular structure. The proton precesses not in the applied field but in a net resultant magnetic field which exists at the nucleus. The gyromagnetic ratio quoted in the preceding paragraph is the radian frequency of precession per externally applied

[33] H. A. Thomas, R. L. Driscoll, and J. A. Hipple, *Phys. Rev.* **78,** 787 (1950).

gauss, and hence, if the purpose of the measurement is the determination of the atomic constant (m_pc/e), no correction for this diamagnetic effect need be made since ω_p/γ_p is independent of such effects.

Sommer, Thomas, and Hipple[34] at the U.S. National Bureau of Standards obtained

$$\mu = 2.792685 \pm 0.000030 \tag{2.4.3a}$$

and Collington, Dellis, Sanders, and Turberfield at Oxford[35] found

$$\mu = 2.792730 \pm 0.000040. \tag{2.4.3b}$$

Bloch and Jeffries (after correction for orbit distortion effects[36]) obtained

$$\mu = 2.792670 \pm 0.000100. \tag{2.4.3c}$$

2.4.1.3. Magnetic Moment of the Electron. The cyclotron frequency of electrons in a magnetic field can be measured. Because of the difference in magnetic moment of the electron and proton the cyclotron frequency of the former is approximately 650 times larger than that of the latter, and hence quite different experimental techniques are required. Gardner and Purcell[37] have measured the ratio of the electron-cyclotron frequency to the proton-precession frequency (and hence the ratio of Bohr magneton to the absolute proton magnetic moment). For the experimental details reference should be made to the original papers; the result obtained was

$$\omega_e/\omega_p = \mu_0/\mu_p' = 657.475 \pm 0.008. \tag{2.4.4}$$

A closely related quantity was measured by Koenig, Prodell, and Kusch[38] and by Beringer and Heald.[39] This is a measurement of the ratio of the electronic g-value to the proton g-value, $g_e/g_p' = \mu_e/\mu_p'$. The important feature of this experiment is that it measured directly the magnetic moment of the electron rather than the Bohr magneton. The distinction between the two (the anomalous magnetic moment of the electron) is an important test of the newer theories of quantum electrodynamics.

Koenig, Prodell, and Kusch utilized the well-known molecular beam method[40] for measuring nuclear magnetic moments and obtained a value

[34] H. Sommer, H. A. Thomas, and J. A. Hipple, *Phys. Rev.* **82,** 697 (1955).

[35] D. J. Collington, A. N. Dellis, J. H. Sanders, and K. C. Turberfield, *Phys. Rev.* **99,** 1622 (1955).

[36] F. Bloch and C. D. Jeffries, *Phys. Rev.* **80,** 305 (1950); C. D. Jeffries, *ibid.* **81,** 1040 (1951); K. R. Trigger, *Bull. Am. Phys. Soc.* **1,** 220 (1956).

[37] J. H. Gardner and E. M. Purcell, *Phys. Rev.* **76,** 1262 (1946); J. H. Gardner, *ibid.* **83,** 996 (1951).

[38] S. H. Koenig, A. G. Prodell, and P. Kusch, *Phys. Rev.* **88,** 191 (1952).

[39] R. Beringer and M. A. Heald, *Phys. Rev.* **95,** 1474 (1954).

[40] I. I. Rabi, S. Millman, P. Kusch, and J. R. Zacharias, *Phys. Rev.* **55,** 526 (1939).

(uncorrected for diamagnetism of the proton resonance sample)

$$g_s/g_p' = \mu_e/\mu_p' = (\mu_0/\mu_p')(\mu_e/\mu_0) = 658.2288 \pm 0.0004. \quad (2.4.5a)$$

Beringer and Heald used a completely different method by making use of a microwave absorption resonance of the Zeeman-split energy levels in hydrogen. The good agreement between the two results is a strong indication of the absence of any unsuspected systematic error in the experiments. Beringer and Heald's result is

$$g_s/g_p' = 658.2298 \pm 0.003. \quad (2.4.5b)$$

The measurement of μ_e/μ_p' may be combined with Gardner and Purcell's measurement of μ_0/μ_p' with the result

$$\mu_e/\mu_0 = 1.001147 \pm 0.000012. \quad (2.4.6a)$$

The error here is completely that due to the uncertainty of Gardner and Purcell's measurement, and this error masks any difference between the two values for g_s/g_p'. This experimental value of μ_e/μ_0 may be compared with the theoretical expression of Karplus and Kroll[40a]

$$\mu_e/\mu_0 = 1 + \frac{\alpha}{2\pi} - 2.973\,\frac{\alpha^2}{\pi^2} = 1.00114536. \quad (2.4.6b)$$

Petermann[40b] has shown that the theoretical expression for μ_e/μ_0 computed by Karplus and Kroll is in error. The calculation of the magnetic moment has been repeated, correct to terms in α^2, independently by Sommerfield,[40c] Petermann,[40d] and Kroll.[40d] The corrected expression is

$$\mu_e/\mu_0 = 1 + \frac{\alpha}{2\pi} - 0.328\,\frac{\alpha^2}{\pi^2} = 1.00115961. \quad (2.4.6c)$$

2.4.1.4. Experiments Involving Planck's Constant. There are several precision experiments which measure quantities involving Planck's constant h. All of these involve a conversion of energy between gamma rays (or X-rays) and other forms of energy.

When electrons are accelerated through a potential difference V, as in an X-ray tube, this energy is available for conversion into radiation when the electron strikes the target and is brought to rest. The shortest wavelength of the emitted radiation corresponds to the conversion of all of the available electron energy into a single photon. If V_a is the total

[40a] R. Karplus and N. M. Kroll, *Phys. Rev.* **77**, 536 (1950).

[40b] A. Petermann, *Nuclear Phys.* **3**, 689 (1957); **5**, 677 (1958).

[40c] C. M. Sommerfield, *Phys. Rev.* **107**, 328 (1957).

[40d] A. Petermann, *Helv. Phys. Acta* **30**, 407 (1957). (Mention of Kroll's recalculation is included in this reference.)

accelerating voltage, λ_g the short wavelength limit of the X-ray spectrum in cgs units, and λ_s the same wavelength on the Siegbahn scale, we have

$$(e/c)V_a = h\nu = hc/\lambda_g = hc/\Lambda\lambda_s \tag{2.4.7}$$

and hence the voltage-wavelength product is directly related to the ratio h/e

$$V_a\lambda_s = hc^2/e\Lambda. \tag{2.4.8}$$

Cohen and DuMond[41] have recently given an extensive discussion of the experimental determinations of this quantity and conclude that the best experimental value is

$$hc^2/e\Lambda = V_a\lambda_s = 12370.8 \pm 0.2 \text{ emu cm.} \tag{2.4.9}$$

A closely related class of experiments involves measurement of the excitation potentials of atoms and a comparison of this value with the wavelength of the corresponding optical transition. Dunnington and his co-workers[42] have measured the excitation potential of the $2p\,^1P_1{}^0$ state of helium by bombarding helium with low energy electrons and noting the energy loss of these electrons when they have enough energy to produce an inelastic scattering in the helium gas. The accuracy of the experiment does not quite approach the accuracy of the short wavelength limit experiments but its theoretical interpretation appears at the moment to be clearer. The experimental result is

$$hc^2/e = 12393.7 \pm 1.8 \text{ emu cm.} \tag{2.4.10a}$$

Nilsson[43] has measured the intensity of K-series X-ray lines as a function of the exciting voltage. There are experimental difficulties in this procedure associated with the problem of accurately identifying points on the excitation curve with "corresponding" features of the K-series X-ray absorption edge. The mean result obtained from a study of four lines was

$$hc^2/e\Lambda = 12372 \pm 5 \text{ emu cm.} \tag{2.4.10b}$$

The de Broglie wavelengths of electrons can be measured by diffraction methods from crystals whose lattice constant is known on the Siegbahn scale. If the accelerating voltage of these electrons is also accurately measured, one can obtain a measurement of $h^2c/em\Lambda^2$, as has been done by von Friesen[44] and with more accuracy by Rymer and Wright.[45] The

[41] E. R. Cohen and J. W. M. DuMond, *in* "Handbuch der Physik—Encyclopedia of Physics" (S. Flügge, ed.), Vol. 35, p. 1. Springer, Berlin, 1957.

[42] F. G. Dunnington, C. L. Hemenway, and J. D. Rough, *Phys. Rev.* **94,** 592 (1954).

[43] A. Nilsson, *Arkiv Fysik* **6,** 544 (1953).

[44] S. von Friesen, *Proc. Roy. Soc.* **A160,** 424 (1937).

[45] T. B. Rymer and K. H. R. Wright, *Proc. Roy. Soc.* **A215,** 550 (1952).

latter result is

$$h^2c/em\Lambda^2 = (2.99515 \pm 0.00060) \times 10^{-6} \text{ emu cm}^2. \qquad (2.4.11)$$

When an electron and positron recombine at rest, two gamma rays are emitted in opposite directions. DuMond, Lind, and Watson[46] measured the radiation emitted following the β-decay of Cu^{64} with a curved crystal spectrometer. These measurements yielded the value

$$h/mc\Lambda = (2.4213 \pm 0.0003) \times 10^{-10} \text{ cm}. \qquad (2.4.12)$$

2.4.2. Evaluation of Experimental Data

In the previous paragraphs, we have listed almost two dozen different experimental results having a bearing on only about a half-dozen independent physical quantities. The problem which remains is how to evolve from these data a consistent set of numbers which will represent the best values of these physical constants. An exact and unequivocal definition[47] of what is meant by the word "best" in this case can be given mathematically in terms of the theory of least-squares, which will not be discussed in detail here.[48] The essence of the mathematical procedure is simple: A weight is assigned to each experimental result; this weight reflects the accuracy with which the experimental number is known. The weight assignable to a linear combination of numbers is readily written in terms of the weights of the individual numbers. If we combine the experimental data in order to extract from it the value of a given fundamental constant we can assign to that number a certain weight. In general, there are infinitely many ways in which a value of a variable may be found as different linear combinations of the given experimental data; to each of these is assigned a weight. That numerical result whose assigned weight is the largest is considered to be the "best" numerical value of the physical quantity. In this way, DuMond and Cohen obtained the values[49] given in Table IV for the fundamental atomic constants.

In Tables III and IV a modified exponent notation has been used. Except in those cases where it is convenient to write the number without an exponent, the exponent has been adjusted so that the standard deviation of the quantity is a decimal with no leading zeros. This notation not only saves space, but makes it easier to evaluate the accuracy of the

[46] J. W. M. DuMond, D. A. Lind, and B. B. Watson, *Phys. Rev.* **75,** 1226 (1949).

[47] E. R. Cohen, *Revs. Modern Phys.* **25,** 709 (1953).

[48] E. Whittaker and G. Robinson, "Calculus of Observations," 4th ed., p. 224. Van Nostrand, Princeton, New Jersey, 1944.

[49] E. R. Cohen and J. W. M. DuMond, *in* "Handbuch der Physik—Encyclopedia of Physics" (S. Flügge, ed.), Vol. 35, p. 83. Springer, Berlin, 1957.

TABLE III. Conversion Factors

A. Mass-Energy	
1 gm	$= 89875.8 \pm 0.2 \times 10^{16}$ erg
	$= 5610.01 \pm 0.11 \times 10^{23}$ Mev
1 electron mass	$= 0.510976 \pm 0.000007$ Mev
1 atomic mass unit	$= 931.143 \pm 0.010$ Mev
1 proton mass	$= 938.213 \pm 0.010$ Mev
1 neutron mass	$= 939.505 \pm 0.010$ Mev
B. Energy-Temperature	
k	$= 13804.4 \pm 0.7 \times 10^{-20}$ erg deg^{-1}
	$= 8616.7 \pm 0.4 \times 10^{-8}$ ev deg^{-1}
R_0	$= 1.98780 \pm 0.00008$ cal mole^{-1} deg^{-1} (thermochemical)
1 ev	$= 11605.4 \pm 0.5$ deg
	$= 23069.2 \pm 0.3$ cal mole^{-1} (thermochemical)
C. Photon-Energy	
1 ev	$= 16020.6 \pm 0.3 \times 10^{-16}$ erg
$E/\bar{\nu} = E\lambda_g$	$= 19861.8 \pm 0.7 \times 10^{-20}$ erg cm
	$= 12397.67 \pm 0.22$ ev-Å
$E\lambda_s$	$= 12372.44 \pm 0.16$ kv-X units
E/ν	$= 6625.17 \pm 0.23 \times 10^{-30}$ erg sec
	$= 6625.17 \pm 0.23 \times 10^{-24}$ erg Mc^{-1}
	$= 41354.1 \pm 0.7 \times 10^{-19}$ ev sec
$\bar{\nu}/E$	$= 5034.79 \pm 0.17 \times 10^{12}$ cm^{-1} erg^{-1}
	$= 8066.03 \pm 0.14$ cm^{-1} ev^{-1}
ν/E	$= 15094.0 \pm 0.5 \times 10^{22}$ sec^{-1}
	$= 24181.4 \pm 0.4 \times 10^{10}$ sec^{-1} ev^{-1}
	$= 24181.4 \pm 0.4 \times 10^{4}$ Mc ev^{-1}

tabulated numbers. It does this at the expense of some familiar exponents, but the decimal point may be easily readjusted by the user.

In all of the tables in this chapter the errors have been quoted as standard errors or standard deviations rather than probable errors. The standard deviation has a more useful analytical significance than the probable error and has the psychological advantage of being larger.

TABLE IV. Fundamental Physical Constants * †

Quantity	Symbol	Value
Velocity of light	c	299793.0 ± 0.3 km sec^{-1}
Gravitational constant	G	$6.670 \pm 0.008 \times 10^{-8}$ dyne cm^2 gm^{-2}
Electronic charge	e	$48028.6 \pm 0.9 \times 10^{-14}$ esu
	$e' = e/c$	$16020.6 \pm 0.3 \times 10^{-24}$ emu
Electron rest mass	m	$9108.3 \pm 0.3 \times 10^{-31}$ gm
Planck's constant	h	$6625.17 \pm 0.23 \times 10^{-30}$ erg sec
	$\hbar = h/2\pi$	$10544.3 \pm 0.4 \times 10^{-31}$ erg sec
Avogadro's constant	N	$6024.86 \pm 0.16 \times 10^{20}$ (gm mole)$^{-1}$
Loschmidt's constant		$2687.19 \pm 0.10 \times 10^{16}$ cm^{-3}
Fine structure constant	$1/\alpha$	137.0373 ± 0.0006
	α	$72972.9 \pm 0.3 \times 10^{-7}$
Rydberg constant	R_∞	109737.309 ± 0.012 cm^{-1}
	$R_\infty c$	$328984.8 \pm 0.3 \times 10^{10}$ sec^{-1}
	$R_\infty hc$	$21795.8 \pm 0.7 \times 10^{-15}$ erg
	$R_\infty hc^2 e^{-1} \times 10^8$	13.60488 ± 0.00022 ev
Universal gas constant	R_0	$8316.96 \pm 0.34 \times 10^4$ erg mole^{-1} deg^{-1}
		82.0820 ± 0.0034 cm^3 atmos mole^{-1} deg^{-1}
		$8207.97 \pm 0.34 \times 10^{-5}$ liter atmos mole^{-1} deg^{-1}
Standard volume of perfect gas	V_0	22420.7 ± 0.6 cm^3 atmos mole^{-1}
		22.4201 ± 0.0006 liter atmos mole^{-1}
Faraday constant	F	$28936.6 \pm 0.3 \times 10^{10}$ esu mole^{-1}
		9652.19 ± 0.11 emu mole^{-1}
Charge to mass ratio of the electron	e/m	$52730.5 \pm 0.7 \times 10^{13}$ esu gm^{-1}
		$17589.0 \pm 0.2 \times 10^{3}$ emu gm^{-1}
Ratio h/e	h/e	$13794.2 \pm 0.2 \times 10^{-21}$ erg sec esu^{-1}
Ratio of proton to electron mass	m_p/m_e	1836.12 ± 0.02
Bohr radius	a_0	$52917.2 \pm 0.2 \times 10^{-13}$ cm
Compton wavelengths		
Electron	$\lambda_e = h/mc = \frac{1}{2}\alpha^2/R_\infty$	$2462.6 \pm 0.2 \times 10^{-14}$ cm
	$\bar{\lambda}_e = \frac{1}{2\pi}\lambda_e$	$38615.1 \pm 0.4 \times 10^{-15}$ cm
Proton	$\lambda_p = h/m_p c$	$13214.1 \pm 0.2 \times 10^{-17}$ cm
	$\bar{\lambda}_p$	$21030.8 \pm 0.3 \times 10^{-18}$ cm
Neutron	λ_n	$13195.9 \pm 0.2 \times 10^{-17}$ cm
	$\bar{\lambda}_n$	$21001.9 \pm 0.3 \times 10^{-18}$ cm
Classical electron radius	$r_0 = e^2/mc^2 = \alpha^3/4\pi R_\infty$	$28178.5 \pm 0.4 \times 10^{-17}$ cm
	r_0^2	$7940.30 \pm 0.21 \times 10^{-29}$ cm^2
	$\frac{8\pi}{3} r_0^2$	$6652.05 \pm 0.18 \times 10^{-28}$ cm^2
		0.665205 ± 0.000018 barn
Zeeman displacement per gauss	$e/4\pi mc^2$	$46688.5 \pm 0.6 \times 10^{-9}$ cm^{-1} gauss^{-1}

TABLE IV (*Continued*)

Zeeman displacement per megacycle (proton resonance response)	$e/4\pi\gamma' mc^2$	$109655.04 \pm 0.10 \times 10^{-7}$ cm^{-1} Mc^{-1}
First radiation constant	$c_1 = 8\pi hc$	$4991.8 \pm 0.2 \times 10^{-18}$ erg cm
Second radiation constant	$c_2 = hc/k$	$14388.0 \pm 0.7 \times 10^{-4}$ cm deg
Atomic specific heat constant	c_2/c	$4799.31 \pm 0.23 \times 10^{-14}$ sec deg
Wien displacement constant	$\lambda_{max}T$	0.289782 ± 0.000013 cm deg
Stefan–Boltzmann constant $= \dfrac{2\pi^5}{15}\dfrac{k^4}{h^3c^2}$	σ	$566.89 \pm 0.10 \times 10^{-7}$ erg cm^{-2} sec^{-1} deg^{-4}
Bohr magneton	μ_0	$9273.1 \pm 0.2 \times 10^{-24}$ erg gauss
Electron moment	μ_e	$9283.7 \pm 0.2 \times 10^{-24}$ erg gauss
Nuclear magneton	μ_n	$5050.4 \pm 0.2 \times 10^{-27}$ erg gauss
Proton moment	μ	2.79275 ± 0.00003 nm
Gyromagnetic ratio for proton in hydrogen	γ'	26752.3 ± 0.4 rad sec^{-1} gauss^{-1} (uncorrected)
	γ	26753.0 ± 0.4 rad sec^{-1} gauss^{-1} (corrected)

* All values are given on the physical scale of atomic weights.

† The corrected calculation of the anomalous magnetic moment of the electron, μ_e/μ_0, appeared in the literature after the manuscript for Part 2 was submitted. It was not possible to incorporate this change into the numerical results of Table IV without extensive revision of the entire section. Table IV, therefore, represents the best values of constants as of 1957. As of August 1, 1958, the following partial list may be considered as the most reliable current values. (The errors assigned to these values, and the correlations which exist between them remains unchanged from those given in Table IV.[49,50])

$$\begin{aligned} c &= 299792.5 \text{ km sec}^{-1} \\ e &= 48027.34 \times 10^{-14} \text{ esu} \\ m &= 9108.21 \times 10^{-31} \text{ gm} \\ h &= 6624.92 \times 10^{-30} \text{ erg sec} \\ N &= 6025.00 \times 10^{20} \text{ (gm mole)}^{-1} \\ 1/\alpha &= 137.0391 \\ \mu_e/\mu_0 &= 1.00115961 \end{aligned}$$

From these, the numerical values of the other quantities listed in Table IV may be found.

It is unfortunate that a final and complete list of the values of the physical constants cannot be written down, but it is a reflection of the vitality of physics that this is so; it is only for a dead science that the final answers can be given.

[50] E. R. Cohen, K. M. Crowe, J. W. M. DuMond, "Fundamental Constants of Physics," pp. 247–271, Interscience, New York (1957).

Note Added in Proof

A joint redefinition of the inch by the United States and Great Britain[51] has eliminated the difference between the United States and British inch. The inch is now defined as

$$1 \text{ in.} = 2.54 \text{ cm exactly.}$$

This defines the foot as exactly 30.48 cm and the yard, 91.44 cm. The meter is then

$$\begin{aligned} 1 \text{ m} &= 39.37007874 \text{ in.} \\ &= 3.280839895 \text{ ft} \\ &= 1.0936132983 \text{ yd.} \end{aligned}$$

[51] *Natl. Bur. Standards (U.S.), Tech. News Bull.* **43**, 1–2 (1959).

3. MECHANICS OF SOLIDS*

3.1. Length and Related Quantities

3.1.1. Length

The measurement of length is of fundamental importance in all experimental work. The selection of units is, of course, arbitrary, and those which are internationally accepted are listed in the preceding chapter. Measurement techniques will be discussed in this section in a classical context because relativistic effects are negligible in all terrestrial observations except those involving atomic or nuclear particles where velocities comparable to that of light are encountered. It should be borne in mind, however, that if two observers moving with relative velocity v measure an interval of length oriented along the relative velocity and at rest with respect to one observer, the interval as measured by the moving observer is shortened by the factor $(1 - v^2/c^2)^{\frac{1}{2}}$. For a detailed discussion of relativistic effects on measurements, reference should be made to any of the excellent treatises on the subject.[1]

3.1.1.1. Measurement Techniques. Most of the length-measurement techniques reduce to a direct comparison of the unknown with a standard and differ only in the varying degrees of complexity of the apparatus which makes this comparison possible. Excellent detailed descriptions of the instruments devised for very precise standard comparisons are available in the literature[2] and will not be discussed further here.

In any such comparison work, correction must be made for temperature effects on the lengths of any solids involved. If T is the temperature, the length of a solid bar is given by $L_T = L_0(1 + \Sigma_n a_n T^n)$, where the a_n are constants determined empirically for any substance and, in general, all terms higher than quadratic in T are neglected. The a_1 and a_2 coefficients are listed in various reference compilations.[3] Length measurements

[1] A. Einstein, "The Meaning of Relativity," 3rd ed. Princeton Univ. Press. Princeton, New Jersey, 1950; P. G. Bergmann, "Introduction to the Theory of Relativity." Prentice-Hall, New York, 1942; L. R. Lieber, "The Einstein Theory of Relativity." Rinehart, New York, 1945.

[2] H. Geiger and K. Scheel, eds., "Handbuch der Physik," Vol. II. Springer, Berlin, 1926; R. T. Glazebrook, "Dictionary of Applied Physics," Vol. 3. P. Smith, New York, 1950; S. Flügge, ed., "Handbuch der Physik—Encyclopedia of Physics," Vol. 24, p. 171. Springer, Berlin, 1956.

[3] "International Critical Tables of Numerical Data: Physics, Chemistry, and Technology." National Research Council (U.S.). McGraw-Hill, New York, 1926–1933.

* Part 3 is by **D. A. Bromley.**

are assumed to be performed at normal atmospheric pressures; in any case, the extremely small compressibility of solids makes pressure corrections unnecessary at all but very high, artificially produced pressures.

Standards of length are conventionally grouped into two types, end and line standards. As these names imply, the length of interest is established in one case as the interval between polished parallel faces of the standard and, in the other, as that between two engraved marks on the surface of the standard.

End standards have been developed to a high degree of perfection, first by the C. E. Johansson Co. in Sweden and later by the National Bureau of Standards in the U.S.A. and the National Physical Laboratory in England, such that the surfaces of any two, if placed together, adhere firmly by "wringing." These "gage" blocks are widely used as secondary standards in precision industrial practice. They have the advantage that standards of any convenient magnitude may be constructed by wringing together an appropriate series of gage blocks; their optically polished surfaces make possible more direct comparisons using optical interferometric studies (see Section 7.2.2).

The international standards of length are of the line variety, as are all the conventional scales and tapes in common use. Intercomparison of these requires the use of more elaborate comparators than in the case of the end standards. In essence, all of these employ optical systems which allow observation of the fiducial marks on the specimen and a direct comparison of the distance traversed by the perpendicular axis of the optical system between fiducial marks, with either a standard or a secondary standard derived therefrom. One of the most familiar of the comparators is that due to Abbe in which a traveling microscope with axis perpendicular to the length being measured moves on a graduated scale with microscopically observed vernier indication.

Much of the instrumental development in this field has been motivated by the requirements of optical spectroscopy, where the comparators are used in the determination of wavelengths from the line positions on spectra obtained either prismatically or with diffraction gratings. The work of Harrison and his associates at Massachusetts Institute of Technology as given in the M.I.T. Wavelength Tables (Wiley, New York, 1939) is representative of this work.

Extremely elaborate precautions have been devised to maximize the precision of all these comparison measurements and are available in the literature.[2]

Under controlled conditions in a given laboratory, comparison of either two end or two line meter standards is possible with a precision of one part in 10^7. Intercomparison of a line with an end standard is less accurate by an order of magnitude. By the use of comparators of the types mentioned previously, calibration of the subdivisions of a meter scale in terms of its whole length is possible to an accuracy of about 10^{-4} mm.

In the most precise applications, the methods of interferometry are used to determine lengths in terms of the wavelengths of certain selected atomic lines (see Section 7.2.3). The classical work of Michelson,[4] who related the length of the standard meter to the wavelength of the cadmium red line, and the confirmation of this measurement by Benoit, Fabry, and Perot[5] establish the entire system of length units in terms of a natural atomic standard. It was found that the number of wavelengths of the cadmium red line in the standard meter was 1,553,164.13 with a precision of one part in 1.6×10^7. It is almost certain that in time the wavelength of a standard line will be accepted as the fundamental unit of length. The cadmium red line is the best primary standard naturally available. It has recently been suggested that the green line (5461 Å) from the radioisotope Hg^{198} is probably somewhat more suitable;[6] this, however, is only one of several possible new standard lines. A definite selection of one of these will not be made before 1960.

3.1.1.2. Triangulation. When direct comparison of known and unknown lengths is impractical, indirect comparison of considerably inferior accuracy is possible by selecting a convenient base line of known length and determining the angles which lines drawn from the ends of the unknown interval to selected points in the base line make with this line. It is then possible to solve the resultant triangles trigonometrically for the desired interval. Clearly, the determination of these angles can be performed using either optical,* radio,† or acoustic methods, with decreasing accuracy. These techniques are basic in all modern navigation and range finding.

3.1.1.3. Time of Flight. Provided that the velocity of propagation of a wave is known, as is the case for electromagnetic and acoustic waves in many media, measurement of the propagation time over a given interval provides a measurement of the interval length. As normally used, radar, for example, is based on the determination of range from the antenna from a measurement of the time of flight of a pulse of electromagnetic radiation to an object and back (see Section 3.4.2). Although the method

* See this volume, Part 7.

† See Vol. 2.

[4] A. A. Michelson, *Trav. mém. bur. intern. poids et mesures, Paris* **11,** S1–237 (1895).

[5] J. R. Benoit, C. Fabry, and A. Perot, *Compt. rend.* **144,** 1082 (1907); *Trav. mém. bur. intern. poids et mesures, Paris* **15,** S1500 (1913).

[6] J. Wiens and L. Alvarez, *Phys. Rev.* **58,** 1005 (1940).

is primarily suited to measurement of long intervals, present timing techniques can measure intervals of the order of 10^{-9} sec with relative ease and so measure distances of fractions of meters. The accuracy of these measurements is in fact limited by the uncertainties in the accepted value for the propagation velocity of light,[7] namely, 299,793.0 $\pm$ 0.3 km/sec (see Section 2.3.2).

3.1.1.4. Extensometers and Strain Gages. These are devices specifically designed to measure small changes in length and are, therefore, widely used in strain measurements (see Section 3.7.1.1.1).

3.1.2. Angle

The determination of angular magnitudes in all instances involves either a direct or an indirect measurement of length, as is suggested by the definition of angle in terms of the dimensionless radian, as the ratio of a subtended circular arc to the corresponding radius.

The simplest such determination involves only the measurement of lengths and their conversion to angle using the standard plane or spherical trigonometric functions.

Perhaps the most widely used method of plane angular measurement involves the comparison of the unknown angle with the standard divisions on a graduated circle, as used in transits and other surveying instruments, for example. Using vernier scales and microscopic observation, angles can be determined conveniently to within from 1 to 10 sec of arc with portable instruments. Perhaps an additional order of magnitude in precision can be gained by using standard circles of large radius, but the method is extremely inconvenient.

Although rarely used for quantitative angular measurements, the ordinary surveyor's level can be calibrated and used for measurement of small angles about the horizontal with an accuracy of the order of 1 sec of arc.

The autocollimator, which is essentially a telescope with a graduated angular scale in its eyepiece and a movable index which can be superimposed on the image and located relative to the graduation, is useful for the determination of small angular separations between objects visible in the field of the telescope.

A familiar instrument specifically designed for angular measurements is the sextant; the name derives from the fact that the graduated scale on the instrument comprises a circular arc of one-sixth of a circle. In use, the angle between the lines of sight to two different reference objects is determined by bringing the image of one, as observed telescopically,

[7] K. D. Froome, *Proc. Roy. Soc.* **A223,** 195 (1954); S. Flügge, ed., "Handbuch der Physik—Encyclopedia of Physics," Vol. 24, p. 1. Springer, Berlin, 1956.

into alignment with that of the other, observed by means of an adjustable mirror system on a half-silvered plate. When this is accomplished, the angular separation may be read directly from the graduations. Accuracies of 30 sec of arc are attainable.

For measurements of very small angular separation, interferometric methods are useful. If a telescope is fitted with a double slit before its objective, the familiar interference fringes are observed for a point object. A second point object subtending a small angle ϕ with the first also produces a fringe pattern, but displaced from it by this same angle. Since the angular separation of the fringes in each pattern is $\phi = \lambda/b$, where λ is the wavelength and b the slit separation, it is clear that by varying b the fringe patterns can be arranged to cancel, and that the condition for this disappearance of the fringes is $\phi = \lambda/2b$, whence ϕ can be determined. This technique has been used to measure angles down to 0.05 sec of arc.[8]

Several widely used angular-measurement devices operate by converting the unknown angle to an appropriate electrical signal. There are obviously many ways of doing this; among the more common are: (1) those in which the angular displacement is converted to a resistance change or analog voltage using a calibrated slidewire; (2) those in which the angular displacement produces a phase shift between balanced ac circuits as in the Selsyn system; and (3) those in which the angular displacement is converted by a cam or rack arrangement into a change in capacitance or inductance in an oscillating circuit with corresponding change in frequency.

All three of these techniques are of an analog nature. In situations where precision greater than 0.1 to 1% is required it is becoming standard practice to change from analog to digital systems; there are many commercially available digitometers, all of which have in common the conversion of the input information in the form of mechanical motion, force, pressure, electrical signals, etc. into a quantized form, usually in a standard code. This has the very great advantage that the information so obtained may be channeled directly to a digital computer for further analysis or processing. Similarly, with standard computer techniques it is simple to arrange to have the computer output automatically set the parameter under study using the same digitometer in a servo linkage. Clearly these same operations are possible using analog signals throughout but with inherently lower accuracy and more limited computational facilities.

The solid angle measures the opening between surfaces, either planar

[8] A. A. Michelson, "Studies in Optics," Univ. of Chicago Press. Chicago, Illinois, 1927.

or nonplanar, which meet in a common point, and is numerically equal to the area on a sphere of unit radius cut out by the defining surfaces. The unit of solid angle is the steradian and equals the solid angle subtended by an area r^2 on a sphere of radius r. Measurement techniques will not be detailed here but are, in general, extensions of the methods given.

3.1.3. Area

If the equations of the bounding curves and of the surface on which these curves are located are known, the ordinary methods of integration are applicable to the determination of the enclosed area, and reference should be made to any standard mathematical text for these methods. In the majority of cases of practical interest, the areas of interest are those bounded by a straight line and a curve, and several approximate formulas have been developed for these situations:

(a) Trapezoidal Rule: If the curve is made up of line segments joining ordinates $y_0, y_1, \ldots y_n$ at corresponding abscissas $x_0, x_1, \ldots x_n$, then the area under this curve is

$$A = \tfrac{1}{2} \sum_{r=1}^{n} (y_r + y_{r-1})(x_r - x_{r-1}). \tag{3.1.1}$$

(b) Simpson's Rule: If the base of the area is divided into an even number of segments of length h, and ordinates erected to the curve from these segment ends are $y_1, \ldots y_n$, then the area is given approximately by

$$A = \frac{h}{3} [y_1 + y_n + 2(y_3 + y_5 + \cdots y_{n-2}) + 4(y_2 + y_4 + \cdots y_{n-1})]. \tag{3.1.2}$$

(c) Gauss' Rule: Assuming a reasonably continuous curve and a base from $x = 0$ to $x = d$, Gauss showed that accuracy approximately double that possible using Simpson's rule could be obtained by using ordinates $y_0, y_1, y_1', y_2, y_2', y_3, y_3'$ at abscissas given by $x_0 = 0.5000d$, $x_1 = 0.9745d$, $x_2 = 0.8707d$, $x_3 = 0.7209d$, $x_1' = 0.0255d$, $x_2' = 0.1293d$, and

$$x_3' = 0.2791d$$

and the equation

$$A = d[0.2090y_0 + 0.0647(y_1 + y_1') + 0.1399(y_2 + y_2') + 0.1909(y_3 + y_3')]. \tag{3.1.3}$$

Many mechanical devices are available for the direct determination of plane areas. The most common and useful of these is the planimeter, one version of which, the Amsler polar planimeter, is shown schematically in Fig. 1. Detailed discussions of the basic theory of this and similar instruments may be found in the literature.[9] In use, O is fixed at any convenient point outside of the area to be measured and, with the rolling wheel and vernier at P set to zero, the tracing point Q is moved completely over the bounding curve. The wheel and vernier reading, when multiplied

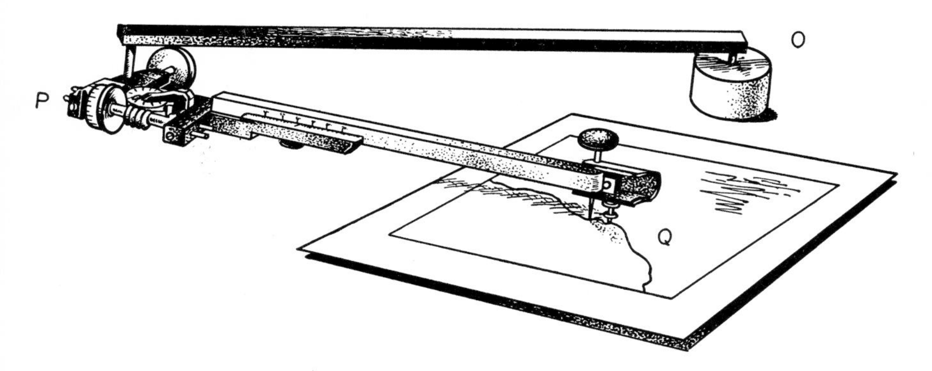

FIG. 1. Schematic drawing of a polar planimeter. O is the fixed point, Q the tracing point, and P the wheel and vernier assembly.

by an appropriate scale factor, which can readily be obtained by repeating the measurement over a rectangular area of convenient dimensions, is the desired area. The maximum error occurs when the direction of motion is at 45° to the wheel axis and, in even relatively crude instruments, is less than 0.1%. For this reason the accuracy of the planimeter reading is almost always limited by the accuracy with which the bounding curves can be drawn.

3.1.4. Volume

Practical difficulties preclude the precision measurement of volume in terms of direct measurements of the linear dimensions involved for all but the simplest geometries. The volume of an irregular solid can be determined by comparing its mass with that of a regular solid of the same material whose volume can be computed from its linear dimensions, or by determining the volume of gas or liquid which it displaces. Thus, measurements of volume are reduced, in general, to those of mass or pressure.

For all regular figures where boundaries are conveniently expressed

[9] See R. T. Glazebrook, "Dictionary of Applied Physics," Vol. 3, p. 450. P. Smith, New York, 1950.

in analytic form, the standard methods of volume integration are applicable. Details are available again in any standard calculus text. There are, however, several general theorems relating to volume determination which will be quoted here for reference.

(a) Cavalier's Theorem: "If in two solids of equal altitude the sections made by planes parallel to and at the same distance from their respective bases are always equal, the volumes of the solids are equal." This theorem is often useful in relating the volume of an irregular solid to that of a regular solid whose volume is calculable.

(b) Theorem of Pappus: "The volume of any solid generated by the revolution of a plane area about an external axis in its plane is equal to the product of the area of the generating figure and the distance its centroid moves."

(c) Prismatoid Theorem: "The volume of a general prismatoid of second order is $V = (h/6)\,(U + L + 4M)$, where h is the altitude, U the area of the upper base, L that of the lower, and M that of the section at altitude $h/2$." By definition, a general second-order prismatoid is a solid such that the area of a section at altitude x above L is given by

$$A_x = ax^2 + bx + L \tag{3.1.4}$$

where a and b are constants.

In fluid-displacement measurements of volume, the fluid normally used is water, and the measurement consists of weighing the volume of water equivalent to the solid. Extensive tables are available for correcting these measurements for buoyancy and temperature effects.[10]

3.2. Mass and Force

3.2.1. Mass

Classically, the concept of mass requires three fundamental experiments: (1) the operational definition of gravitational mass, (2) the operational definition of inertial mass, and (3) the demonstration that the ratios of the inertial and gravitational masses are equal for any two bodies. If, then, unit inertial and unit gravitational mass are assigned to a given international standard, the statement that an object has M units of gravitational mass implies that it has M units of inertial mass and vice versa. No attempt has been made as yet to define a standard of mass in terms of any natural standard such as the mass of a selected atom.

[10] R. T. Glazebrook, "Dictionary of Applied Physics," Vol. 3, p. 784. P. Smith, New York, 1950.

Relativistically, the mass of an object is no longer an invariant but is a function of its velocity relative to the observer's inertial frame of reference and is given by $m = m_0/(1 - v^2/c^2)^{\frac{1}{2}}$, where m_0 is the so-called "rest mass," v is the velocity with respect to the observer, and c is the velocity of light. In ordinary terrestrial measurements on a macroscopic scale, relativistic effects are negligible; in measurements on a microscopic scale, in atomic and nuclear experiments, for example, the velocities encountered are sufficiently great that these effects must always be included.

3.2.1.1. Gravitational Mass. The most familiar instrument for the measurement of gravitational mass is, of course, the equal-arm balance. In this instrument the gravitational attraction on the unknown mass is balanced by an appropriate selection of multiples or submultiples of the internationally accepted unit of mass.

Various techniques have been developed for minimizing the various sources of error which may be present including, as the most prevalent, nonequal length of the arms, friction in the pivots, buoyancy effects due to unequal densities of the standard and unknown masses, and temperature effects.

To minimize temperature effects in precision work, the balance beam itself may be constructed from a material such as Invar with a negligible temperature coefficient. This, unfortunately, introduces the possibility of magnetic disturbances, since an Invar beam may have a slight magnetic moment. Usually, the balance is placed in a constant temperature enclosure and manipulated by remote-control devices.

Since residual friction is always present, the rest point of the balance is always determined by the so-called "method of swings." This consists of noting the pointer reading at reversal for a number of swings. It can readily be shown that the number of such readings must be odd in order to eliminate frictional effects. The rest point is found by taking the arithmetic mean of the means of the readings on each side of the rest point.

To correct for unequal arm lengths, the unknown mass M is placed on the left arm and balanced by a standard mass G_1 on the right; then M is placed on the right and balanced by G_2 on the left. The true value for M is then the geometric mean of G_1 and G_2.

If the density of the medium in which the balance operates (usually air) is a, that of the unknown mass m, and that of the standard mass g, then, at balance

$$M - \frac{aM}{m} = G - \frac{aG}{g}, \qquad \text{or} \qquad M = G\frac{1 - a/g}{1 - a/m}. \tag{3.2.1}$$

Usually $a \ll g$ or m, whence $M \approx G(1 - a/g + a/m)$.

To avoid the necessity for such buoyancy correction, many experimenters have devised methods for carrying out the weighing *in vacuo*.[11]

With a superior quality analytical balance the limit of accuracy in comparing masses is about one part in 10^6. Under very carefully controlled conditions, in standards laboratories, for example, accuracies of one part in 10^8 are possible in comparing kilogram standards.

3.2.1.2. Spring Balances. All spring balances depend for their operation on the balancing of the gravitational attraction on a mass with the elastic restoring forces set up in the deformed support. Provided that Hooke's law holds for the support, i.e. that the restoring force is directly proportional to the deformation (see Section 3.7.1.), the equilibrium deformation is a direct measure of the gravitational mass and may be calibrated to read directly. Since Hooke's law is rigorously valid only for infinitesimal deformations and since creep and hysteresis are always present in the elastic element, the spring balance is inherently not a precision instrument. Its wide use derives entirely from its convenience.

In the measurement of small masses, the quartz torsion microbalance has proved extremely useful. Extremely delicate craftsmanship is required in the construction of these balances, and the reader is referred to the extensive literature for details.[12] When used by a skilled operator, such balances are capable of determining 1 μg of material with 1% accuracy on scale pans weighing 100 mg, corresponding to an accuracy of one part in 10^5.

3.2.1.3. Inertial Mass. An operational definition of inertial mass results from the study of two-body collisions in an inertial reference frame.[13] If the two bodies in question initially have velocities v_1 and v_2 and masses represented by m_1 and m_2, then experiment has shown that, independent of the magnitudes of the velocities used,

$$\frac{\Delta v_1}{\Delta v_2} = -m_2/m_1. \tag{3.2.2}$$

Similarly, using a third object with v_3 and m_3,

$$\frac{\Delta v_1}{\Delta v_3} = -m_3/m_1 \tag{3.2.3}$$

where Δv represents the change in velocity during the collision. In theory, then, having once defined m_1 to be a standard, it is possible by observing

[11] G. W. Monk, *J. Appl. Phys.* **19,** 485 (1948).

[12] H. Carmichael, *Can. J. Phys.* **30,** 524 (1952); B. B. Cunningham, *Nucleonics* **5,** 62 (1949).

[13] E. Mach, "The Science of Mechanics" (translated by T. J. McCormack), 5th English ed., Open Court, La Salle, Illinois, 1942.

such collisions to relate any two inertial masses. This method of relating masses has the advantage that it carries over intact into relativistic theory, whereas gravitational methods do not.

The above equations are, of course, merely statements of the general principle of conservation of linear momentum in binary collisions. It is clear that, from an experimental point of view, the use of a balance is much more convenient than the study of collisions and, as mentioned previously, one of the major emphases in the study of inertial mass lay

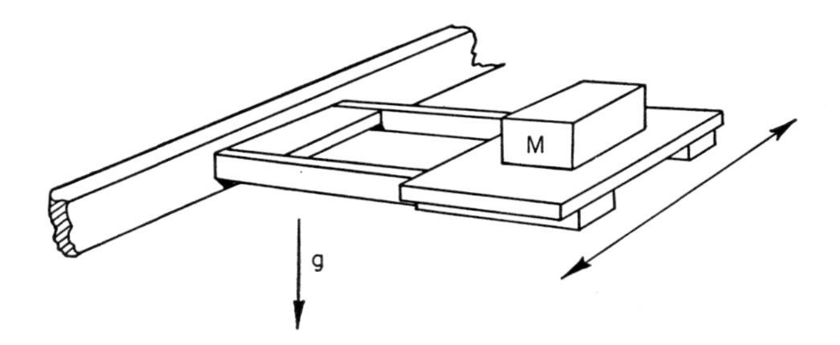

FIG. 2. Schematic drawing of the Schriever inertial balance.

in demonstrating that the ratio of two inertial masses, determined in this way, is identical with the ratio of the corresponding gravitational masses. There has been no experimentally measured difference in these ratios. Consequently, the use of gravitational methods for the determination of both inertial and gravitational masses is justified.

Any device which measures inertial mass directly depends for its operation on the acceleration of the body under study. As an example of such a device, Fig. 2 shows a schematic drawing of the inertial balance used by Schriever[14] consisting of a platform supported for horizontal oscillatory motion on two steel reeds. The masses under study are placed on the platform and the period of horizontal oscillations measured. Empirically, it is found that the inertial mass is given by $M = aT^2 + b$, where T is the period and a and b are constants for the particular balance which must be determined by experiment.

Another type of inertial balance consists of a horizontal platform which is capable of rotation in its plane. The centrifugal acceleration is given by ω^2/r, where ω is the angular velocity and r is measured in the plane from the axis of rotation at equilibrium. Consequently, the elastic restoring forces in the spring balance the product of the inertial mass and this acceleration, and the spring deformation gives the inertial mass, since ω and r can be measured directly.

[14] W. Schriever, *Am. J. Phys.* **5**, 202 (1937).

3.2.1.4. Electromagnetic Methods. These methods[15] are applicable in the region of atomic and nuclear masses and are thus complementary to those already described.*

3.2.1.4.1. CONVENTIONAL MASS SPECTROSCOPES. In general, these are instruments designed to measure the charge-to-mass ratio of ions by noting their deflection in a fixed electric or magnetic field; since the charges are at most small integral multiples of the electronic charge, ze, the masses m can readily be deduced. The first of these instruments,

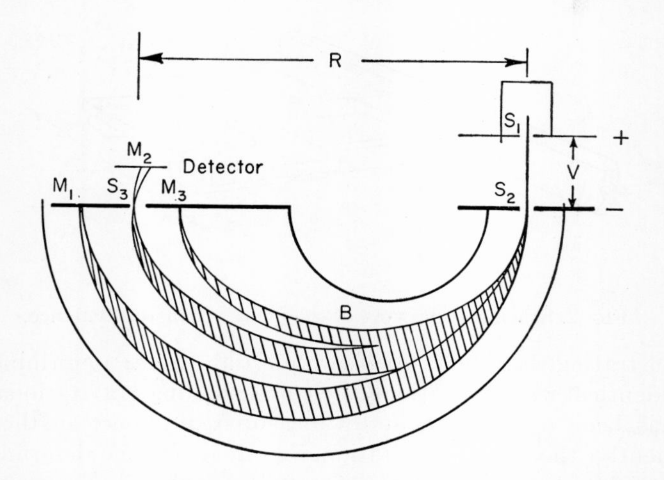

FIG. 3. The Dempster mass spectrograph. The magnetic field B is directed into the plane of the figure. S_1, S_2, and S_3 are slits defining the radius R. As shown, ions of mass M_2 traverse the slits to the detector; those of mass M_1 and M_3 are rejected. Photographic plates are often used to record all masses simultaneously.

using electrostatic deflection, was that of Thompson, with which he was able to separate the neon isotopes of mass 20 and 22 atomic mass units (1.66×10^{-24} gm). Dempster devised one of the first instruments using a 180° magnetic field as shown in Fig. 3. The positively charged ions were produced by electron bombardment of various salts and were then accelerated by a high voltage V to an energy

$$E = \tfrac{1}{2}mv^2 = zVe. \tag{3.2.4}$$

These ions are bent by the magnetic field B into circular paths of radius R given by $Bzev = mv^2/R$, whence

$$m = zeB^2R^2/2V. \tag{3.2.5}$$

* See also Vol. 4.

[15] V. H. Dibeler, *Anal. Chem.* **26**, 58 (1954).

Much effort has been devoted to the design of more precise mass spectrometers with larger solid angles and various special features.[16]

3.2.1.4.2. Time-of-Flight Spectrometers. The major advantage of the time-of-flight spectrometer over the conventional one is that the accuracy depends ultimately on electronic timing circuitry rather than on delicate mechanical alignment and the maintenance of precise magnetic fields.

Goudsmit has suggested[17] a spectrometer which measures the time of flight of an ion traversing a helical path in a magnetic field. In such an instrument, $m = TB/KN$, where T is the time of flight, B the uniform magnetic field, N the number of revolutions, and K an instrumental constant. This instrument has the tremendous advantage that its resolution does not decrease with increasing mass as is the case with conventional spectrometers. An accuracy of one part in 2×10^5 has been attained for heavy masses.

A second variant of the time-of-flight method[18] has the ion source at one end and the collector electrode at the other end of an evacuated tube. High voltage pulses applied to appropriate grids inject a pulse of ions into the tube. The time of flight to the collector is a function of the ion mass, and the entire mass spectrum is displayed for each pulse by connecting the collector electrode to the vertical deflection plates of an oscilloscope.

This is by far the most rapid technique available. Unfortunately, its resolution is still somewhat inferior to that of the previously mentioned methods.

3.2.1.4.3. Resonance Methods. The essential attractive feature of the resonance methods is that they reduce the measurement of mass to the measurement of frequency, which inherently has the highest precision of any mechanical measurement.

An instrument called the "omegatron" has been devised by Hipple, Sommer, and Thomas,[19] which is essentially a miniature cyclotron. At resonance in such a device the applied radio-frequency equals the rotation frequency of the ion in its expanding spiral orbit which, in turn, is a function of the ion mass so that a determination of the resonant frequency for a given ion also gives its mass. Inherently, this type of device would

[16] K. T. Bainbridge, *in* "Experimental Nuclear Physics" (E. Segre, ed.), Vol. I. Wiley, New York, 1953.

[17] S. Goudsmit, *Phys. Rev.* **74,** 622 (1948).

[18] R. E. Fox, W. M. Hickam, T. Kjeldaas, Jr., and D. J. Grove, *Phys. Rev.* **84,** 859 (1951).

[19] J. A. Hipple, H. Sommer, and H. A. Thomas, *Phys. Rev.* **76,** 1877 (1949).

seem to have the possibility of attaining the greatest precision in mass measurements.

As a final example, the study of microwave spectroscopy[20] in many instances allows the determination of mass data with a precision approximately equal to that attainable with conventional techniques. This is particularly important since the method is applicable to elements which are not amenable to study with any other techniques.

3.2.2. Force

Force is here considered as a secondary concept defined as the time rate of change of momentum which, in the case of constant mass, reduces to the product of inertial mass and acceleration. In principle, force may be measured in one of three ways: (1) by comparison with the gravitational acceleration of known masses; (2) by elastic deformation; and (3) by conversion to hydrostatic pressure, which can then be measured directly.

In usual terrestrial laboratory measurements, the acceleration due to gravity, $g = GM_1/R^2$, is assumed constant $= 978.049(1 + 0.0052884 \sin^2 \phi - 0.0000059 \sin^2 2\phi)$ cm/sec^2, with M_1 the earth's mass, R its radius, and ϕ the latitude of the laboratory. A rule of thumb is that this acceleration decreases by about three parts in 10^7 for each meter of additional altitude and that the contribution to the apparent gravitational acceleration due to the earth's rotation is at most 3% at the equator (see Section 2.3.1.1).

3.2.2.1. Force Weighing. This section requires little comment beyond that of the previous section. The beam balance and platform scale are used in a wide variety of types in the measurement of forces produced by dynamometers (see Section 3.5.5).

3.2.2.2. Elastic Deformation Force Measurements. For the operation of all these devices it is assumed that Hooke's law applies. All have in common that they must be calibrated by comparison with standard weights or forces, but beyond that the number of types has been limited only by the imagination of the experimenter. Only a few representative examples will be given here.

The spring balance discussed previously is widely used in dynamometer measurements for force determination and for determination of tension in cables. In the measurement of large values of tension, the slight elongation of a standard metal rod is often used, requiring extensometers or strain gages to detect the deformation. These will be discussed in detail in Section 3.7.1.

[20] W. Gordy, W. V. Smith, and R. Trambarulo, "Microwave Spectroscopy." Wiley, New York, 1953.

In the measurement of tensile or compressive forces, use is made either of the change in length of a standard specimen or, more conveniently, in the diameter of a metal ring to which the force is applied. A standard device is the proving ring, a circular ring provided with projecting lugs for loading and some method of measuring the resultant diametral deformation. In the usual design, a micrometer screw with a large-diameter graduated wheel is used, but many modifications of this for greater sensitivity have been used, including the substitution of wire strain gages (see Section 3.7.1.1.2). By adjusting the radial thickness, diameter, and width of these rings, they may be used for a very wide range of forces from 10 to 10^6 kg.

Gibson *et al.*[21] have described an interesting deformation device for the measurement of small forces in the range 10–12,000 mg. The force is applied to the center of a diaphragm which forms one plate of a parallel-plate condenser in one arm of a high-frequency bridge circuit. It can be shown that, if the diaphragm material has Young's modulus E, Poisson ratio ν, thickness t, and radius a, the deformation D produced by a force F is for small deformations

$$D = \tfrac{3}{4}\frac{Fa^2(1 - \nu^2)}{Et^3}. \tag{3.2.6}$$

This deformation is linearly related to the capacitance and hence to the bridge output. The advantage of methods such as this, and those using resistance strain gages, is that they are applicable to alternating-force cycles up to 100 kc/sec and beyond.

3.2.2.3. Hydrostatic Methods. These methods are all based on the fundamental fact that whereas solids transmit force, fluids transmit only pressure, which is by definition, force per unit area. Consequently, if force F is applied uniformly to an area A of the fluid, a hydrostatic pressure $p = F/A$ is developed in the fluid and may be measured by any of the techniques discussed in Part 4. The coupling between force and pressure may be a diaphragm, bellows, or piston; the latter is preferable because it allows uniform force application over a sharply defined area. The effective area of a corrugated bellows is ill-defined, as is that of a diaphragm.

The familiar hydraulic jack or Brahma press, consisting of a pair of pistons of area A_1 and A_2 ($A_1 \gg A_2$), separated by a confined volume of fluid, is a step-up force transformer acting on these principles. Neglecting friction, a force F_1 applied to the smaller piston results in a force $F_2 = F_1 \times A_1/A_2$ being exerted by the larger. Force amplifications of several orders of magnitude are possible in this way. Conversely, very

[21] R. Gibson, J. Ingham, and L. J. Postle, *J. Sci. Instr.* **30,** 159 (1953).

large forces may be measured by using this same device as a step-down transformer to couple them to lower range measuring devices.

3.2.3. Density

The density of a substance is, by definition, its mass per unit volume. Since, in general, the volume is a function of both temperature and pressure, the density of solids also has appreciable temperature dependence; owing to the very small compressibility of solids, however, their density is essentially independent of the applied pressure.

3.2.3.1 Mass and Volume Measurement. The most obvious method for the determination of density is the direct measurement of both mass and volume. As noted previously, however, the practical considerations which normally reduce a volume determination to one of mass make this a circular process of limited applicability to all but simple geometric shapes. It should be noted, though, that each of the following three methods gives only the density relative to the fluid used. In the original determinations of this density, direct mass and volume measurements were, of course, involved.

3.2.3.2. Hydrostatic Method. This is a simple application of the familiar Archimedes' principle and is applicable to the determination of density greater than that of the fluid (usually water) used in the procedure.

If the solid under study has a measured weight W_a in air and W_W when suspended under water, the density, neglecting air buoyancy and assuming unit density for water, is given by

$$\delta = W_a(W_a - W_W)^{-1}. \tag{3.2.7}$$

If D is the true density of the solid, δ_W the density of water, and σ that of air at the measurement temperature, then it can be readily shown that $D = \delta(\delta_W - \sigma) + \sigma$.

Extensive tabulations of these corrections are available.[22] The major sources of error in this method result from surface tension effects on the suspension used for the weighing in liquid, and from trapped air films on the specimen. The relative importance of these depends on the sample size. Roughly, a minimum volume of 5 ml is required for accuracy of one part in 10^4 in the corrected density.

For solids lighter than water, the measurement can be carried out using an additional weight which is weighed without and then with the sample. In special cases, use of liquids other than water may be advantageous.

[22] R. T. Glazebrook, "Dictionary of Applied Physics," Vol. 3, p. 134. P. Smith, New York, 1950.

3.2.3.3. Fluid Substitution Method. For small samples, a simple method of determining density involves the weighing of a container first filled to a predetermined level with fluid and, second, with the sample immersed in the fluid. The difference of these weights, corrected as above, and the displaced volume give the density.

3.2.3.4. Flotation. This method is particularly applicable to the determination of the density of very small samples, but is limited by the availability of flotation media of appropriate density. The sample density is taken as equal to the density of the liquid which will just float it. A refinement of this technique which is particularly suited to the measurement of small density changes makes use of the fact that the thermal expansion coefficients of solids are, in general, much smaller than those of liquids, which may have the same densities at a particular temperature.[23] The difference in density between two samples may be determined by placing both in an appropriate liquid, which approximates their densities at a convenient temperature, and observing the characteristic temperature at which each sample will remain suspended in the liquid.[24] For a discussion of the techniques applicable to the determination of the density of the flotation media, reference should be made to Chapter 4.1.

3.3. Time and Frequency

3.3.1. Time*

Counting the recurrences of any natural, recurrent phenomenon constitutes a measurement of time, and the criteria which select the phenomena of most use in time measurement are availability and invariability. Up to the present, time standards have been based on astronomical observations of the period of the earth's rotation. Two separate standards, the solar and sidereal, in which the units are the mean solar second and the sidereal second, defined as $\frac{1}{86400}$ of the corresponding days, are in use. The true solar day is the interval between successive transits of the center of the sun's disk over a given meridian; since this varies throughout the year, the mean solar day is referred to an assumed "mean sun" which transits at intervals equal to the average length of the true solar day. The sidereal day is similarly defined as the interval between successive transits of the first point in Aries and is thus a

* See also Vol. 2, Chapter 9.3.

[23] D. A. Hutchison and H. L. Johnston, *J. Am. Chem. Soc.* **62,** 3165 (1940); *Phys. Rev.* **62,** 32 (1942); D. A. Hutchison, *ibid.* **66,** 144 (1944); *J. Chem. Phys.* **10,** 383 (1945).

[24] I. Estermann, W. J. Leivo, and O. Stern, *Phys. Rev.* **75,** 627 (1949).

measure of the earth's rotation period relative to the "fixed" stars. The solar and sidereal units are in the ratio of 1.002783 to 1.

It is now known that the earth's rotation is not sufficiently invariable to allow its use as a primary standard; variations of the order of three parts in 10^8 are known to have occurred during the past 50 years. In addition to variations due to seasonal motions of the earth's air masses, changing mass of polar ice caps, etc., there is a known very small deceleration due to tidal friction. In consequence, intensive research has been devoted to the development of time standards depending only on the natural constants of crystals, molecules, and atoms, which are less subject to small perturbations. Some of these will be described briefly in this section.

No section on time measurement would be complete without reference to relativistic effects. In Newtonian theory time is assumed to be absolute, independent of space and of the state of motion of the observer. Relativistic theories show that this is not the case and that, relative to an interval of time measured in a "fixed" system, the interval in a "moving" system is reduced by the factor $(1 - v^2/c^2)^{\frac{1}{2}}$, where v is the relative velocity of the systems, and c is the velocity of light. The reader is referred to any of the many excellent texts on relativity for detailed discussion of this point.[1]

3.3.1.1. Mechanical Oscillation Measurements. The science of horology has been devoted to the construction of secondary time standards based on oscillating mechanical systems since the time of Galileo, and an extensive literature exists in this field.[25] Familiar examples include the pendulum, balance wheel, escapement devices, etc. The simpler of these depend for their operation on the fact that, in simple harmonic oscillation of infinitesimal amplitude, the period is strictly constant and a function of the restoring forces involved. More elaborate devices such as the cycloidal pendulum have been devised to extend this exact isochronous behavior to finite amplitudes. The practical limit of accuracy of such techniques appears to be about one part in 10^5.

3.3.1.2. Piezoelectric Techniques. During the past 25 years pendulum clocks have been replaced in accurate applications by the quartz-crystal clocks. These utilize the fact that quartz, as well as other naturally occurring crystals such as Rochelle salt and tourmaline, and ceramics such as barium titanate, exhibit piezoelectric behavior in that electrical potential differences develop across the specimen when strained and vice versa. By appropriate selection of the orientation from which the sample

[25] H. Geiger and K. Scheel, eds., "Handbuch der Physik," Vol. II, Chapter 6. Springer, Berlin, 1926; H. M. Smith, *J. Sci. Instr.* **32**, 199 (1955); F. D. Lewis, *Proc. I. R. E.* (*Inst. Radio Engrs.*) **43**, 1046 (1955); G. M. Clemence, *Science* **123**, 567 (1956).

is removed from the original crystal, it is possible to get a wide range of natural mechanical resonant frequencies as well as positive, negative, or zero temperature coefficients. These specimens are incorporated in an electronic oscillating circuit and automatically stabilize the circuit frequency at the resonant frequency of the specimen. The output of this circuit is used to drive a synchronous motor connected to clock dials in the usual way. In normal laboratory work, the frequency of a quartz-crystal clock can be expected to be accurate to about one part in 10^6 at best. In very precise installations, accuracies of the order of one part in 10^7 to one in 10^8 have been maintained.[26] Unfortunately, these clocks are unsuited to measurements over very long periods since the resonant frequency increases uniformly as the crystal ages, necessitating frequent checks which, up to the present, have been against astronomical observations.

3.3.1.3. Microwave Methods.* The use of atomic or molecular systems as the basis for a time standard has always been particularly attractive because the periodicity of their characteristic motions should not be subject to small perturbations. Only recently has radio-frequency technology advanced to the stage where the advantages of these techniques can be exploited, and at present two general types of "atomic clocks" are under development. In one type, molecular rotational and other frequencies are detected by microwave absorption and used to stabilize an oscillating system and thus a clock. In the other, hyperfine structure or resonance effects are detected by atomic and molecular beam techniques and provide a stabilizing signal. For details of circuits and relevant theoretical treatment, reference should be made to review articles in the literature.[27]

The practical limit of stability of the absorption clock at present appears to be about one part in 10^{12}. In the case of ammonia absorption, for example, one of the standard lines used is that corresponding to inversion of the molecule at 24×10^9 cycles/sec. A 10% change in pressure of the absorbing gas would produce a frequency shift of three parts in 10^{11} at 10^{-2} mm Hg pressure. An electric field of $\frac{1}{30}$ volt/cm produces a Stark shift of one part in 10^{12}.

The beam type of atomic clock has the highest possible theoretical accuracy and at present generally utilizes the 9.2×10^9 cycles/sec resonance for a beam of cesium atoms. Essen and Parry[28] have measured

* See also Vol. 2, Part 10.

[26] J. M. Shaull and J. H. Shoaf, *Proc. I. R. E.* (*Inst. Radio Engrs.*) **42,** 1300 (1954).

[27] C. H. Townes, *J. Appl. Phys.* **22,** 1365 (1951).

[28] L. Essen and J. V. L. Parry, *Nature* **176,** 280 (1955).

this resonance frequency to be 9192.631830 Mc (± 10 cycles/sec) relative to a smoothed mean solar time. The National Physical Laboratories[29] in England now employ a cesium standard with a quoted accuracy of one part in 10^{10}. Zacharias at M.I.T. is now developing a cesium-beam standard which is expected to have an accuracy of a few parts in 10^{14}, which will be the most precise of the physical standards by many orders of magnitude. It is certain that, once developed, these clocks will replace astronomical observations in fixing the primary standards of time.

Very recently, a primary frequency standard utilizing cesium-beam techniques has become available commercially.* A minimum stability of five parts in 10^{10} is claimed; this represents an improvement by an order of magnitude over any previously available commercial instrument.

3.3.1.4. Digital Counting Techniques.† The recent development of fast electronic scaling circuits has made possible precise determinations of time intervals in terms of the frequencies of the standard oscillators just described in that the number of oscillations occurring in the interval is counted exactly. The accuracy available with present techniques is about one part in 10^8.

3.3.1.5. Electrical Techniques. One of the most widely used time measurements utilizes the well-known exponential build-up of charge on the capacitance in a resistance-capacitance-series circuit. The charge on, or potential across the capacitor is clearly a measure of the time interval after application of a potential difference to the circuit terminals. This technique is not well suited to time intervals less than about 10^{-6} sec because of inaccuracy introduced by stray circuit capacitances, nor to intervals greater than about 10^2 sec because of inaccuracy due to leakage in the capacitances used.

The conventional oscilloscope is, of course, a most convenient instrument for the measurement of time intervals. The intervals measurable in this way are limited by the sweep speeds available and, in popular commercial models, this range is from 12 sec/cm to 2×10^{-8} sec/cm.[30]

The fact that the propagation velocity of electromagnetic signals along coaxial cables is determined by the geometry of the cable and is considerably less than the velocity *in vacuo* permits the use of so-called delay lines, either of the coaxial distributed type or of the lumped-constant approximation to this. Time intervals are then determined by comparison with the time required for signals to traverse a given delay cable. For

* Available from The National Co. Inc., Malden and Melrose, Massachusetts.

† See also Vol. 2, Chapter 9.1.

[29] J. M. Steele, *Proc. Inst. Elect. Engrs.* (*London*) **B102**, 155 (1956).

[30] O. S. Puckle, "Time-Bases," 2nd ed. Wiley, New York, 1951.

details of these techniques reference should be made to the many excellent reports on the subject.[31]

One of the most widely used modern techniques in short time-interval measurements involves the electronic conversion of time intervals into voltage pulses of proportional amplitude, which are then analyzed using conventional pulse-height analyzers.[32] All these devices use the "chronotron" principle originally used by the Los Alamos group.[33]

3.3.2. Frequency*

It is clear that all the techniques of the preceding section on time measurements are equally applicable here. In addition, many instruments have been developed specifically for frequency measurements. The simplest of these is useful over a restricted frequency range and comprises a selection of mechanical vibrators with resonant frequencies at appropriate intervals in this range. When coupled to a vibrating system, the relative amplitudes of forced oscillation of these vibrators permit a determination of the applied frequency. In the usual form these vibrators are tuned reeds excited either by an electromagnet or by mechanical coupling.

Many commercial instruments are available for measurement of frequency by heterodyne comparison with a standard oscillator. When the beat frequency is reduced to zero, the frequencies are identical. Direct wavelength measurements using slotted transmission lines and absorption measurements of radio-frequency power in a tunable-circuit type of wave meter are particularly useful for frequencies in the range above 10^7 cycles/sec.

In essentially all practical frequency and time measurements, the individual standards are calibrated by comparison with standard frequencies and time markers broadcast by the various governmental agencies. The National Physical Laboratories in England broadcast over WCBR at many frequencies, as does the National Bureau of Standards in the United States over WWV. The time signals broadcast daily by all standards laboratories are intercompared daily by the International Time Bureau in Paris.

* See also Vol. 2, Chapter 9.2.

[31] R. E. Bell, *in* "Beta- and Gamma-Ray Spectroscopy" (K. Siegbahn, ed.), Chapter 19. Interscience, New York, 1955; Z. Bay, *IRE Trans. on Nuclear Sci.* **NS-3,** 12 (1956).

[32] P. G. Koontz, C. W. Johnstone, G. R. Keepin, and J. D. Gallagher, *Rev. Sci. Instr.* **26,** 546 (1955); G. von Dardel, *Appl. Sci. Research* **B3,** 209 (1953); A. B. Van Rennes, *Nucleonics* **10**(7), 20 (1952); **10**(8), 22 (1952); **10**(9), 32 (1952); **10**(10), 50 (1952).

[33] S. H. Neddermeyer, E. J. Althaus, W. Allison, and E. R. Schatz, *Rev. Sci. Instr.* **18,** 488 (1947).

3.4. Point Kinematics

3.4.1. Displacement

The measurement of displacement is essentially that of length, which has been discussed under that heading in Section 3.1.1 and under "Strain Measurement" in Section 3.7.1.1.2. Some additional representative techniques used in displacement measurement will be discussed in this and in the following two sections.

It should be noted that quantum mechanically the ultimate accuracy in the measurement of displacement is related to that of momentum by the familiar uncertainty principle, $\Delta x \cdot \Delta p \gtrsim h/2\pi$, where h is Planck's constant.[34] In all macroscopic measurements this uncertainty is many orders of magnitude smaller than the accuracy of the measurements.

One of the most versatile of the displacement-measuring instruments is the unbonded strain gage. Here the gage wire is supported only at its ends rather than being bonded to a surface as described in Section 3.7.1.1.2; hence the change in resistance is directly proportional to the displacement of these ends, rather than to strain over the interval. These have the advantages of simplicity, slight interaction with the system under study, and wide frequency response.

The differential transformer is one of a number of electrical devices which give an output proportional to displacement.[35] This consists of a single primary winding and two opposed secondary windings on a movable iron core. The position of this core may be adjusted until the secondary voltage is zero and, if it is then connected to the system under study, the secondary voltage will be proportional to the core displacement over a restricted interval.

Many of the commercially available phonographic pick-ups give electrical outputs proportional to displacement. These include (1) piezoelectric units; (2) capacitance units which essentially couple the displacement to be measured to one plate of a capacitor and use appropriate circuitry to measure the resultant capacitance change; and (3) eddy current units which have a high-resistance vane which varies the inductance of a resonant circuit depending upon its position. Displacement of this vane produces a proportional change in the amplitude of electrical oscillation in the circuit.

All of the devices to be described in the following sections for measurement of velocity and acceleration, which give an electrical output, can be

[34] D. Bohm, "Quantum Theory," Part VI. Prentice-Hall, New York, 1951.

[35] A. Miller, *in* "The Right Angle," Vol. 4, No. 1, p. 4. Sanborn Company, August 1956.

directly converted for displacement measurement by the addition of one or two simple resistance-capacitance integrating circuits on the output, just as any of the displacement-measuring devices may be converted to measurement of velocity or acceleration by addition of one or two stages of resistance-capacitance differentiating circuits.[36] It should be noted, however, that after insertion of these circuits static calibration is no longer possible.

The generic title of *transducer* has been applied to measuring devices such as these whose conveniently measurable output changes in proportion to the changes in the parameter under study. As discussed in Section 3.1.2, depending upon the precision required and the possible computations or automatic control applications which may involve the transducer output signal, it may be desirable to arrange for either an analog or digital output and in some cases for both.

3.4.2. Velocity

From its definition, average velocity may be measured by combining independent measurements of length and time. These measurements have ranged from the most simple, using meter sticks and stop watches, to the very elaborate methods used in ballistic research.[37] One of the classic methods involves the use of an Atwood machine, in which a sheet of smoked or sensitized paper is attached to the moving object and is drawn under a sharp point executing uniform harmonic motion perpendicular to the paper motion or under a spark gap which fires at uniform intervals. A similar technique involves high-speed photography of the moving system at a known sequence rate. Displacement during the sequence interval can then be determined directly from the record, and the velocity averaged over this interval can be obtained.

The most widely used ballistic chronograph is that devised by Le Boulengé, which measures the flight time over a known distance in terms of the distance through which a mass falls freely during the same interval. Classically, the interval is defined by having the bullet cut suspending wires; many variations of this device have been used. By maintaining an electrical brush discharge across the gun muzzle, for example, the bullet acquires an electrostatic charge which is sufficient to give a triggering pulse on passing through a wire loop. This pulse can then be used to turn on and off electronic timing devices described previously.

The familiar ballistic pendulum is also widely used for velocity measurements. By requiring conservation of momentum during the impact

[36] F. E. Terman, "Radio Engineering," 3rd ed., Chapter 12. McGraw-Hill, New York, 1947.

[37] R. H. Bacon and W. J. Kroeger, *Am. J. Phys.* **12,** 269 (1944).

and conservation of energy thereafter, it follows that the projectile velocity is given by

$$v = \frac{M + m}{m} \times (2gy)^{\frac{1}{2}}$$

where m and M are the masses of the projectile (with initial velocity v) and of the suspended target mass, and y is the maximum vertical displacement of the center of mass of the combined mass after impact. It should be remembered in this device that, for $M \approx 1000m$ as is usual, only about 0.1% of the projectile kinetic energy is converted to measurable potential energy and 99.9% is lost to thermal energy in the block.

One of the most direct mechanical velocity measurements may be made in terms of the familiar Doppler effect.[38] As is well-known, if a source emitting a wave train either of sound or electromagnetic waves is in motion relative to an observer, the wavelength (or frequency) of the detected waves is a function of the relative velocity of source and observer and of the wave propagation velocity. If the emitted frequency is f_s, a general nonrelativistic expression for the detected frequency, assuming collinear motion, is the following:

$$f_D = \frac{C + v_R}{C - v_R} \cdot \frac{C + v_D}{C + v_s} \cdot f_s \tag{3.4.1}$$

where C, v_R, v_D, and v_s are the velocities of propagation, of a reflector if one is present, of the detector, and of the source, respectively. In this expression v_R is to be taken as intrinsically positive if the reflector is approaching the generator and detector, and negative if retreating. Similarly, v_s is positive if the source moves in a direction opposite to that of the wave train being sent to the detector, and v_D is positive if the detector moves toward the oncoming wave train. The reader is referred to any of the standard treatises for a full treatment of the relativistic Doppler effect. It can readily be shown that if the observer's velocity, v, makes an angle α with the line of sight to the source, the observed frequency is given by

$$f_D = f_s(1 - \beta \cos \alpha + \beta^2/2 + \text{higher terms}); \qquad \beta = v/c. \tag{3.4.2}$$

The $\beta^2/2$ term appears only in the relativistic treatment and has been observed experimentally by Ives at the Bell Telephone Laboratories in a study of the H_β line from hydrogen ions accelerated to 18 kev energy. If the frequencies are known and all but one of the velocities, as is usually

[38] J. O. Perrine, *Am. J. Phys.* **12,** 23 (1944); G. F. Herrenden-Harker, *Am. J. Phys.* **12,** 175 (1944); L. Landau and E. Lifshitz, "The Classical Theory of Fields" (translated by M. Hamermesh). Addison-Wesley, Cambridge, Massachusetts, 1951.

the case, the unknown velocity may be obtained directly from these expressions. This is the basis for the method used in astronomy for determining stellar velocities in terms of apparent shifts in known spectral-line frequencies.

In the so-called CW radar, use is made of the Doppler effect in determining the radial velocities of the targets relative to the radiating antenna. In this system, the phase of the pulse frequency relative to a continuous wave transmission is noted in both transmission and reception, and the phase shift gives the velocity directly. These systems are capable of high accuracy and have been widely used for velocity determinations in contemporary missile research. Such measurements do not give information on the target range since in general the "transmitter on" time is much greater than for conventional pulse radar, and echo time measurements are not feasible. If, however, two separate frequencies are transmitted simultaneously, the difference in phase between the two received, Doppler-shifted signals is directly proportional to the target range.

Katz and his associates at the Westinghouse Research Laboratories developed the first electromagnetic-wave, Doppler-shift, velocity-measuring system for use in military ballistic studies; the familiar traffic control "radar" is a commercial adaptation of this device.

In nuclear research, the Doppler effect for electromagnetic waves again becomes important and has been used for the determination of nuclear velocities in terms of emitted gamma radiation energy (i.e., frequency).[39]

The electromagnetic and magnetostriction phonograph pickups commercially available give an output proportional to the stylus velocity. The three most common types are: (1) electromagnetic (moving iron) units, consisting of a pick-up coil wound on the yoke of a small permanent magnet and a small iron slug arranged to move in the gap of this magnet, giving a coil output proportional to the velocity of this slug; (2) dynamic (moving coil) units in which a coil is arranged to move in a fixed magnetic field, again giving an output proportional to velocity; (3) magnetostriction units using the property of certain ferromagnetic metals such as nickel, iron, cobalt, and manganese alloys, which results in a change of the magnetic reluctance of these metals when strained; hence it is possible to use them to obtain a voltage output proportional to velocity. In the pickup, a section of nickel wire is put under slight tension and mounted rigidly in a magnetic field. Two pickup coils are wound on this wire, and the stylus is fastened at right angles between them.

[39] P. B. Moon and W. Davey, Proc. Birmingham Conf. Nuclear Phys. University Press, Birmingham, 1953.

Velocity meters can be simply and economically constructed by arranging a long permanent bar magnet so that one end of it moves axially in a simple solenoid. Perls and Buchmann[40] have used a $\frac{1}{4}$-in. diam. Alnico V bar magnet 6 in. long to obtain a sensitivity

$$\epsilon/v = 5.10 \times 10^{-4} \times n$$

where n is the number of turns on the solenoid, ϵ the output in volts, and v the magnet velocity in centimeters per second.

As noted in the previous section, any velocity-measuring device with an electrical output can readily be converted for measurement of displacement or acceleration by addition of an integrating or differentiating circuit respectively.

3.4.3. Acceleration

Average acceleration is readily measurable in terms of measurements of length and time, and the Atwood machine measurements mentioned earlier are widely used, as are the methods of high-speed photography. Many refinements have been devised for improving the accuracy of such methods; Pugh, for example, has described a series of measurements using least-squares fitting to the spark record.[41]

One of the most direct measurements of acceleration is essentially a comparison with the gravitational acceleration of a suspended mass. As shown in Fig. 4, if the mass is suspended by a light, rigid rod from a support which is accelerated horizontally, the equilibrium angle θ is a direct measure of this acceleration, since $a = g \tan \theta$.

An equivalent manometer accelerometer is shown schematically in Fig. 5. Since the fluid surface automatically aligns itself normal to the total acceleration vector, the acceleration, $a = 2gd/b = Dg/b$, may be obtained simply from the manometer reading. For precise work, a detailed study of the theory of the instrument is required, incorporating effects due to viscosity and turbulence.[42]

A common piezoelectric accelerometer is constructed by cementing two properly oriented square slabs of piezoelectric material with electrodes between and on either side of the resultant unit, thus giving twice the single-crystal output. This is clamped rigidly at three of the four corners, and the inertial reaction of the unsupported corner develops a force, proportional to the acceleration and normal to the crystal plane. This, in turn, strains the crystal, producing a voltage output proportional to acceleration. The very high natural frequency of the crystal greatly

[40] T. A. Perls and C. Buchmann, *Rev. Sci. Instr.* **22,** 465 (1951).

[41] P. Pugh, *Am. J. Phys.* **4,** 217 (1926).

[42] D. Steen and D. Casey, *Rev. Sci. Instr.* **24,** 1021 (1953).

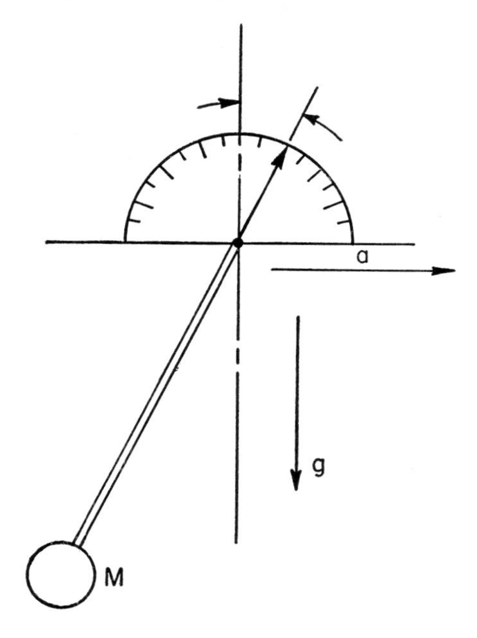

FIG. 4. The pendulum accelerometer. The applied acceleration a is applied horizontally and in the direction shown.

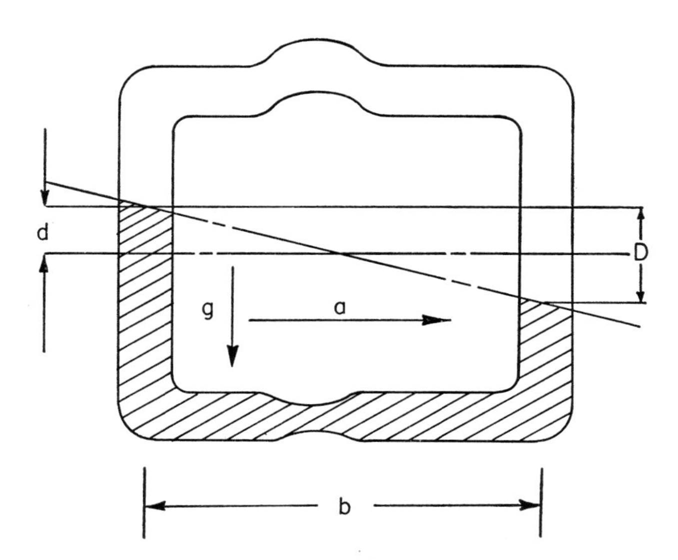

FIG. 5. The manometer accelerometer. d is the fluid displacement in one of the vertical legs corresponding to a horizontal acceleration a applied in the plane of the accelerometer.

extends the useful range of this accelerometer since, as will be shown in Section 3.5.4.3, this natural frequency must be considerably greater than the highest frequency of interest in the measurement. To increase the sensitivity of the device, the free corner is sometimes loaded with an additional mass, but this sacrifices some of the inherent frequency range.

Piezoelectric crystals are available both as natural crystals (quartz and tourmaline) and as synthetics (Rochelle salts, ammonium dihydrogen phosphate and ethylenediamine tartarate). The natural crystals have much greater elastic moduli and are thus suited to high force situations; they have low leakage, are suitable for temperatures below 300°C, and the single crystal sensitivity is typically 100 mv per 1000 kg of applied force.

Synthetic crystals have sensitivities roughly 10^3 times greater than this. Rochelle salts and ammonium dihydrogen phosphate are adversely affected both by high humidity and by temperatures above 100°C. Barium titanate with a small lead titanate admixture, bonded into a ceramic, is widely used because of its high sensitivity and physical stability. Accelerations in the range from 10 to 10^6 cm/sec^2 are conveniently measured at frequencies up to 10^5 to 10^6 cycles/sec; sensitivities as high as 1 millivolt/gram force have been obtained.

A similar device uses a cantilever beam loaded at its outer end with an adjustable mass to permit adjustment of the sensitivity. Bonded wire strain gages are applied to the top and bottom of this beam near its root, where the strains are essentially uniform. Since when one of these gages is in tension the other is in compression, they are connected in adjacent arms of a bridge circuit to double the output and minimize ambient temperature effects. By mounting three of these (or any of the other units which do not depend on the gravitational acceleration for their operation) to measure the component accelerations along three mutually perpendicular axes, accelerations of arbitrary magnitude and direction can be completely determined.

A wide variety of electron tube transducers have been developed in which one or more of the tube electrodes are free to move under the applied acceleration forces. This change in electrode geometry appears as a change in the tube output proportional to the applied acceleration. Large output and high sensitivity are available because of the high internal gain of the tube, and the frequency response is essentially unlimited.

An ingenious optical accelerometer has been described by Lindsay and Masket,[43] which is applicable to the measurement of extremely

[43] J. C. Lindsay and A. V. Masket, *Rev. Sci. Instr.* **25**, 704 (1954).

high accelerations over short displacements such as are encountered in ballistic studies when projectiles are penetrating a surface. In this device, a light beam is projected parallel to the surface to a photocell detector. As the projectile reaches the surface it intercepts this beam but as soon as penetration begins, the back surface of the projectile, which is arranged to have a sharp edge, begins to uncover a fraction of the light beam. The acceleration of the projectile can then be deduced from the shape of the photocell output as a function of time displayed on an oscillograph. This device is operable for accelerations up to the order of 10^8 meters sec^{-2}.

The seismic unit described in Section 3.5.4.3 can be used to measure displacement, velocity, or acceleration, depending upon the damping and frequency range used, and is the basis for many of the commercial transducers.

A very simple mechanical accelerometer, useful in the frequency range below 20 kc/sec, is obtained by connecting a tuned wire to the seismic mass and setting it into vibration by electromagnetic means. Movement of the seismic mass under applied acceleration loads the wire and changes the natural resonant frequency. This results in a frequency-modulated, nonlinear output.

3.5. Rigid-Body Mechanics

For purposes of calculation, a "rigid" body is always considered as an idealized array of mass points held in a completely nonextensible force lattice. The study of rigid-body mechanics is presented with great formal elegance in many excellent treatises on the subject.[44]

3.5.1. Moments of Mass

Several useful concepts are introduced by a consideration of the different moments of this array of mass points relative to some axis such as PQ in Fig. 6. These moments, for a continuous specimen, are then given by $\int r^n \, dm$, where n is integral and is the order of the moment considered. The zeroth moment is simply the algebraic sum of the component masses and is the mass of the body.

[44] E. T. Whittaker, "Treatise on the Analytical Dynamics of Particles and Rigid Bodies: With an Introduction to the Problem of Three Bodies," 4th ed. Dover, New York, 1944; H. Goldstein, "Classical Mechanics." Addison-Wesley, Cambridge, 1951; H. Geiger and K. Scheel, eds., "Handbuch der Physik," Vol. V, Chapters 2, 3, 4. Springer, Berlin, 1926; E. J. Routh, "Treatise on the Dynamics of a System of Rigid Bodies." Macmillan, London, 1905.

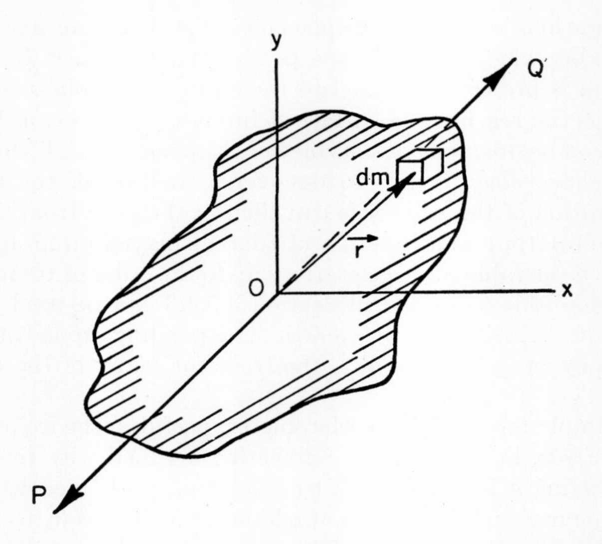

FIG. 6. PQ is the axis to which the various moments of the mass M are referred.

3.5.1.1. Center of Mass. For $n = 1$, $\int \mathbf{r}\, dm = \bar{\mathbf{r}}M$, where $\bar{\mathbf{r}}$ is the position vector of the center of mass of the body. For objects of regular geometry the position of this point can be calculated directly from this definition. For irregular objects, use is made of the fact that in any body the centers of mass and of gravity coincide. For a two-dimensional object, this point is readily located merely by suspending it from different points and noting that, from the definition of center of gravity, the plumb lines dropped from each of these points must all cross at that point. For a three-dimensional object, the same process may be applied by balancing the object on knife edges in different orientations. The point common to the vertical planes through the knife edges is then the center of gravity of the object.

3.5.1.2. Moment of Inertia. For $n = 2$, the moment of inertia of the body *relative to the axis PQ* is obtained: $I = \int r^2\, dm = k^2M$, where k is by definition the "radius of gyration" of the body relative to PQ. Here again, for simple geometries the moment of inertia is directly calculable from the definition. In such calculations, a useful, easily derived theorem originally due to Steiner states that, if the moment of inertia relative to an axis passing through the center of mass of an object is I_0, the moment of inertia relative to any parallel axis displaced a distance d from the first is given by $I_d = I_0 + d^2M$, where M is the mass of the body.

One of the most direct methods for the experimental determination of moments of inertia utilizes the torsion pendulum. In this device, the rigid body is supported by a cylindrical rod, clamped at its upper end so that the axis of interest in the body is coaxial with the support. If the body is rotated slightly from equilibrium and then released, the period of the subsequent oscillations is

$$T = 2\pi \sqrt{\frac{2L}{\pi M R^4}} \times \sqrt{I + \tfrac{2}{3} I_R} \tag{3.5.1}$$

where M is the shear modulus, L the length, R the radius of the support rod, I_R its axial moment of inertia, and I and M the desired moment of

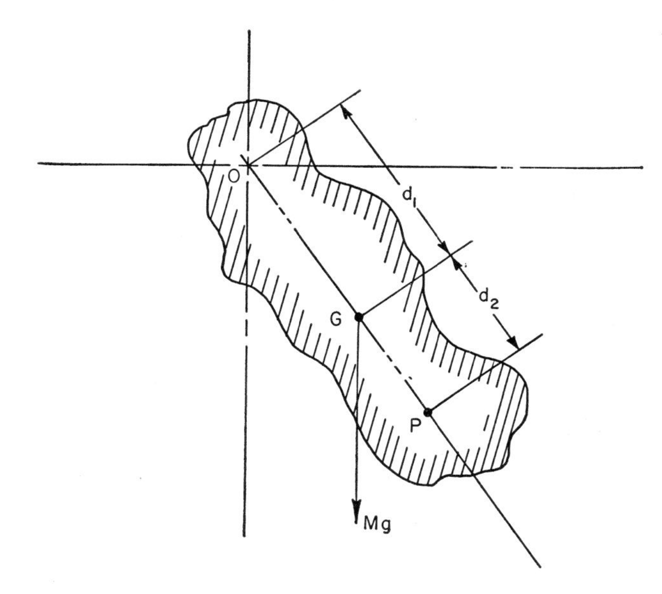

FIG. 7. The compound pendulum. O, G, and P are the centers of suspension, gravity, and oscillation or percussion, respectively.

inertia and the mass of the suspended object. Usually, a comparison method is used wherein the period T_s is determined with the unknown I replaced by a known I_s. This can then be substituted in the above equation to eliminate the support constants. For normal equipment $I \gg I_R$, and

$$I = I_s \frac{MT^2}{M_s T_s^2} \qquad \text{or} \qquad k = k_s \frac{T}{T_s}. \tag{3.5.2}$$

Still another useful method involves using the specimen as a compound pendulum. It is first suspended from knife edges so that it is free to oscillate in the plane of Fig. 7 about a perpendicular axis through O, and the period of oscillation is determined. A second point P is found experimentally such that an axis parallel to that at O, through P, defines a plane which contains the center of mass of the specimen and such that the period of oscillation of the specimen, if suspended about this axis, is identical with that originally determined. The radius of gyration of the specimen about a parallel axis passing through the center of mass can then be simply shown to be $k_c^2 = d_1 \cdot d_2$. The points O and P, related in this way, are the so-called "centers of suspension and of oscillation." It can be readily demonstrated that the center of suspension and the center of percussion of the body coincide. Consequently, if an impulse is delivered to the body at P in a direction perpendicular to OP, no reaction is required at the suspension point O. The location of this point, which often is of paramount importance in the design of physical apparatus, can be obtained simply by the method outlined.

3.5.2. Moments of Force

Only the first moment of force, or torque, is widely used in mechanics and is by definition $\boldsymbol{\tau} = \mathbf{r} \times \mathbf{F}$, where $\mathbf{r}$ is the radius vector from the point of application of the force to the reference point about which τ is defined. The most direct measurement of static torque consists of independent measurements of the applied force and the torque arm length.

3.5.2.1. Torsion Balance. The torsion balance is a device which provides a measurement of applied torque by balancing it against the restoring elastic shear stresses in a rod to which the torque is applied (see Section 3.2.1.2). Assuming that the distortions produced do not exceed the elastic limit of this rod, the deflection produced is linearly proportional to the applied torque, and the balance can be conveniently calibrated by applying a known torque and noting the deflection produced.

3.5.2.2. Dynamic Torque Measurements. It can be shown that at any point on the surface of a cylindrical shaft under torsion the stresses can be decomposed into equal compressive and tensile components directed along 45° helices coaxial with the shaft.[45] Then, if wire strain gages are arranged as shown in Fig. 8, one is subjected to pure compression and the other to pure tension. If these are connected in adjacent arms of a bridge network, the bridge output will be proportional to the torque, assuming that Hooke's law is operable, and may be calibrated by application of a known torque. This arrangement has the advantage that thrust loading, bending, and temperature effects subject both gages to tension or to compression and, consequently, do not affect the bridge output.

[45] A. E. H. Love, "Treatise on The Mathematical Theory of Elasticity," 4th ed. 1927. Macmillan, New York, 1927.

There is, of course, the additional very great advantage that the output of the strain gages may be applied to slip rings, and the torque determined under dynamic conditions as the shaft rotates under load.

An ingenious device which also provides dynamic torque measurements has been reported by Field and Towns.[46] This consists of two disk collars attached to opposite ends of the load-transmitting shaft. Each bears a

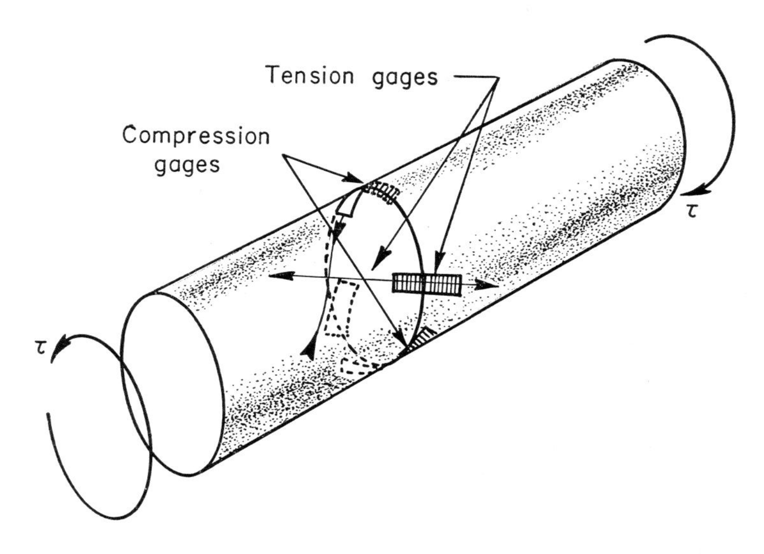

FIG. 8. Application of bonded strain gages to the determination of torque transmitted by a cylindrical shaft.

large number of holes near its periphery, and a beam of light is directed to pass through these holes to a photocell detector as each hole passes the projector. In the unstressed condition the angular positions of these collars are aligned so that pulses from the detectors are in exact time coincidence. As torque is applied to the shaft, the collars rotate relative to one another, producing a phase difference between the output pulses which is proportional to the torsion and, therefore, to the torque. This method has the considerable advantage of requiring no physical connection to the rotating shaft. Accuracies of 0.25% are claimed for this type of device.

A further method which shares the advantage of requiring no physical connection to the shaft utilizes the change in permeability in a ferromagnetic shaft under torsion.[47] In iron, for example, for relatively low

[46] J. F. Field and D. H. Towns, *Electronic Eng.* **26,** 486 and 529 (1954).

[47] R. A. Beth and W. W. Meeks, *Rev. Sci. Instr.* **25,** 603 (1954).

values of stress and magnetic flux the permeability is increased in the direction of applied tension and decreased in the direction of applied compression. By connecting two sets of coils, one with flux return along the tension helix and one along the compression helix previously mentioned, and connecting these in adjacent arms of an inductance bridge, the bridge unbalance due to changing permeability can be calibrated to give an accurate measure of the applied torque. The method has the disadvantage of being applicable only to ferromagnetic shafts.

Additional torque measuring devices will be discussed in Section 3.5.6.

3.5.3. Rotation

Under this general heading, the important measurements are those of angular displacement θ, angular velocity ω, and angular acceleration α. The first of these has essentially been covered in Section 3.1.2, whereas the latter is rarely measured directly and is usually deduced from the measurement of some appropriate linear acceleration, as discussed in Section 3.4.3.

The measurement of angular velocity is often required, and a variety of instruments has been developed for this purpose under the general title of tachometers.

The most direct measurement of average angular velocity requires a measurement of the angular displacement, that is, number of revolutions, and the time. The revolution counter may range from a simple mechanical counter directly coupled to the rotating shaft, to very elaborate devices such as those which reflect light from polished facets on the shaft to photocells, so that the resultant electrical pulse frequency may be determined using the methods of Section 3.3.2. Similar devices use projections on the shaft to vary the inductance or capacitance of adjacent electrical circuits to generate these pulse trains.

3.5.3.1. Tachometers. The simplest and oldest of these instruments is the familiar flyball governor mechanism which is based on the simple conical pendulum. The sensitivity, $dh/d\omega$, is inversely proportional to the cube of the angular velocity ω, where h is the equilibrium height of the conical pendulum.

Magnetic and viscous drag tachometers operate on somewhat similar principles in that in both, the rotational displacement of a disk against a retarding spring load is proportional to the angular velocity measured. In the magnetic type, a bar magnet is mounted to rotate in a plane parallel to that of the pivoted disk, and the drag, proportional to ω, arises from magnetic interaction with eddy currents induced in the pivoted disk. In the viscous drag instrument, both the driven disk and pivoted disk are enclosed in a case filled with fluid. This has the advantages of constancy of calibration and simplicity of construction but has the major disadvantage of a quadratic scale, since the drag is proportional to ω^2. The magnetic instrument is that normally used in automotive speedometers.

As will be shown in Chapter 8.3, the voltage output of an alternator is directly proportional to the angular velocity of the armature and thus provides the so-called "generating tachometer," which has the considerable advantage that the sensing and indicating units need not be adjacent. A similar device of greater inherent accuracy, but considerably greater cost, uses an electronic frequency meter to measure the frequency of the alternator output.

For measurements of the angular velocity of inaccessible units, or where mechanical connection is not permissible, stroboscopic methods are extremely convenient.[48] Stroboscopic devices produce intense, very short light pulses ($\sim$1 μsec duration) at known, variable frequencies. If the pulse frequency is synchronized with the angular frequency to be measured, the rotating member appears stationary and can be made to *appear* to rotate in either direction by slightly changing the pulse frequency.

3.5.4. Oscillation

In a great many of the experimental applications which involve the study or measurement of oscillation, to a first approximation at least, the motion can be considered as simple harmonic, characterized by an acceleration which is a linear function of the displacement. Nonharmonic oscillations with accelerations which are functions of higher powers of the displacement will not be considered here since the majority require mathematical treatment beyond the scope of this chapter. All the equations to follow will pertain, for convenience, to a linear oscillator.

3.5.4.1. Linear Oscillator. The differential equation for the motion represented in Fig. 9(a) is

$$m \frac{d^2x}{dt^2} + kx = 0 \qquad (3.5.3)$$

where k is the force constant of the restoring mechanism.[49] The solution of this equation is the familiar $x = x_0 \sin \omega_0 t$, $v = +\omega_0 x_0 \cos \omega_0 t$, and $a = -\omega_0{}^2 x_0 \sin \omega_0 t = -\omega_0{}^2 x$, and the period of oscillation is

$$T = \frac{2\pi}{\omega_0} = 2\pi \sqrt{\frac{k}{m}}. \qquad (3.5.4)$$

[48] C. W. McLeish and D. H. Rumble, *Proc. I. R. E. (Inst. Radio Engrs.)* **42,** 594 (1954).

[49] N. W. McLachlan, "Theory of Vibrations." Dover, New York, 1951; W. T. Thompson, "Mechanical Vibrations," 2nd ed. Prentice-Hall, New York, 1953.

A more realistic model is that shown in Fig. 9(b) where damping proportional to the velocity has been included. The corresponding differential equation is

$$m \frac{d^2x}{dt^2} + \mu \frac{dx}{dt} + kx = 0 \tag{3.5.5}$$

and the solution is of the form

$$x = Ae^{\alpha+t} + Be^{\alpha-t} \tag{3.5.6}$$

where

$$\begin{aligned} \alpha\pm &= \frac{-\mu}{2m} \pm \sqrt{\frac{\mu^2}{4m^2} - \frac{k}{m}} \\ &= -\gamma \pm \sqrt{\gamma^2 - \omega_0^2} \\ &= -\gamma \pm \omega \end{aligned} \tag{3.5.7}$$

by definition of γ and ω. This motion will be considered in somewhat greater detail since many of the concepts introduced will be required

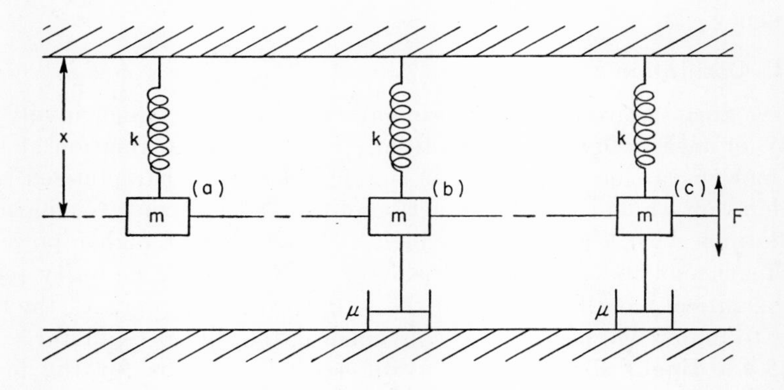

FIG. 9. Schematic representations of the three oscillators discussed in Section 3.5.4.

later in this chapter. Three types of motion are possible depending on the relative magnitude of γ^2 and ω_0^2, that is, the relative magnitudes of resisting and restoring forces. These are the following:

(a) Underdamped Motion: When $\omega_0^2 > \gamma^2$, the foregoing general solution reduces to $x = Ce^{-\mu t} \cos(\omega t + \phi)$, and the motion is illustrated in Fig. 10(a); μ and ϕ are the damping and phase constants, respectively. A convenient measure of the damping present is the "logarithmic decrement" δ, which is defined as the logarithm of the ratio of successive maximum or minimum amplitudes in the oscillation. It can be shown by direct substitution that $\delta = \gamma T$; hence measurement of δ and T suffices to determine μ.

(b) Overdamped Motion: When $\gamma^2 > \omega_0^2$, the general solution reduces to $x = Ae^{-(\gamma-\omega)t} + Be^{-(\gamma+\omega)t}$; the amplitude decays exponentially with time, without oscillation, as shown in Fig. 10(b).

(c) Critically Damped Motion: When $\gamma^2 = \omega_0^2$, the solution reduces to $x = (A + Bt)e^{-\gamma t}$; the amplitude again decays exponentially but more

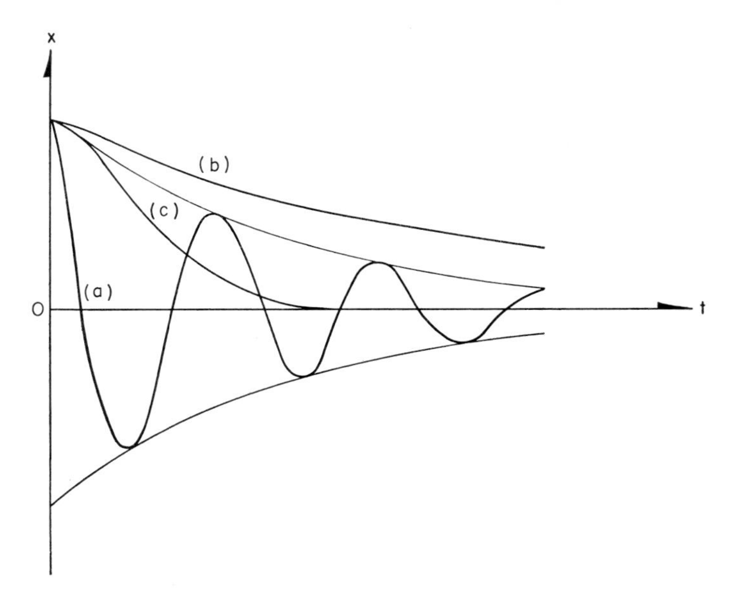

FIG. 10. Amplitude versus time plot for a harmonic oscillator as a function of the applied damping: (a) underdamped, (b) overdamped, and (c) critically damped.

rapidly than in case (b). The critically damped motion is that which is desired in any mechanical system, such as a galvanometer, whose return to equilibrium is governed by frictional forces. If the system is critically damped, the time required to re-establish equilibrium is a minimum, as shown in Fig. 10(c).

3.5.4.2. Forced Harmonic Motion. If, as in Fig. 9(c), a sinusoidally oscillating external force of angular frequency ω' is applied to the oscillating mass, the equation of motion becomes

$$m\frac{d^2x}{dt^2} + \mu\frac{dx}{dt} + kx = F_0 \cos(\omega' t + \theta_0) \tag{3.5.8}$$

and the solution is of the form

$$x = Ae^{-\gamma t}\cos(\omega t + \phi_0) + \frac{F_0}{m}\left\{\frac{1}{(\omega_0^2 - \omega'^2)^2 + 4\gamma^2\omega'^2}\right\}^{\frac{1}{2}} \sin(\omega' t + \theta_0 - \beta)$$

where

$$\beta = \tan^{-1}\frac{2\gamma\omega'}{\omega_0^2 - \omega'^2}. \tag{3.5.9}$$

A and ϕ_0 are the two arbitrary constants characteristic of the solution of a second-order differential equation and are determined by the initial conditions of the system. The first term, which decays exponentially with time, is the "transient" term, and the second is the "steady state" solution. The amplitude of this term is a function of the applied frequency and is a maximum when the applied frequency ω' equals $\sqrt{\omega_0^2 - 2\gamma^2}$. This is the condition of "resonance," and the amplitude at resonance is limited only by the damping.

The total mechanical energy of the damped harmonic oscillator can be obtained by substitution, as $E = E_0 e^{-2t\gamma}$, decreasing exponentially with time at twice the rate of decrease of the amplitude, and the fractional rate of energy decrease is

$$\frac{1}{E}\frac{dE}{dt} = \frac{d(\ln E)}{dt} = -2\gamma.$$

A quantity which characterizes the behavior of an oscillatory system is defined as

$$Q = 2\pi \times \frac{\text{energy stored per cycle at resonance}}{\text{energy dissipated per cycle at resonance}}$$

and from the differential equation this can be shown to be equal to $m\omega/\mu$, as compared to the more familiar $Q = \omega/R$ for resonant electrical circuits. It can easily be shown that $\delta = \pi/Q$ and that $Q \simeq f_0/\Delta f$ for $\delta > 4 \times 10^{-2}$, where f_0 is the resonant frequency of the system and Δf is the frequency interval between frequencies at which the amplitude is 70.7% of its peak resonant value; Δf is the so-called *bandwidth* of the system.

In forced damped harmonic motion, substitution again gives the total mechanical energy as

$$E = \frac{F_0^2}{2m}\frac{\omega'^2}{(\omega_0^2 - \omega'^2)^2 + 4\gamma^2\omega'^2} \tag{3.5.10}$$

which is a maximum when the applied force oscillates at the natural frequency of the system.

3.5.4.3. Oscillation Measurements. The quantities of most general interest in an oscillatory system are the frequency, amplitude, and logarithmic decrement. A method for the determination of the logarithmic decrement has been given; the methods previously described for measurement of frequency and displacement suffice for the other two quantities.

Many of the techniques described in Chapter 3.4 are directly applicable to vibration measurements.

The basic element in almost all vibration measuring instruments is the seismic unit shown schematically in Fig. 11. Depending on the damping and frequency range used, the motion of the suspended mass relative to the base is proportional to the displacement, velocity, or acceleration of the base.

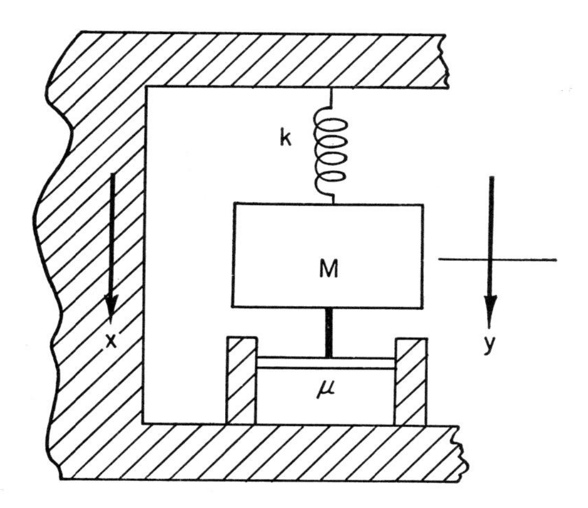

FIG. 11. Schematic drawing of a seismic unit. x and y represent the collinear displacements of the frame and suspended mass, respectively.

If the relative displacement of the mass and base is denoted by $z = y - x$, viscous damping assumed, and the motion of the base assumed to be simple harmonic with $x = X \sin(\omega' t + \theta_0)$, the differential equation for z is

$$m \frac{d^2 z}{dt^2} + \mu \frac{dz}{dt} + kz = m\omega'^2 \sin(\omega' t + \theta_0). \tag{3.5.11}$$

Since viscous damping is built into the instrument, only the steady-state solution of Eq. (3.5.11) is required. For convenience, this solution can be written in the form $z = Z \sin(\omega' t + \theta_0 - \beta)$, and if the dimensionless ratio ω'/ω_0 is represented by r,

$$Z = \frac{r^2 X}{[(1 - r^2)^2 + 4\gamma^2 r^2]^{\frac{1}{2}}} \quad \text{and} \quad \beta = \tan^{-1}\left[\frac{2\gamma r}{1 - r^2}\right]. \tag{3.5.12}$$

Figure 12 shows the behavior of the phase angle β and the amplitude ratio Z/X as a function of r. Since $x = X \sin \omega' t$,

$$v = \omega' X \cos \omega' t = \omega' X \sin(\omega' t + 90°)$$

and $a = -\omega'^2 X \sin \omega' t = \omega'^2 X \sin(\omega' t + 180°)$, the requirement to measure displacement is that Z/X be constant and that there be no phase distortion. From these equations and Fig. 12, if the resonant frequency of the instrument, ω_0, is low compared to ω', r is large, $Z/X \approx 1$ independent of γ, $\beta \approx 180°$ and essentially constant. Such an instrument is called a *vibrometer*, and the measurable displacements are limited by the dimensions of the device.

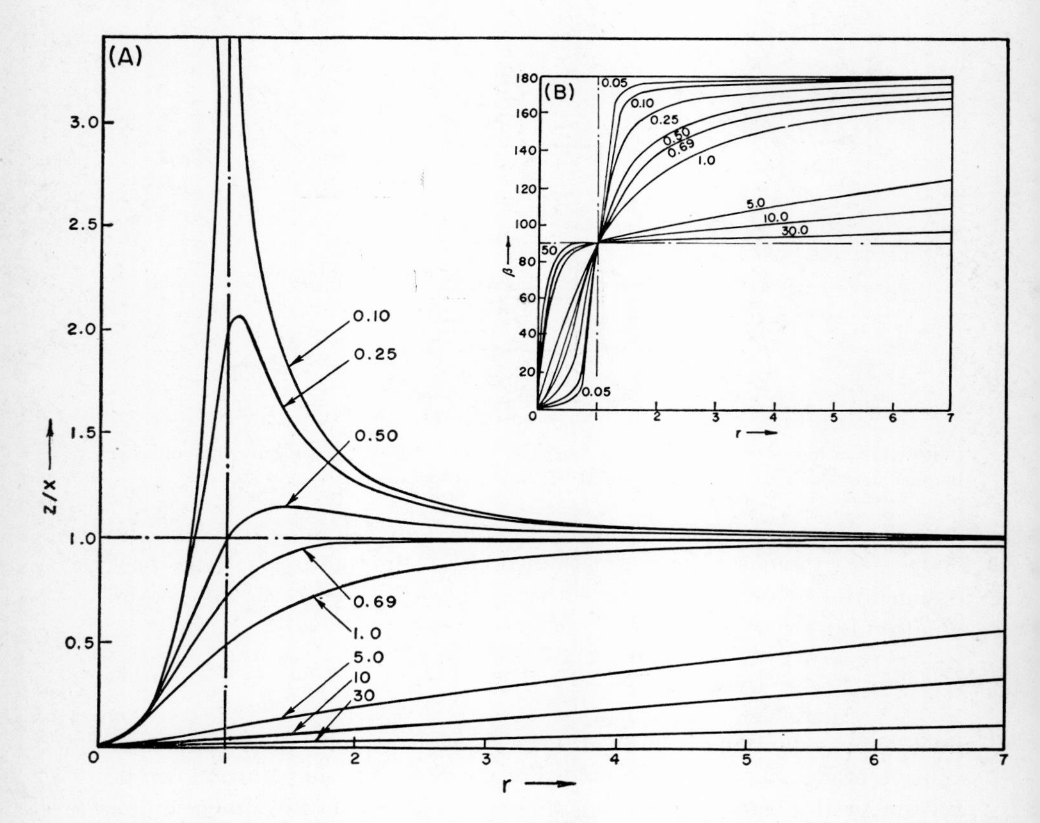

FIG. 12. (A) Amplitude response of the seismic unit as a function of the applied frequency, and (B) phase response of the seismic unit as a function of the applied frequency.

When $\omega_0 \gg \omega'$, the instrument constitutes an *accelerometer*, since the requirement here is that the ratio $Z/\omega'^2 X$ be constant and that the phase be 180° different from that in the vibrometer. From the above equation it can readily be shown that, for $r < 0.4$, this ratio is almost exactly unity for $\gamma = 0.69$, and from Fig. 12 this is seen to give $\beta \approx 0$

over this range as required. This is why the high natural resonant frequencies of the piezoelectric crystals are of particular advantage in constructing the accelerometer described in Section 3.4.3.

Finally, for very large damping, the instrument constitutes a *velocity meter,* since here the requirement is that $Z/\omega X$ be a constant and $\beta \sim 90°$ with respect to either the vibrometer or accelerometer. Reference to Fig. 12 shows that, whereas for the accelerometer range the Z/X curve has the required quadratic shape for $\gamma \sim 0.69$, for very large γ the curve is essentially linear over a wide range of r about $r = 1$, and $\beta \sim 90°$ over the same range.

In many instruments based on this seismic unit, the relative motion of the mass and base is converted to an electric output by making the mass a permanent magnet which moves in a coil attached to the base. In this case the output is proportional to dz/dt. To obtain the displacement, an integrating circuit is required and to obtain the acceleration, a differentiating circuit.

Since the output of all these instruments drops off rapidly with increasing γ, the lowest acceptable damping for the particular application should be used.

3.5.5. Work and Power

Work is defined as the scalar product of force and resultant displacement or torque and resultant angular displacement, and represents the increase in the total mechanical energy of the system to which the force or torque is applied, plus any energy converted to thermal energy and lost to the system as a result of friction. The units of work are, therefore, dyne-centimeters or ergs in the cgs system (or foot-pounds in the English system). Since the erg is impractically small, the joule $= 10^7$ ergs is more widely used.

Power is defined as the rate of doing work and is, therefore, given by the scalar product of force and velocity or of torque and angular velocity. The units of power are joules per second = watts in the cgs system, and horsepower in the English system where, by definition, 1 hp = 500 ft-lb/sec.

Before the equivalence of work and heat were established, the two units of heat defined were (a) the calorie, which was defined as the quantity of heat required to raise the temperature of one gram of pure water from 14.5°C to 15.5°C at atmospheric pressure and (b) the British thermal unit (Btu), defined as the quantity of heat required to raise the temperature of 1 lb of pure water from 59.5°F to 60.5°F, again at atmospheric pressure.

The internationally accepted "mechanical equivalents" are that 1 calorie = 4.1860 joules and 1 Btu = 778.26 ft-lb = 251.996 calories (see Section 6.2.1.2).

3.5.5.1. Calorimetry. Perhaps the oldest method of measuring work and power is calorimetry, in which the work done is converted to heat and is measured by observing the temperature rise of a given volume of

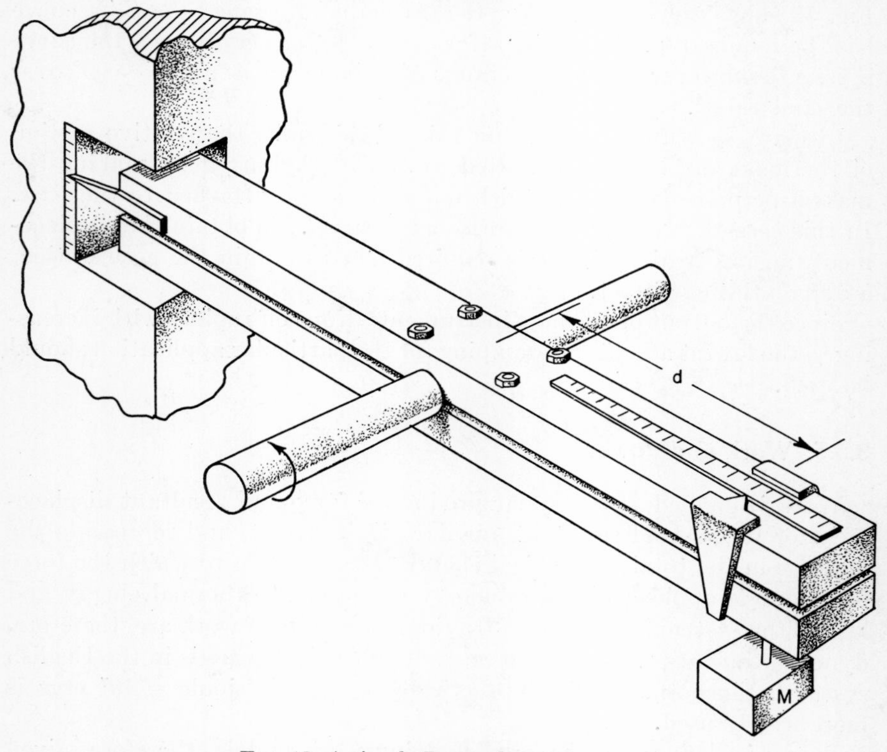

FIG. 13. A simple Prony dynamometer.

matter, usually water. Calorimeters are roughly classifiable as either isolated or steady state; in the former a fixed volume of water and in the latter a steady flow of water absorbs the energy and experiences a measurable temperature rise. Extremely elaborate devices have been developed to minimize errors due to energy loss from the system by radiation or convection, and for details reference should be made to Chapter 6.2 and to any of the standard texts in the field.[50] Such a measurement of work and a time measurement give the average power directly.

[50] M. Saha and B. N. Srivastava, "Treatise on Heat," Indian Press, Calcutta, India, 1950.

The usual methods of work and power measurement in electrical circuits will be discussed in a later chapter.

3.5.5.2. Dynamometer Measurements. One of the most commonly required measurements of power and work is that of the output of various types of motors and engines where the output is a torque applied to a shaft, and the measurement consists of a determination either of the product $\tau\theta$ or $\tau\omega$ to give either work or power.

As noted earlier in Section 3.5.2.3, dynamometer instruments are widely used in these measurements.[51]

3.5.5.2.1. ABSORPTION DYNAMOMETERS. A large number of these instruments are adaptations of, or improvements on, the original Prony dynamometer, which consisted simply of two symmetric wooden beams clamped to the rotating shaft, with the clamping friction adjustable by means of screws as shown in Fig. 13. A mass M is moved along one of the beams until the exerted torque is just sufficient to hold the beams stationary at some index point midway between two stops which prevent rotation of the beam with the shaft. If the beam system is initially balanced relative to the rotating shaft, and the distance from the shaft axis to the point of suspension of M at balance is d, then the torque τ exerted by the shaft is Mgd. The power is then given in ergs by the product $Mgd\omega$. Many adaptations have been directed at controlling the constancy of the clamping friction which is the limiting factor in this device. A somewhat similar device due to Kelvin is shown in Fig. 14, in which ropes or cables are wound on a drum as shown and connected to a spring balance and suspended mass. Here the brake horsepower in English units is given by

$$\frac{(Mg - P)R\omega}{33{,}000}$$

where P is the balance reading in pounds, R the radians in feet, and M is the mass in slugs (see Chapter 2.2). Since Mg/P is $e^{\mu\theta}$, where μ is the coefficient of friction and θ is the arc of contact measured in radians, P can be made much smaller than Mg by increasing θ. This results in a very accurate device, since errors in P are then negligible; additional advantages are those of simplicity and economy. In the metric system, the power would be $(Mg - P)R\omega \cdot 10^{-7}$ watts, where M is given in grams, P in dynes, and R in centimeters.

Hydraulic dynamometers are constructed in much the same way as the now familiar automotive torque converter in that two opposed

[51] R. J. Sweeney, "Measurement Techniques in Mechanical Engineering." Wiley, New York, 1953.

turbine wheels are arranged to run in a common fluid bath. One wheel is attached to the rotating shaft and the other fixed to a casing which is free to turn relative to the shaft, but whose rotation is prevented by some adaptation of the lever and mass system just described. For a given

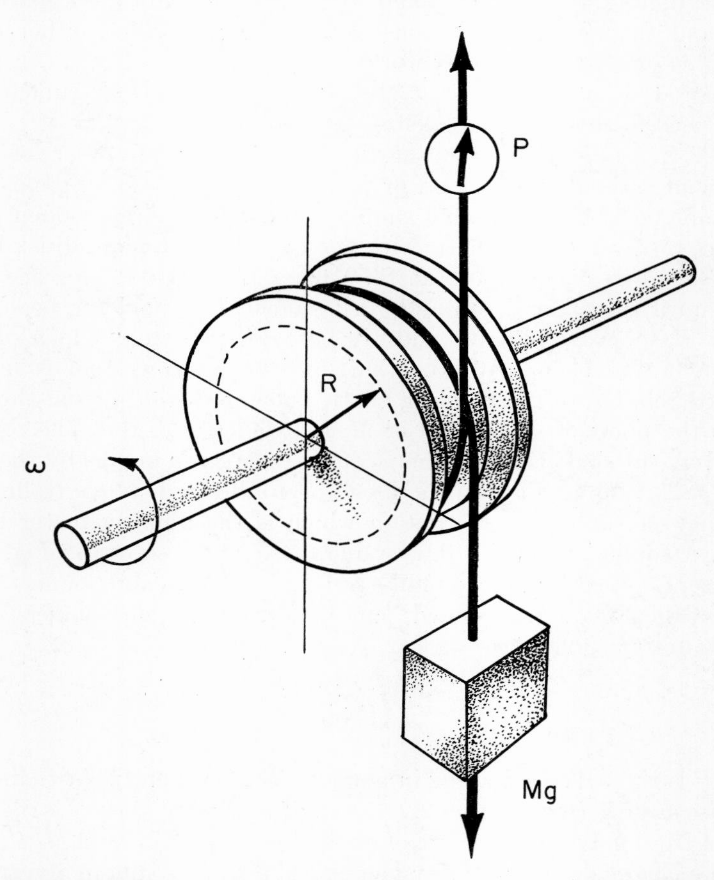

FIG. 14. The Kelvin rope dynamometer. P is a spring balance inserted in the rope to indicate the applied upward tension.

angular velocity and instrument, the torque transmitted to the casing is proportional to the amount of fluid present. This torque is also proportional to the square of the angular velocity over a very wide range. Devices of this type have been used in the testing of marine engines rated up to 2000 hp at 90 rpm.

In the eddy-current dynamometer, a metallic disk coupled to the shaft rotates in a magnetic field established by a field structure which again is free to rotate relative to the shaft but is constrained by a "torque weighing" arm. The rotation of the disk induces eddy currents in the disk and thus dissipates energy which subsequently is converted to heat and removed by appropriate water cooling. The power absorption of this device is proportional to the square of the magnetic-flux density through which the disk rotates and is directly proportional to the angular velocity.

3.5.5.2.2. TRANSMISSION DYNAMOMETERS. Any of the dynamic methods for torque measurement described in Section 3.5.3 may be used in a transmission dynamometer. If the torque and the angular velocity are known, the power delivered by the shaft in question is given by their product.

The most versatile of the modern dynamometers is the "electric cradled" type. This utilizes the fact that the torque acting on the armature of an electric motor is necessarily equal to that acting on the stator and frame. If the frame is mounted so that it would be free to rotate about the armature axis except for the usual "torque weighing" arm and restraints, then the torque indicated by the arm and the angular velocity give the power transmitted by the armature.

These instruments are widely used in the automotive industry.

3.6. Friction

The study of friction is one of the oldest in physics, and only recently has anything like an adequate understanding of the phenomenon emerged.[52] The earliest theory of friction is that of surface roughness developed by Coulomb, who assumed that the work expended against friction is due to the lifting of surface irregularities of one of the sliding surfaces over those of the other. It has been shown that this process can account for at most a few per cent of the observed friction.[53] The molecular theory of friction developed by Tomlinson,[54] which attributes friction to irreversible disturbance of the surface molecules, has never received convincing experimental support.

The most successful theory is that of localized adhesion or welding due to Bowden and Tabor[55] and others. In this theory it is assumed that, at

[52] J. T. Burwell and E. Rabinowicz, *J. Appl. Phys.* **24,** 136 (1953).

[53] C. R. Lewis and C. D. Strang, *J. Appl. Phys.* **20,** 1164 (1949).

[54] G. A. Tomlinson, *Phil. Mag.* [7] **7,** 905 (1929).

[55] F. P. Bowden and D. Tabor, "The Friction and Lubrication of Solids." Oxford Univ. Press, London, 1950.

the actual contact points between the surfaces, such high pressures develop that cold welding occurs. When a tangential load is applied to the interface, at first elastic deformation of these junctions and of the supporting media occurs until the shear strength of the junctions is reached. Shearing of the junctions follows, and the coefficient of friction drops from its high static value, corresponding to the shear strength of the strong junctions formed during static loading, to its lower dynamic value, corresponding to the weaker junctions continuously being formed and sheared.

A further effect, first noted by Ernst and Merchant,[56] known as "snowballing" or "galling," results from the deformation of the junction substrate, in that areas of the surface which otherwise would have remained separate are brought into contact and can greatly increase the coefficient of friction.

The function of a lubricant is thus twofold: first, it must prevent the formation of large metallic junctions and, secondly, it must prevent galling of the surfaces. With adequate lubrication there is little difference between the static and dynamic coefficients; with very slight lubrication the reduction in friction is almost entirely due to the elimination of galling.

3.6.1. Measurement Techniques

Almost all measurements of the coefficient of friction consist of simple determinations of the tangential force required across the interface of the surfaces in question, either to initiate sliding motion or to maintain this motion with uniform velocity once started, depending upon whether the static or dynamic coefficient is desired. From elementary mechanics there are a variety of detailed methods for performing the measurements. In that used by Eagleson,[57] for example, shown schematically in Fig. 15, the relative velocity of the surfaces is automatically averaged. If M_2 falls a distance h from rest before striking a support, and M_1 moves an additional distance d before coming to rest, it follows that

$$\mu_k = \frac{M_2 h}{M_1 h + (M_1 + M_2) d}$$

or, if $M_1 = M_2$,

$$\mu_k = \frac{h}{h + 2d} \tag{3.6.1}$$

The familiar inclined-plane experiment is widely used for these meas-

[56] H. Ernst and M. E. Merchant, Proc. M.I.T. Sum. Conf. Surf. Finish, 1940.

[57] H. V. Eagleson, *Am. J. Phys.* **13**, 43 (1945).

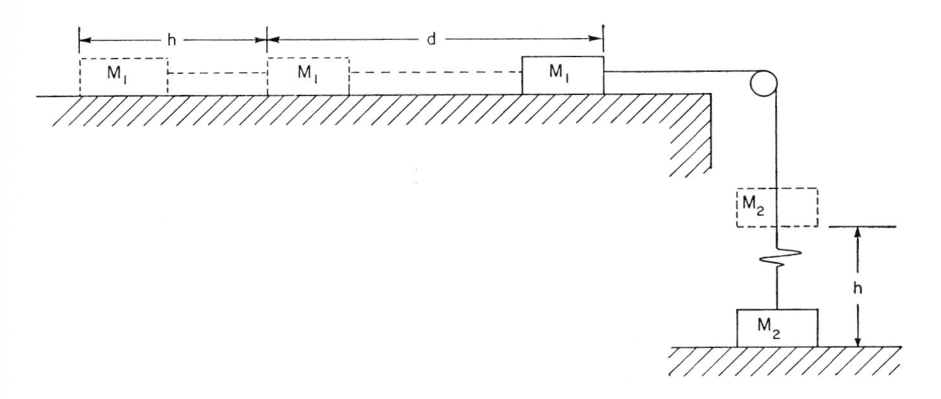

FIG. 15. Schematic representation of the Eagleson technique for determination of the dynamic coefficient of friction.

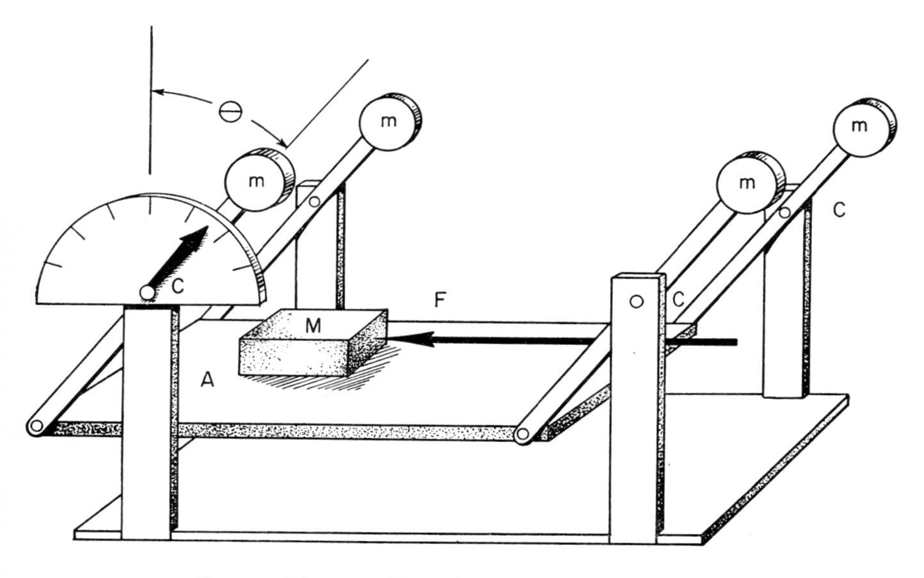

FIG. 16. Direct-reading friction meter due to Gough.

urements since, for a mass sliding on the plane in dynamic equilibrium, $\mu_k = \tan\theta$, where θ is the plane angle. Šimon *et al.*[58] have used this method for measurement of μ_s by noting the angle at which slight vibration of the plane would initiate sliding motion. Rabinowicz[59] has used the inclined plane to study the transition from μ_s to μ_k as a function of

[58] I. Šimon, H. O. McMahon, and R. J. Bowen, *J. Appl. Phys.* **22,** 177 (1951).
[59] E. Rabinowicz, *J. Appl. Phys.* **22,** 1373 (1951).

displacement by using a plane angle θ, such that $\mu_s \gg \tan \theta > u_k$, and by noting the distance which a small spherical mass must roll down the plane to initiate sliding of a larger mass initially at rest.

An ingenious direct-reading friction meter is described by Gough[60] and is shown schematically in Fig. 16. A is a platform carried by four identical levers, counterweighted by masses m so that the system is in neutral equilibrium and the center of mass always lies in the plane of the pivots. The test surface is placed on the platform and a horizontal force applied. Neglecting pivot friction, the levers give the direction of the resultant force acting at the interface, and the coefficients of friction μ_s and μ_k are given by the tangent of the angle between the vertical and the levers at the moment when slipping starts and when the surfaces are in uniform relative motion respectively.

A number of tabulations of coefficients of friction for various surfaces exist in the literature.[61]

3.7. Deformation*

3.7.1. Elastic Deformation

The criterion of elasticity is the ability of a sample, which has been deformed by an external force, to return exactly to its original dimensions upon removal of applied forces. In the limit of small deformations, almost all solids satisfy this criterion, and in this section this will be assumed. The external forces may be one of, or a combination of, three general types: tensile, compressive, or shearing forces. In an elastic sample, internal restoring forces develop to balance the applied forces at the equilibrium deformation, and the stress is defined to be the elastic restoring force per unit area at any cross section within the sample. The strain is a measure of the magnitude of the deformation produced and is usually defined as change in length per unit length and thus is dimensionless. In a perfectly elastic sample, stress and strain are linearly related, and the constant of proportionality which characterizes the sample is the elastic modulus. As shown in Fig. 17, there is a differentiation between systems involving simple tensile and compressive stresses and shear stresses. If the applied force is F and the area of section A, then in both cases the stress $\sigma = F/A$. In Fig. 17(a), if the original length is L and the change in

* See also Vol. 6, Chapter 4.2.

[60] V. E. Gough, *J. Sci. Instr.* **30,** 345 (1953).

[61] E. C. Pike, *J. Roy. Aeronaut. Soc.* **27,** 1085 (1949); O. W. Eshbach, "Handbook of Engineering Fundamentals," 2nd ed., pp. 2–52, 4–54. Wiley, New York, 1952.

length is ΔL, then the strain ϵ_L is by definition $\Delta L/L$. The modulus of elasticity, or Young's modulus, is by definition $E = \sigma/\epsilon_L$ and has the units of force/unit area. In Fig. 17(b) the shearing strain is, by definition, $\epsilon_s = \Delta L/H \approx \phi$ (in radians), where ϕ is the shear angle and the shear modulus of elasticity, or modulus of rigidity G, is by definition σ/ϵ_s.

In analogous fashion, the bulk modulus B is defined as the negative of the ratio of applied pressure to the relative volume change produced.

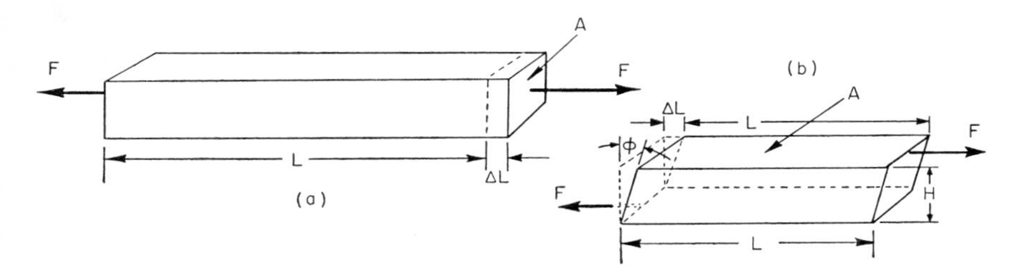

FIG. 17. Solid prism under (a) tension and (b) shear.

Thus, $B = -pV/\Delta V$, where the negative sign is included to allow the tabulation of positive quantities. The compressibility k is defined as the reciprocal of the bulk modulus and is a more widely used constant.

The final constant customarily used to define the properties of an elastic body relates to the experimental observation that, if the linear dimensions are deformed in one direction, those in the transverse directions are also changed. It has been empirically observed that the ratio of the transverse to the longitudinal changes in length is constant and is the so-called Poisson ratio ν.

It will be noted that for simplicity in the above paragraphs linear systems have been assumed throughout in that stress and strain have been assumed to be collinear. In anisotropic substances this is not the case and, in a general treatment, both stress and strain are represented by tensors of rank three.[62] In this treatment it follows directly that the four quantities defined above are not independent and that the following relationships hold

$$\begin{aligned} G &= E/2(\nu + 1) \\ k &= 3(1 - 2\nu)/E. \end{aligned} \tag{3.7.1}$$

It is obvious from this equation that, since k must be positive, $\nu \lesssim 0.5$

[62] S. Timoshenko, "Strength of Materials," Parts I and II. Van Nostrand, Princeton, New Jersey, 1955; R. V. Southwell, "Theory of Elasticity," Oxford Univ. Press, London and New York, 1941; J. C. Lindsay and A. V. Masket, *Rev. Sci. Instr.* **25,** 704 (1954).

and that $\nu = 0.5$ corresponds to an incompressible substance. For most solids ν is found to lie close to 0.3.

3.7.1.1. Strain and Stress Measurements.* Stress as such is rarely measured and instead is inferred from strain measurements or calculations. The number of experimental devices which have been invented for the measurement of strain is legion, and what is hoped to be a representative sampling of these will be included here.

3.7.1.1.1. Extensometers. This is a generic term covering a wide variety of devices for determining the change in length of a given sample.[63] For use with tension testing machines, the sample is in the form of a cylindrical rod; in the unstrained state, gage points are selected on this rod some convenient distance apart, and the extensometer is used to measure the change in the spacing of these gage points under tension. In one of the most common forms, collars are attached rigidly to the test rod at the gage points, one of which carries either a fine screw micrometer or dial micrometer which just contacts a fixed rod attached to the other. Using electrical circuit continuity to indicate the point of contact, direct measurements of elongation are possible to an accuracy of perhaps one part in 5×10^4 at best. In a more sensitive version of this type, a lever is pivoted at one of the clamps; the short end is pivoted a fixed distance from the second clamp, and the long end carries a hairline indicator in the field of a microscope fitted with an optical micrometer scale in its eyepiece. The mechanical magnification of the sample elongation can be computed directly from the dimensions of the lever system. The most sensitive variant of this device uses an optical lever to increase the mechanical magnification of the sample elongation. A common model consists of a rigid bar with a fixed knife edge at one end held to one gage point by spring tension. The other end of the bar is supported away from the gage point by a small member with two knife edges, one against the test rod and the other against the bar. As the distance between gage points changes, this member rotates, as does a small mirror attached to it. This rotation, and hence the elongation, is determined by noting the motion of a light beam reflected from the mirror. By choosing the light paths appropriately, any required sensitivity may be obtained.

The important point to note is that all extensometers give an average strain, averaged over the distance between gage points.

3.7.1.1.2. Strain Gages. In almost all practical applications the dimensional changes occurring at or near the surfaces of the sample are those of most significance. Wire strain gages are particularly suited to

* See also Vol. 6, Section 4.2.1.

[63] A. H. Sully, "Metallic Creep and Creep Resistant Alloys," Chapter 2. Interscience, New York, 1949.

measurement of such surface strains. All are based on the well known strain-resistance characteristics of long wires, and for details of construction reference should be made to the extensive literature in the field.[64] If ρ is the resistivity, L the length, and A the cross sectional area of the wire, then the resistance $R = \rho L/A$, and, by differentiating this expression and noting that $\delta A/A = -2(\delta L/L)\nu$, where ν is Poisson's ratio for the wire material, the fractional resistance change is given as

$$\delta R/R = \epsilon(1 + 2\nu) + \delta\rho/\rho$$

where ϵ is the strain.

This would indicate that all materials should have the same strain sensitivity, defined as $\delta R/\epsilon R$. This is found not to be the case and indicates that resistivity changes with strain are more important than dimensional changes.

In use, the resistance wire is first bonded to a paper backing in some appropriate folded pattern to obtain a greater, more easily measured total resistance change than would be possible with a single strand, and this in turn is bonded to the surface under study. The terminals are connected to a bridge circuit, usually a simple Wheatstone bridge and, after calibration, the bridge output gives the strain directly. Calibration is usually carried out by mounting the gages near the foot of a cantilever beam and calculating the strain corresponding to given distortions of the free end or by direct comparison with an extensometer of the type described above.

The electrical strain gage has many noteworthy advantages: (1) in contrast to extensometers, true strain rather than average strain is determined; (2) because of their low inertia, inductance, and capacity, they will measure alternating strains at frequencies as high as 50 kc/sec without special effort; (3) the fractional resistance change is linear with strain up to about 0.2% strain; (4) the outputs from several gages can have various arithmetic operations performed electrically and automatically in complex systems; (5) the lowest strain observable with ordinary gages is about 5×10^{-5}, although with special care this limit can be reduced to the order of 10^{-7}.

The practical limit on accuracy is at present about 1%.

3.7.1.1.3. PHOTOELASTIC STUDIES. It has been found experimentally that normally isotropic media such as glass and methyl-methacrylate plastics become birefringent or "double refracting" under stress (see Chapter 7.4). Furthermore, it has been found that the relative retardation of the ordinary and extraordinary rays transmitted in the media is

[64] J. Yarnell, "Resistance Strain Gauges: Their Construction and Use," Morgan Bros., London, 1951; J. L. Thompson, *J. Brit. Inst. Radio Engrs.* **14,** 583 (1954).

directly proportional to the stress and thus provides an indirect method for its measurement. To obtain stress patterns in solid objects, models are constructed from appropriate transparent media and subjected to known stresses in the same fashion as the original object. If circularly polarized light (produced with a polaroid sheet and mica quarter-wave plate) is incident on the model, the transmission velocities of the mutually perpendicular-plane polarized components of the incident beam will

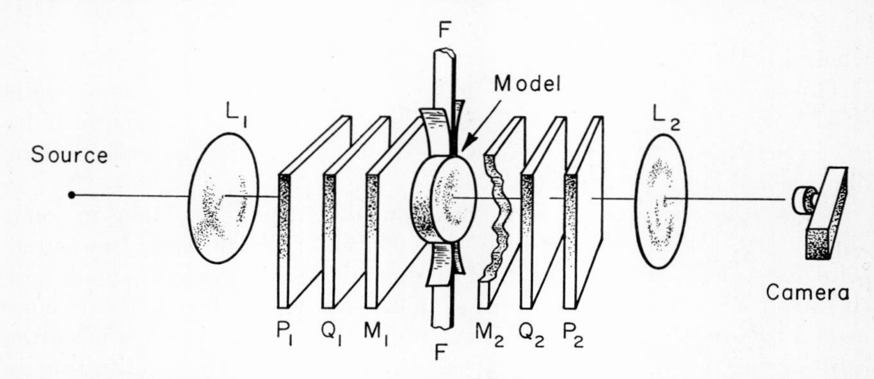

FIG. 18. Experimental arrangement for photography of photoelastic stress patterns. L_1 and L_2 are lenses to produce and focus the parallel light beam through the model; P_1 and P_2 are polaroid sheets, Q_1 and Q_2 quarter-wave plates, and M_1 and M_2 are half-silvered mirrors (if these are used). If, in addition to the isochromatic lines which indicate the magnitude of stress by their concentration, the isoclinic lines which give the stress directions are required, the quarter-wave plates are removed, and several exposures are made with the crossed polaroids at different angular orientations relative to the model.

differ by an amount depending directly upon the stress present. A second quarter-wave plate and polaroid combination placed beyond the model will then show interference patterns due to varying phase differences between the two components transmitted through the model. Figure 18 is a schematic diagram of the photographic equipment required to record these interference patterns. Qualitatively, the higher stress regions are shown by fringe concentrations; quantitatively, the model material is calibrated by examining some simple structure such as a loaded beam, where the stresses are calculable, and determining the stress equivalent to a single fringe. In further models constructed from the same material, the stress at any point may then be determined merely by counting fringes.[65] Post[66] has reported a simple method for increasing the resolu-

[65] H. T. Jessop and F. C. Harris, "Photoelasticity: Principles and Methods." Dover, New York, 1949; E. G. Coker and L. N. G. Filon, "Treatise on Photoelasticity." Cambridge Univ. Press, London and New York, 1931; see also Proc. Semi-Ann. Photoelasticity Conf. M.I.T., Cambridge, Massachusetts.

[66] D. Post, *J. Appl. Phys.* **25**, 1060 (1954).

tion of this technique by sharpening the fringes produced. This is done by inserting partial mirrors of high reflectivity on either side of the specimen. The Haidinger fringes (see Section 7.2.2) thus produced are an order of magnitude sharper than those normally obtained.

It should be noted that, as described, this method gives only a two-dimensional section of the stress pattern.

One of the most important recent developments in photoelasticity is the so-called "stress freezing" process which avoids this restriction and allows stress determination within the solid. It has been shown that many of the transparent, thermosetting plastics, when maintained at a temperature such that they are thermoplastic, obey Hooke's law accurately and in short periods, 15 min, reach equilibrium under static loading. If the material is then cooled under load, the strains are almost completely frozen in, and slices carefully cut from the solid can be polished and examined to show the strain pattern in the original solid in the plane of the slice. O'Rourke[67] has given a detailed treatment of the theory of three-dimensional photoelastic studies.

3.7.1.1.4. X-Ray Diffraction. In the diffraction of X-rays from crystalline material, constructive interference is observed at diffraction angles θ given by the Bragg equation $n\lambda = 2d \cdot \sin \theta$, where n is an integer, λ is the wavelength of the incident radiation, and d is the interplanar spacing in the lattice. If the lattice is strained, d changes microscopically, consequently shifting the diffraction pattern by measurable amounts. The advantages of this method of strain determination are: (1) strains can be measured in a very restricted area on the surface, of the order of a square millimeter or less; and (2) it does not require indirect calibration. The disadvantages are: (1) the equipment is complex and the process slow; (2) it is applicable only to static stresses; and (3) it is not applicable to samples with pronounced internal strains, since these then broaden the diffraction pattern lines to such an extent that accurate measurement of their position is impossible.

For details of the experimental method and interpretation, reference should be made to several detailed treatments in the literature[68] and to the discussion in Section 7.10.2.

3.7.1.1.5. Strain-Sensitive Lacquers. If a brittle lacquer is sprayed on the surface of a sample which is subsequently deformed, the location, number, and orientation of the resultant cracks in the lacquer film indicate qualitatively the location, magnitude, and direction of the surface strains in the sample. The lacquer films may be calibrated for semiquantitative work by applying identical films to the sample and to the surface

[67] R. C. O'Rourke, *J. Appl. Phys.* **22**, 872 (1956).

[68] D. E. Thomas, *J. Appl. Phys.* **19**, 190 (1948); W. T. Sproull, "X-Rays in Practice," Chapter 21. McGraw-Hill, New York, 1946.

of a standard beam which is then stressed in a known fashion and the resultant crack pattern related to the calculated strain.

3.7.1.2. Modulus Measurements. The variety of testing machines which have been developed for the measurement of the elastic properties of materials is well known. For detailed descriptions of these, reference should be made to the publications of the American Society for Testing Materials (ASTM) and to any of the many reference works on the subject.

3.7.1.2.1. YOUNG'S MODULUS. Almost all the testing machines have in common the fact that test samples of standard geometry are mounted so that a measurable stress can be applied to them and the resultant strain measured using any of the methods described in the preceding section. The stress is determined from the measured applied force and the specimen cross section. From these data the static Young's modulus E is obtained from its definition as $E = \sigma/\epsilon$. A common method used for the determination of the corresponding dynamic modulus involves observation of vertical oscillations of a mass suspended on the end of a long wire specimen. It can be readily shown that $E = 4\pi^2 LM/AT^2$, where M is the suspended mass, L and A the suspension length and cross section, and T the observed period of vertical oscillation. A more elegant version of this method consists of the study of the longitudinal vibrations of a specimen rod, most conveniently induced and detected by piezoelectric or magnetostriction techniques previously mentioned. It can be readily shown that the resonant frequency of such vibrations, which is detected by noting the frequency for maximum amplitude, is given by

$$f_R = \frac{n}{2L}\sqrt{E/\rho}, \qquad \text{whence} \qquad E = \frac{4L^2 f_R}{n^2}\,\rho \tag{3.7.2}$$

where ρ is the rod density, L its length, and n an integer which characterizes the order of the mode of oscillations considered.

3.7.1.2.2. SHEAR MODULUS. Measurements of the static shear modulus G are normally carried out by clamping one end and applying a known torque τ to the other end of a cylindrical sample of radius R and length L and noting the angle of twist θ produced. The modulus G is then given by $G = 2L\tau/\pi\theta R^4$. In analogy with the measurement of E, the simplest dynamic measurement of G makes use of the torsion pendulum. The specimen, in the form of a rod or wire, is clamped at one end, and a massive body of known moment of inertia is attached to the other. This mass is rotated slightly from equilibrium, and the period T of the oscillations following its release is determined. The modulus is then obtained from

$$G = 8\pi L/T^2R^4(I + \tfrac{2}{3}I_s) \tag{3.7.3}$$

where I is the moment of inertia of the suspended mass about the support axis, and I_s is that of the support itself about its own axis. In the usual apparatus $I \gg I_s$, and I_s is neglected.

3.7.1.2.3. ULTRASONIC MEASUREMENTS. From a study of the conditions for equilibrium in a rigid body, it can be shown that the propagation velocities of transverse and longitudinal waves in elastic solids can be expressed in terms of E, G, and ν as follows[62]

$$V_t = \left\{\frac{E}{2\rho(1+\nu)}\right\}^{\frac{1}{2}} = \left\{\frac{G}{\rho}\right\}^{\frac{1}{2}}$$
$$V_e = \left\{\frac{E(1-\nu)}{\rho(1+\nu)(1-2\nu)}\right\}^{\frac{1}{2}}. \quad (3.7.4)$$

Furthermore, if the ratio of these velocities is defined to be $\alpha = V_e/V_t$

$$\nu = \frac{2-\alpha^2}{2(1-\alpha^2)}$$
$$G = \rho V_t^2$$
$$E = \frac{\rho(1+\nu)(1-2\nu)V_e^2}{1-\nu}. \quad (3.7.5)$$

Consequently, a measurement of V_e and V_t suffices to determine all the elastic constants of the sample.[69] The measurement of these velocities will be considered in Section 3.7.4.

3.7.2. Nonelastic Deformation

Figure 19 is an idealized stress–strain curve for most solids. As the stress is increased from zero up to a value σ_1, the strain is linearly proportional to stress, and the slope of the curve OA is the elastic modulus E for the material; if the stress is removed, the strain returns to zero. As the stress is increased beyond σ_1, the behavior is no longer perfectly elastic, and the curve slope decreases. The stress σ_1 is termed the elastic limit or proportional limit for the material, and its value clearly depends on the sensitivity of the extensometer used. For stresses beyond σ_1, the deformation is partly elastic and partly plastic since, if the sample is stressed to σ_2 and then released, the stress–strain curve during the unloading is BHE. On reloading, the curve is essentially parallel to OA and follows BE elastically, but a permanent extension or set OE remains in the sample. The curve $CKFJ$ is a repetition of this cycle at higher stress. It will be noted that after being stressed cyclically beyond the original elastic limit σ_1, this limit appears to increase to σ_2, σ_3, etc. This is the phenomenon of work hardening.

[69] W. C. Schneider and C. J. Burton, *J. Appl. Phys.* **20,** 48 (1949).

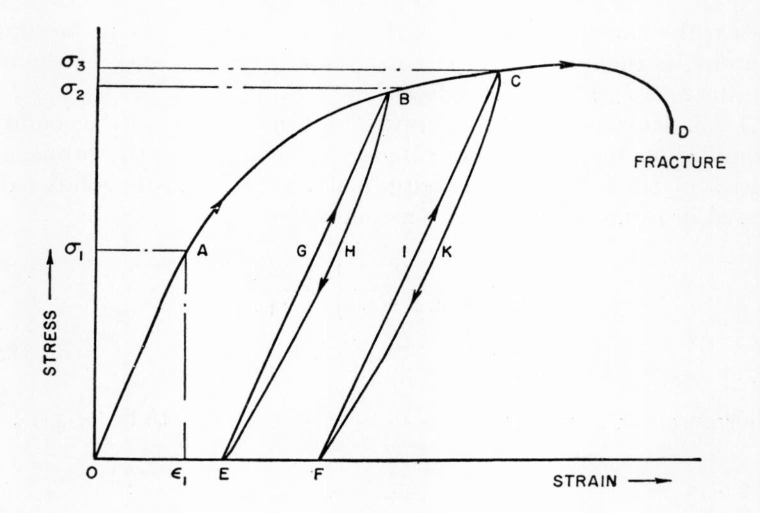

FIG. 19. Idealized stress–strain curve for a solid.

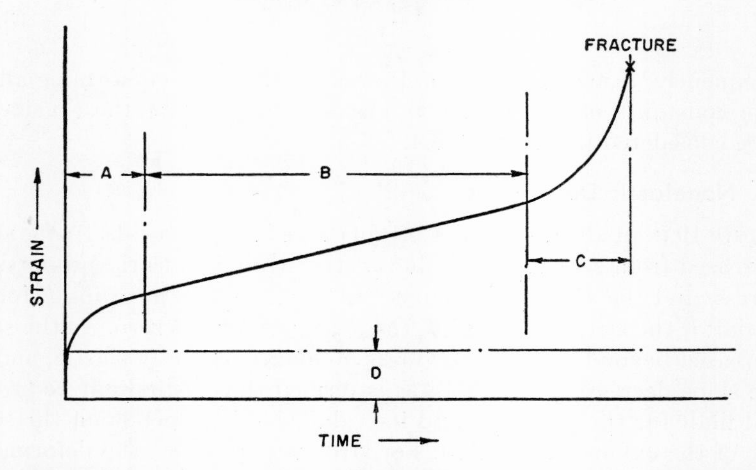

FIG. 20. Idealized creep curve for a solid. Regions labelled A, B, and C represent the so-called primary, secondary, and tertiary stages of creep, respectively. D represents the initial deformation.

The areas enclosed by the loops $BHEG$ and $CKFJ$ represent energy loss during the cycle. The phenomenon is known as elastic hysteresis, and the area of the loops representing the energy dissipated per cycle is closely related to the concept of internal friction to be discussed below.

The fall-off of the curve at D results from a localizing of the deformation with consequent decrease in diameter and fracture.

Unfortunately, the characteristics of this curve are functions of the sample temperature and the duration of the stress application. If the applied stress is maintained constant at other than very low values, a detectable increase in strain with time is observed. This phenomenon is known as creep. It has never been conclusively shown that there exists, for any material, a limiting stress below which creep does not occur.

Figure 20 is an idealized creep curve for solids and is arbitrarily divided into three sections, although the relative duration of each of these is strongly temperature dependent. In mechanical design the entire useful lifetime of the structure lies in the secondary creep section. Consequently, in principle, all designs must be based on the assumption of a certain duration of service. Working stresses are then calculated to insure that creep does not exceed set limits during this period. Creep rates of up to 1% in 10^4 hr are standard for steam piping and boiler tubes, whereas in steam turbines, for example, creep must be kept well below 1% in 10^5 hr.[63]

Andrade[70] has shown that creep curves of all pure metals could be represented by $L = L_0(1 + \beta t^{\frac{1}{3}})e^{kt}$, where β, k, and L_0 are constant; L_0 is the specimen length immediately after initial loading. This equation is customarily broken into contributions from two types of creep. If $k = 0$, the creep rate $dL/dt = \frac{1}{3}L_0\beta t^{-\frac{2}{3}}$, representing the so-called "transient flow," which damps out with time. Similarly, if $\beta = 0$, $dL/dt = L_0 k e^{kt} = Lk$, which is essentially constant, representing the so-called "quasi-viscous flow." The combination of these two gives the observed creep curve, the transient part being responsible for primary creep and the quasi-viscous for the secondary. Experimentally, it is found that k increases with increasing stress, whereas β is essentially constant; with increasing temperature β increases and k remains essentially constant. An extensive literature has developed on the interpretation of creep phenomena in terms of dislocation theory of metals.[71]*

A related concept is that of stress relaxation. If a test sample is strained by a fixed amount and maintained at that strain, it is found that the stress, to first approximation, decreases exponentially with time due to creep. The stress may be written $\sigma(t) = \sigma_0 e^{-t/\tau}$, where τ is defined as the relaxation time. The methods of measurement of creep and relaxation require no further discussion since they are merely extensions of methods previously considered.[72]

* See also Vol. 6, Chapter 4.3.

[70] E. N. da C. Andrade, *Proc. Roy. Soc.* **A84,** 1 (1910); **A90,** 329 (1914).

[71] W. T. Read, Jr., "Dislocations in Crystals." McGraw-Hill, New York, 1953; W. P. Mason, *Phys. Rev.* **98,** 1136 (1955); *J. Acoust. Soc. Am.* **27,** 643 (1955).

[72] For further discussion of creep and relaxation see B. Gross, *J. Appl. Phys.* **18,** 212 (1947); **19,** 257 (1948); **22,** 1034 (1951).

3.7.3. Viscoelasticity

The term viscoelastic is applied to rubberlike solids which have relatively large energy-absorbing abilities in addition to normal elasticity. The property of solids which acts to damp out elastic vibrations, once established, is called internal friction. Many different processes can contribute to this damping; in ferromagnetic solids, energy losses due to the permeability changes with stress previously mentioned occur; thermal energy flow occurs in specimens from regions under compression to those in tension, etc. The major component of internal friction is, however, due to the quasi-viscous or plastic flow just discussed.

In order to retain some of the earlier formalism from normal elasticity, the elastic moduli are redefined as complex quantities such that $E = E_1 + iE_2$ and $G = G_1 + iG_2$. Here E_1 and G_1 represent the normal elastic behavior and are determined using the methods described previously, while E_2 and G_2 are the imaginary moduli connected with the energy losses in the solid and are determined by various methods to be described, all of which require the establishment and subsequent study of elastic oscillations in the solid.

A simple method for the determination of the complex shear modulus uses a sample of the material as the support member of a torsion pendulum having either rectangular or circular cross section.[73] The equation of motion of the pendulum for small oscillations is $I(d^2\theta/dt^2) + kG_2\theta = 0$, where k is a shape constant $= CD^3\mu/16L$ for a support of rectangular cross section and $= \pi R^4/2L$ for one of circular cross section (μ is a shape factor tabulated, for example, by Nielsen[74]). The solution is then $\theta = \theta_0 e^{-\gamma t} e^{j\omega t}$, and, since the logarithmic derivative $\delta = \gamma T$, where T is the period of oscillation, it can be shown readily that

$$G_1 = I(4\pi^2 - \delta^2)/kT^2 \qquad \text{and} \qquad G_2 = 4\pi I\delta/kT^2. \qquad (3.7.6)$$

This method has been used for shear moduli up to 10^{11} dynes/cm² and for δ between 0.01 and 5.

There are a variety of methods for determining E_1, as discussed in the previous section. The simplest of these is the static stress–strain test, where the slope of the initial section of the curve gives E_1 and the area of the hysteresis loops gives the energy dissipated per cycle which, as will be shown presently, can be used to compute E_2. This is a tedious and little used method.

A further method, also of limited applicability, requires a measurement of the energy input required to maintain the specimen in oscillation of

[73] L. E. Nielsen, *Rev. Sci. Instr.* **22,** 690 (1951).
[74] L. E. Nielsen, *ASTM Bull.* **No. 165,** 48 (1950).

constant amplitude and gives a measure of the damping and hence of E_2.

The methods which are almost always used are: (1) the measurement of the resonant frequency and bandwidth of the specimen under forced oscillation; and (2) measurement of the decay in amplitude of the elastic oscillations after the driving force is removed.[75]

These methods are complementary since, when the damping is small, the resonance curve is extremely sharp, requiring precise measurement, while the decay is slow and easily measured; when the damping is large, the decay is too short to measure conveniently, but the bandwidth is large and easily determined.

The range of damping over which the bandwidth measurements are useful depends directly on the accuracy required; to obtain accuracies of 1%, problems of stability in the driving oscillator require that δ be greater than about 3×10^{-4}. The fact that the approximate expression $\delta = \pi/Q = (\Delta f/f_0)\pi$, normally used, breaks down for $\delta \leq 4 \times 10^{-2}$ imposes a lower limit. In the decay method, the apparatus is not normally sufficiently sensitive to measure individual oscillation amplitudes. Therefore, the time T required for the amplitude envelope to decrease by a factor of two is determined, whence $\delta = 1/(f_0 T)\ln 2$. Cottell *et al.*[76] have determined that in direct observation the limit of accuracy using this method appears to be about 4%. By displaying the oscillation envelope on an oscilloscope together with timing markers, an accuracy of about 1% is possible. This technique is useful for all $\delta < 4 \times 10^{-2}$. Having determined δ, E_2 is given by $E_2 = \gamma\omega = \delta\omega^2/2\pi$.

One of the most widely used bandwidth methods measures f_0 and Δf for the specimen undergoing forced oscillations as a cantilevered reed.[77] For this system it can be shown directly that

$$\begin{aligned} E_1 &= \rho(4\pi^2/a_0^2k^2)L^4(f_0^2 + \tfrac{1}{8}(\Delta f)^2) \\ E_2 &= \rho(2\pi/a_0^4k^2)L^4\,\Delta f\omega \end{aligned} \tag{3.7.7}$$

where k is the radius of gyration of the reed cross section, equal to $d/2\sqrt{3}$ for a rectangular section where d is the width, and r/L for a circular cross section where r is the radius. L is the length, ρ the density, ω the angular frequency, a_0 a constant $= 1.875$ for the fundamental mode of vibration, 4.694 for the first harmonic, etc.

A quantity often used to characterize the internal friction is the "loss tangent" defined as $\tan\phi = E_2/E_1 = \omega\tau$, where τ is the relaxation time

[75] J. R. Pattison, *Rev. Sci. Instr.* **25,** 490 (1954).

[76] G. A. Cottell, K. M. Entwistle, and F. C. Thompson, *J. Inst. Metals* **74,** 373 (1948).

[77] A. W. Nolle, *J. Appl. Phys.* **19,** 753 (1948); M. Horio and S. Onogi, *J. Appl. Phys.* **22,** 977 (1951).

and ϕ is the phase angle by which the strain lags behind the applied stress.

In essentially all measurements of the complex moduli, piezoelectric or magnetostriction devices are used both to induce and to detect the elastic oscillations in the sample.

3.7.4. Wave Propagation

The propagation velocities for transverse and longitudinal elastic waves in solids are given in terms of the elastic constants in Section 3.7.1.2.3.[78]

One of the most direct methods of measurement of these velocities and of the damping of such waves in solids is, unfortunately, limited to transparent specimens and is widely used for gels and concentrated polymer solutions by placing them in a rectangular transparent cell. Transverse oscillations are induced in the sample by maintaining a thin plate at one end of the tank in sinusoidal oscillation in its own plane, whereas longitudinal waves are generated by forming one end of the cell of a piston driven by a piezoelectric or magnetostriction device. The amplitude, wavelength, velocity, and damping of the wave are then determined by examining the sample in polarized light as discussed in Section 3.7.1.1.3.[79]

Solid acoustic delay lines are examples of the practical applications of the fixed propagation velocities of elastic waves in solids. In many commercially available multichannel pulse-height analyzers and elementary computers, the memory element consists of a nickel wire along which acoustic waves are transmitted, the delay time being fixed by the propagation time for a longitudinal wave down the wire. Magnetostriction drive and detection is normally used in that a magnetic pulse applied to one end of the wire generates a compression pulse which then travels down the line at a fixed velocity determined only by the elastic constants of the material. Quartz delay lines are also widely used and have the very considerable advantage of greater bandwidth, that is, less internal friction, and are able to operate satisfactorily with pulse repetition rates in the megacycle range, whereas the nickel lines are normally not used for repetition rates over 500 kc.

Richardson[80] has given a detailed treatment of the application of infra- and ultrasonic techniques to the determination of the complex moduli of viscoelastic solids.

[78] H. Kolsky, "Stress Waves in Solids." Oxford Univ. Press, London and New York, 1953; E. H. Lee and J. A. Morrison, *J. Polymer Sci.* **14,** 93 (1956).

[79] F. T. Adler, W. M. Sawyer, and J. D. Ferry, *J. Appl. Phys.* **20,** 1036 (1949).

[80] E. G. Richardson, "Relaxation Spectrometry." Interscience, New York, 1957.

It is possible to study the propagation of elastic waves in crystalline substances by investigating the diffuse scattering of X-rays from the samples. Theoretically, for a perfect crystal at absolute zero, the intensity of scattered X-rays would be zero between the Bragg maxima. At ordinary temperatures, the nonzero intensity of this diffuse scattering between the maxima is attributed to the thermal motion of the crystal atoms relative to their equilibrium lattice locations. Elastic waves propagating in the crystal produce similar effects. Detailed discussions of this technique for determining these velocities are given by James and by Laval.[81]

3.7.4.1. Seismology. Seismology consists, to a large extent, of the study of the propagation, through the earth, of elastic waves produced either by natural phenomena such as earthquakes or, more recently, by chemical or nuclear explosions. In all, six different types of wave motions have been observed; of these, the longitudinal P waves with periods of 1 to 5 sec, and the transverse S waves with periods of 5 to 10 sec are most common. Both are body waves transmitted through the interior of the earth.

In addition to these, the R and Q waves are surface waves transmitted in the earth's crust. Each individual particle in the path of an R or Rayleigh wave moves in an elliptic orbit, whereas the Q wave is effectively a polarized transverse vibration in which the individual particle amplitudes are perpendicular to the direction of propagation with no component normal to the surface in which the wave is propagating. These have been investigated in detail by Rayleigh[82] and Love,[83] respectively. Finally, the two remaining types of waves, C and H, have not been observed as yet in studies of natural phenomena but only following nuclear explosions.[84] The H or hydrodynamic wave has a transmission mechanism similar to that observed in liquids and the C, or coupled wave is a combination of compressional and shear displacements.

Seismographs of extreme sensitivity have been produced for recording the characteristics of the longitudinal and transverse wave components as described, for example, by Clewell *et al.*[85] In essence all these instruments are adaptations of the seismic unit described in Section 3.5.4.

[81] R. W. James, "The Optical Principles of the Diffraction of X-Rays." Bell, London, 1948; J. Laval, *Bull. soc. franc. minéral.* **64,** 1 (1941); H. Cole, *J. Appl. Phys.* **24,** 482 (1953).

[82] J. W. Strutt, Baron Rayleigh, *Proc. London Math. Soc.* **17,** 4 (1885).

[83] A. E. H. Love, "Some Problems of Geodynamics." Cambridge Univ. Press, London and New York, 1911.

[84] L. D. Leets, *Am. Scientist* **34,** 198 (1946); **45,** 114 (1957).

[85] D. H. Clewell, R. A. Broding, G. B. Loper, S. N. Heaps, R. F. Simon, R. L. Mills, and M. B. Dobrin, *Rev. Sci. Instr.* **24,** 243 (1953).

With natural wave sources, the velocities of propagation, and consequently the elastic and geological constants of the intervening media are determined by combining the observations at several different locations and thus fixing the location and time of the initial disturbance by triangulation. With artificial sources, propagation times are measured directly and, by appropriately arranging a grid of such sources, geological fine structure is easily determined.

4. MECHANICS OF FLUIDS

4.1. Static Properties*

The current interest in the mechanics of fluids is strongly concentrated on fluids in motion which are treated extensively in Chapter 4.2. Measurements in fluid dynamics involve, however, the measurement of certain static properties of fluids which are discussed in this chapter.

4.1.1. Mass, Volume, Pressure, and Density

4.1.1.1. Mass. For the determination of the mass of fluids, the same weighing methods that are used for solids (Chapter 3.2) are generally applicable, except for the need of enclosing the fluid in a suitable vessel. For liquids, an ordinary weighing glass or flask with ground stopper can be used. Since atmospheric gases are soluble in most liquids, it is necessary, for precision measurements, to degas the liquid before weighing. This is usually done through heating under reduced pressure in a vessel connected to a reflux condenser. A very effective degassing procedure is to subject the liquid to a high-frequency sound field.[1] It is also frequently necessary to dry the liquid in order to remove dissolved water. For this purpose, a suitable, nonreacting drying agent (calcium chloride, phosphorus pentoxide, metallic sodium, or silica gel) is added and allowed to remain with the liquid for several hours. For storage of such liquids, sealed-off vessels are strongly recommended. Care has to be taken, of course, that these storage vessels, as well as those subsequently used for weighing and other operations, are thoroughly dry; glass vessels normally carry substantial quantities of adsorbed water which is difficult to remove. A water layer on the external surface may also influence the weight of a glass vessel; it is, therefore, frequently desirable to create reproducible conditions by saturating the outside of the glass vessels before each weighing operation by wetting, and subsequent drying with a clean cloth.

For the direct mass determination of gases it is necessary to enclose them in a sealed-off vessel. Since their density is of the same order as that of air, corrections for buoyancy (see Section 3.2.1.1) are absolutely necessary.

If the composition (and the molecular weight M) of the gas is known, its mass m can be determined from measurement of its pressure p,

[1] I. Rudnick, personal communication (1953).

* Chapter 4.1 is by **I. Estermann.**

its volume v and its temperature T by the application of the gas law, $pv = (m/M)RT$, or by the use of a more accurate equation of state.

4.1.1.2. Volume. Most of the conventional methods for volume measurements on fluids are relative and based on the mass and density of a

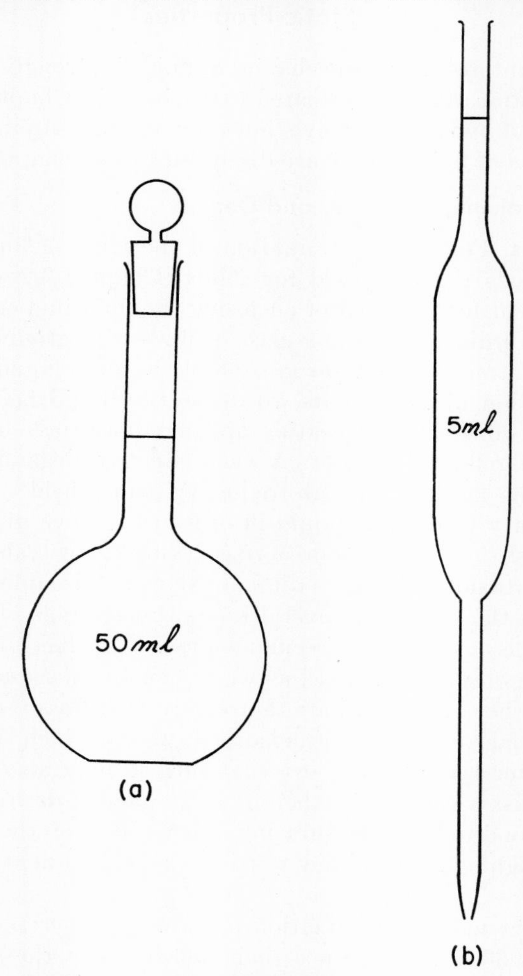

FIG. 1. (a) Measuring flask; (b) pipette.

"standard liquid" (mercury or water). According to the amount and the nature of the fluid, various tools are used. For liquids, measuring flasks, pipettes, or burettes (Fig. 1) are used. They are usually calibrated as indicated above, either by the supplier or the user. Since the volume of these vessels depends on the ambient temperature t, which is normally different from the calibration temperature t_0, a correction according to

$$V_t = V_0[1 + 3\alpha(t - t_0)] \quad (4.1.1)$$

where V_t is the volume at t, V_0 the volume at t_0, usually 15°C, and α the linear thermal expansion coefficient of the material of the vessel.

The volume of gases is measured with gas pipettes or burettes. Since the volume of a given quantity of a gas is indeterminate unless its pressure is known, volume measurements on gases are always taken together with a measurement of pressure, and frequently together with a measurement of the mass in order to obtain the density.

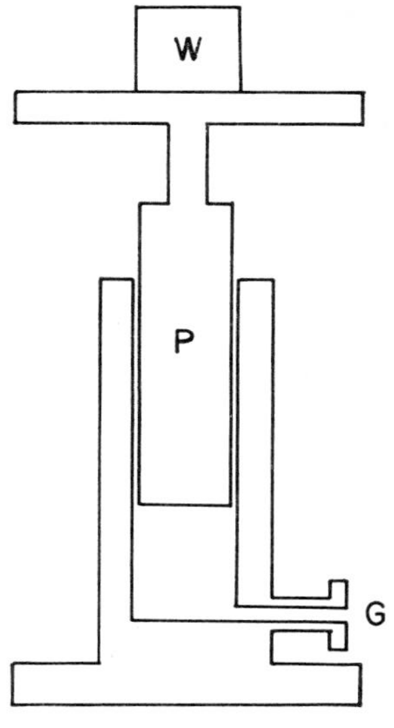

FIG. 2. Dead-weight manometer.

4.1.1.3. Pressure. The standard method of determining gas pressures is the dead-weight manometer (Fig. 2) in which the (gravitational) force on a piston of known area is balanced by the gas pressure.[2] To prevent sticking, the piston is slowly rotated. For pressures from about 1 mm Hg up to a few atmospheres, the hydrostatic pressure of a column of mercury, oil, or water is used to provide this balance. These columns are usually contained in a U-tube, of which one end is connected to the gas, the other is either open to the atmosphere, closed and evacuated, or closed and filled with a gas (air) of a certain pressure at a certain calibration volume. The use of these instruments and the calculation of the readings is obvious.

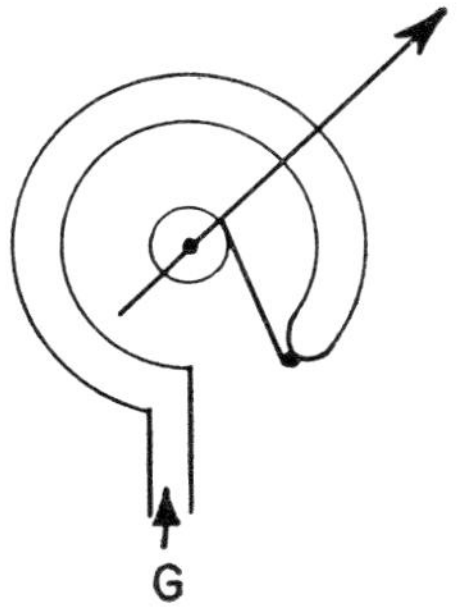

For measurements of pressures of less than 1 mm Hg see Sections 4.2.3–4.2.4. Mechanical manometers, based on deformation of hollow tubes of various shapes (Bourdon tubes, Fig. 3) are now available with precision and accuracy of about 1 part per thousand, but most instruments of this kind lack the necessary reliability for laboratory measurements.

FIG. 3. Bourdon tube manometer.

By indirect measurement, volumes of fluids can be obtained by measuring the flow rate and time. The flowmeters most commonly used in the laboratory consist

[2] A. Michels, *Koninkl. Ned. Akad. Wetenschap. Proc.* **35**, 994 (1932).

of a slightly tapered vertical tube in which a steel ball is lifted by the gas stream to a height indicating the rate of flow. Another type, the Venturi meter (Fig. 4) consists of a differential manometer attached to both sides of a constriction in the gas line. According to Bernouilli's principle, the kinetic energy of the fluid per unit volume

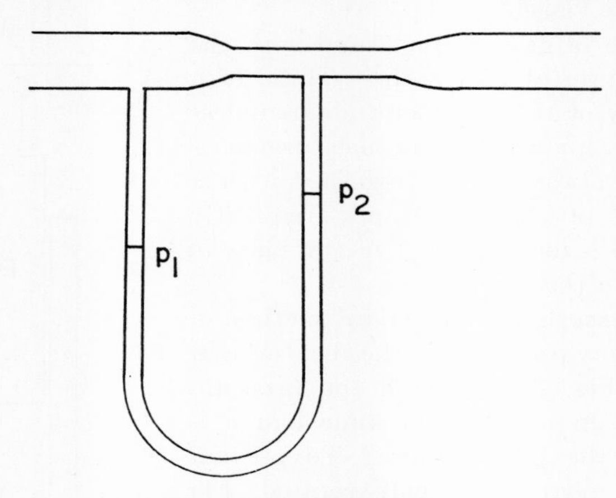

FIG. 4. Venturi flowmeter.

which has been accelerated across the constriction in order to satisfy the continuity relation (Eqs. (4.2.1) and (4.2.7)) is equal to the pressure difference, or

$$p_1 - p_2 = \tfrac{1}{2}\rho(v_2^2 - v_1^2). \tag{4.1.2}$$

The volumetric flow rate is

$$Q = A_1 v_1 = A_2 v_2 \tag{4.1.3}$$

where A_1 and A_2 are the areas of the unconstricted and the constricted pipe (neglecting the compressibility of the fluid). Hence

$$Q = A_1 A_2 \sqrt{\frac{2(p_1 - p_2)}{\rho(A_1^2 + A_2^2)}}. \tag{4.1.4}$$

4.1.1.4. Density. The density of fluids may be determined by a combination of mass and volume measurement $\rho = m/V$. For most liquids a pycnometer designed by Ostwald (Fig. 5) is very convenient. For filling, the capillary tube a is connected to a suction apparatus, while b is immersed into the liquid whose density is to be measured. Before weighing, the instrument is immersed in a temperature bath, and after equilibrium is reached, the meniscus in a is brought to the calibration mark by touching the end of tube b with filter paper, leaving b completely filled. Corrections for thermal expansion and air buoyancy have to be applied as required. The volume of the pycnometer is usually determined by weighing it when filled with mercury, thus the density obtained is a relative value. Accuracy of 10 ppm can be reached. The major limitations of accuracy are caused by adsorbed water layers on the glass, and by thermal hysteresis of the volume of the vessel if measurements are taken far from room temperature.

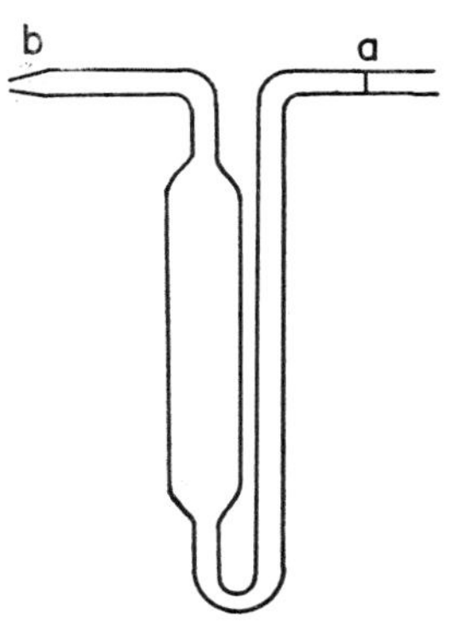

FIG. 5. Pycnometer.

Indirect measurements are based on Archimedes' principle according to which a submerged body is buoyed up by the liquid with a force equal to the weight of the displaced liquid. If W_a is the weight of the submerged body of volume V_0 in air (corrected to vacuum if necessary) and W_e the weight when submerged, the density of the liquid is given by

$$\rho = \frac{W_a - W_e}{V_0}. \tag{4.1.5}$$

Volume V_0 is usually determined by measuring the buoyancy in a standard liquid (water). Various "hydrostatic balances" have been designed which make this procedure quick and easy. Since the thermal expansion of most liquids is of the order of 10^{-5}/°C, accurate density measurements have to be made at or reduced to a standard temperature, usually 15°C.

In the case of gases (or vapors) the density depends even stronger on the temperature and also on the pressure. Here it is customary to give the results reduced to standard temperature and pressure (stp) viz., 0°C and a "standard atmosphere," 760 mm Hg or 1,0131250 dyne/cm² (see Section 2.3.3). Measurements at other conditions are reduced to stp with the aid of an appropriate equation of state. In most practical cases, the "ideal" gas law

$$pV = p_0 V_0[1 + 0.003662(t - t_0)] = nRT = \frac{m}{M} RT \tag{4.1.6}$$

is adequate. The density at stp is obtained from

$$\rho = \frac{m}{V_0} = \frac{mp_0[1 + 0.00366(t - t_0)]}{pV}. \tag{4.1.7}$$

Another way of expressing the density of a gas is to give the ratio of the weight of a certain volume of gas to that of an equal volume of dry air at the same pressure and temperature. As long as ideal gas conditions prevail this ratio is independent of temperature and pressure.

For the experimental determination of gas densities at room temperature and ambient pressure, a flask of known V equipped with stopcocks is weighed, first evacuated, then filled with the gas at ambient pressure. As counterweight, one uses for accurate measurements a glass flask of equal volume and surface area which has been treated in the same way as the weighing flask, thus compensating for air buoyancy and water adsorption. Corrections to stp are made as stated before.

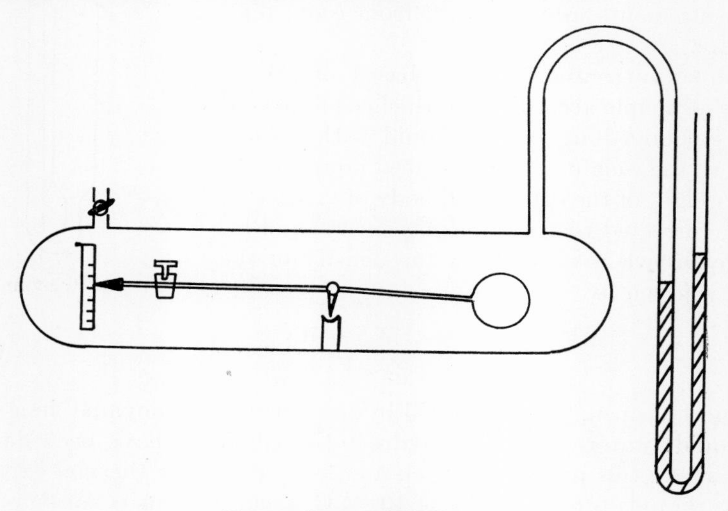

FIG. 6. Aerostatic balance.

Another suitable instrument for relative measurements is the aerostatic balance[3,4] (Fig. 6). It consists of a glass chamber with attached manometer surrounding a small balance, carrying a hollow weight on one end and a small counterweight on the other. By filling this chamber alternately with air and with the gas, and noting the pressures at which the balance reaches equilibrium, one obtains the relative density of the gas from the equation

$$\rho_{\text{gas}}/\rho_{\text{air}} = p_{\text{air}}/p_{\text{gas}}. \tag{4.1.8}$$

Multiplication with the density of air at stp gives the desired gas density.

4.1.2. Compressibility

The compressibility of a fluid is the reciprocal of the bulk modulus B

$$\kappa = \frac{1}{B} = -\frac{1}{V_0}\frac{dV}{dp} = \frac{1}{\rho_0}\frac{d\rho}{dp}. \tag{4.1.9}$$

[3] A. Stock and G. Ritter, *Z. physik. Chem. (Leipzig)* **119**, 333 (1926); **124**, 204 (1926); **126**, 172 (1927).

[4] T. S. Taylor, *Phys. Rev.* **10**, 653, 1917; E. Lehrer and E. Kuss, *Z. physik. Chem. (Leipzig)* **163**, 73 (1931).

If the temperature is kept constant during compression (as in most direct measurements) we deal with the isothermal compressibility

$$\kappa_T = \frac{1}{\rho_0}\left(\frac{\partial \rho}{\partial p}\right)_T. \tag{4.1.10}$$

If the compressed body is thermally insulated (or if the compression is so fast that the heat of compression is not removed, as in the velocity of sound method) we deal with the adiabatic or isentropic compressibility

$$\kappa_S = \kappa_T + \frac{T\alpha^2}{\rho c_p} \tag{4.1.11}$$

where T is the absolute temperature, α the thermal volume expansion coefficient, and c_p the specific heat at constant pressure.

The compressibility is related to the thermal expansion and the pressure coefficient by the thermodynamic relation

$$\left(\frac{\partial V}{\partial p}\right)_T \left(\frac{\partial p}{\partial T}\right)_V \left(\frac{\partial T}{\partial V}\right)_p = -1. \tag{4.1.12}$$

Measurements of the compressibility of liquids are usually carried out at large pressures since the absolute values of κ are rather small, of the order of 10^{-4}–10^{-6}. The liquid is enclosed in a glass vessel (piezometer) (Fig. 7) which is equipped with a U-type neck holding two electrical contacts. A mercury column serves as piston and indicates its position by making or breaking contacts. The whole instrument is enclosed in a pressure bomb which is filled with a suitable liquid and connected to a hydraulic pump. When pressure is applied, the mercury level sinks and the electrical contact is broken. The experiment is repeated after the addition of a known mass of mercury. The pressure necessary to break contact is determined again, and the compressibility is obtained by the ratio of the additional pressure and the volume of the added mercury. Corrections for the compressibility of the mercury, which can be obtained in the same way, and for that of the glass have to be applied. The procedure is repeated up to the limits set by the yield strength of the bomb and the range of the manometer. Leak-free closure of the bomb at high pressures or elevated temperatures requires rather complicated construction details.[5]

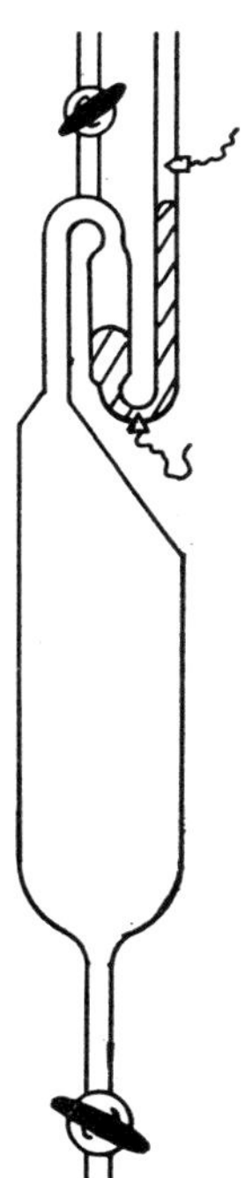

FIG. 7. Piezometer.

[5] P. W. Bridgman, "Physics of High Pressures." Macmillan, New York, 1952.

The compressibility of gases is usually obtained from the equation of state and Eq. (4.1.9). For ideal gases,

$$V = RT/p \qquad \text{and} \qquad \kappa_T = 1/p. \tag{4.1.13}$$

For a Van der Waals gas, one obtains

$$\kappa_T = \frac{V^2(V - b)^2}{RTV^3 - 2a(V - b)^2}. \tag{4.1.14}$$

The measurement of the compressibility of gases at high pressures is carried out in similar fashion as in the case of liquids.

The most common indirect method for the determination of compressibility is based on the measurement of the velocity of sound c (see Section 5.1.4), from which the adiabatic compressibility can be obtained by use of the equation

$$c = (\rho\kappa_S)^{-\frac{1}{2}}. \tag{4.1.15}$$

The use of ultrasonic frequencies and corresponding short wavelengths is often desirable.

4.1.3. Surface Tension

The most important physical property which distinguishes a liquid or solid from a gas or vapor is the existence of a phase boundary. When this boundary separates the condensed phase from a gas or its own vapor, it is referred to as surface; if two condensed phases are separated by a boundary, it is called interface.

The work dW necessary to enlarge the surface or interface by an increment dA is proportional to the additional area,

$$dW = \sigma\, dA. \tag{4.1.16}$$

The proportionality constant σ is called surface tension and has the dimension force per unit length; it is, therefore, measured in dyne/cm. Direct measurements of σ are possible only for liquids, for solids σ has to be calculated on the basis of molecular theory.

A liquid surface bordering on a solid wall will form a contact angle θ which is zero for a completely wetted surface like water on clean glass, or 180° for a completely nonwetting surface like mercury on clean glass. For $90° < \theta < 180°$ the liquid forms a convex meniscus and is depressed by a capillary, for $\theta < 90°$, it forms a concave meniscus and rises in a capillary. A common method for the measurement of σ (Fig. 8) is based on the observation of the difference in height h between the liquid in a capillary of radius r and that of a large "flat" surface with which it is in contact.[6] The force exerted by the surface tension is $2\pi r\sigma \cos \theta$. This force is balanced by the force of gravity acting on the column of the liquid, $\pi r^2 \rho g h$, where ρ is the density of the liquid and g the acceleration of gravity. Hence

$$\sigma = \tfrac{1}{2} r \rho g h / \cos \theta. \qquad (4.1.17)$$

(If h' is the height measured from the apex of the meniscus to the flat surface, the effective height, $h = h' + (r/3) - 0.129/(r^3/h')$.) The contact angle θ is difficult to measure, and this method is reliable only for $\cos \theta = \pm 1$, i.e., for completely wetting or nonwetting surfaces. Minute impurities have a strong influence on θ.

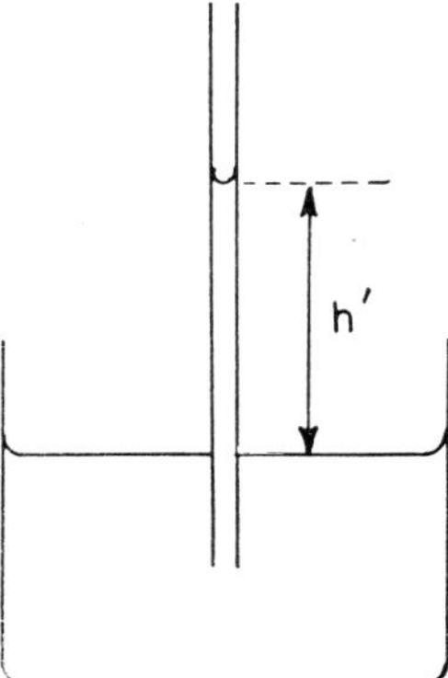

FIG. 8. Rise of liquid in capillary tube.

Other methods are based on the measurement of the force required to tear a wetted ring from a surface,[7] on the size of drops of a liquid flowing from a capillary,[8] on the vapor pressure of small droplets, and on the oscillations in suspended drops.

4.1.4. Viscosity

The viscosity is not strictly speaking a static property, but because of its close relation to the other properties of liquids, it is briefly treated here. Further discussion is given in Chapter 4.2.

Although fluids do not resist infinitely slow deformations, a finite rate of deformation is opposed by a force. For many substances (Newtonian fluids), the shear stress in a moving fluid is proportional to the velocity gradient normal to the direction of the flow

$$\tau = \mu(\partial v/\partial n). \qquad (4.1.18)$$

The proportionality factor μ is known as the absolute viscosity and has the dimension $ML^{-1}T^{-1}$ and its unit gm cm^{-1} sec^{-1} is called Poise. (In many problems in fluid dynamics the important quantity is $\nu = \mu/\rho$ called the kinematic viscosity. It is measured in units cm^2 sec^{-1} called Stokes, see Section 4.2.1.)

[6] P. Volkman, *Ann. Physik* [3] **53**, 633 (1894); [3] **66**, 194 (1898); J. R. Ligenza and R. B. Bernstein, *J. Am. Chem. Soc.* **73**, 4636 (1951).

[7] H. Moser, *Ann. Physik* [4] **82**, 993 (1927); H. W. Fox and C. H. Chrisman, *J. Phys. Chem.* **56**, 284 (1952); W. D. Harkins and H. F. Jordan, *J. Am. Chem. Soc.* **52**, 1751 (1930).

[8] Lord Rayleigh, *Phil. Mag.* [5] **48**, 321 (1899); W. D. Harkins and F. E. Brown, *J. Am. Chem. Soc.* **41**, 499 (1919).

For a liquid flowing through a cylindrical tube of radius r and length l, the volumetric rate of flow is given by Poiseuille's law

$$\frac{dV}{dt} = \frac{\pi}{8\mu} r^4(p_1 - p_2) \tag{4.1.19}$$

where $p_1 - p_2$ is the pressure difference at the ends of the tube. This law is valid only if the flow is laminar, see Section 4.2.1.

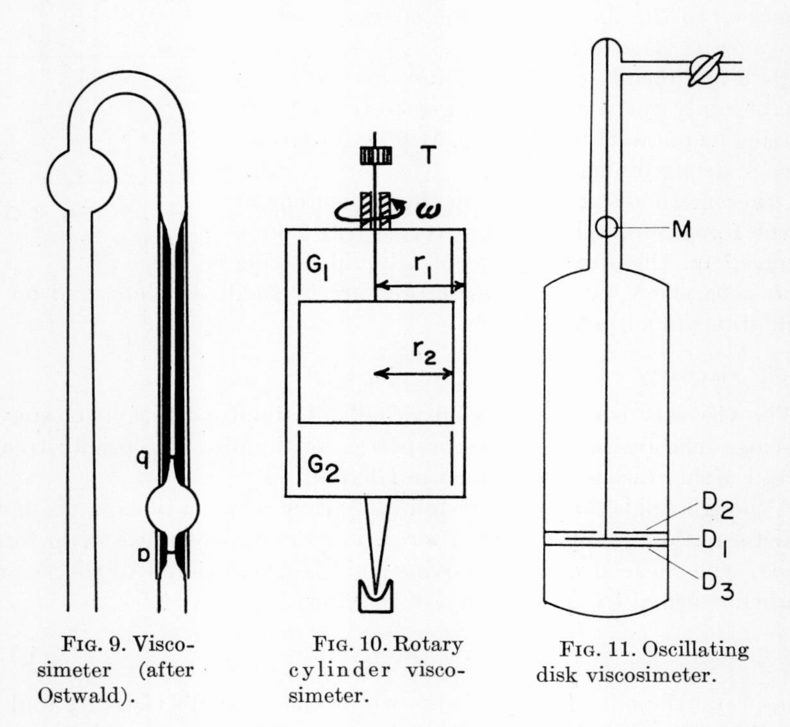

FIG. 9. Viscosimeter (after Ostwald).

FIG. 10. Rotary cylinder viscosimeter.

FIG. 11. Oscillating disk viscosimeter.

The absolute determination of μ from Poiseuille's law[9] is rather difficult and requires corrections for the kinetic energy of the flow, deviations from the ideal flow pattern at the end of the tube, etc. For most practical measurements one uses instruments by which the unknown viscosity is compared with that of a standard liquid, e.g., water. Such an instrument is Ostwald's viscosimeter (Fig. 9). The times t and t_0 required for equal volumes of the two liquids to flow through the capillary are measured.

[9] J. F. Swindells, J. R. Coe, Jr., and T. B. Godfrey, *J. Research Natl. Bur. Standards* **48**, 1 (1952).

Then

$$\mu/\mu_0 = (\rho/\rho_0)(t/t_0) \tag{4.1.20}$$

or

$$\nu/\nu_0 = t/t_0 \tag{4.1.21}$$

where the index 0 refers to the standard liquid.

Another procedure utilizes the laws of motion of a sphere in a viscous medium under the influence of a constant force.[10] According to Stokes' law, this motion proceeds with a constant velocity v which is related to the absolute viscosity of the fluid μ by the equation

$$v = \frac{2}{q} \frac{\rho_s - \rho_f}{\mu} gr^2 \tag{4.1.22}$$

where r is the radius of the sphere, ρ_s its density, ρ_f the density of the fluid, and g the acceleration of gravity. This method permits the determination of μ if r is known, but also of r if μ is known. As example, we refer to Millikan's oil drop measurement of the electronic charge (see Section 1.2.1).

A third method is that of the rotating cylinder[11] (Couette flow, see Section 4.2.1.4). The instrument (Fig. 10) consists of two concentric cylinders of radii r_1 and r_2. The gap between the cylinders is filled with the fluid. The external cylinder is rotated with a constant angular velocity ω and the torque transmitted by the fluid to the inner cylinder is balanced by a torsion wire suspension. For laminar flow conditions this torque is proportional to the viscosity.

The methods described above are applicable for gases as well as for liquids, always provided that laminar flow persists. An additional technique frequently used for the measurement of the viscosity of gases is that of the oscillating disk[12] (Fig. 11). A thin circular plate is suspended from a torsion fiber between two stationary plates and set into oscillatory motion. If the space between the plates is filled with a gas, the oscillations are damped and the logarithmic decrement Λ of the motion is proportional to the viscosity μ. Because of various corrections, this instrument is mostly used for relative measurements. The equation

$$\mu/\mu_0 = (\Lambda - \Lambda_K)/(\Lambda_0 - \Lambda_{0K}) \tag{4.1.23}$$

is used for the determination of the unknown viscosity if that of the

[10] G. G. Stokes, "Mathematical and Physical Papers," Vol. 1, p. 75. Cambridge Univ. Press, London, 1880; R. A. Millikan, *Phys. Rev.* **21,** 217 (1923).

[11] A. R. Kuhltau, *J. Appl. Phys.* **20,** 217 (1949).

[12] J. C. Maxwell, *Phil. Trans. Roy. Soc. London, Ser. A* **156,** 249 (1866); P. Günther, *Z. physik. Chem. (Leipzig)* **110,** 626 (1924).

standard gas μ_0 is known. The terms Λ_K and Λ_{0K} represent the damping of the various parts of the apparatus other than the oscillating disk.

The viscosity is strongly dependent on the temperature.[13] For air between $t = 10°C$ and 30°C, the viscosity is given by

$$\mu_t = [181.8 + 0.495(t - t_0)]10^{-6} \text{ poise.}$$

For normal pressures, the viscosity of gases is, in accordance with the kinetic theory, independent of the pressure. For rarefied gases, i.e., under conditions where the mean free path is comparable to the dimensions of the apparatus, the considerations given above are no longer valid (see Section 4.2.3).

4.2. Dynamic Phenomena*

4.2.1. Fundamental Considerations of Fluid Dynamics[1,2]

4.2.1.1. Definitions—The Nature of Fluids. The phenomena of fluid dynamics of greatest interest to the physicist are those which deal with gases rather than with liquids. Although many of the fundamental principles discussed in this section are equally applicable to gases and to liquids, the presentation is based on the concepts and formulations of the kinetic theory of gases. In accordance with these concepts, a gas is considered to be an aggregation of unconnected molecules in rapid thermal motion where the mean thermal energy is sufficiently great to prevent condensation. As a fundamental consequence of the theory, the fluid properties of a gas are found to result from the propagation of disturbances by the mechanisms of intermolecular and molecule-surface collisions.†

Although the molecular nature of a gas ultimately determines its macroscopic properties, one must necessarily deal experimentally and theoretically with fluid elements in which the molecular motions may be averaged. By a consideration of the time variation of molecule density in an element of phase space one may establish Boltzmann's transport

[13] A. Sutherland, *Phil. Mag.* [5] **36,** 507 (1893); M. Trautz, *Ann. Physik* [5] **11,** 190 (1931); [5] **18,** 816 (1933).

† See also Vol. 3, Part 9.

[1] S. Chapman and T. G. Cowling, "The Mathematical Theory of Non-Uniform Gases." Cambridge Univ. Press, London and New York, 1939.

[2] S. Goldstein, "Modern Developments in Fluid Dynamics." Oxford Univ. Press, London and New York, 1943.

* Chapter 4.2 is by **F. C. Hurlbut.**

equation and from it derive the general fluid mechanical equations. Thus the theory provides structural support for the known results from experiment, i.e., that a gas at normal density may be treated as a viscous compressible fluid. Certain restrictions are put on this view; the gas cannot be too rarefied, and the viscous and thermal stresses at the boundary cannot be too great.

The problems of gas dynamic experiment are in large part the problems of measurement in the flow field. The quantities of interest, the local heat rate to a body immersed in the flow, for example, or the aerodynamic forces on it or the distribution of normal pressure about it, are determined for the most part through the performance of elementary measurements. While the measurements in themselves are elementary, the interpretations of them may not be, and in fact must rest upon a certain minimum structure of theory. It is with these matters that we shall be concerned in the present section.

4.2.1.2. The General Equations of Gas Dynamics.[3] A system of five conservation equations and an equation of state constitute the fundamental framework of gas dynamics. The principle of conservation of mass leads to the equation*

$$\frac{D\rho}{Dt} + \rho\Delta = 0 \tag{4.2.1}$$

where

$$\frac{D}{Dt} \equiv \frac{\partial}{\partial t} + u_i \frac{\partial}{\partial x_i} \quad \text{(the material derivative)} \tag{4.2.2}$$

$$\Delta = \frac{\partial u_i}{\partial x_i} \quad \text{(the dilatation)} \tag{4.2.3}$$

where ρ is the density and u_i the components of velocity.

Linear relations are assumed to exist between the rate of strain components and the stress components. The static pressure p is defined to be independent of the rates of strain, so that the gas law holds for a gas in motion. The momentum equations may then be written:

$$\rho \frac{Du_i}{Dt} = \rho X_i - \frac{\partial p}{\partial x_i} + \mu\nabla^2 u_i + \frac{\mu}{3}\frac{\partial}{\partial x_i}\left(\frac{\partial u_j}{\partial x_j}\right) - \frac{2}{3}\frac{\partial \mu}{\partial x_i}\left(\frac{\partial u_j}{\partial x_j}\right) + \frac{\partial \mu}{\partial x_j}\left(\frac{\partial u_i}{\partial x_j} + \frac{\partial u_j}{\partial x_i}\right) \quad \text{(Navier–Stokes equations)} \tag{4.2.4}$$

Here X is the force due to external fields and μ is the absolute viscosity.

* The present statement of the equation is somewhat less general than that to be found in reference 3. The summation convention has been employed throughout Section 4.2.

[3] H. W. Liepmann and A. Roshko, "Elements of Gasdynamics." Wiley, New York, 1957.

If E is the internal energy of the gas per unit mass, the energy equation may be written

$$\rho \frac{DE}{Dt} = \Phi_1 + J \frac{\partial}{\partial x_i}\left(K \frac{\partial T}{\partial x_i}\right)$$

with

$$\Phi_1 = \tfrac{1}{2} p_{ij} e_{ij}$$

where

$$p_{ij} = -(p + \tfrac{2}{3}\mu\Delta)\delta_{ij} + \mu e_{ij}$$

and

$$e_{ij} = \left(\frac{\partial u_i}{\partial x_j} + \frac{\partial u_j}{\partial x_i}\right) \tag{4.2.5}$$

and where T is the absolute temperature and J is the mechanical equivalent of heat.

The equation of state is normally taken to be the perfect gas law

$$p = \rho \frac{R}{M} T \tag{4.2.6}$$

where R is the universal gas constant and M is the molecular weight.

Solutions of the complete system lie beyond the range of present analytical techniques. In consequence, solutions to incomplete equations or incomplete systems are ordinarily sought, often through approximation and linearization. Each of these simplified systems constitutes a field of fluid mechanics, in the most important of which there exists a substantial literature and a significant area of applicability. Thus we have incompressible inviscid flow (classical hydrodynamics), compressible inviscid flow, incompressible viscous flow, and compressible viscous flow; each in one, two, or three dimensions.

4.2.1.3. Effects of Compressibility.[4] Many problems of flows in ducts, or of low speed flight, for example, may be treated within the realm of incompressible flow theory. However, when the fluid speed becomes an appreciable fraction of the speed of sound, proper cognizance must be taken of the effects of compressibility. Many features of compressible fluid flow can be illustrated in a one-dimensional system, and it is with such a system that we shall deal here. It will also be considered that the effects of viscosity can be neglected for this discussion, an approximation which is good in most external flow except in regions of the boundary layer (see Section 4.2.1.4).

From the conservation equations and the equations of state one may derive the compressible Bernoulli equation

[4] A. Shapiro, "The Dynamics and Thermodynamics of Compressible Fluid Flow," Vol. 2. Ronald, New York, 1954.

$$\frac{u^2}{2} + \frac{\gamma}{\gamma - 1} \frac{p_0}{\rho_0} \left(\frac{p}{p_0}\right)^{(\gamma-1)/\gamma} = \frac{\gamma}{\gamma - 1} \frac{p_0}{\rho_0} \tag{4.2.7}$$

which relates the stream flow variables u, p for an isentropic acceleration of gas from a reservoir at stagnation conditions ρ_0, p_0. The words "stagnation conditions" are used here and elsewhere in this presentation to signify the state of the gas essentially at rest, immediately preceding the convergent section of a nozzle. In such an acceleration through a simple orifice u cannot exceed the local velocity of sound a where

$$a = \sqrt{\frac{\gamma RT}{M}} \tag{4.2.8}$$

and T is the local temperature. A critical speed, a^*, may be defined for flow through a convergent nozzle such that $u = a = a^*$. It may be represented in terms of a_0, the speed of sound at the stagnation temperature, as

$$a^* = a_0 \sqrt{\frac{2}{\gamma + 1}} \tag{4.2.9}$$

One may define the Mach number M as the ratio

$$\frac{u}{a} = M \tag{4.2.10}$$

and use it together with the energy equation and the relation for isentropic flow

$$\left(\frac{a_0}{a}\right)^2 = \frac{T_0}{T} = \left(\frac{p_0}{p}\right)^{(\gamma-1)/\gamma} \tag{4.2.11}$$

to provide a multiplicity of relationships for compressible flows.[5]

If the acceleration to sonic speed takes place in a nozzle in which there exists a divergent exit, the flow may continue to accelerate, and thus becomes supersonic. If A^* is the throat area, and A is the area of any section downstream, the Mach number of an accelerated flow at this latter section is given by the area ratio equation

$$\left(\frac{A}{A^*}\right)^2 = \frac{1}{M^2}\left[\frac{2}{\gamma + 1}\left(1 + \frac{\gamma - 1}{2} M^2\right)\right]^{(\gamma+1)(\gamma-1)}. \tag{4.2.12}$$

A normal shock wave may arise in a supersonic flow as a discontinuity across which an abrupt change in the flow variables u, p, T occurs. One finds the pressure increase across such a wave to be

[5] See, for example, "Tables of Compressible Flow," National Advisory Committee for Aeronautics, Technical Report No. 1135 (1953).

$$\frac{p_2 - p_1}{p_1} = \frac{2\gamma}{\gamma + 1}(M^2 - 1) \tag{4.2.13}$$

while the speed before and after the shock wave is related by

$$u_1 u_2 = a^{*2}. \tag{4.2.14}$$

The speed u_2 is required to be the lesser by the direction of entropy change, while ρ_2, p_2, and T_2 must be greater than ρ_1, p_1, T_1.

An obstacle placed in a supersonic flow creates a shock wave as the only available means of adjusting flow-direction to that required. The shock wave in this case is inclined at an angle to the stream, and where the body is axisymmetric, lies as a conical envelope about the nose. If the obstacle in question is a supersonic pitot tube, the measured impact pressure is related to the stagnation pressure of the flow behind the shock. The ratio of the free stream static pressure p_1 to the measured pressure p_0' may then be expressed in terms of the Mach number as

$$\frac{p_1}{p_0'} = \frac{\{[2\gamma/(\gamma + 1)]M_1^2 - (\gamma - 1)/(\gamma + 1)\}^{1/(\gamma-1)}}{\{[(\gamma + 1)/2]M^2\}^{\gamma/(\gamma-1)}}$$

(Rayleigh's formula). (4.2.15)

4.2.1.4. Effects of Viscosity.[6,7] We shall return to a remark of the preceding section in which it was stated that compressible flow theory served well except in regions near the boundary. In a more precise formulation Prandtl[8] in 1904 postulated that the fluid viscosity acts to alter the flow about a streamline body only in a small region or layer immediately adjacent to the surface. Thus the flow outside this *boundary layer* may be treated by any suitable theory of inviscid flow. Within the thickness of the boundary layer, however, the local velocity must decrease from that of the inviscid flow of the outer edge to zero at the surface, and it is within this region that the attendant viscous stresses must be established.

The present discussion is not limited, however, to cases of the boundary layer approximation. In very slow flows, for example, where the concept of a boundary layer is inapplicable, the viscous stresses are known to propagate through the fluid to distances very far from the body. It is our first task, therefore, to find relations suitable for an expression of the viscous stresses in the most general formulation. To this end it is assumed that the rates of strain are linearly connected with the components of shearing stress, the assumption of a Newtonian fluid. The most general

[6] S. Goldstein, see reference 2, p. 90.

[7] H. Schlichting, "Boundary Layer Theory." McGraw-Hill, New York, 1955.

[8] L. Prandtl, *in* "Aerodynamic Theory" (W. F. Durand, ed.), Vol. 3, Section 20, pp. 34–208. Springer, Berlin, 1943.

relations of this nature consistent with the assumption of isotropy are

$$p_{xx} = -p' + \lambda\Delta + \mu e_{xx} \tag{4.2.16}$$

plus similar expressions yy and zz and $p_{yz} = \mu e_{yz}$ etc., where p_{ij} are the stresses, e_{ij} are the rates of strain, and λ and μ are coefficients. The static pressure is defined as the negative of the mean of the principal stresses, i.e.,

$$p = -\tfrac{1}{3}p_{ii} = p' - (\lambda + \tfrac{2}{3}\mu)\Delta. \tag{4.2.17}$$

For the static pressure to be independent of the rates of strains, we must take $\lambda + \frac{2}{3}\mu = 0$. Then

$$p_{ij} = -(p + \tfrac{2}{3}\mu\Delta)\delta_{ij} + \mu e_{ij} \tag{4.2.18}$$

where δ_{ij} is the Kronecker delta, and the coefficient μ is the absolute viscosity. It is the above formulation of the p_{ij} which was assumed in the development of the Navier–Stokes equations (Eq. (4.2.4)). More generally, one can define a "bulk" or "second" viscosity coefficient, κ, equal to the quantity $3\lambda + 2\mu$.

In the metric system, values of absolute viscosity are expressed in poises (dyne sec/cm^2). The ratio

$$\nu = \frac{\mu}{\rho} \tag{4.2.19}$$

is termed the kinematic viscosity and expressed in Stokes (cm^2/sec). Absolute viscosities of common gases may be found in Table I.

TABLE I. Absolute Viscosity of Common Gases[a]

Gas	Symbol	Viscosity $\mu \times 10^7$ poises/0°C	Temperature increment $(\Delta\mu)_T \times 10^7$ poises/°C
Air	—	1,813	4.78
Argon	A	2,225	6.37
Carbon dioxide	CO_2	1,463	4.50
Chlorine	Cl_2	1,330	4.51
Ethylene	C_2H_4	1,000	8.20
Helium	He	1,953	4.64
Hydrogen	H_2	882	2.00
Krypton	Kr	2,474	7.35
Methane	CH_4	1,090	3.30
Neon	Ne	3,112	6.97
Nitric oxide	NO	1,898	5.38
Nitrogen	N_2	1,750	4.55
Oxygen	O_2	2,030	5,87
Xenon	Xe	2,246	7.25

[a] Abstracted from J. Kestin *in* "American Institute of Physics Handbook," pp. 2–207, Table 2V-4. McGraw-Hill, New York, 1957.

The role of viscosity in determining the shear stress at a surface is most simply visualized in connection with the laminar motion of a gas between parallel plates in relative translation as in Fig. 1 (Couette flow). In this case a linear velocity gradient du/dy is established such that the viscous stress at the surface is given by

$$p_{yx} = \mu \frac{du}{dy}. \tag{4.2.20}$$

It is instructive to view the case of Couette flow in light of the kinetic theory.[9,10] One may see that a molecule moving from region A to region B, Fig. 1, carries with it tangential momentum appropriate to the region A. This extra momentum is transferred to a molecule in region B, while

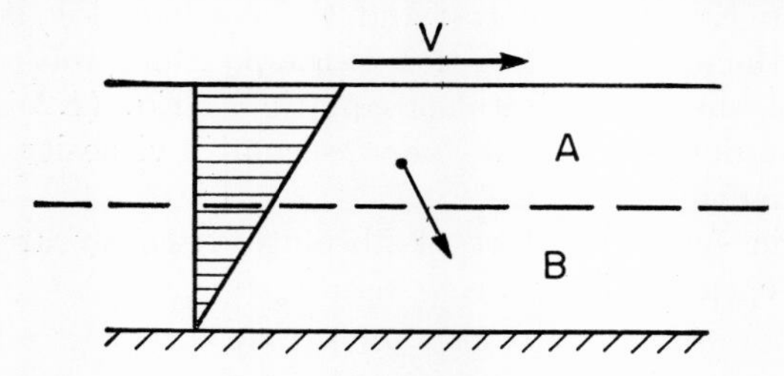

FIG. 1. Gas in Couette flow.

molecules of region B are similarly effecting a reverse transfer of momentum to region A. Thus, the transfer of tangential momentum in a gas is effected by mechanisms of molecular collision. The most exact theory gives the absolute viscosity as

$$\mu = 0.499\rho\bar{c}\lambda \tag{4.2.21}$$

where $\bar{c}$ is the mean random speed of the molecules due to their thermal motion and λ is the mean path length between molecular collisions. In this expression the common mean free path is used, given for hard spheres as

$$\lambda = \frac{1}{\sqrt{2}\, n\pi\sigma^2} \tag{4.2.22}$$

where σ is the molecular diameter and n is the molecule number density. Since the molecule thermal speed $\bar{c}$ is proportional to $(T)^{\frac{1}{2}}$ where T is the absolute temperature, the viscosity of gases in the case of this elementary model is seen from Eqs. (4.2.21) and (4.2.22) to be independent of the pressure, but proportional to the square root of the absolute temperature.

[9] E. H. Kennard, "Kinetic Theory of Gases." McGraw-Hill, New York, 1938.
[10] L. B. Loeb, "Kinetic Theory of Gases," 2nd ed. McGraw-Hill, New York, 1934.

Experiment shows that the hard sphere model does not lead to quite the correct result. The most recent determinations of viscosity are well correlated by the Keyes[11] formula

$$\mu = \frac{a_0 T^{\frac{1}{2}}}{1 + (a/T) + 10^{-a_1/T}} \tag{4.2.23}$$

where a_0, a, and a_1 are empirical constants. For air the values are

$$a_0 = 1.488 \times 10^{-5}$$

$a = 122.1$, $a_1 = 5$ in the temperature range 79–1845°K. Other values may be obtained from the tables of Reference 11.

The problem of boundary layer growth on a semi-infinite flat plate immersed in a laminar flow has been the subject of many investigations in the past half century. The considerations of Prandtl had led to a simplification of the Navier–Stokes equations, but it remained for Blasius[12] in 1908 to supply an approximate solution for the flat plate. It is a consequence of the solution that the value of u, the local velocity, reaches that of U, the free stream velocity, only as y, the transverse distance from the plate, approaches infinity. However, the ratio u/U reaches the value 0.994 when

$$y = \delta = \frac{5.2x}{\sqrt{\mathrm{Re}_x}} \tag{4.2.24}$$

where δ is defined as the boundary layer thickness, and where Re_x is the Reynolds number based on the distance x from the leading edge of the plate.

The Reynolds number is a dimensionless parameter first expressed by Reynolds[13] in connection with studies of laminar and turbulent pipe flow. It is written

$$\mathrm{Re} = u\rho d/\mu \tag{4.2.25}$$

where u is the mass velocity, μ is the absolute viscosity, ρ is the density, and d is a characteristic dimension of the flow. In the case of flow in pipes, the characteristic dimension is the pipe diameter; in the case of the boundary layer thickness as expressed above, the dimension of significance for the prediction and correlation of the phenomena is the dis-

[11] F. G. Keyes, "The Heat Conductivity, Viscosity, Specific Heat and Prandtl Number for Thirteen Gases." Project Squid, Mass. Inst. Technol. Tech. Rept. 37 (1952).

[12] H. Blasius, *Z. Math. Physik* **1,** 56 (1908). Also *Natl. Advisory Comm. Aeronaut. Tech. Mem.* **1256** (1950).

[13] O. Reynolds, *Phil. Trans. Roy. Soc. London* **174** (3), 935 (1883).

tance x. For a review of the more recent work in the field of boundary layer theory the reader is referred to References 7 and 9.

As the relative speed of the surfaces of Fig. 1 is increased, the flow becomes turbulent. Momentum transfer between regions of the flow field is no longer accomplished solely by the travel of individual molecules, but by the random movement of fluid elements of larger size. Thus the simple picture of the development of shear stresses at the surface in laminar flow must be replaced by a more complicated model in which the statistical fluctuation of fluid elements are also considered.

It is found that turbulent flows are developed under ordinary circumstances at Reynolds numbers between 2300 and 4000, irrespective of the absolute value of pipe size, density, viscosity, or speed of flow. Thus the Reynolds number becomes the significant parameter for correlating the phenomena of turbulence, one of the most interesting and far reaching discoveries in the whole of fluid mechanics. The Reynolds number attending the onset of turbulent flow is referred to as *critical.* The consequences of turbulence to the flow of gases in general cannot be explored in a paper of the present scope, although certain specific ramifications will be discussed in later sections. No complete theory of turbulent flows exists because of the complexity of the fundamental equations, but there are many semi-empirical and phenomenological treatments which find application in various areas of gas mechanics, particularly in the aeronautical sciences.[14,15]

4.2.1.5. Heat Transfer.* [16,17] The transfer of thermal energy from one region of a gas to another, or from the gas to a wall, proceeds by a complex of means which are ordinarily grouped under conduction, convection, and radiation. The mechanisms of conduction and convection are completely, although obscurely, stated for ideal gases in the terms of the energy equation, Eq. (4.2.5). However, for real gases where radiation also plays an important role in thermal energy transfer, it is essential to conduct investigations along lines which may be less general while being physically more complete. Thus, for this reason and because of the immense complexity of the general equation of fluid flow, there exist numerous special theories of heat transfer in gases, each applicable to a limited range of conditions. The measurement of the transfer of heat proceeds by the measurement of temperatures at surfaces, or within the bodies of

* See also Chapter 6.3.

[14] H. L. Dryden, *Advances in Applied Mechanics* **1,** 2 (1948).

[15] B. A. Bakhmeteff, "The Mechanics of Turbulent Flow." Princeton Univ. Press, Princeton, New Jersey, 1936.

[16] W. H. McAdams, "Heat Transmission," 3rd ed. McGraw-Hill, New York, 1954.

[17] W. H. Giedt, "Principles of Engineering Heat Transfer." Van Nostrand, Princeton, New Jersey, 1957.

probes exposed to the stream. Exceptions may be found in gaseous radiation measurement, which may proceed by photometric or spectrographic techniques. Although the surface temperature determinations are accomplished by the ordinary techniques of thermometry, resistance wires, thermisters, thermocouples, etc., the interpretation of these measurements is dependent upon the applicable theory.

It is appropriate to begin with the concept of thermal conductivity. Conductivity in a gas occurs by virtue of the random thermal motion of

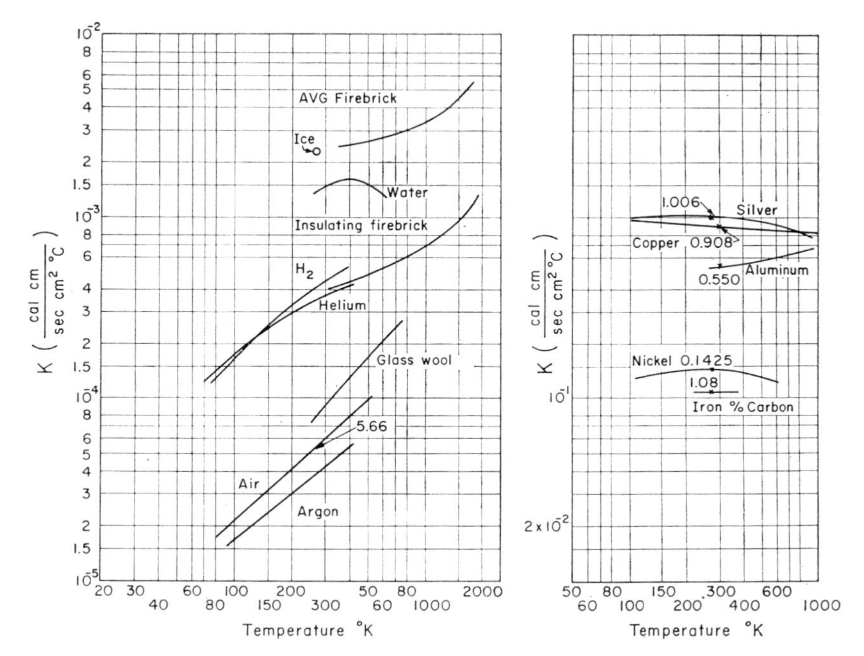

FIG. 2. Thermal conductivities of gases, liquids, and solids.

the molecules, and is thereby distinguished from convection which occurs through the motion of fluid elements. The basic equation of conductivity in one dimension states that the energy flow per unit time (heat rate) per unit area is proportional to the temperature gradient. Thus,

$$\frac{dQ}{dt} = q = -kA\,\frac{dT}{dx} \tag{4.2.26}$$

where Q is the quantity of heat, A is the area, T is the temperature, and k is a coefficient termed the *thermal conductivity*. In the cgs system the units of k are cal/sec cm °K. Values of thermal conductivity for several gases, liquids, and solids may be found in Fig. 2. It is occasionally useful to relate the thermal conductivity to the thermal capacity for a substance and to consider this combination of properties a single property. The ratio

$$\frac{k}{\rho C_v} = \alpha \tag{4.2.27}$$

is termed the *thermal diffusivity* where ρ is the density and C_v is the specific heat.

The thermal conductivity may be discussed in terms of the kinetic theory in a manner quite analogous to that employed in connection with viscosity, Section 4.2.1.4. Where a temperature gradient is established

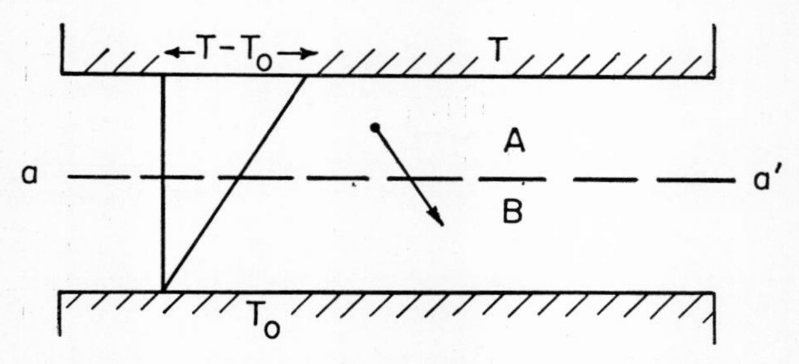

FIG. 3. Temperature gradient established in a gas.

in a gas as in Fig. 3, molecules from region A transport an excess of thermal energy to region B. It can be shown that the net transport of energy per unit of time across unit area of a boundary such as aa' is given by

$$q = \tfrac{1}{4}(9\gamma - 5)\mu C_v \frac{dT}{dx} \tag{4.2.28}$$

where γ is the ratio of specific heats, C_v is the specific heat of the gas at constant volume, and μ is the absolute viscosity as given in Eq. (4.2.21). Equation (4.2.28) follows a suggestion made by Eucken[18] in 1913 whereby it was proposed that changes in the state of a gas are accomplished at one rate in the translational energy of the molecules, and at other rates in their various internal energies. Some progress has recently been made toward refinement of the concepts, notably as a result of information from observations of relaxation times in shock wave experiments,[19] and the dispersion and absorption of ultrasonic vibrations in gases.[20] Here the

[18] A. Eucken, *Physik. Z.* **14,** 324 (1913).

[19] J. C. Logan, "Relaxation Phenomena in Hypersonic Aerodynamics." Reprint No. 728, Institute of Aeronautical Sciences, New York, 1957.

[20] J. O. Hirschfelder, C. F. Curtiss, and R. B. Bird, "Molecular Theory of Gases and Liquids." Wiley, New York, 1954.

times required for the interchange of energy among its various forms may be studied in detail.

It is central to the field of heat transfer to discuss what occurs in the neighborhood of the boundaries. Consider a heated vertical wall adjacent to a stationary body of gas. Since the gas must be at rest at the wall, heat transfer through the first layer is governed by the conduction equation. A short distance out from the wall the mechanisms of convection become effective in creating a current of gas upward. As we move further out, the velocity gradient continues to increase, then decreases as the temperature gradient diminishes and as the local temperature approaches that of the main body of the gas. The *thermal boundary layer* thickness is defined as the point at which the temperature gradient has effectively returned to zero. It should be noted that heat transfer is not *out* through the boundary layer to the body of gas, but rather in the vertical direction by reason of the convective flow. If the wall is high enough, turbulent motion ensues.

In a partial avoidance of the manifest complexities, much of the work of heat transfer is formulated in terms of the heat transfer coefficient h, proposed by Newton in 1701 and defined by the relation

$$q = hA(T_w - T_\infty) \tag{4.2.29}$$

in which T_w is the surface temperature and T_∞ is the remote temperature of the gas. An inspection of Eq. (4.2.29) in relation to the conduction equation shows that h is a measure of the slope of the temperature gradient immediately at the wall.

The problem of the velocity boundary layer on a flat plate in laminar flow has a close analog in the problem of the thermal boundary layer for the same configuration. Where the temperature and velocity gradients are low, equations of the same form govern the formation of both boundary layers where α, the thermal diffusivity, is interchanged with ν, the kinematic viscosity, and where T is interchanged with the tangential velocity u. The temperature and velocity distributions would be identical where $\nu = \alpha$, a condition which obtains for most gases. This fact has suggested the dimensionless ratio

$$\mathrm{Pr} = \nu/\alpha = \mu C_p/k \tag{4.2.30}$$

where Pr is known as the Prandtl number, as a measure of the similarity of the velocity and thermal boundary layers. Where $\mathrm{Pr} < 1$, the thermal boundary layer is the thicker, while for $\mathrm{Pr} > 1$ the reverse is true.

The local heat transfer coefficient h_x per unit width for a plate oriented parallel to a laminar flow is found to be

$$h_x = 0.332k\text{Pr}^{\frac{1}{3}}\sqrt{\frac{U_\infty}{x\nu}} \tag{4.2.31}$$

where x is the distance downstream from the leading edge, and from this, where the plate is of length l the average heat transfer coefficient h is obtained as

$$h = 0.664k\text{Pr}^{\frac{1}{3}}\sqrt{\frac{U_\infty}{l\nu}}. \tag{4.2.32}$$

It should be noted that a singularity exists at the leading edge where the local heat transfer coefficient becomes infinite, while downstream the values are finite and vary inversely as root x.

As a further step toward the generalization of theoretical and experimental findings, the dimensionless grouping known as the Nusselt number Nu may be written as

$$\text{Nu}_x = h_x x/k. \tag{4.2.33}$$

Thus, for the thermal boundary layer, when the Nusselt number is to be based on the local heat transfer coefficient, we have from Eq. (4.2.31),

$$\text{Nu}_x = 0.332\text{Pr}^{\frac{1}{3}}\text{Re}_x^{\frac{1}{2}}. \tag{4.2.34}$$

A similar equation results from Eq. (4.2.32), where Nu is based on the average heat transfer coefficient. Where the convective heat transfer takes place in tubes and occurs in connection with a laminar flow, the problem is more complex and the solution generally of more restricted applicability, i.e., limited to cases of restrictive assumptions of initial flow velocity and thermal boundary condition. Even so, much has been accomplished in the correlation of theory and experiment.

Where the flows are turbulent, effort in analysis has centered on a determination of the significant parameters for the correlation of the experimental data, rather than on solutions of the general governing equations. Results of such efforts are summarized in detail in the literature.[21]

In high speed flows, and particularly where compressibility is important, substantial quantities of heat may be generated as a result of viscous dissipation and fluid compression in the boundary layer. These effects cause the temperature in gas adjacent to the leading surface to approach its *stagnation* value, defined as the temperature resulting in an isentropic slowing of the stream from velocity U_∞ to zero, and expressed as

$$T_0 = \frac{U_\infty^2}{2JC_p} + T_\infty. \tag{4.2.35}$$

[21] H. A. Johnson and M. W. Rubesin, *Trans. Am. Soc. Mech. Engrs.* **71,** 447 (1949). A summary of the literature on Aerodynamic Heating.

If a probe of simple geometry, for example a sphere, is instrumented by means of a thermocouple placed within the probe body, the equilibrium temperature may be found when the probe is placed within the high speed gas stream.

A *recovery factor* R may be defined such that

$$R = \frac{T_p - T_\infty}{T_0 - T_\infty} \tag{4.2.36}$$

where T_p is the measured probe temperature, T_∞ is the static temperature of the stream, and T_0 is the stagnation temperature. Values of R are found to be of the order unity.

4.2.1.6. Diffusion. The general fluid dynamical equations as they are formulated in Section 4.2.1.2 apply only to the fluid motions of a gas which is homogeneous with respect to composition. No general formulation exists which accommodates variable relative concentration where more than one molecular species is present, nor which allows for the attendant variation in specific heat or viscosity. However, some cases of one-dimensional flow have been studied extensively, where the central interest lies in the changing molecular concentrations rather than in the fluid mechanical effects. The molecular processes by which these concentration changes are effected are those of diffusion.

The subject of diffusion may be approached by an examination of the mechanism of what is termed *pure* diffusion.[22] In pure diffusion the center of mass of a binary mixture is assumed to remain at rest while the relative spatial distribution of the two molecular species becomes modified with increasing time. In the simplest case we may envisage a tubular reservoir having a gate valve at the center which divides the reservoir into regions A and B, containing molecules of mass 1 and mass 2 respectively. We assume no initial pressure difference between the two regions. When the gate is withdrawn, molecules of mass 1 move by a succession of random motions and collisions into region B, while molecules of mass 2 proceed in similar fashion into region A. If Γ_1 designates the net number per second of molecules of type 1 passing into B across a unit of area normal to the concentration gradient, and if Γ_2 is similarly defined with respect to molecules of type 2, the coefficient of diffusion D may be defined by the following equations:

$$\Gamma_1 = -D\frac{dn_1}{dx}; \qquad \Gamma_2 = -D\frac{dn_2}{dx} \tag{4.2.37}$$

where n_1 and n_2 are the molecule number densities. The process continues until each species fills the volume uniformly.

[22] E. H. Kennard, "Kinetic Theory of Gases." McGraw-Hill, New York, 1938.

The process is quite analogous to the flow of heat in an adiabatic reservoir. In consequence, it may be shown that the governing equation is mathematically identical to the Fourier heat equation, and may be written in the form

$$\frac{\partial n_1}{\partial t} = D\nabla^2 n_1 \tag{4.2.38}$$

where D is the coefficient of diffusion defined above, and n_1 is the local number density of molecules of mass 1. Because of this mathematical identity, the results of many investigations in the field of heat transfer in solids may be applied directly to problems of diffusion.[23,24]

It should be noted that the assumed absence of net mass transfer requires a volumetric counterflow in the direction of the region initially containing the more rapidly diffusing molecules. By application of this consideration to an elementary flux balance in which we write $\bar{c}$, the mean molecular speed, and λ_1, the mean free path of species 1 in the mixed gas, we find the Meyer[25] formula for the coefficient of diffusion, in which D is represented as

$$D = \tfrac{1}{3}\left(\frac{n_2}{n}\bar{c}_1\lambda_1 + \frac{n_1}{n}\bar{c}_2\lambda_2\right) \tag{4.2.39}$$

where $n = n_1 + n_2$. This result shows D to be strongly dependent on the relative concentration, a conclusion at substantial variance with the experimental findings. The more elaborate work by Chapman[26] and Enskog[27] in which account is taken of the nonequilibrium nature of the velocity distribution function, predicts correctly that the diffusion coefficient is slightly dependent on relative concentration for dissimilar molecules. The Chapman–Enskog equation may be written

$$D_{12} = (1 + \lambda_{12})(\tfrac{3}{8})\sqrt{\frac{\pi}{2}}\frac{1}{nS_d}\left(\frac{m_1 + m_2}{m_1 m_2}kT\right)^{\frac{1}{2}} \tag{4.2.40}$$

where S_d is the collision cross section ($S_d = \pi\sigma^2$ for hard spheres), and λ_{12} is a constant which varies in value from zero for molecules repelling each other as r^{-5} where r is the radial separation of the molecules to 0.132 for

[23] J. Crank, "The Mathematics of Diffusion." Oxford Univ. Press, London and New York, 1956.

[24] H. S. Carslaw and J. C. Jaeger, "Conduction of Heat in Solids." Oxford Univ. Press, London and New York, 1947.

[25] O. E. Meyer, "Kinetic Theory of Gases." Translation published by Longmans, Green, New York, 1899.

[26] S. Chapman, *Phil. Trans. Roy. Soc. London, Ser. A* **217**, 115 (1918).

[27] D. Enskog, "Kinetische Theories der Vorgänge in mässig verdünnten Gases." Ph.D. Dissertation, Uppsala University, Sweden, 1917. See also reference 1.

hard spheres of extremely unequal mass. Values of D_{12} for several gas pairs at room temperature are presented in Table II. A useful, although inexact, formula for the prediction of the temperature and pressure dependence of D_{12} may be written

$$(D_{12})_{T,p} = (D_{12})_{T_0,p_0} \left(\frac{T}{T_0}\right)^n \frac{p_0}{p} \tag{4.2.41}$$

where n is a constant having the value 1.75 for the permanent gases and 2 for the more condensible gases.

TABLE II. Diffusion Coefficients D_{12} for Common Gases

Gases	D_{12} (observed) cm²/sec
H_2–Air	0.661
H_2–O_2	0.679
O_2–Air	0.1775
O_2–N_2	0.174
CO–H_2	0.642
CO–O_2	0.183
CO_2–H_2	0.538
CO_2–Air	0.138
CO_2–CO	0.136
N_2O–H_2	0.535
N_2O–CO_2	0.0983
H_2O–Air	0.219
H_2–D_2	1.20

The discussion thus far has considered a special case of *concentration* diffusion in which, in finite systems, inequalities in molecular concentration became erased with the passage of time. The most general theories predict two further processes, *pressure* and *thermal* diffusion, each proceeding toward a *separation* of constituents in the presence of strong pressure or thermal gradients. The possibility of thermal diffusion was first suggested by Enskog[28] and later predicted, and then experimentally verified, by Chapman.[29] For temperature differences of about 200°C, measured changes in concentration of mixtures of hydrogen and sulfur dioxide, for example, were of the order of 3–4%. An interesting application of pressure diffusion has been made recently by Becker,[30] in which he has demonstrated the possibility of economical isotope separation by means of rarefied gas jets in which strong pressure gradients occur.

[28] D. Enskog, *Physik. Z.* **12,** 533 (1911).

[29] S. Chapman and F. W. Dootson, *Phil. Mag.* [6] **33,** 248 (1917).

[30] E. W. Becker, K. Bier, and H. Burghoff, *Z. Naturforsch.* **10a,** 565 (1955).

One further and somewhat better-known diffusion process is that of *forced* diffusion, in which the constituents are selectively influenced by an external field. In the most important case, that of an ionized gas in an electric field, studies of forced diffusion are complicated by the multitude of processes which may occur at or below energies sufficient to produce ionization. Some or all of these may affect the momentum, energy, quantum state, and number density of the particles. For further information the reader is referred to references in the fields of gaseous electronics[31] and plasma dynamics.[32,33]

4.2.1.7. Similarity. It is a consequence of the great complexity of the most interesting problems of gaseous flow that much of the work of gas dynamics has in the final analysis related to questions of *similarity*. With each experimental investigation there exists the accompanying problem: what are the significant parameters of the flow; how shall the flow be characterized so that the present findings have predictive value in a flow which is similar but not identical?

For purposes of discussion, the experimental flow shall be designated as the *model* flow, while the one with which it is to be compared shall be termed the *prototype*. It would seem evident at the outset that *geometrical* similarity between model and prototype must exist; that is, that the ratios between corresponding lengths must be the same. A possible exception to this first requirement should be noted, in which an analog of the flow is created, either by means of a transformation of coordinates, or by the use of other physical phenomena. It may be reasonably argued from another view that such analogs do not in fact violate the requirement of geometrical similarity.

It is also necessary that the ratios of all corresponding velocities and accelerations be equal, the requirement of *kinematic* similarity. It follows immediately that the ratios of all corresponding forces on the fluid masses must be the same. This is the requirement of *dynamic* similarity. Of all possible forces acting on fluid elements, only a few are of interest in the characterization of gaseous flows. For example, forces due to gravity are usually not of interest except in meteorological studies and forces of surface tension are nonexistent. The forces of interest which remain are those of pressure, viscosity, inertia, and elasticity, and with them three independent ratio equations between model and prototype may be written. If the forces in question are written

[31] L. B. Loeb, "Basic Processes in Gaseous Electronics." University of California Press, Berkeley, California, 1955.

[32] W. P. Allis, "Handbuch der Physik—Encyclopedia of Physics" (S. Flügge, ed.), Vol. 21, 383–444. Springer, Berlin, 1955.

[33] R. F. Post, *Rev. Modern Phys.* **28**, 338 (1956).

$$
\begin{aligned}
F \text{ inertia} &= \rho V^2 l \\
F \text{ pressure} &= p l^2 \\
F \text{ viscosity} &= \mu l V \\
F \text{ elasticity} &= l^2 a^2 \rho
\end{aligned} \tag{4.2.42}
$$

where a is the speed of sound, and use has been made of the relation $a = \sqrt{(\gamma p)/\rho}$, the similarity equations may be written

$$\left(\frac{\rho V^2}{p}\right)_P = \left(\frac{\rho V^2}{p}\right)_M \tag{4.2.43}$$

$$\left(\frac{Vl\rho}{\mu}\right)_P = \left(\frac{Vl\rho}{\mu}\right)_M \quad \text{(Reynolds number)} \tag{4.2.44}$$

$$\left(\frac{V}{a}\right)_P = \left(\frac{V}{a}\right)_M \quad \text{(Mach number).} \tag{4.2.45}$$

The requirement of Eq. (4.2.45) will ordinarily be satisfied, since approximate thermal similarity occurs under most circumstances and ρ and p are connected by the equation of state. Thus we are led again to the Mach and Reynolds numbers as the significant parameters of the flow for aerodynamics interpretations and correlations. Additional requirements may be applied and other dimensionless groupings may be found useful as well. The reader is referred, for example, to Section 4.2.1.5.

4.2.2. Flow Phenomena—Normal Pressures

The flow phenomena to be considered here are those encountered in the everyday range of engineering experience, extending from pressures of a few standard atmospheres to a lower limit of perhaps one thousandth of an atmosphere. These are flows of the continuum regime for which lower limits are more precisely stated in Section 4.2.3 and upper limits fixed for each gas by consideration of the utility of the equation of state of an ideal gas. In general, the gas may be treated as a compressible viscous fluid or by means of a simplified fluid model, inviscid, incompressible, or both, where reason and experience permit. Here at normal pressures are to be found all ordinary phenomena of aerodynamics, as well as those associated with the great majority of flows in air heating, cooling, and ventilating systems, over architectural or other structures, in many varieties of power plant, and in process equipments of the chemical and metallurgical industries. Although the diversity of the subject is immense, the practical interest is such that many excellent books of reference and handbooks exist.[34] Accordingly, the present section will confine itself to

[34] See, for example, L. S. Marks, "Mechanical Engineers' Handbook." McGraw-Hill, New York, 1955.

the presentation of a few of the methods and experimental findings which are of most general applicability.

4.2.2.1. Ducted Flows—Flows in Pipes and Wind Tunnels.[35,36] Gaseous flows may be roughly divided into two varieties, internal and external. In the study of internal flows, the interest lies in the whole motion of the gas as it moves in a duct, in distributions in pressure, temperature, and velocity, in dissipative effects, in volumetric and mass flow rates, and in the power expended in maintaining the flow. In external flows the interest is centered on a body immersed in the fluid, on the aerodynamic forces acting on the body, on the heat transfer between the body and the surrounding stream, and on the field of flow in immediate interaction with the body.

A great many of the problems of gaseous flows in ducts may be treated by the methods suggested in Section 4.2.1 in connection with the discussion of one-dimensional compressible flow. Where the gas velocity is low the incompressible Bernoulli equation and the equation of continuity may serve. These may be written

$$\frac{u^2}{2} + \frac{p}{\rho} = \text{const} \quad \text{(Bernoulli's equation)} \tag{4.2.46}$$

$$\rho u A = \text{const} \quad \text{(Equation of continuity).} \tag{4.2.47}$$

Where the flow velocity is an appreciable fraction of the velocity of sound the equations of compressible flow must be employed. It should be noted that no account is taken of dissipative effects in these expressions. Hence, the range of applicability is limited to short, large diameter systems in which there are inappreciable temperature gradients in the gas or along the boundaries.

If a gas which is initially at room temperature flows at moderate speeds in a long metal pipe or duct standing exposed to the ambient air, a good approximation may be made by the assumption of isothermal conditions. If the Reynolds number is below critical, the flow will be laminar, and in a straight circular pipe will have shear stresses and velocities symmetrically distributed about the center line. The law of Poiseuille flow applies, and may be expressed as

$$Q = \frac{\pi a^4}{16\mu l} \frac{p_1{}^2 - p_2{}^2}{RT} \tag{4.2.48}$$

where Q is the flow rate in g/sec, a is the pipe radius, and l its length,

[35] A. Shapiro, "The Dynamics and Thermodynamics of Compressible Fluid Flow," Vols. 1 and 2. Ronald, New York, 1954.

[36] H. W. Liepmann and A. Roshko, "Elements of Gasdynamics," Wiley, New York, 1957.

T is the temperature of the gas, and μ is its viscosity, p_1 and p_2 are the pressures at entrance and exit of the pipe.

If the flow is turbulent, the "friction factor" of Darcy and others may be employed to supply a connection between the shearing stress at the wall, τ_0, and the velocity V of the mass flow. Thus

$$\tau_0 = f\rho \frac{V^2}{8} \tag{4.2.49}$$

where f is the friction factor and ρ is the density. The pressure drop for a given mass flow rate Q may then be calculated in terms of f, as

$$p_1^2 - p_2^2 = \frac{QRT}{\pi^2 a^4}\left[2 \ln \frac{V_2}{V_1} + f\frac{l}{d}\right] \tag{4.2.50}$$

where T is the temperature and R is the universal gas constant. The term $\ln V_2/V_1$ is ordinarily small by comparison with $f(l/d)$; otherwise

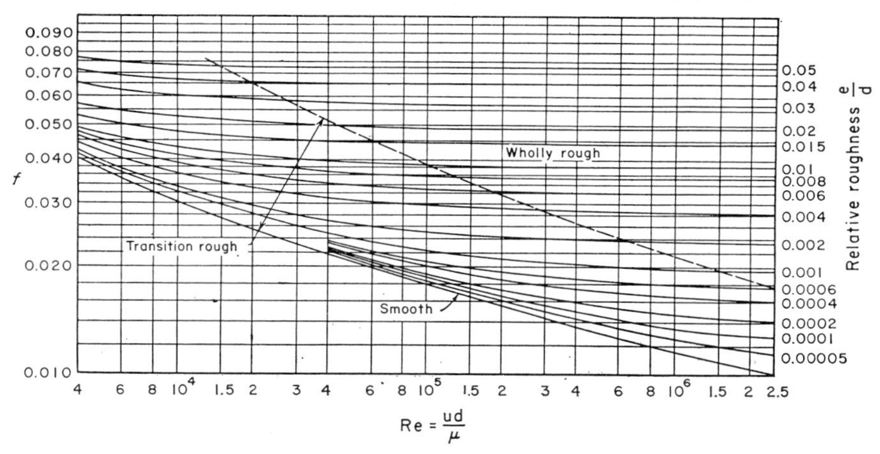

FIG. 4. Relation of friction factor, Reynolds number, and roughness for commercial pipes.

the solution must be found by trial. Values of f for various conditions of Reynolds number and pipe roughness may be found from the diagram, Fig. 4.

For flows of greater thermodynamic complexity recourse must be taken in the semi-empirical formulations of the appropriate field of mechanical engineering.

The wind tunnel as a special case of a ducted flow system presents a circumscribed problem of considerable technical interest. It is the function of all wind tunnels to produce a prescribed stream at the tunnel test section, often with the additional requirement that this be accomplished with the greatest possible economy. The flow is accelerated and formed in the nozzle, which acts to transform the thermal energy of the gas upstream of the nozzle throat into kinetic energy of flow at the nozzle exit. Directly preceding the nozzle is the stagnation region which, in turn, follows the compressor, while immediately downstream of the nozzle are the tunnel test section, a transition region connecting with the diffuser, and the return circuit to the compressor in that order. It is the function of the diffuser to effect a partial recovery of the thermal energy transformed in the expansion, and to guide the flow into the return circuit.

Heat exchangers before and after the pump may be employed to adjust the temperature, and in particular to adjust the stagnation temperature

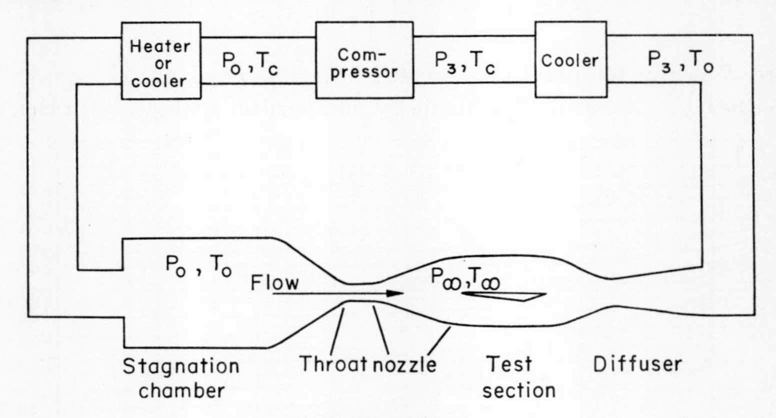

FIG. 5. Wind tunnel schematic.

after recompression. The configuration of the present discussion is schematically illustrated, Fig. 5. The majority of wind tunnels operating at ordinary pressures are of the return type for economic reasons, and most are totally enclosed in the fashion illustrated, so that test section pressures can be fixed in design to meet specified aerodynamic conditions.

The pump must compress the air through the ratio p_0/p_3 where p_0 is the stagnation pressure and p_3 is that at the pump intake. In an analysis by Crocco[37] it is assumed that a shock wave is formed at the exit of the test section and that the rest of the diffuser pressure recovery occurs subsonically. An average diffuser efficiency may be assumed and with it the recovery in p_0 due to the diffuser may be calculated. Since the loss in pressure recovery due to entropy change across the shock wave may also be calculated, the over-all pressure loss may be found as a function of Mach number. One such estimated curve[38] is presented in Fig. 6. From the

[37] L. Crocco, *Aerotecnica* **15,** 237 (1935) and **15,** 735 (1935).

[38] J. Lukasiewicz, *J. Aeronaut. Sci.* **20,** 617 (1953).

information of Fig. 6, the one-dimensional continuity equation and performance characteristics of the pump, tunnel test section sizes, or desired pump characteristics may be found.

4.2.2.2. Flows about Immersed Bodies. The subject of fluid forces and heat transfer on bodies totally immersed in the stream is of substantial interest and importance in many fields of study in addition to those of the aeronautical sciences. Effort will be directed in this section toward the presentation of those concepts of the most general interest and applicability.

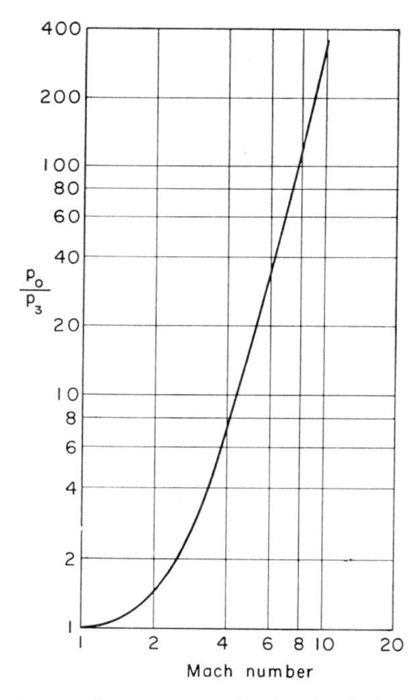

FIG. 6. Wind tunnel pressure ratios for fixed throat diffusers.

In the ordinary convention, forces on the body parallel to the incident stream are designated as those of drag, while those normal to the stream are termed lift. Moments about the three coordinate axes may also be induced on the body, termed the moments of pitch, yaw, and roll. In the case of a nonspinning sphere only drag is present. For other configurations some or all of the forces or moments may be in evidence, depending on the condition of motion with respect to the stream. It should be appreciated that even in the case of a relatively simple body, a cone, cylinder, or flat plate, for example, the origin of the various gross forces and moments may be quite complex in detail. In spite of this difficulty, the results of fluid-mechanic experiment can be correlated rather well and many results predicted where the fluid force or moment is equated to the free stream dynamic pressure multiplied by a coefficient specific to the shape. Thus the drag is written

$$D = C_d A \frac{\rho U^2}{2} \tag{4.2.51}$$

where D is the drag force, A is the projected area of the body, ρ is the density of the stream, U is the relative velocity of body and stream, and C_d is the coefficient of drag. A moment coefficient C_M is similarly defined by the expression

$$M = C_M l A \frac{\rho U^2}{2} \tag{4.2.52}$$

where M is the moment and l is a characteristic length.

Once again it is found desirable to correlate values of the force coefficients by means of the Reynolds number. The drag coefficient, for example, is ordinarily found to be a slowly varying function of Reynolds number, but in some instances rather sharp changes may occur. The sphere is a shape which exhibits the latter behavior, and at the same time serves in illustration of many details of the interaction of fluid and stream.[2] At very low free stream velocities the streamlines about the sphere resemble those of an ideal fluid flow. It should be remarked, however, that the resemblance is superficial, since the properties of the real fluid flow are largely determined by the viscous effects. In the very slow flow limit the Stokes relation shows the drag coefficient to be given by the expression

$$C_d = 24/\text{Re} \tag{4.2.53}$$

where Re is based on sphere diameter.

As the velocity of the flow increases, the energy loss in fluid elements nearest the surface is sufficient to prevent these elements from returning along the adverse pressure gradient to the full free stream static pressure at the rear stagnation point. In this circumstance, the flow breaks from the surface of the sphere at the *separation point,* in this case located in the forward hemisphere about 83° from the forward stagnation point. The streamlines are diverted from their closing configuration and encompass instead a large diameter turbulent wake. This condition and that of the slow flow are illustrated in Fig. 7.

At Reynolds numbers of 1×10^5 to about 4×10^5, transition occurs in the boundary layer forward of the separation point, accompanied by a sudden shift of the separation point to the rear hemisphere and an attendant abrupt decrease in the drag coefficient. The coefficient of drag

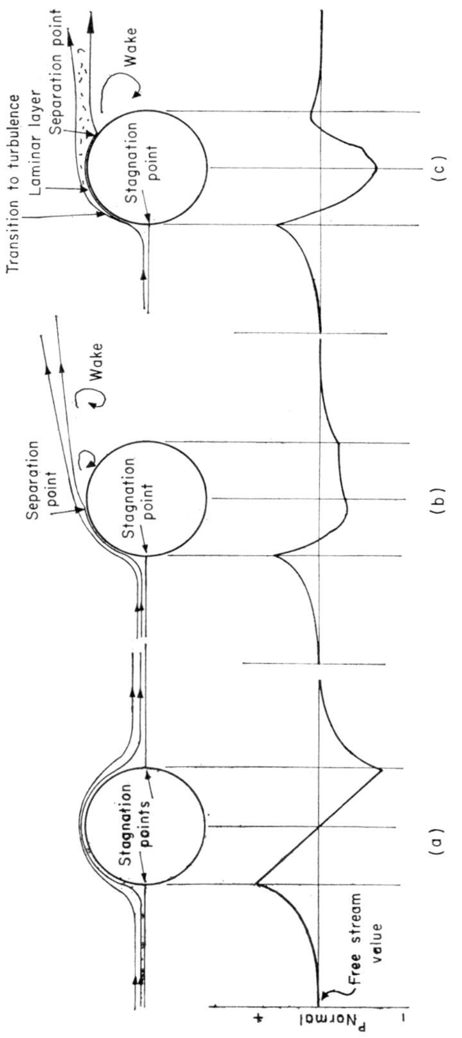

FIG. 7. Diagrams of the flow about spheres for three conditions of Reynolds number: (a) Stokes regime Re < 0.1; (b) laminar separation $10^3 <$ Re $< 2.5 \times 10^5$; and (c) turbulent separation Re $> 2.5 \times 10^5$.

for a sphere is plotted as a function of Reynolds number in Fig. 8, and with this are similar plots of drag coefficients for other simple aerodynamic shapes.

Although the fluid mechanical sources of drag are intimately connected, it is convenient to consider the total drag as being composed of two parts, that due to skin friction, and that due to form. The form drag is obtained by integrating the axial component of normal pressure over all elements of the surface, while the skin friction drag is obtained by a similar summation of the tangential stresses. The distinction is illustrated in the case of a flat plate which, when oriented normal to the flow, is influenced almost entirely by form drag, but when oriented along the flow is influenced largely by skin friction drag. For most bodies in the very slow flows the skin friction drag is much larger than the form drag, while for Reynolds numbers of a thousand and greater the skin friction has dropped to a small fraction of the total.

If a symmetrical body is inclined to the stream, there will occur an unbalance of normal pressures on corresponding elements of its surface. If the body is not symmetrical a similar unbalance may occur at zero angle of attack, i.e., where some principal reference line of the body lies parallel to the stream. The component transverse to the stream is designated as the lift, the lift coefficient being defined in the same fashion as that of the drag. The lift coefficient is evidently of greatest interest when its value is known as a function of angle of attack. Representative lift curves for simple aerodynamic shapes and for a typical airfoil are presented in Fig. 9. For more detailed information the reader is referred to the Index of NACA Technical Publications.[39]

As the relative speed of stream and body increases toward that of sound, the description of the flow and of its interaction with the body grows in complexity. At a critical speed the flow becomes sonic as it is accelerated over the body with the resultant appearance of shock waves at points of return to subsonic flow. Under these conditions the flow is said to be transonic. It is interesting to note that coefficients of drag increase to a maximum in the transonic regime and return again to lower values as the flow becomes entirely supersonic. For spheres, for example, the coefficient rises from approximately 0.5 to 1.0 through the transonic regime, and then falls again to less than 0.95.[40] When the body passes through the gas at greater than sonic speed, a shock wave is formed

[39] Index of NACA Technical Publications, published annually by the National Advisory Committee for Aeronautics, U.S. Government Printing Office, Washington, D.C. The 1915–1949 combined index, in particular, contains many references of value in the field of aerodynamic forces in subsonic flows.

[40] A. C. Charters and R. N. Thomas, *J. Aeronaut. Sci.* **12**, 468 (1945).

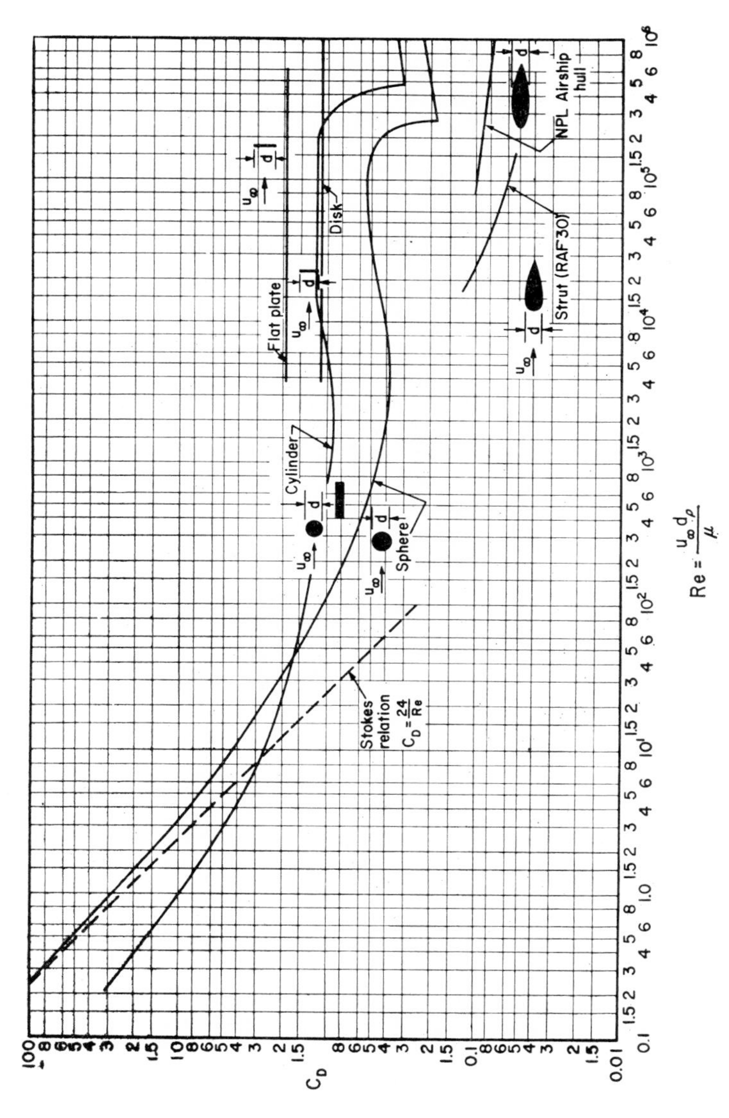

FIG. 8. Representative drag coefficients in subsonic flows.

which precedes the body at the stagnation point. The flow behind the shock is subsonic in the region of the stagnation point, but remains supersonic behind regions of the wave removed from the body. From this discussion it may be imagined that there exist many theoretical and experimental obstacles to a rigorous understanding of the supersonic flow field; however, much work of practical value has been done where aerodynamic force coefficients in supersonic flows are determined for specific classes of bodies and defined in terms of the free stream conditions.

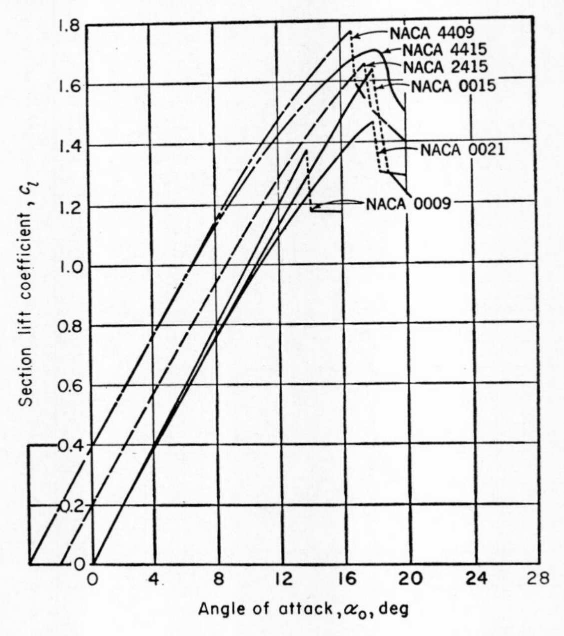

FIG. 9. Aerodynamic characteristics of some four-digit airfoils; $R \sim 8 \times 10^6$ (from "Principles of Aerodynamics," by James H. Dwinnell).

The production of flows at normal densities is accomplished by propellers or fans, Roots type or centrifugal blowers, or multistage axial flow compressors. Detailed information is best supplied by engineering references in these fields or, in the case of the largest equipment, by the manufacturers. It should be noted that the power requirement for the production of large scale supersonic flows becomes very large. At least two[41] supersonic wind tunnels within the United States require electrical inputs of the order of 2×10^5 kw.

[41] National Advisory Committee for Aeronautics Unitary Tunnel Facility, Ames Laboratories, Sunnyvale, California. Propulsion Wind Tunnel, Arnold Engineering Development Center, Tullahoma, Tennessee.

4.2.2.3. Measurement in Gas Flows. In the present section we shall be more directly concerned with those methods of measurement which relate to the flowing gas, rather than with those which might be used, for example, in a precise determination of the viscosity or the ratio of specific heats. The variables of interest in the stream are the pressure, density, temperature, local velocity, flow rate, and turbulence level. If the stream is in interaction with a boundary, measurements of aerodynamic forces and heat transfer are also of importance.

The measurement of pressure and temperature in a static gas presents no special problems and may be accomplished by a great variety of familiar means. Even the density may be determined easily and directly by weighing. Unless otherwise indicated, it will be assumed in this section that a differential manometer with either oil or mercury as a working fluid will suffice for pressure measurement, and that the temperatures of bodies can be read with a thermometer or thermocouple.

Conditions are substantially altered when the gas is in motion, since the pressure within an open ended tube depends on the location of the tube in the flow and on its orientation. Similarly, a thermometer placed in the stream does not indicate the stream temperature, but instead a recovery temperature which depends on many variables of the flow and of the geometry. Thus the various pressure measurements are differentiated along the following lines: that at the point on the forward surface of a probe where the flow is brought to rest (*stagnation point*) is termed *impact* pressure, that measured along a wall normal to adjacent streamlines is known as *static*, and that measured on the slant surface of a cone may be termed *cone* pressure.

The simultaneous measurement of impact and static pressures with a pitot-static probe, Fig. 10(a), provides information for a calculation of free stream Mach number. For compressible fluids one can solve immediately for U_1/a_1 = Mach in the expression

$$\frac{p_i}{p_1'} = \left[1 + \frac{\gamma - 1}{2}\left(\frac{U_1}{a_1}\right)^2\right]^{\gamma/(\gamma-1)} \tag{4.2.54}$$

where p_1' is the static pressure and p_i is the impact pressure, and a_1 is the speed of sound at the temperature of the stream. Equation (4.2.54) extends only to sonic speed; above that the Rayleigh formula, Eq. (4.2.15) must be employed since the assumption of an isentropic deceleration from the free stream to the stagnation point no longer applies. Note that Eq. (4.2.54) does not give the velocity directly unless the temperature and hence a_1, the sound speed, is known.

If the flow has been produced by an isentropic expansion from a reservoir to the free stream condition, the free stream temperature may be calculated from the isentropic relations; otherwise the temperature must be determined by means of a total temperature probe. One form of the total temperature probe, Fig. 10(b), is constructed by mounting a thermocouple junction within an impact tube. In order to supply energy lost to the thermocouple by radiation and conduction, some flow through the tube is permitted by small holes drilled near the base of the thermocouple. Recovery factors close to unity are achieved in some designs. It should be remarked that these are calibrated devices, and so have only limited superiority over unshielded cylinders or spheres used as total temperature probes.

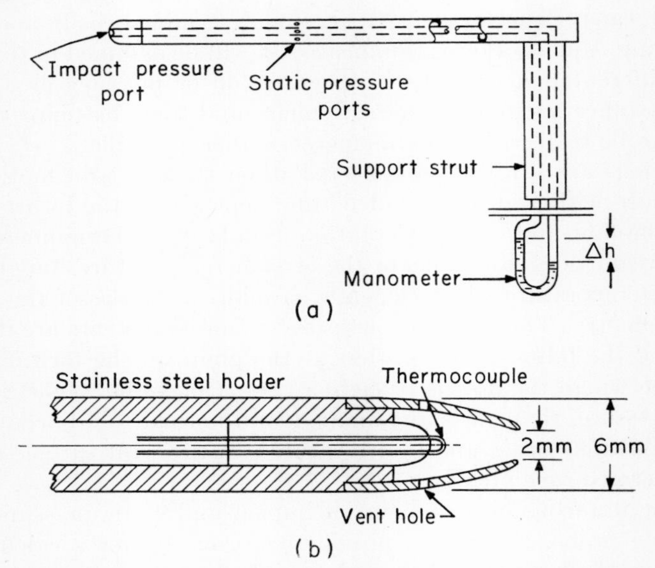

FIG. 10. (a) Pitot static tube; and (b) total temperature probe.

The determination of local values of the density presents some difficulties, since it would seem to require measurement of both the pressure and the temperature. Other means exist, namely those depending on the interaction of the gas and electromagnetic radiation. Both the speed of the radiation through the medium and the cross section for absorption depend on the density of the gas. The first effect may be used in connection with an interferometer to produce a visual picture of the flow field in which the local shift of the interference fringes is interpretable in terms of the local density of the flow field, and in which abrupt changes in the flow are made visible as discontinuities in the fringe pattern. The first effect is also used in the Schlieren system, in which changes in the refractive index are made visible by placing a knife edge in the image plane of the source. The interferometer technique yields information of more direct quantitative value except where distances or angles of the shock wave structure are in themselves of quantitative importance; in which case the Schlieren results are equally informative. Both suffer from problems of interpretation when the flow field is not two-dimensional.

The direct absorption of X-radiation along an optical path across the test section has been studied by Winkler[42] and also by Dimeff,[43] and mass absorption coefficients of approximately 790 cm^2/gm have been measured. The loss in radiant intensity at the detector is a measure of the integrated absorption along the path in accordance with the relation

$$\ln (I/I_0) = - \int_0^l \rho\mu \, dx \tag{4.2.55}$$

where μ is the mass absorption coefficient and ρ is the local density. Thus the total absorption is an integrated effect, just as is the total phase shift in the interferometric technique. In either case where circular symmetry exists a systematic study of one transverse scan of the flow field can lead in principle to complete knowledge of the local density distribution within the plane traversed by the beam.[44]

In subsonic flows local stream direction may be studied by yaw tubes, a special form of pitot tube in which pressures at two or more pairs of ports around the head are compared. To some extent this may be done in supersonic flows. It must be remembered that such devices may create disturbances which have far reaching effects on the entire flow field.

In many gas handling systems it is necessary to measure the flow rate in terms of total mass or volumetric flow per unit time. This may be accomplished quite exactly by use of the variable volume gasometer or the float rotometer, or by use of the positive displacement utility meter. In many instances the use of one of these devices may be infeasible, for example, where the volumetric flow is very large. In such cases use may be made of a Venturi meter, a convergent–divergent region of the duct in which pressures are measured upstream of the constriction and at the throat, as indicated in Fig. 11. The mass rate of flow for compressible fluids is given by the relation

$$Q = \frac{CA_2Y\rho_1}{\sqrt{1 - (A_2/A_1)^2}} \sqrt{\frac{2}{\rho_1}(p_1 - p_2)} \tag{4.2.56}$$

[42] Eva Winkler, U.S. Naval Ordnance Laboratory, White Oak, Maryland, Memo 10118 (1949).

[43] J. Dimeff, R. K. Hallet, and C. F. Hansen, *Natl. Advisory Comm. Aeronaut. Tech.* TN-2845 (1952).

[44] P. M. Sherman, *J. Aeronaut. Sci.* **24,** 93 (1957).

where C is the loss coefficient, Y is the expansion factor, A_2 is the throat area, and ρ_1 is the density upstream of the convergent section of the meter. The value of the loss coefficient is about 0.98 for nozzles of 10–40 cm diameter at Reynolds numbers of 10^5 or greater, and the value of Y may be found from tables of compressible flow or from Table III for gases with a γ of 1.40.

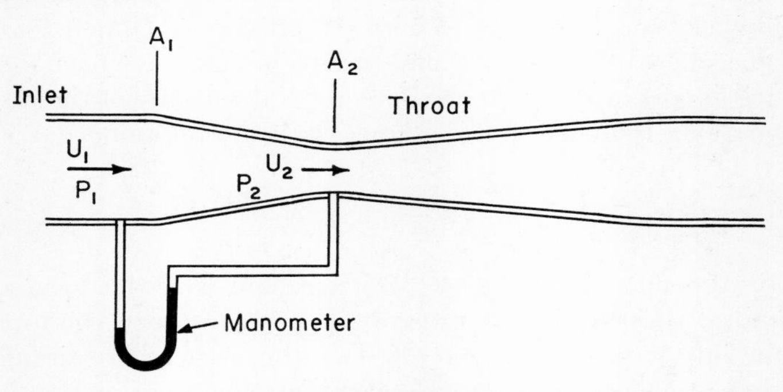

FIG. 11. Venturi meter (schematic).

In many instances, particularly where air is to be admitted to a large low velocity duct system, it is most convenient to mount an aperture plate at the system entrance. The mass rate of flow through such a meter is given approximately by

$$Q = CA\sqrt{2\rho(p_1 - p_2)} \tag{4.2.57}$$

where C is an empirical coefficient to be determined for each condition but having a value of about 0.6 where the Reynolds number based on orifice diameter is greater than 2×10^4.

Where the necessary pressure ratio between reservoir and flow system exists it is sometimes convenient to use a critical flow nozzle. This is a nozzle of known throat area in which critical flow is established (see

TABLE III. Loss Coefficients for Gases of $\gamma = 1.40$

A_2/A_1	p_2/p_1				
	0.95	0.90	0.85	0.80	0.75
0	0.975	0.947	0.918	0.888	0.855
0.2	0.973	0.945	0.915	0.883	0.850
0.3	0.972	0.941	0.910	0.878	0.842
0.4	0.967	0.935	0.900	0.866	0.830
0.5	0.962	0.925	0.886	0.850	0.810
0.6	0.956	0.913	0.870	0.828	0.785

Section 4.2.1.3), and through which the mass flow rate is determined from the flow velocity and density at the throat and from the throat area.

The finite volume of manometric measuring systems imposes a severe limitation on the response time of the measurement. The hot wire anemometer[45] minimizes this difficulty and possesses other advantages as well, most of these being related to the very small size possible for the measuring element. In the most usual application the temperature of an electrically heated wire is determined from the known thermal coefficient of resistivity (see Section 4.2.3.4). The cooling of the wire is largely a measure of the product of the local velocity and the density. Empirical data for various length to diameter ratios, end conditions, and flow conditions, are found to correlate well in terms of the Nusselt number. Since the response time can be made quite short, the measurement of rapidly fluctuating local velocities becomes possible. For example, fluctuation frequencies in turbulence studies have been measured to 10^5 cycles/sec.

4.2.3. Flows in Rarefied Gases[46,47]

4.2.3.1. The Regimes of Rarefied Gas Flows. A flow of a rarefied gas is one in which the length of the molecule mean free path λ is comparable to some significant dimension L of the flow field. The choice of a characteristic length L may sometimes constitute a problem of the investigation; more often L is readily found to be, for example, the diameter of a pipe or orifice, the width or thickness of a body in the stream, the boundary layer thickness or the thickness of any region in which there exists a sharp gradient in velocity, temperature, or density of the gas flow.

A parameter useful in characterizing rarefied gas flows is the Knudsen number K, equal to the dimensionless ratio λ/L. From Eq. (4.2.21), and from the connection between $\bar{c}$, the mean thermal speed of the molecules of a gas, and a, the speed of sound in the gas, we may devise an expression for λ in terms of the absolute viscosity μ and the speed of sound. Thus,

$$\lambda = 1.26 \sqrt{\gamma} \frac{\mu}{\rho a} \tag{4.2.58}$$

[45] M. V. Morkoven, "Fluctuations and Hot-Wire Anemometry in Compressible Flows." Wind Tunnel Agardograph Series, North Atlantic Treaty Organization Advisory Group for Aeronautical Research and Development, Palais de Chaillot, Paris, November, 1956.

[46] S. Chapman and T. G. Cowling, "The Mathematical Theory of Non-Uniform Gases." Cambridge Univ. Press, London and New York, 1939.

[47] S. Schaaf and P. Chambre, *in* "High Speed Aerodynamics and Jet Propulsion" (J. V. Charyk and M. Summerfield, eds.), Vol. 4, Part G, pp. 687–739. Princeton Univ. Press, Princeton, New Jersey, 1957.

where γ is the ratio of specific heats and ρ is the density. From the definition of Mach and Reynolds numbers (Eqs. (4.2.10) and (4.2.23)) and Eq. (4.2.58), one may obtain

$$K = 1.26\sqrt{\gamma}\,\frac{M}{\mathrm{Re}}. \tag{4.2.59}$$

Regimes of rarefied gas flows are often specified in terms of K as formulated in Eq. (4.2.59). However, since rarefaction effects in flows at $\mathrm{Re} \gg 1$ first make themselves known in the boundary layer where the characteristic length is the boundary layer thickness δ, it is also important to formulate K in terms of λ/δ. Thus, since

$$\frac{\delta}{L} \sim \frac{1}{\sqrt{\mathrm{Re}}} \tag{4.2.60}$$

we have

$$K \sim \frac{M}{\sqrt{\mathrm{Re}}}. \tag{4.2.61}$$

Although some disagreement among investigators remains, the designation of flow regimes as *continuum*, *slip*, *transition*, and *free molecule* has gained widespread acceptance. The laws of ordinary continuum gas dynamics will prevail wherever $M/\sqrt{\mathrm{Re}} \ll 1$ or $M/\mathrm{Re} \ll 1$, whichever is appropriate. While there is some arbitrariness in the assignment of the lower boundary of the continuum regime, experimental evidence would fix it by placing the *slip* regime within the following interval:

$$\begin{aligned} &0.01 < M/\sqrt{\mathrm{Re}} < 0.1, \qquad \mathrm{Re} > 1 \\ &0.01 < M/\mathrm{Re} < 0.1, \qquad \mathrm{Re} < 1. \end{aligned} \tag{4.2.62}$$

Slip flow effects might thus be expected where λ is from 1 to 10% of the characteristic length.

In the transition regime the mean free path is of the order of the characteristic dimension of the body. Molecule–surface collisions and molecule–molecule collisions are both of importance in determining the flow. Analytical investigations have so far yielded few results; most information concerning the transition flow regime is empirical.

The regime of free molecule flow, on the other hand, is moderately well understood. Here the flow is extremely rarefied and λ is substantially longer than the characteristic length. Experimental evidence would suggest

$$\frac{M}{\mathrm{Re}} > 3 \qquad \text{or} \qquad K > 5 \tag{4.2.63}$$

as a suitable definition of the free molecule flow limit, but in some circumstances it would appear that K should be greater than 10.

Phenomena of the slip flow regime will be treated at greater length elsewhere in the present section, while transition and free molecule flows in some of their aspects will be considered in Section 4.2.4. The flow regimes are schematically summarized in Fig. 12 in terms of Mach and Reynolds numbers where the characteristic length has been taken to be 1 meter. Corresponding altitudes have also been included, where atmospheric density has been taken from the NACA proposed standard atmosphere tables.[48]

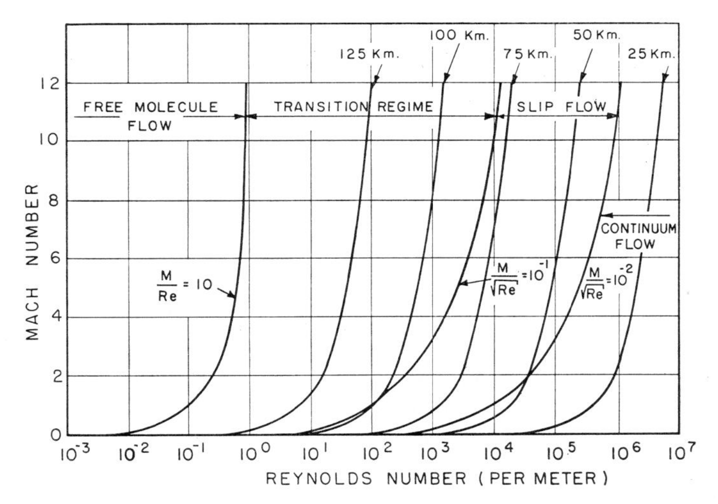

FIG. 12. The regimes of gas dynamics.

4.2.3.2. Slip, Temperature Jump, and Flows in the Slip Regime.[49,50] In continuum viscous flow theory the gas is assumed to be at rest at the boundary. However, it has been recognized from the time of Maxwell[51] that there is a finite velocity of "slip" at the boundary of a rarefied gas flow. A similar failure of continuum boundary conditions occurs in respect to energy transfer at the gas–surface interface. It was shown by Knudsen,[52] 1911, that a temperature jump occurs at the boundary of a rarefied gas flow. Both slip and temperature jump reflect the fact that near the

[48] C. N. Warfield, "Tentative Tables for the Properties of the Upper Atmosphere." *Natl. Advisory, Comm. Aeronaut. Tech. Notes* **1200** (1947).

[49] E. H. Kennard, "Kinetic Theory of Gases." McGraw-Hill, New York, 1938.

[50] L. B. Loeb, "Kinetic Theory of Gases," 2nd ed. McGraw-Hill, New York, 1934.

[51] J. C. Maxwell, "Scientific Papers," Vol. 2, p. 708. Cambridge Univ. Press, London and New York, 1890.

[52] M. Knudsen, *Ann. Physik* [4] **34,** 593 (1911).

wall the gas is constituted half by molecules carrying the property of the wall (average momentum zero or $T = T_w$), and half carrying the gas property one mean free path away.

A slight complication is added by the imperfect adjustment of the molecule tangential momentum and energy to that of the wall in the interaction. The interaction model almost universally adopted assumes that a fraction f of the molecules is diffusely reflected (average tangential momentum is zero), while the remainder, $1 - f$, are specularly reflected, thus retaining their incident tangential momentum. Similarly, an exchange of energy occurs in the encounter such that one may define an accommodation coefficient α as

$$\alpha = \frac{E_i - E_r}{E_i - E_w} \tag{4.2.64}$$

where i, r, and w refer to incident, reflected, and wall, respectively. E_i is the mean energy of the incident molecules, while E_w is the energy these would have possessed had they issued from a gas in equilibrium at the wall temperature. As a special case Eq. (4.2.64) is often formulated in terms of molecular "temperatures."

Much theoretical effort has been directed toward the proper formulation of the boundary conditions with slip and temperature jump. All solution to order λ agree with the result given by Kennard,

$$u_0 = \frac{2 - f}{f} \lambda \left(\frac{\partial u}{\partial Y}\right)_0 + \frac{3}{4} \frac{\mu}{\rho T} \left(\frac{\partial T}{\partial x}\right)_0 \tag{4.2.65}$$

$$T_0 - T_w = \frac{2 - \alpha}{\alpha} \frac{2\gamma}{\gamma + 1} \frac{\lambda}{\mathrm{Pr}} \left(\frac{\partial T}{\partial Y}\right)_0 \tag{4.2.66}$$

where u_0 is the velocity of the gas at the wall, μ is the absolute viscosity, γ is the ratio of specific heats, ρ is the gas density, and Pr is the Prandtl number. The first section of Eq. (4.2.65) is the isothermal slip term, and the second portion is the "thermal creep" term.

It would appear from studies of the dispersion of sound waves and from studies of the internal structure of weak normal shock waves that the Navier–Stokes solutions yield better agreement with the experimental results than those available from the Burnett[53] or thirteen moment[54] equations, "higher order" approximations to solution of the Maxwell–Boltzmann equations. Thus it seems best to apply the Navier–Stokes equations throughout the slip regime and to use, where appropriate, the

[53] D. Burnett, *Proc. London Math. Soc.* **40,** 382 (1935); also see reference 1.

[54] H. Grad, *Comm. on Pure and Appl. Math.* **2,** 331 (1949).

slip and temperature jump boundary conditions of Eqs. (4.2.65) and (4.2.66).

Slip at the surface is readily exhibited in two classical experiments; the measurement of the flow of gases through tubes, and the measurement of drag in a rotating cylinder device. A discussion of the first is more appropriate to the next section; the second will be considered briefly here.

The Navier–Stokes equation for compressible viscous flow in the annulus between two cylinders can be solved exactly to yield

$$F_D = \tfrac{1}{2}\rho U^2 \frac{2}{\mathrm{Re}\{1 + 2[(2 - f)/f](\lambda/h)\}} \tag{4.2.67}$$

where U is the relative velocity of the surfaces, and Re is based on h, the annular gap. Experimental results agree nicely with the theory and afford at the same time a means of determining f, the coefficient of specular reflection. Values of f and α are summarized in Table IV and in standard texts on the kinetic theory.

Slip regime phenomena are characterized not only by the noncontinuum effects at the boundary, but by the highly viscous character of the flow at the low Reynolds numbers. Viscous effects in the case of

TABLE IV. Values of f, Coefficient of Specular Reflection, and of α, Coefficient of Thermal Accommodation

Gas	Surface	Values of f	Reference
air	Machined brass	1.00	*a*
CO_2	Machined brass	1.00	*a*
air	Old shellac	1.00	*a*
CO_2	Old shellac	1.00	*a*
air	Hg	1.00	*a*
air	Oil	0.90	*a*
CO_2	Oil	0.92	*a*
H_2	Oil	0.93	*a*
air	Glass	0.89	*a*
He	Oil	0.87	*a*
air	Fresh shellac	0.79	*a*
air	Ag_2O	0.98	*b*
He	Ag_2O	1.00	*b*
H_2	Ag_2O	1.00	*b*
O_2	Ag_2O	0.99	*b*
air	Oil on machined aluminum	0.90	*c*
N_2	Glass	0.97	*d*
N_2 or air	Mild steel polished or etched	1.00	*d*
N_2 or air	Aluminum polished or etched	1.00	*d*

TABLE IV (*Continued*)

Gas	Surface	Values of α	Reference
H_2	Bright Pt	0.32	*e*
H_2	Black Pt	0.74	*e*
O_2	Bright Pt	0.81	*e*
O_2	Black Pt	0.93	*e*
N_2	Pt	0.50	*f*
N_2	Tungsten	0.35	*f*
Air	Flat lacquer on bronze	0.88–0.89	*g*
Air	Polished bronze	0.91–0.94	*g*
Air	Machined bronze	0.89–0.93	*g*
Air	Etched bronze	0.93–0.95	*g*
Air	Polished cast iron	0.87–0.93	*g*
Air	Machined cast iron	0.87–0.88	*g*
Air	Etched cast iron	0.89–0.96	*g*
Air	Polished aluminum	0.87–0.95	*g*
Air	Machined aluminum	0.95–0.97	*g*
Air	Etched aluminum	0.89–0.97	*g*
He	Tungsten	0.025–0.057	*h*
He	Nickel not flashed	0.20	*h*
He	Nickel flashed	0.085	*h*
He	Tungsten flashed	0.17	*i*
		0.12	*f*
He	Tungsten not flashed	0.53	*i*
Argon	Tungsten flashed	0.82	*i*
		0.46	*i*
Argon	Tungsten not flashed	1.00	*i*

[a] R. A. Millikan, "Coefficients of Slip in Gases and the Law of Reflection of Molecules from the Surfaces of Solids and Liquids." *Phys. Rev.* **21**, 217 (1923).

[b] E. Blankenstein, "Coefficient of Slip and Momentum Transfer in H_2, He, air and O_2." *Phys. Rev.* **22**, 582 (1923).

[c] S. F. Chiang, "Drag on a Rotating Cylinder at Low Pressures." Inst. Engr. Research, Univ. California Eng. Proj. Rept. HE-150-85 (1951); also Ph.D. Thesis. University of California, Berkeley, California, 1951.

[d] F. C. Hurlbut, "Studies of Molecular Scattering at the Solid Surface." *J. Appl. Phys.* **28** (8), 844–850 (1957).

[e] M. Knudsen, "Kinetic Theory of Gases." Methuen, London, 1934.

[f] R. N. Oliver and M. Farber, "Experimental Determination of Accommodation Coefficients as Function of Temperature for Several Metals on Gases." Jet Propulsion Lab., California, Inst. Technol. Memo No. 9–19 (1950).

[g] M. L. Weidmann, "Thermal Accommodation Coefficients." *Trans. Am. Soc. Mech. Engrs.* **68,** 57 (1946).

[h] J. K. Roberts, "The Exchange of Energy Between Gas Atoms and Solid Surfaces," *Proc. Roy. Soc.* **A129,** 146 (1930); **A135,** 135 (1932).

[i] W. C. Michels, "Accommodation Coefficients of Helium and Argon against Tungsten." *Phys. Rev.* **49,** 472 (1932).

impact probes in subsonic and supersonic streams have been studied by Sherman.[55] It is important to note that interpretation of pressure measurements may require substantial correction for these effects at sufficiently low Reynolds numbers.

4.2.3.3. Ducted Flows. Internal flows in the slip and transition regimes always occur at Reynolds numbers less than critical (Re $<$ 2400), and hence are laminar. Poiseuille's equation may be corrected to include the slip boundary condition giving, for an isothermal tube,

$$Q_{pv} = \frac{\pi a^4}{8\mu L}\left[1 + 4\left(\frac{2-f}{f}\right)\frac{\lambda}{a}\right](p_1{}^2 - p_2{}^2) \qquad (4.2.68)$$

where Q_{pv} is the flow rate in microbar cm^3 per second at the temperature of the tube, p_1 and p_2 are in microbars, a is the radius of the tube, L its length, and μ is the viscosity. The flow rate in grams may be obtained by dividing Q_{pv} by RT where R is the gas constant per gram and T is the absolute temperature.

It may be noted that as the ratio λ/a increases or as f decreases, the value of the quantity in the brackets becomes large. Brown *et al.*[56] conducted an extensive review of the data on the flow of gases in pipes. The correlation among many observations was excellent, with the value of f being found to be 0.77 for glass tubes and the value of the quantity in brackets being as high as 40,000. Values of f were found to be close to unity for iron pipe.

If we divide Q_{pv} by the pressure difference we obtain the conductance F, the flow rate per unit pressure difference. For air at 20°C, Eq. (4.2.68) yields the following expression for the conductance of a long tube in viscous flow with f assumed to be unity:

$$F = 0.182\frac{d^4}{L}p_m \text{ liters/sec.} \qquad (4.2.69)$$

In the above expression p_m is the mean pressure in microns.

The conductance of an aperture is expressed as

$$F = 76.6\gamma^{0.712}\sqrt{1-\gamma^{0.288}}\frac{A}{1-\gamma} \text{ liters/sec.} \qquad (4.2.70)$$

[55] F. S. Sherman, "New Experiments on Impact Pressure Interpretation in Supersonic and Subsonic Rarefied Air Streams." *Natl. Advisory Comm. Aeronaut. Tech. Notes* **2995** (1953).

[56] G. P. Brown, A. DiNardo, G. K. Cherry, and T. K. Sherwood, *J. Appl. Phys.* **17,** 802 (1946).

Guthrie and Wakerling[57] give the conductance of a conduit of rectangular section as

$$F = 0.26Y \frac{a^2b^2}{L} p_m \text{ liters/sec} \tag{4.2.71}$$

where Y is given in Table V:

TABLE V. Conductance Coefficients for Conduits of Rectangular Section

a/b	1.0	0.9	0.8	0.7	0.6	0.5	0.4	0.3	0.2	0.1
Y	1.00	0.99	0.09	0.05	0.90	0.82	0.71	0.58	0.42	0.23

4.2.3.4. Measurements in Rarefied Gases. The three regimes of rarefied gas flows as defined in the first portions of Section 4.2.3 may well be joined within a single flow field. For example, the interpretation of pressures behind a small orifice may require a consideration of free molecule effects although the body in which the orifice is placed lies in continuum flow. It should be appreciated that such a mergence of regimes complicates the development of theories of the flow and the interpretation of measurements of all kinds therein. Occasionally, however, this state of affairs may be turned to advantage. It will be our present task to consider the instruments which must be associated with measurements in rarefied gas flows and the problems relevant to their use.

In rarefied gas flows, as in flows of ordinary density, the measurement of pressure is fundamental. The pressures under consideration here lie between one micron and a few millimeters of mercury at temperatures below the melting point of the common metals. Instruments for pressure measurement in this range, as at higher pressures, may be divided into two general varieties, those which depend ultimately on the change of molecular momentum at the gas–surface interface, and those which depend on some other property of the gas, its thermal conductivity, for example, or its response to electromagnetic radiation. However, many of the instruments which are of value at higher pressures must be substantially transformed in order to meet the requirements of the new regime, while some instruments useful in the rarefied gas regime have no counterpart of significant utility elsewhere.

An example of such an instrument is to be found in the McLeod gage. This device measures pressure by trapping a sample of gas at p_1, and then compressing it isothermally until a readily measurable pressure, p_2, is reached. The schematic diagram, Fig. 13, shows a McLeod in the compressed position. Mercury (occasionally oil), serving as the compressing

[57] A. Guthrie and R. K. Wakerling, "Vacuum Equipment and Techniques." McGraw-Hill, New York, 1949.

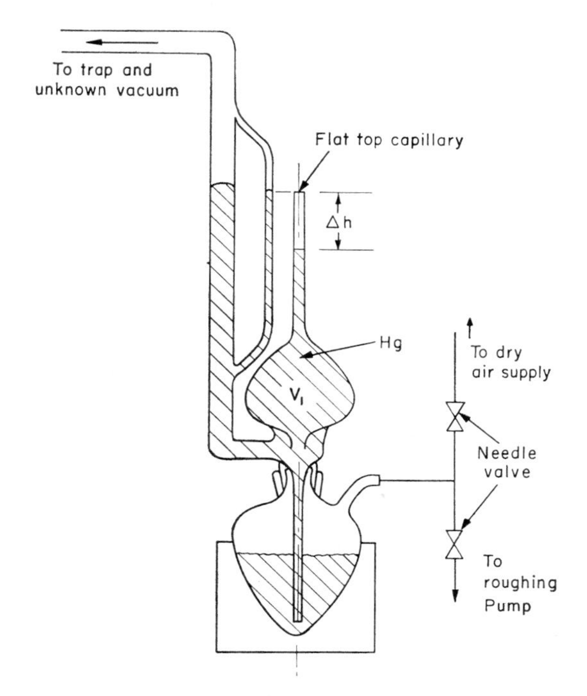

FIG. 13. McLeod gage.

fluid, is forced into volume V_1 by air pressure. The pressure in volume V_2 is measured as Δh, the differential height of the mercury columns in the measuring and comparison legs, while the volume of V_2 is known by the product ΔhA where A is the cross sectional area of the capillary. Since, by Charles' law

$$p_1V_1 = p_2V_2 \tag{4.2.72}$$

we have

$$p_1 = \frac{(\Delta h^2)A}{V_2}. \tag{4.2.73}$$

The McLeod gage is almost universally applied as the reference instrument in the calibration of continuously indicating gages, except at pressures where the precision mercury manometer is applicable. Two serious limitations attend the use of the McLeod; (1) the trapping and reading operation is delicate and time consuming, and (2) the large volume V_1 terminating the connecting tube makes application under nonsteady state conditions infeasible. With careful design and precise evaluation of the capillary area and compression volume the instrument can be made of great accuracy and reliability for use within the pressure range cited above.[58]

The U-tube manometer can also be made an instrument of almost universal applicability in the present range of pressures if the reference leg (Fig. 14) is connected to a region of high vacuum. Mercury and oil are the most popular working fluids, a low vapor pressure oil being chosen for use with pressures between a few microns and a few millimeters Hg. For application in this range the manometer tubing must be a centimeter or more in inside diameter in order to avoid capillarity effects, and optical

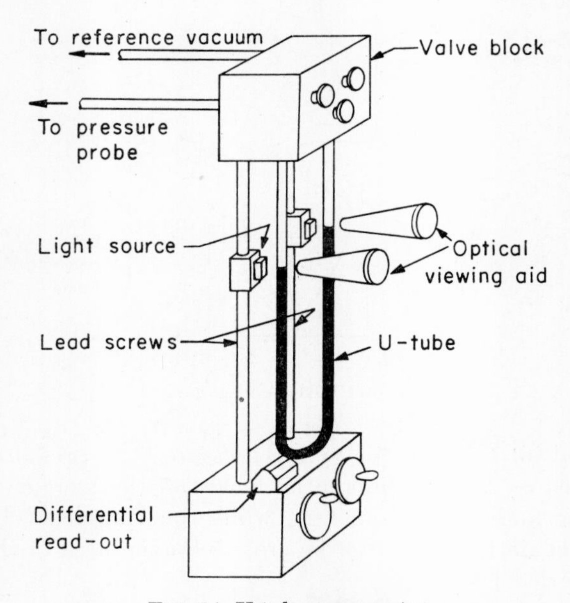

FIG. 14. U-tube manometer.

aids must be employed in viewing the meniscus. Maslach[59] has described a differential manometer with which pressure measurements to $\pm 1\,\mu$ can be made, and which is suitable for rapid routine use in connection with a rarefied gas wind tunnel.

The manometer may also suffer from problems of response time in some measuring applications. In any event it is not an indicating instrument or one readily adapted to the production of continuous chart records. Instruments much more suitable on both counts lie in the second group, and of these the most universally employed are the heat conduction

[58] F. S. Sherman, *Natl. Advisory Comm. Aeronaut. Tech. Notes* **2995** (1953).
[59] G. J. Maslach, *Rev. Sci. Instr.* **23**, 367 (1953).

gages, frequently now labeled gages of the Pirani type.[60] Heat is transferred from a heated element to the gage walls by gas conduction in the space within the envelope. At ordinary pressures the heat transfer is independent of pressure, but for Knudsen numbers of 10^{-3} and larger based on the radius of the envelope, it becomes pressure dependent. An elementary heat balance yields the following equation

$$E = E_R + E_c = \{\epsilon\sigma(T_w{}^4 - T^4) + \frac{n\bar{c}}{4}\alpha\frac{C_v}{N_A}(T_w - T)\}A \quad (4.2.74)$$

where E, the total heating supplied is equated to that lost by radiation plus that lost by gas conduction. End losses at the ends of the filaments are neglected. The coefficient ϵ is the emissive power, σ the radiation constant, n the number density of molecules in the volume, $\bar{c}$ the molecule mean speed, α the accommodation coefficient, C_v the molar specific heat, N_A Avogadro's number, and T and T_w the temperature of filament and wall respectively.

Some distinction is made among the various Pirani gages on the basis of the material of the heated element and the method of sensing the temperature change in it. The common Pirani uses a wire filament, as in Fig. 14, as both heated element and resistance thermometer. The "Thermistor" Pirani employs a glass coated bead of semiconducting material[61] in the same fashion. Thermistors possess sharp, negative, temperature-resistance characteristics and present certain advantages thereby. The thermocouple gage is a four-terminal Pirani employing one wire as the heated element and using a thermocouple for the temperature measurement.

All Pirani gages suffer in some degree from a sensitivity to changes in envelope temperature. The difficulty is partially remedied by the inclusion of a second Pirani within a balanced measuring circuit (Fig. 15). Since the emissivity and accommodation coefficient of the sensitive element influence the response of the Pirani, the aging or contamination of these critical surfaces modifies the calibration. Further, the resistance wire of the common Pirani changes its temperature-resistance characteristic with time. It may be thus appreciated that the Pirani must be repeatedly calibrated if it is to be used in measurements of precision. Kavanau[62] has described a thermistor Pirani system of remarkable

[60] S. Dushman, "Scientific Foundations of Vacuum Technique." Wiley, New York, 1949.

[61] J. A. Becker, C. B. Green, and G. L. Pearson, *Trans. Am. Inst. Elec. Engrs.* **65**, 711 (1946).

[62] L. L. Kavanau, "Base Pressure Studies in Rarefied Supersonic Flows." Ph.D. Dissertation, University of California, Berkeley, California, 1954, also University of California Eng. Proj. Rept. HE-150-125 (1954).

refinement. The thermistor was chosen in this case for the presumed internal and surface stability of the element. Great care was exercised in the thermal regulation of the envelopes of the metering and comparison gages. In the range 10–300 μ Hg the reproducibility of the indication was of the order of 0.1%, the sensibility was about 0.1 μ, while the absolute accuracy was determined by that of the primary standard, in this case about 1%. On the other hand, because of the simplicity of the metering circuit, the Pirani is frequently used as a rough and ready pressure indicator in the apparatus of rarefied gas investigation at points where high

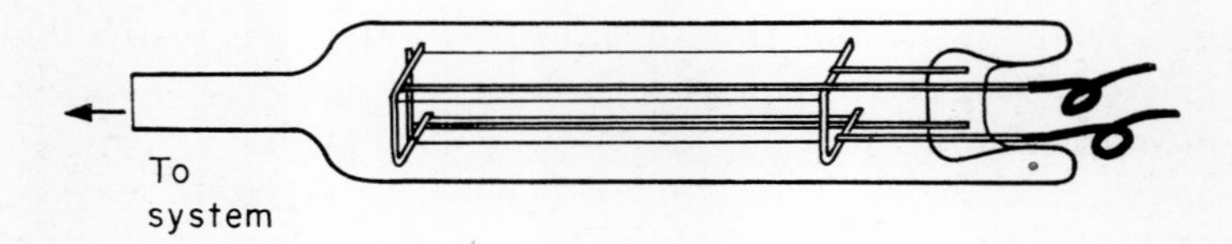

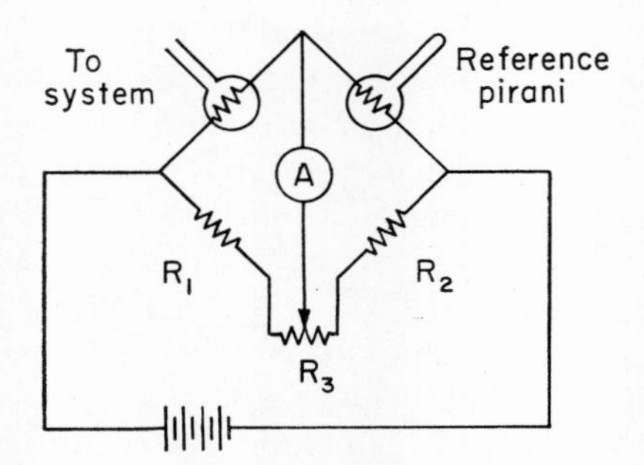

FIG. 15. Pirani gage and metering circuit.

accuracy is of little importance and in consequence, where a calibration is required infrequently.

A second instrument of very general applicability is the "Alphatron"[63] gage, a device which takes its name from the radioactive alpha particle source in the sensitive volume. Ions of the specimen gas formed in collision with alpha particles are collected in an ionization chamber which is associated with a metering circuit. The range of the instrument extends from 1000 mm to 10 μ full scale. It is a suitable instrument for measurement in the rarefied gas range where higher pressures may also be encountered in the course of an experiment.

[63] J. R. Downing and G. Mellen, *Rev. Sci. Instr.* **17,** 218 (1946).

Of somewhat limited utility at the low end of the rarefied gas range, but very useful at pressures of 1 mm Hg and up are those devices which involve the measurement of the deflection of a diaphragm, or the extension of a bellows, or measure the null position where a balancing force is applied. A number of such devices have been developed, and some are commercially available. Attempts to extend the range to the lowest pressures are made difficult by thermal and hysteresis effects in the sensitive membrane, and by the response times imposed by the relatively large diaphragm areas.[64] Even so, many fields of investigation are served well by these devices.

No general theory of response times in gaging systems for rarefied gases exists at the present where account is taken of the variation of flow regime which may be encountered within each element of the system in the course of a measurement. The reader is referred to Section 4.2.4.3 for a résumé of response times where free molecule flow conditions obtain throughout, and to References 65 and 66 for a consideration of response times in other regimes. It should be recognized as a general principle of such systems that the gage volume and that of the connecting tubing must be filled through an orifice of limited conductance. In the rarefied gas regime response time constants may be of the order of 1–100 sec or more, even after some attention has been directed to limiting the gage volume.

Entrance orifices of small area and connecting tubing of small diameter may permit systematic errors in pressure measurement due to thermal transpiration or thermal creep,[67,68] where temperature gradients exist in the system. Thermal transpiration is a free molecule phenomenon whereby steady state pressures in two volumes at T_1 and T_2 connected by a tube of diameter such that p_1 and p_2 are related by the expression

$$p_1/p_2 = \sqrt{T_1/T_2}. \tag{4.2.75}$$

Thermal creep, on the other hand, occurs at the boundary of a surface, and a gas in slip or transition flow. This surface layer, approximately one mean free path in thickness, is found to move along the surface in the direction of the thermal gradient. The velocity is given by the kinetic

[64] D. Alpert, C. G. Matland, and A. O. McCoubrey, *Rev. Sci. Instr.* **22** (6), 370–371 (1951).

[65] J. M. Kendall, "Time Lags Due to Compressible-Poiseuille Flow Resistance in Pressure Measuring Systems." Naval Ordnance Laboratory, White Oak, Maryland, Memo 10677 (1950).

[66] A. L. Ducoffe, *J. Appl. Phys.* **24,** 1343 (1953).

[67] E. H. Kennard, "Kinetic Theory of Gases." McGraw-Hill, New York, 1938.

[68] L. B. Loeb, "Kinetic Theory of Gases," 2nd ed. McGraw-Hill, New York, 1934.

theory as

$$u = \frac{3\mu R}{4pM} \cdot \frac{\partial T}{\partial x} \tag{4.2.76}$$

where R is the gas constant, M is the molecular weight, μ is the absolute viscosity, and p is the local pressure of the gas. In the slip regime, if we solve the equations of Poiseuille flow for the creep velocity boundary condition at the wall, we have for zero net mass transfer

$$\frac{\Delta p}{\bar{p}} = \frac{p_1 - p_2}{\frac{1}{2}(p_1 + p_2)} = \frac{3.8K^2}{1 + 4K}\left(\frac{T_1 - T_2}{\frac{1}{2}(T_1 + T_2)}\right) \tag{4.2.77}$$

where p_1 is the pressure to be measured, p_2 is the pressure within the gage shell, T_1 and T_2 are the corresponding temperatures, and K is the Knudsen number. At larger Knudsen numbers, i.e., in the transition regime, one must employ the semiempirical relation of Knudsen[69,70]

$$\frac{\Delta p}{\bar{p}} = \frac{1}{2}\left[1 + \frac{2.46(K + 3.15)}{K(K + 24.6)}\right]^{-2} \frac{\Delta T}{\bar{T}} \tag{4.2.78}$$

where empirical constants have been set down here for hydrogen. It is not expected that results for other gases would differ appreciably. It should be remarked, however, that these expressions are better employed in estimating experimental errors so that they may be minimized by experimental technique than in correcting measurements where large temperature gradients exist.

Further error in rarefied gas pressure measurement may be introduced by *drift*, residual pressures resulting from system outgassing or ingassing. In engineering systems the phenomena of drift may always be present in some degree. Schaaf and Cyr[71] have considered the problems of drift in the free molecule regime, while Kavanau[72] has developed semiempirical expressions for drift in the slip and transition regimes.

Measurement of other properties of rarefied gas flows, of mass flow rate, Mach number, or temperature, for example, may proceed by means similar to those discussed in Section 4.2.2.3. Proper cognizance of the effect of viscosity must be taken, however, a process which in itself may require a considerable effort. The reader is referred to the work of Sherman[73] on impact pressure interpretations, or of Talbot[74] on interpretation of pressures on cone probes.

[69] M. Knudsen, *Ann. Physik* [4] **31,** 205 (1910).

[70] M. Knudsen, *Ann. Physik* [4] **83,** 797 (1927).

[71] S. A. Schaaf and R. R. Cyr, *J. Appl. Phys.* **20,** 860 (1949).

[72] L. L. Kavanau, Ph.D. Dissertation, University of California, Berkeley, California, 1954.

[73] F. S. Sherman, *Natl. Advisory Comm. Aeronaut. Tech. Notes* **2995** (1953).

[74] L. Talbot, T. Koga, and P. M. Sherman, "Hypersonic Viscous Flow Over Slender Cones." University of California Eng. Proj. Rept. HE-150-147 (1957).

Other techniques of measurement in rarefied gas streams avoid the problem of system response times or those which might result from the introduction of a probe into the flow field. Of greatest importance among these are the free molecule wire probe techniques and that of glow flow visualization.

The free molecule wire probe was first applied by Stalder[75] and later refined by Sherman[76] in connection with studies of shock wave structures. If a wire is of sufficiently small diameter by comparison with the mean free path it is presumed, and has been demonstrated, that it may be stretched across the flow field without creating a disturbance in that field. Further, the wire, being in free molecule flow, achieves an equilibrium temperature, neglecting end effects, which can be predicted from the local molecular distribution function.[77,78] The equilibrium temperature wire and a later development, the heat transfer free molecule probe[79] have been used to survey two-dimensional flow fields in regions of the boundary layer and shock wave about flat plates and other simple aerodynamic shapes. The model is fixed in the stream, while the wire is moved relative to it.

At densities often encountered in rarefied gas flows the ordinary techniques of flow visualization, the Schlieren and interferometric techniques, fail by reason of insufficient optical thickness in the flow field. The flow may be made visible, however, by the mechanisms of afterglows,[80] so that shock waves, boundary layers, and other structures are revealed by changes in luminous intensity within the flow field. The technique is of value for qualitative studies supplementing other techniques of measurement, but appears infeasible for quantitative applications.

A direct measure of gas density may be provided by the electron beam technique[81] whereby the reduction of beam intensity by scattering is a function of the integrated gas density[82] along the beam path. The beam is projected from gun to detector across the flow field, and may be moved relative to the model, as in the case of the free molecule wire probe.

[75] J. Stalder, G. Goodwin, and M. O. Creager, *Natl. Advisory Comm. Aeronaut. Rept.* **1052** (1951).

[76] F. S. Sherman, *Natl. Advisory Comm. Aeronaut. Tech. Notes* **3298** (1955).

[77] J. S. Stalder and D. Jukoff, *Natl. Advisory Comm. Aeronaut. Rept.* **944** (1949).

[78] S. Bell and S. A. Schaaf, *Jet Propulsion* **25**, 168 (1955).

[79] J. A. Laurmann and D. C. Ipsen, "Use of a Free Molecule Probe in High Speed Rarefied Gas Studies." University of California Eng. Proj. Rept. HE-150-146 (1957).

[80] W. B. Kunkel and F. C. Hurlbut, *J. Appl. Phys.* **28**, 827 (1957).

[81] F. C. Hurlbut, "An Electron Beam Density Probe for Measurements in Rarefied Gas Flows." WADC TR-57-644 (1957).

[82] The analysis discussed by P. M. Sherman, reference 44, Section 4.2.2.3, in connection with radiation absorption techniques is applicable here.

Measurements made by means of an electron optical analog of the Schlieren system also lead to density determination for the rarefied gas flow field.[83]

4.2.4. High Vacua* [84,85]

The term "high vacuum" applies in its strictest sense to regions of pressure of 10^{-6} atmos or lower. However, while the region of highest vacuum is often the site of a physical experiment, the vacuum system is an interdependent complex of components and regions, all of which must be discussed. Thus the present section will not be limited to discussions of high vacua but will be directed toward summarizing all those methods and techniques useful in the production, maintenance, and instrumentation of regions of high vacuum that have not been discussed in Section 4.2.3.

4.2.4.1. The Production of High Vacua. Pumps for the production of high vacua are ordinarily characterized by empirical data relating volumetric flow and pressure as measured at the pump intake flange. Pressures are ordinarily expressed in terms of microns (μ) of Hg (10^{-3} mm) if in the range of 0.001 to 1 mm Hg, and in terms of mm Hg at higher and lower values, e.g., 10^{-6} mm Hg. Volumetric flow rate is commonly termed "speed" and is most frequently expressed in liters/sec. Units and conversion factors may be found in Table VI. The ultimate vacuum or blank-off pressure is also a characterizing feature and may be determined from the manufacturer's performance data or from separate statements in the literature. Pumps have not only a maximum compression ratio over which operation may be maintained, but may have an upper limit to the forepressure beyond which there is a fundamental failure of the flow within the pump. In this instance the limiting forepressure becomes a further performance characteristic of importance. As a result, it is usually necessary to employ pumps of various capabilities in a matched cascade arrangement in order to bridge the range from atmospheric pressure to the desired high vacuum.

4.2.4.1.1. MECHANICAL PUMPS. Mechanical pumps for vacuum application are found in five classes: axial compressors, rotary vane water-sealed pumps, reciprocating piston compressors, rotary vane oil sealed pumps, the so-called Roots pumps and "molecular" pumps. Only the

* See also Vol. 4, Chapter 12.1.

[83] L. Marton, D. C. Schubert, and S. R. Mielczarek, *J. Appl. Phys.* **27**, 4 (1956).

[84] S. Dushman, "Scientific Foundations of Vacuum Technique." Wiley, New York, 1949.

[85] A. Guthrie and R. K. Wakerling, "Vacuum Equipment and Techniques." McGraw-Hill, New York, 1949.

Table VI. Units and Conversion Factors

Pressure
- 1 microbar = 1 dyne/cm^2 = 10^{-6} bar
- 1 standard atmosphere (A_0) = 760 mm Hg at 0°C = 1.01325×10^6 microbar
- 1 mm Hg = 1,333.22 microbar = 10^3 micron (μ) Hg = 1 Torr (German)
- 1 A_0 = 29.921 in Hg at 32°F = 14.696 lb/in^2

Temperature
- $T = t°\text{C} + 273.15$ absolute Centigrade (Kelvin)
- $T_F = t°\text{F} + 459.7$ absolute Fahrenheit (Rankine)
- $T = \frac{5}{9}(t°\text{F} - 32)°\text{C}$

Volume
- 1 liter = 0.035316 ft^3

Volumetric Rates
- 1 liter/sec = 2.119 ft^3/min = 3.6 m^3/hr
- 1 μ liter/sec = 2.119 μ ft^3/min = 4.738 atm cm^3/hr

Mass
- 1 gm = 2.2046×10^{-3} lb

Miscellaneous

$$m = \frac{\text{number of molecules}}{\text{cm}^3} = 9.656 \times 10^{18}\ Pmm/T$$

$$= 2.687 \times 10^{19} \text{ at } 0°\text{C and } 760 \text{ mm Hg pressure}$$

$$\nu = \text{number of molecules striking } 1 \text{ cm}^2/\text{sec}$$

$$= 3.513 \times 10^{22}\ Pmm/\sqrt{MT}$$

where M is the molecular mass

latter three classes are of value in the production of high vacua, the other three being limited by design to pressures above 10 mm Hg or so.

The rotary vane oil sealed pumps are available in a wide range of sizes and performance characteristics. All employ the principle illustrated in Fig. 16 with minor variations in the details. The eccentrically mounted rotor compresses the gas entering at port A and sweeps it out through poppet or feather valves at B. The separation between intake and exhaust regions is maintained by the sliding vane at C and by the oil seal at point D.

Clearances are small and the quality of workmanship must be excellent for acceptable results. An essential feature of these pumps and one not included in Fig. 16 is the oil reservoir. Low vapor pressure oil supplied to the vacuum region and shaft seals by air pressure becomes ejected with the exhaust air and returned to the oil reservoir through a trap.

Rotary vane oil seal pumps are available with free air displacements as low as 0.16 liters/sec, or as high as 400 liters/sec. Pumps of moderate displacement and of compound design are available with blank-off pressures as low as 0.05 μ, while the larger single stage pumps may be obtained with blank-off pressures of the order of 10 μ. A typical pump with speed of approximately 60 liters/sec blanks off at 10–25 μ in ordinary service, and is supplied with a 5-hp electric motor.

Mechanical pumps of this type are a well perfected device and have been developed by the manufacturers to the point where they possess a high degree of reliability and will give years of satisfactory service. The larger pumps require circulating water for cooling, but aside from occasional vacuum oil changes, need in addition only the maintenance servicing given any rotating machinery. They exhaust to atmospheric pressure, and hence serve almost universally in conventional vacuum systems as the "roughing" pump (that which produces a rough vacuum

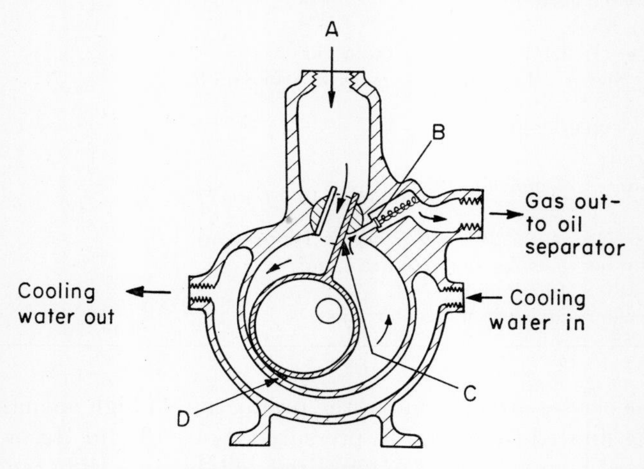

FIG. 16. Rotary oil pump.

throughout the equipment when it is first closed from the atmosphere), and as the fore-vacuum pump preceding the vapor diffusion pumps.

On the debit side, the rotary vane oil seal pumps possess relatively small volumetric speeds for the weight, power consumption, and initial expense by comparison with steam, air, or oil jet ejectors which in some instances perform the same function. The ultimate vacuum in rotary pumps is determined in part by the vapor pressure of the seal oil and this, in turn, is partly a property of contaminants transported from the vacuum system or from moisture condensation in the oil reservoir. In addition, the degree of solubility of the atmosphere in the oil and the completeness with which the gas is eliminated at each cycle play an important role. Thus, the rotary pump cannot be used at all to provide vacua where there is a high moisture content or a high percentage of any condensable gas. Even in the best of circumstances the reservoir oil requires periodic checking and occasional changing.

The Roots pump, of which one variety is illustrated in Fig. 17, is essentially a rotary displacement pump, but with such small clearances between moving and stationary parts that the need for an oil seal is eliminated. At low pressures, the exchange of momentum between the gas molecules and the rotating impeller, which characterizes the operation

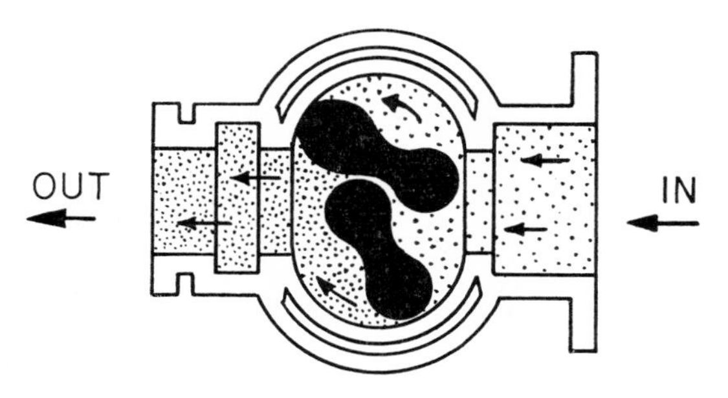

FIG. 17. Roots pump.

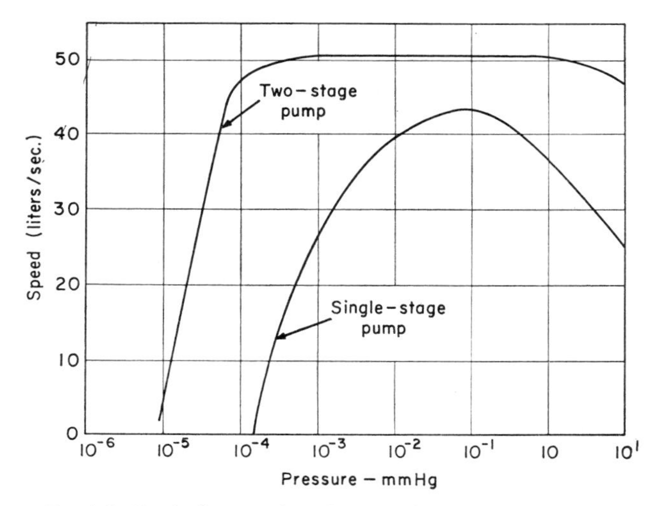

FIG. 18. Typical operating characteristics—Roots pumps.

of the molecular pump, is also utilized. Efficient operation can only occur when the fore pressure is suitably low, and the character of the speed-pressure function is largely determined by the magnitude of this pressure. Typical operating characteristics are illustrated in Fig. 18 for single- and double-stage pumps. The single-stage pump with the maximum fore pressure (10–20 mm Hg) exhibits a decidedly peaked performance with maximum pumping speed at approximately 0.1 mm Hg. The double-stage pump, on the other hand, has a characteristic which is flat to 10^{-4} mm Hg.

The molecular pump,[86–88] which was originally developed by Gaede, is now commercially available in a form designed by Siegbahn. It is constructed in the manner of a centrifugal pump with a smooth disk rotating at high speed serving as impeller. The clearances between the disk and the casing are of the order of a molecular mean free path and are usually arranged to require a fore vacuum of 0.1 to 1 mm Hg. The direction of the molecular flow is toward the center along a spiral groove cut in the casing, and the sense of rotation is such as to impart momentum to the gas molecules in the inward spiraling direction. These molecules are removed from the center region by a fore pump.

Molecular pumps merit consideration in many applications because of their freedom from back-diffusing pump fluids and because of their ability to handle vapor-laden or even dust-bearing atmospheres. They do not require a cooled trap and are, therefore, usable where liquid nitrogen or similar coolants are not available. These devices are available in a wide range of capacities from a few liters/sec to 6000 liters/sec. They are quiet, compact, and the power and cooling requirements are slight.

4.2.4.1.2. VAPOR PUMPS. A second large class of pumps is to be found in the molecular entrainment devices, in which momentum exchange occurs between the molecules of the gas to be pumped and those of a directed stream, rather than between these and a moving solid surface. In these pumps there are no moving parts. The details of the momentum exchange differ among the various types of pump, as does the language commonly used to describe the phenomena, yet there are many points of similarity. Pumps of this class may be found for any inlet pressure from atmospheric to high vacuum, with capacities from a fraction of a liter per second in the case of a small diffusion pump to 10^5 liters/sec or more in the case of a large jet ejector.

A résumé of operating ranges and performance characteristics of the various vapor pumps may be found in Fig. 19. Vapor pumps operating in the high vacuum range are of the diffusion type and employ either mercury or low vapor pressure esters or purified hydrocarbons as the working fluid. Single stage "booster" pumps of the jet aspirator type may operate to inlet pressures of a few microns and against forepressures of up to a few mm Hg. In the pressure range $10\ \mu$–1 mm, where peak speed is desired in the neighborhood of $100\ \mu$, the oil jet ejector is most frequently used. It is in this range too that the five-stage steam jet

[86] W. Gaede, *Ann. Physik* [4] **41,** 337 (1913).
[87] F. Holweck, *Compt. rend.* **177,** 43 (1923).
[88] S. von Friesen, *Rev. Sci. Instr.* **11,** 362 (1940).

ejector may be applied, although the peak speed of the best available units appears to occur at several hundred microns. Each stage of a steam ejector maintains across it a pressure ratio of about ten. Thus, a four-stage ejector can be designed for maximum volumetric flow at 1 mm Hg or so, with ejectors of fewer stages operating at correspondingly higher pressures.

The diffusion principle as applied to pumping was first investigated by Gaede[89] in 1915, and soon after by Langmuir.[90] Most contemporary diffusion pumps employ the Langmuir configuration with modifications

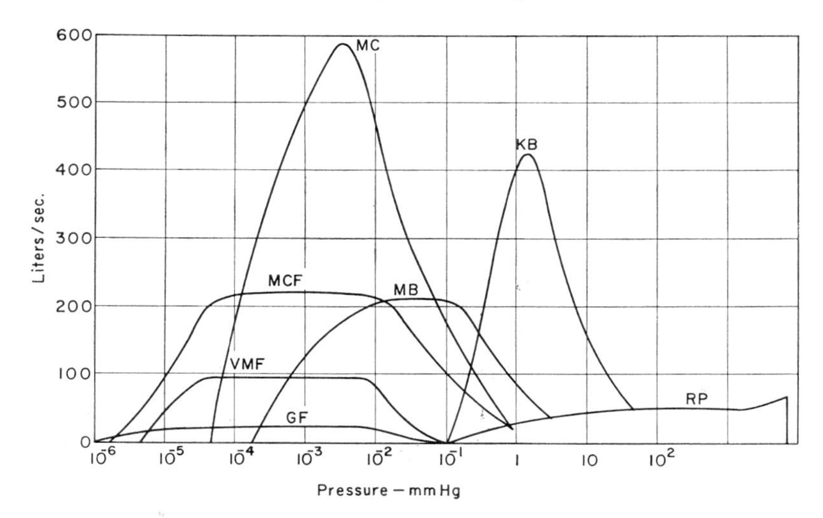

FIG. 19. Typical operating characteristics—vapor pumps. KEY: RP: rotary oil-sealed mechanical pump; KB: oil ejector pump; MB: metal booster type vapor diffusion pump; MC: high speed metal diffusion pump—nonfractionating; MCF: high speed metal diffusion pump—fractionating; VMF: small metal diffusion pump—fractionating; GF: glass diffusion pump—fractionating.

introduced by Hickmann[91] and later investigators. The schematic diagram, Fig. 20, illustrates a three-stage fractionating oil diffusion pump of a type having speeds (depending on the size) of a few hundred to several thousand liters per second. Oil vapor, formed in the boiler, passes up the central column and then becomes directed down to form vapor screens by action of the nozzles or aprons which surround the column. The cooled walls provide a condensing surface for the vapor (see Langmuir[90]) which then returns in the liquid phase to the boiler. Gas molecules, by reason of

[89] W. Gaede, *Ann. Physik* [4] **46,** 357 (1915); see also *Z. tech. Physik* **4,** 337 (1923).
[90] I. Langmuir, *Gen. Elec. Rev.* **19,** 1060 (1916).
[91] K. D. D. Hickmann, *J. Franklin Inst.* **221,** 383 (1936).

their thermal motion, travel from the pump entrance to the region of the vapor screen where the probability of collision with an oil molecule is high. Compression by vapor stream entrainment is repeated at each stage, with the compressed gas moving on into the forepressure region at the pump outlet.

Back diffusion of the gas is also possible, and would determine the ultimate vacuum if the pump fluid possessed zero vapor pressure at the temperature of the condensing surface nearest the inlet. Pump oil evaporates from this point of contact and from points above to which the

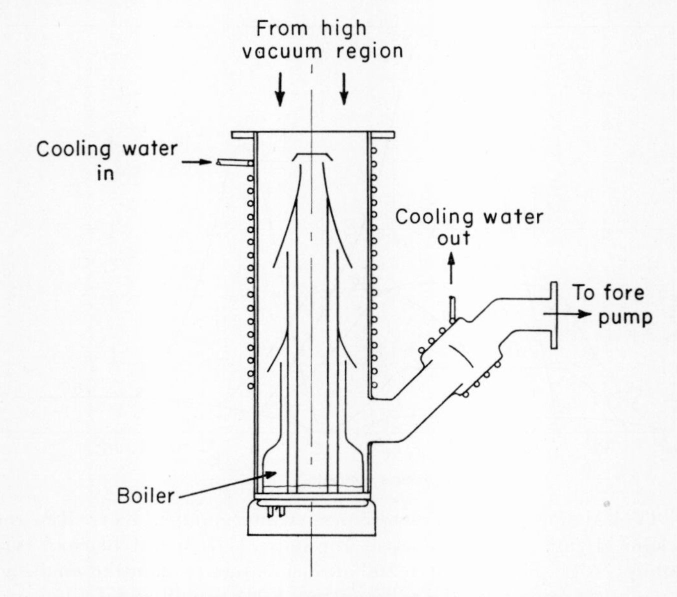

FIG. 20. Schematic diagram of diffusion pump.

oil has moved by surface migration. The test volume becomes filled with vapor at the vapor pressure of the oil. The fractionating type[91] of oil diffusion pump was devised to insure that only the least volatile of the pump oil fractions would be present in the central column, the lighter fractions contributing to vapor screens in later and less critical stages. For the production of the highest vacua and nearly always in the case of the mercury pumps a vapor trap is interposed between the pump and the region to be evacuated. Vapor pressure–temperature curves for various substances are presented in Fig. 21, and vapor pressures of common pump oils are summarized in Table VII.

Ejectors of both the oil and steam jet variety are designed to accomplish the exchange of momentum between the high velocity stream of the

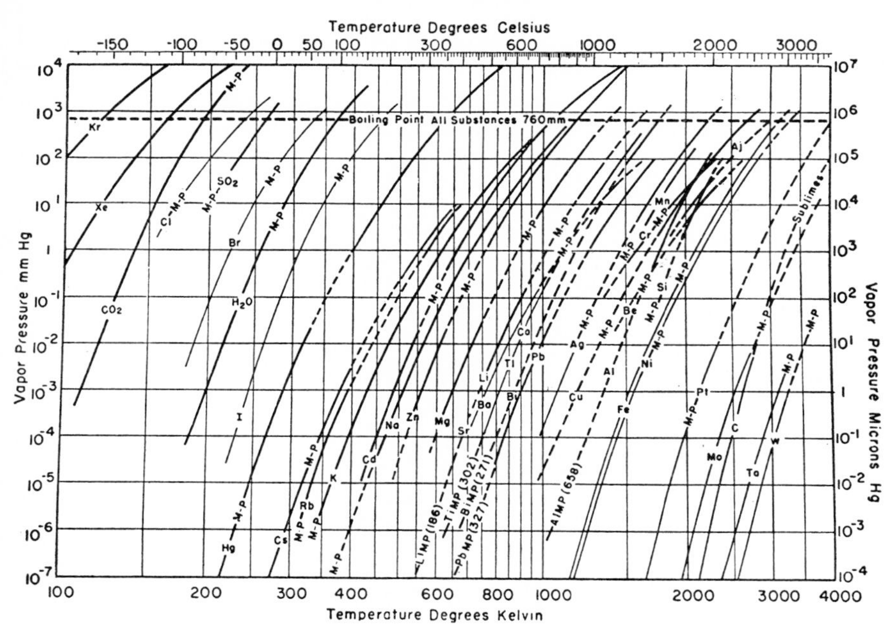

Fig. 21. Vapor pressure—typical curves for various substances.

Table VII. Vapor Pressures of Common Pump Oils[a]

Fluid	Chemical nature	Molecular weight	Vapor pressure, mm Hg	Temperature, °C
Apiezon B	Hydrocarbon	468	10^{-7}	22
Apiezon C	Hydrocarbon	574	10^{-8}	22
Apiezon BW	Hydrocarbon	472	10^{-7}	22
DC-702	Methyl polysiloxanes	530	8×10^{-6}	20
DC-703	Methyl polysiloxanes	570	1.2×10^{-6}	20
Norcoil 20 Octoil-S	Di-2-Ethyl sebacate	426	3×10^{-8}	25
Norcoil 10	Chlorinated diphenyls	326	3×10^{-4}	25
Norcoil 30 Octoil	Di-2-Ethyl hexyl pthalate	390	3×10^{-7}	25

[a] Consolidated Vacuum data given in terms of ultimate vacuum attainable in a given pump. See company data sheet on vapor pump fluids.

jet and the relatively low velocity stream of the entering gas. The flows are of the transition, slip, or continuum nature, and the momentum exchange occurs between molecular aggregations rather than between individual molecules. Typical configurations of an oil ejector and of a single stage of a steam jet ejector system are illustrated, Figs. 22(a) and (b). Note that in the oil ejector as in the diffusion pump the working vapor is removed by condensation at the wall and returned directly to the boiler. The operation of the steam ejector is more complicated. Here, advantage is taken of properties of the supersonic stream to achieve both fluid mixing and recompression in a succession of shockwaves established in region A, Fig. 22(b). A subsonic expansion takes place in B where the gas may then go to the next stage or to a vapor condenser.

Ejectors are of substantial importance in food, chemical, biological, and metallurgical industries because of the high speeds available and because of the long life and low maintenance costs for these devices.

4.2.4.1.3. Ion Pumps and Others. Any surface to which gas molecules adhere is in effect a pump which has, subject to certain qualifications, the speed of an aperture of the same area. Various mechanisms are available to effect this adherence. These fall into the general classes of condensation, sorption (adsorption, absorption, desorption),[92] and "gettering." "Gettering" is a word signifying any of several physical or chemical processes at an active surface or within an active surface layer by which surface atoms hold molecules of the residual gas.

Except in application to vapor trapping, condensation has rarely been employed for the production of high vacua. Exceptions are found in the clean-up of residual gas in vacuum insulated helium transfer tubes and storage flasks. It should be noted, however, that the vapor pressure of CO_2 and the vapors of most volatile fluids condense to low vapor pressure solids at the temperature of liquid N_2. Nitrogen, oxygen, and other atmospheric constituents have extremely low vapor pressures at the temperature of liquid helium, 4.2°K or liquid hydrogen, 20°K. Thus, the possibility of meeting a special pumping problem by condensation pumping should not be overlooked.

Various substances have the property of capturing gas molecules at the surface or in the outer layers of the solid at some temperatures and releasing them at others. In this interaction no true chemical compound is formed, although the binding forces may be high enough to prevent chemical activity of the surface atoms. A striking example of surface adsorption is found in the case of activated charcoal, where the surface area of the solid has been estimated to be as much as 2500 sq meters/gm.

[92] S. Brunauer, "Physical Adsorption," Vol. 1. Princeton Univ. Press, Princeton, New Jersey, 1943.

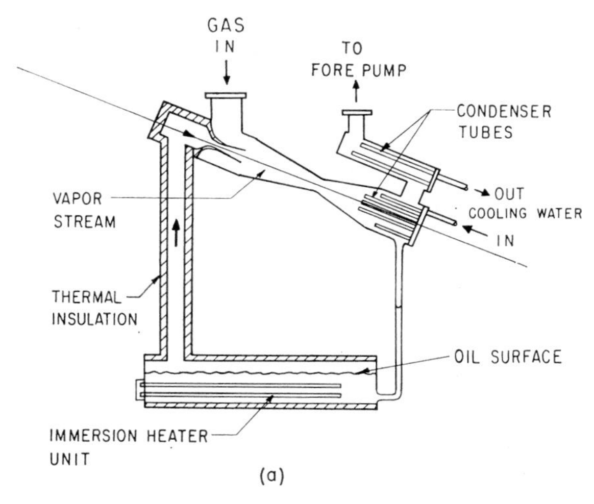

FIG. 22(a). Oil jet ejector system.

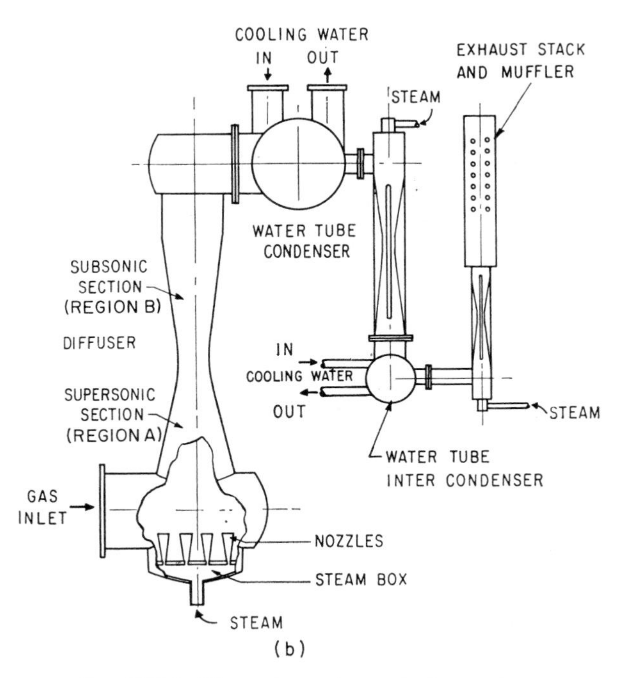

FIG. 22(b). Three-stage steam ejector system.

A popular application is in the maintenance of permanent vacuums in liquid nitrogen storage flasks.

Most fresh metal surfaces are extremely active both chemically and physically and some are well suited for vacuum deposition on the walls of enclosures within which a permanent vacuum is to be maintained. A pellet of magnesium, for example, is degassed in the final bake-out of a vacuum tube and then evaporated by radiofrequency heating at the time of seal-off. The freshly formed metal surface improves and maintains the permanent vacuum within the vacuum envelope. Other common substances found in this application are barium, titanium, calcium, aluminum, sodium, and zirconium. Reduced, oxygen-free copper may be used as a trap[93] and essentially as a pump in the production of very high vacua, 10^{-8} mm Hg or better, an increasingly important technique.

One getter pump of fairly elaborate form is available commercially, the so-called Evapor-Ion pump.[94] In this device titanium wire is continuously volatilized in an arc and allowed to deposit on the walls of the pump. Gaseous ions may also be formed in the discharge and driven into the walls by electrostatic fields. Most of the pumping of the permanent gases at a much reduced speed is attributed to ion pumping action. Speeds of 7000 liters/sec for air are claimed at 10^{-6} mm Hg, with ultimate vacuums of 10^{-7} mm Hg or better. In another type, titanium is volatilized by sputtering in a glow discharge.

One true ion pump deserves mention, the ionization gage.[95] It was discovered in studies of the production of ultra high vacuums that the ion gages attached to baked out systems continued to clean up the residual gas, presumably by driving ions formed in the discharge into the walls of the gage. Alpert[96] reports pressures of 10^{-9} mm Hg or better to be routinely produced in his laboratory, while pressures to 10^{-12} mm Hg have been produced in double shelled systems. The final residual gas is thought to be largely helium, and the permanence of the lowest pressures is found to be limited by the diffusion of atmospheric helium through the Pyrex envelope.

4.2.4.2. Pumping Speed. In addition to the final vacuum obtainable (P_0) and the maximum forepressure permissible (P_f), the most important characteristic of a pump is its speed S. If V is the volume of the recipient,

[93] D. Alpert, *Rev. Sci. Instr.* **24,** 1004 (1953).

[94] R. H. Davis and A. S. Diviatia, "Construction and Performance of Evapor-Ion Pumps." Vacuum Symposium Trans. Comm. on Vacuum Techniques, Boston, Massachusetts, 1954.

[95] D. Alpert and R. T. Bayard, *Rev. Sci. Instr.* **21,** 571 (1950).

[96] D. Alpert, R. S. Buritz, and W. A. Rogers, Research Report R-94436-4-B, Westinghouse Research Lab., East Pittsburgh, Pennsylvania, 1952.

S is defined by

$$-dP/dt = (S/V)(P - P_0) \quad (4.2.79)$$

or, for $P \gg P_0$, S is the volume in which the pressure is reduced by the action of the pump to $1/e$ of its starting value in one second, or to one tenth in 2.3 sec. For the experimental determination of S one uses the integrated form of Eq. (4.2.79)

$$S = (V/t) \ln (p_1 - p_0)/(p_2 - p_0) \quad (4.2.80)$$

where t is the time required for the reduction of the pressure in a volume V from p_1 to p_2. This method is only convenient if $V \gg S$ and if degassing of the walls can be neglected. Otherwise, a dynamic method is preferable. A gas burette is connected to the recipient through a needle valve. If the amount of gas admitted per second is $Q = pv$, and the pressure maintained in the recipient by the pump is P, ($\gg P_0$), then $S = pv/P$. It is also possible to connect the exhaust of the pump to a closed volume V_0, in which the pressure is well below P_f, and to attach a small leak to the recipient. If the pressure increase in V_0 per second is $P_2 - P_1$, the pumping speed is $S = V_0(P_2 - P_1)/P$. The pumping speed is usually expressed in liters per second or in cubic feet per minute. For mechanical pumps, the theoretical maximum value of S is the piston displacement times the number of revolutions per second (or minute). The actual value is lower because of internal leakage, noxious volume, and degassing of the pump oil, and decreases with pressure, dropping to zero for $P = P_0$. For diffusion pumps, the theoretical maximum value of S is determined by the conductance of the diffusion slit, but the actual speeds are considerably smaller. Moreover, the pumping speed of a diffusion pump not only decreases with decreasing pressure dropping to zero for $P = P_0$, but also decreases with increasing pressures, dropping to zero for $P = P_f$ (Fig. 19). Most diffusion pumps have a maximum speed plateau at pressures between 1 and 10^{-3} μ.

4.2.4.3. Free Molecular Flows in Ducts and Pumps. In many regions of a vacuum system rarefied flows may occur in the slip or transition regimes as discussed in Sections 4.2.3.1 through 4.2.3.3. However, wherever the molecule mean free path is of the order of the duct or vessel diameter the concepts and formulations of free molecule flow apply, where all molecular collisions are assumed to be with the walls. The collision process is assumed to effect an erasure of the molecular history, allowing the molecules to become "re-emitted" from the surface with a random or cosine distribution in direction and a Maxwellian distribution in velocities. The utility of this assumption in detail is somewhat in question, as indicated by direct studies of molecular scattering from surfaces.[97]

The number of molecules striking a unit area of the wall is, for a gas in equilibrium within a vessel, equal to $n\bar{c}/4 = p(2\pi mkT)^{-\frac{1}{2}}$ per unit time, where p is the pressure in dynes/cm², m the mass of the molecules, k Boltzmann's constant, and T the absolute temperature. If an aperture of area A cm² leading to a region of negligible pressure is placed in the wall of the vessel, the volumetric flow outward through it per unit time (for air at 20°C) is found from the above relation to be

$$F = 11.6 \times 10^3 A \text{ cm}^3/\text{sec} \tag{4.2.81}$$

where F is known as the conductance of a perfect orifice. It is more appropriate in consideration of pump and duct conductances of common vacuum systems to use the units liters/sec rather than cm³/sec. Thus

$$F_0 = 11.6A \text{ liters/sec.} \tag{4.2.82}$$

The term "speed" is occasionally applied to apertures in the same sense that it is applied to pumps, and signifies the conductance or volumetric flow rate. The lumped conductance of any system of orifices, pipes, valves, and pumps may be found by application of the usual rules for combination of series and parallel conductances. For example, the conductance of a pipe terminated by a valve leading to a fast pump may be expressed as

$$\frac{1}{F} = \frac{1}{F_p} + \frac{1}{F_v} + \frac{1}{F_t} \tag{4.2.83}$$

where the subscripts p, v, t, stand for pump, valve, and tube, respectively. The volumetric flow rate of such a system is independent of the gas density in regions of constant pump speed, the mass rate being determined by the pressure in the volume upstream of the tube t. Where the pump is considered to be a conductance leading to a region of zero pressure, the steady state pressure in a continuously pumped volume is completely prescribed in a balance between mass rate out through known conductances and mass rate in through needle valves or capillaries. For uniform temperature the mass rate Q is expressed as $F\rho \times 10^{-3}$ if F is in liter/sec and ρ is in gm/cm³. Applying the perfect gas law, we may write

$$F_1 p_1 = F_2 p_2 \tag{4.2.84}$$

where F_1 and F_2 are conductances into and out of a volume, and p_1 and p_2 are the pressures upstream of each conductance.

[97] F. C. Hurlbut, *J. Appl. Phys.* **28** (8), 844–850 (1957).

TABLE VIII. Values of K, Clausing Factor for Short Tubes

l/r	K	l/r	K	l/r	K
0	1	1.8	0.5384	10	0.1973
0.1	0.9524	1.9	0.5256	12	0.1719
0.2	0.9092	2.0	0.5136	14	0.1523
0.3	0.8699	2.2	0.4914	16	0.1367
0.4	0.8341	2.4	0.4711	18	0.1240
0.5	0.8013	2.6	0.4527	20	0.1135
0.6	0.7711	2.8	0.4359	30	0.0797
0.7	0.7434	3.0	0.4205	40	0.0613
0.8	0.7177	3.2	0.4062	50	0.0499
0.9	0.6940	3.4	0.3931	60	0.0420
1.0	0.6720	3.6	0.3809	70	0.0363
1.1	0.6514	3.8	0.3695	80	0.0319
1.2	0.6320	4.0	0.3589	90	0.0285
1.3	0.6139	5.0	0.3146	100	0.0258
1.4	0.5970	6.0	0.2807	1000	0.002658
1.5	0.5810	7.0	0.2537		
1.6	0.5659	8.0	0.2316		
1.7	0.5518	9.0	0.2131		

Many useful formulas have been developed by application of the concepts of free molecule flow in ducts. Those of some importance are listed below.

The conductance of a long tube[98]

$$F = 12.1 \frac{d^3}{l} \text{ liters/sec.} \qquad (4.2.85)$$

The conductance of a short circular tube[99]

$$F = K \left(\frac{l}{r}\right) F_0 \text{ liters/sec.} \qquad (4.2.86)$$

The factor $K(l/r)$ due to Clausing, expresses as a function of the ratio l/r the probability that a molecule entering the end of the tube will emerge from the opposite end. Values of K may be found in Table VIII. F_0 is the conductance of a perfect orifice of area A.

The conductance of a short rectangular tube of section $a \cdot b$ where $a \gg b$ and $a \gg l$, a, b, and l are expressed in centimeters, is

$$F = 3.668K'a \cdot b \text{ liters/sec.} \qquad (4.2.87)$$

[98] M. Knudsen, *Ann. Physik* [4] **28,** 75 (1909).

[99] P. Clausing, Ph.D. Dissertation, University of Amsterdam, Amsterdam, 1918; also *Physica* **6,** 48 (1928); *Ann. Physik* [5] 901 (1932).

TABLE IX. Values of K', Clausing Factor for Short Channels

l/b	K'	l/b	K'	l/b	K'
0	1.0000	1.0	0.6848	5.0	0.3582
0.1	0.9525	1.5	0.6024	10.0	0.2457
0.2	0.9096	2.0	0.5417		
0.4	0.8362	3.0	0.4570		
0.8	0.7266	4.0	0.3999		

Representative values of K' may be found in Table IX.

Since conductance in free molecule flow varies as the square root of the temperature, and inversely as the square root of the molecular weight, values of F for conditions other than those for air at 20°C above may be found from the following expression

$$F_{T_1M_1} = \sqrt{\frac{T_1 M_{\text{air}}}{T_{\text{air}} M_1}} F_{\text{air}}. \tag{4.2.88}$$

4.2.4.4. Vacuum Systems. The present section describes the application of certain of the preceding considerations to the design of a specific vacuum system. The system in question, Fig. 23, contains many of the elements of an atomic beam apparatus, but serves equally well as a

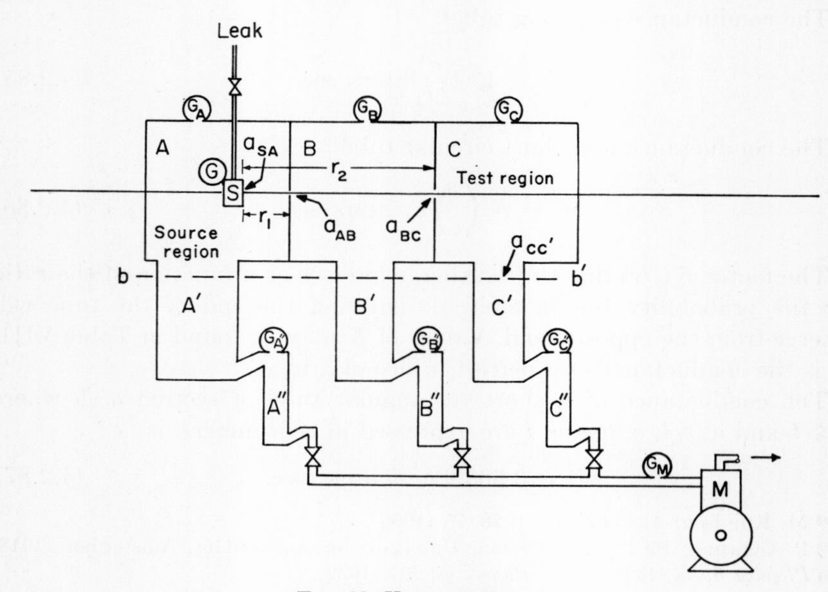

FIG. 23. Vacuum system.

generalization of other equipments in common use. The principal elements are represented here as:

(1) C, the evacuated test region with gaging system G_C

(2) A gas flow, Q_C, into the region C through the aperture A_{BC} (to be calculated)

(3) A high vacuum pump with trap, C', having lumped speed S_C at the entrance flange $a_{CC'}$

(4) A forepump C'' with speed $S_{C''}$ preceded by connecting tubing of conductance $F_{C'C''}$

(5) A mechanical pump M with speed S_M

(6) Conducting piping and valves to the mechanical pump.

In addition, the system of Fig. 23 comprises an intermediate region B, with entrance and exit apertures a_{AB} and a_{BC} and pumping system B', B'', and M, and a source region A which is similarly equipped.

Many systems would include valves at the position of the dotted line bb' for the purpose of isolating the system from the diffusion pumps. This arrangement allows the upper chambers to be opened while the high vacuum pumps remain in operation. A roughing line to each evacuated region would also be included.

We shall first prescribe operating pressures for regions A, B, and C and give aperture areas a_{BC} and a_{AB}, and beam path lengths r_1 and r_2. We shall also prescribe a flow of gas Q_A from the region S and, with these data compute the required pumping speeds for the system A', B', and C'. Let:

$p_C = 10^{-7}$ mm Hg

$p_B = 10^{-6}$ mm Hg

$p_A = 5 \times 10^{-5}$ mm Hg

$Q_A = 10^{18}$ molecules/sec. These have been assumed to issue from slit a_{SA} by thermal effusion where the slit area is 0.0025 cm^2 and the slit width 0.0025 cm. The gas behind the slit is assumed to be nitrogen at $\sim$1 mm Hg pressure and at a temperature of 293°K. Isothermal conditions are assumed throughout.

a_{AB}, $a_{BC} = 0.05$ cm^2

$r_1 = 10$ cm, $r_2 = 50$ cm.

The molecular flow from region S of 10^{18} particles/sec is equivalent to a volumetric flow of 610 liters/sec at the pressure of chamber A, 5×10^{-5} mm Hg. It is reasonable (and also conservative), to consider as negligible both the outflow and inflow through aperture a_{AB} in setting the requirement, in consequence of the known influx above, that the pumping speed of the trap and pump system A' be at least 610 liters/sec at 5×10^{-5}

mm Hg. Now, if the pressure at the forepump connection to A' is to be held at 1 μ Hg, the speed, $S'_{A''}$, of the combined pump A'' and connecting tubing must be 30.5 liters/sec. Thus

$$\frac{1}{S'_{A''}} = \frac{1}{30.5} = \frac{1}{S_{A''}} + \frac{1}{F_{A'A''}} \tag{4.2.89}$$

or

$$S_{A''} = \frac{30.5 F_{A'A''}}{F_{A'A''} - 30.5} \text{ liters/sec.} \tag{4.2.90}$$

If some reasonable value for $F_{A'A''}$ is assumed, say 100 liters/sec, then $S_{A''}$ must be approximately 44 liters/sec. If A'' will work into a forepressure of 50 μ Hg, the conductance of the piping system to the mechanical pump must be $\sim$1 liter/sec at that pressure.

We now consider the gas flowing into region B. For practical purposes only two contributions to the influx need be considered, that from the source region at p_A, and that which comes directly by molecular flow from region S. Here it is necessary to know the distance r_1. The fraction of the total molecular flow from region S passing directly into region B, from elementary kinetic theory, is

$$2 \frac{a_{AB}}{2\pi r_1^2} (610) = \frac{0.10}{628} (610) = 9.70 \times 10^{-2} \text{ liters/sec} \tag{4.2.91}$$

at 5×10^{-5} mm Hg or 4.85 liters/sec at 10^{-6} mm Hg. The other contribution, i.e., that from region A, is

$$50 \times 11.6 \times 0.05 = 29 \text{ liters/sec}$$

at 10^{-6} mm Hg. Thus the total influx to region B' is approximately 34 liters/sec.

Similar considerations apply to the calculation of influx to region C'. The direct contribution from region S amounts to 3.14×10^{-3} liters/sec, while that arriving from region B is 5.8 liters/sec. Thus the pump and trap system must have a speed of 5.8 liters/sec at 10^{-7} mm Hg.

It should be appreciated that the high vacuum required for region C may very well not be achieved in practice, particularly if the system cannot be baked out. In this region of pressure, degassing of walls and gaskets may release large numbers of molecules which have to be added to the volumetric flow requirements and thus play an important role in determining the ultimate steady state value. Most vapor diffusion pumps, even when equipped with excellent traps, exhibit very low pumping speeds in the 10^{-7}–10^{-6} mm range. On the other hand, the pump speed rises sharply as pressures are increased, and in the case of the present system, the mass influx remains fixed. At 2×10^{-7} mm Hg pressure the required pumping speed for C' is 2.9 liters/sec, while at 3×10^{-7} mm Hg the required speed is 1.9 liters/sec. Thus at some point, at a good low pressure it is hoped, the rising speed–pressure characteristic of the pump overtakes the diminishing volumetric flow requirement and the steady state pressure is established.

A glance at the requirements above reveals that the volumetric flow rates required to accommodate pumping systems $B'B''$ and $C'C''$ at the mechanical pump connection are negligibly small. However, if piping of size sufficient only to accommodate this requirement were employed, initial pump-down times would be intolerably long. The conductance of connecting piping and valves is ordinarily determined, therefore, in light of pump-down time considerations. In the present case assume the volume of chamber C to be 30 liters. If the conductance of piping, valves, and mechanical pump together is 5 liters/sec, the pump-down time to 50 μ Hg from 760 mm Hg is, from Eq. (4.2.80)

$$t = \frac{30}{5} \ln \frac{760 \times 10^3}{50} = 58 \text{ sec.} \tag{4.2.92}$$

Experience has shown that times up to 10 min or so in small systems can be tolerated, but times of much greater duration can cause substantial inconveniences, particularly in the trial period of a research apparatus.

Particular attention should be given to the role of the intermediate region in the present system. By its use the source and detector regions are effectively isolated and lower pressures are achieved in the detector region than might otherwise be possible. The technique of differential pumping exhibited here is often used to attenuate pressure fluctuations or to provide a transition from a high pressure to a low pressure in ion or electron guns, molecular beams, mass spectrometers, and quite generally elsewhere in the apparatus of physical research.

4.2.4.5. Pressure Measurements in High Vacua. The most important problem of measurement in vacuum technology is that of the pressure within the vacuum system. Various techniques and devices for pressure measurement in the forepressure range have been discussed in Section 4.2.3. It will be the task of this section to discuss this problem at pressures of the order of 1 μ Hg and lower.

Three types of gages employing ionization of the gas molecules are in current use and commercially available. In the first of these, the so-called ionization gage,[100] electrons emitted from a heated filament produce ions

[100] O. E. Buckley, *Proc. Natl. Acad. Sci. U.S.* **2,** 683 (1916); S. Dushman and C. G. Found, *Phys. Rev.* **17,** 7 (1921).

in collision with gas molecules. The resultant ion current is proportional to the electron emission current, the number density of the gas molecules within the gage volume, and the ionization efficiency of the electrons for the gas molecules present. Damage to the hot cathode fixes the upper pressure limit, while photoelectrons from the anode resulting from X-rays generated by electrons at the grid constitute a zero current which fixes the lower pressure limit. A gage which minimizes this effect[101] is illustrated in Fig. 24(a).

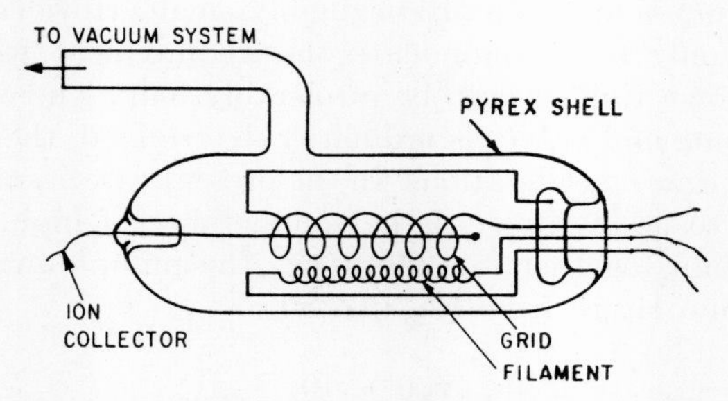

FIG. 24(a). Ionization gage, Bayard–Alpert type.

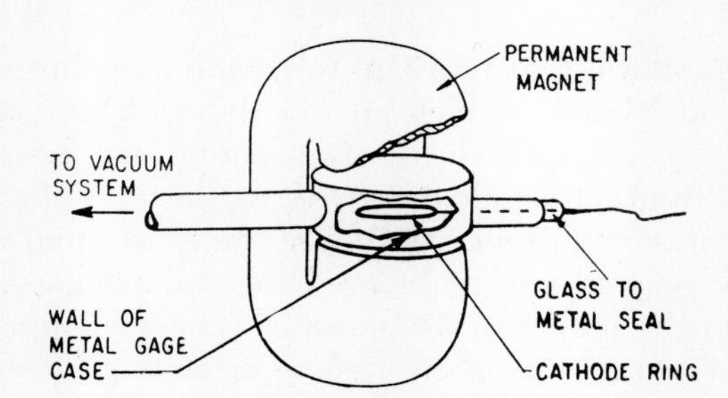

FIG. 24(b). Ionization gage, Philips type.

The second type ionization gage in current use was devised by F. M. Penning[102] and is referred to as the Philips gage (see Fig. 24(b)). This is a cold cathode device in which electrons are held to a circular orbit by a transverse magnetic field. The ion current constitutes the total gage current and is proportional to the density of the gas molecules within the gage volume. The Philips gage may be used to higher pressures than the hot cathode type described above and is normally not damaged by exposure to atmospheric pressure.

The alphatron[103] gage, discussed in Section 4.2.3.4, is usable in some

[101] D. Alpert and R. T. Bayard, *Rev. Sci. Instr.* **21,** 571 (1950).

[102] F. M. Penning, *Physica* **4,** 71 (1937).

[103] J. R. Downing and G. Mellen, *Rev. Sci. Instr.* **17,** 218 (1946).

models to pressures of 0.1 μ and hence may be of value in systems operating in the near high vacuum region. This applies also to the Pirani gage.

While ionization gages are of great practical value, their indications are of significance only after calibration, which is usually carried out by comparison with a McLeod gage (see Section 4.2.3.4). Since the calibration gas is normally air, but the residual gas unknown, the pressure readings of all ionization gases are not absolute, but in terms of equivalent air pressure. It should be pointed out that the range of overlap between a mercury McLeod and an ion gage is at most two decades, namely 1 μ–10^{-5} mm Hg. Lower pressures are measured by linear extrapolation of the known characteristic of the gage. Ionization gages may release adsorbed gases for long periods of time or, when thoroughly degassed, act as getter pumps (see Section 4.2.4.1.3) and thus indicate either higher or lower values than the true pressure value.

The Knudsen[104] gage is one of a large class of gages relying for their operation on the momentum transfer in a rarefied gas. A vane with quartz

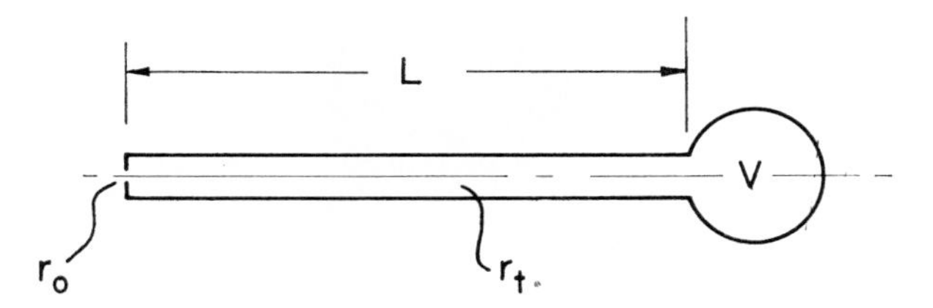

FIG. 25. Elements of a gaging system.

fiber suspension is rotated between two surfaces at different temperatures by the action of the residual molecules coming from different directions with different average momentum. The gage has the advantage of being absolute; its response can be calculated from geometrical factors, surface temperatures, and molecule number density. It has the disadvantage of being a delicate apparatus which requires very careful use if its potential as an absolute gage is to be realized. Other gages utilizing the transfer of momentum are known as "molecular" and involve the damping of oscillatory motion of a fiber-supported disk or the transmission of torque to fiber-supported vanes from an external rotating cage or cylinder. Such gages may be sensitive to pressures as low as 10^{-6} mm Hg.

An important consideration in the design of vacuum gaging systems is that of response time.[105] Where the gaging system consists of the elements of Fig. 25, an orifice between connecting tubing and gage of radius

[104] M. Knudsen, *Ann. Physik* [4] **32,** 809 (1910).
[105] S. A. Schaaf and R. R. Cyr, *J. Appl. Phys.* **20,** 860 (1949).

r_0, connecting tubing of radius r_t, and length L, and a gage of volume V, the response time constant t_i is given by

$$t_i = C\left[\frac{VL}{r_t^3} + \frac{8V}{3r_0^2} + \frac{8\pi V_t^2 L}{3r_0^2} + \frac{4L^2}{\pi r_t}\right] \tag{4.2.93}$$

where C is a constant, $\frac{3}{4}[\frac{1}{2}(2\pi RT)^{\frac{1}{2}}] \approx 10^{-5}$ sec/cm at 20°C, R is the gas constant, and T the temperature in °K. Free molecule flow conditions have been assumed. For a given r_0, L, and V, Eq. (4.2.93) may be optimized with respect to r_t. Thus

$$(r_t)_{\text{opt.}} = [0.18 r_0^2 V]^{\frac{1}{5}}. \tag{4.2.94}$$

4.2.5. Shock Tubes[106–109]

4.2.5.1. Introduction. There exists the strong implication within much of the foregoing material that the measurements discussed or suggested are best made where relatively long measurement periods are available, and where the apparatus has had ample time to come to steady state. Such an interpretation is completely justified by experience. An exception must be made, however, in the case of the shock tube, since the usefulness of this device for the study of certain aerophysical phenomena overshadows the concomitant experimental difficulties.

In its simplest form the shock tube is a pipe-shaped apparatus, separated into two chambers by a diaphragm placed normal to the pipe axis. One chamber, the compression region, is usually filled with gas at a high pressure, while the other chamber, the expansion region, is evacuated or filled with gas at a low pressure. In operation the diaphragm is ruptured, often by a solenoid actuated spike, allowing gas from one chamber to expand rapidly into the other. In this process, driven by the energy of the expanding gas, a normal shock wave travels the length of the expansion tube. Contiguous with the rear face of the shock wave, and traveling with it, is a widening zone of steady flow. The time available for observation is very short; the shock wave passes a model or instrument in the test region in a microsecond or less, while the region of steady flow behind it occupies the same field of study for times of the order of milliseconds at most.

When initial pressure ratios across the diaphragm are of sufficient

[106] W. Payman and W. C. F. Shepherd, *Proc. Roy. Soc.* **A186,** 293 (1946).

[107] W. Bleakney, D. K. Weimer, and C. H. Fletcher, *Rev. Sci. Instr.* **20,** 807 (1949).

[108] I. I. Glass and G. N. Patterson, Proc. 5th Intern. Aeronaut. Conf., pp. 200–240, Inst. Aeronaut. Sci., New York, 1955.

[109] R. Courant and K. O. Friedrichs, "Supersonic Flow and Shock Waves" (Vol. 1 of "Pure and Applied Mathematics," R. Courant, L. Bers, and J. J. Stoker, eds.). Interscience, New York, 1948.

magnitude and other requirements are met, the temperatures in the steady flow region may reach 20,000°K or more, while the shock front itself will be strongly ionized.[110] Thus an environment is created in which the study of important and fundamental problems of molecular physics may occur; problems, for example,[111,112] of relaxation times of vibrational states, electronic excitation, dissociation, ionization, or of chemical kinetics. At the same time, there exists an evident utility of the device

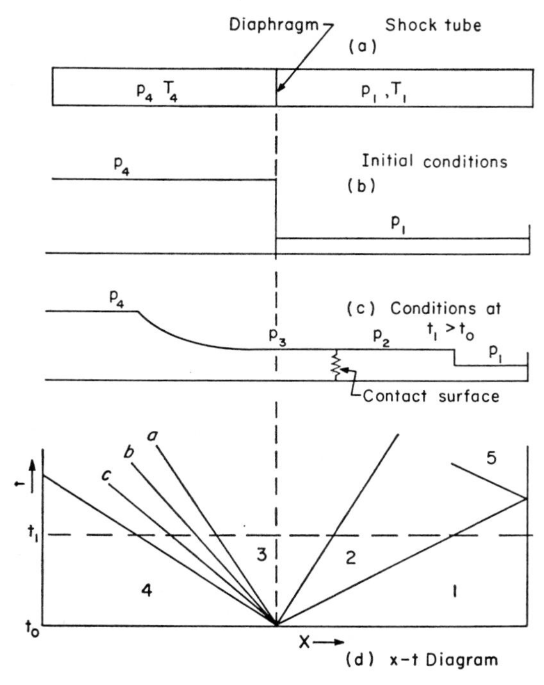

FIG. 26. Schematic of shock tube and shock propagation.

for the study of problems of aerodynamics in hypersonic flows with very large stagnation enthalpies.[113]

4.2.5.2. Theory. The shock tube may be diagrammed as in Fig. 26(a) with initial and subsequent pressures in the various regions of flow as in Figs. 26(b) and (c), respectively. The time history of the shock tube flow may be represented as in the x-t diagram, Fig. 26(d). At time $t = 0$

[110] E. L. Resler, S. C. Lin, and A. Kantrowitz, *J. Appl. Phys.* **23,** 1390 (1952).

[111] E. F. Greene and D. F. Hornig, *J. Chem. Phys.* **21,** 617 (1953).

[112] A. Hertzberg and J. C. Logan, *in* "The Threshold of Space" (M. Zelikoff, ed.), pp. 276–287. Pergamon, London, 1957.

[113] W. Griffith, *J. Aeronaut. Sci.* **19,** 249 (1952).

the gases 1 and 4 have initial states p_1, T_1, and p_4, T_4, and properties γ_1, R_1, γ_4, R_4, etc., where γ is the ratio of specific heats and R is the gas constant. Immediately after the rupture of the diaphragm a shock wave propagates into the expansion chamber with wave speed C_w, while an expansion wave propagates into region 4 with speed a_4 at its front, where a_4 is the speed of sound at temperature T_4. The region of uniform flow at p_2, T_2 is terminated at its upstream boundary by the *contact surface* which marks the point of separation between gas which had been at condition 4 preceding diaphragm rupture, and that which had been at condition 1. Region 3 at the contact surface must be at the same pressure as region 2, but the temperatures and densities in the two regions will not be equal. The centered expansion wave forms the head of region 3. The lines marked a, b, c, Fig. 26(d), are characteristics of the expansion wave, and define the propagation of constant states of the gas. The shock wave of region 1 is reflected from the tube end as shown, and creates thereby a region of gas at condition 5 where T_5 is approximately $2T_2$. The useful portion of the process is terminated when the reflected shock interacts with the contact surface.

The one-dimensional gas dynamic analysis leading to the above idealized picture of the shock tube process is based on assumptions relating to the adiabatic and inviscid character of the flow, to the constancy of specific heat for each molecular species, and to the applicability of the ideal gas law. The *shock strength* p_2/p_1 is given implicitly in terms of initial conditions p_1, p_4, etc., by

$$\frac{p_4}{p_1} = \frac{p_2}{p_1}\left[1 - \frac{\gamma_4 - \gamma_1(a_1/a_4)[(p_2/p_1) - 1]}{\sqrt{2\gamma_1}\sqrt{2\gamma_1 + (\gamma_1 + 1)[(p_2/p_1) - 1]}}\right]^{\gamma_4/(\gamma_4-1)}. \tag{4.2.95}$$

Since $p_3 = p_2$ the *expansion strength* p_3/p_4 is determined from the shock strength and initial conditions. The expansion wave is an isentropic expansion with temperature behind it given by the isentropic relation

$$\frac{T_3}{T_4} = \left(\frac{p_3}{p_4}\right)^{(\gamma_4-1)/\gamma_4}. \tag{4.2.96}$$

If the coefficient μ is defined by the expression

$$\mu = \frac{\gamma + 1}{\gamma - 1} \tag{4.2.97}$$

where $(\mu - 1)$ is the effective number of degrees of freedom of the gas, the temperature behind the shock, T_2, is given by

$$\frac{T_2}{T_1} = \frac{\mu_1 + (p_2/p_1)}{\mu_1 + (p_1/p_2)} \quad \text{(Rankine–Hugoniot relations).} \tag{4.2.98}$$

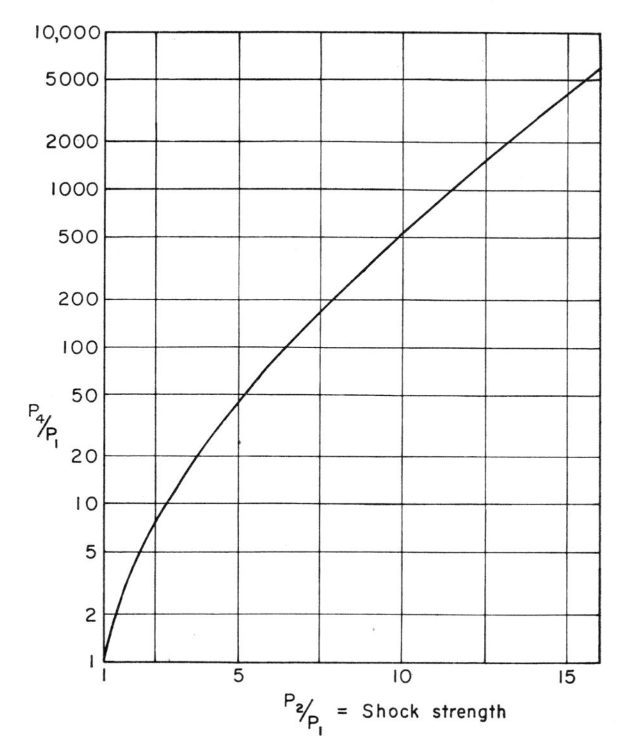

FIG. 27. Initial pressure ratio versus shock strength for air (from R. K. Lobb Report, University of Toronto, Institute of Aerophysics No. 8, May 1950).

Further, one may readily express the Mach number of the shock wave M_1, where $M_1 = (C_w/a_1)$, as

$$M_1 = \left[\frac{\gamma_1 - 1}{2\gamma_1} + \frac{\gamma_1 + 1}{2\gamma_1}\left(\frac{p_2}{p_1}\right)\right]^{\frac{1}{2}} \tag{4.2.99}$$

while that of the fluid behind the shock is given by the expression

$$M_2 = \left(\frac{p_2}{p_1} - 1\right)\left[\frac{2\gamma_1(p_1/p_2)}{(\gamma_1 + 1) + (p_2/p_1)(\gamma_1 - 1)}\right]^{\frac{1}{2}}. \tag{4.2.100}$$

Plots of certain important ideal shock tube results are presented, Figs. 27 through 29. Figure 27 shows the ratio of shock strength to initial pressure ratio for various gases. Figure 28 exhibits the relation between shock strength and fluid Mach number M_2, while Fig. 29 shows temperatures which may be achieved in region 5, both for ideal flow and for ionized gas conditions.

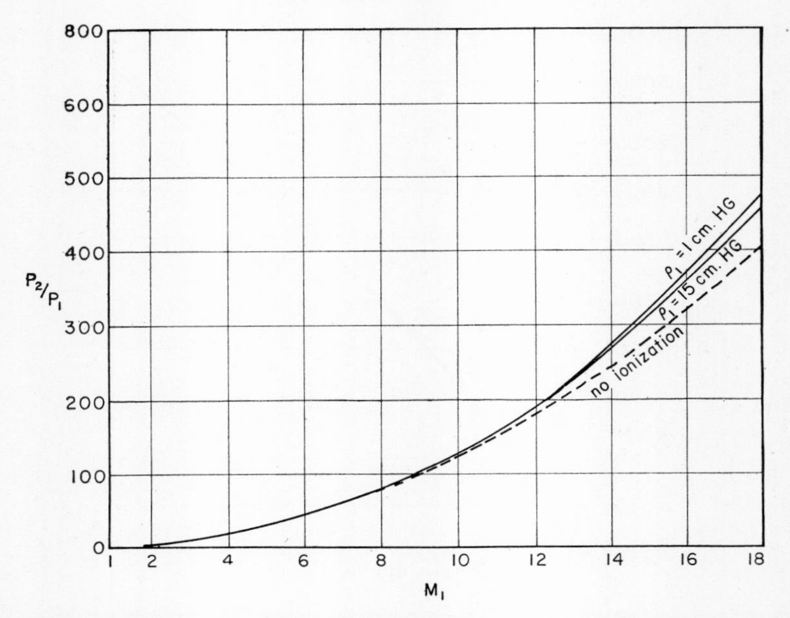

FIG. 28. Pressure ratio across the incident shock in argon (from Shao-Chi Lin, Ph.D. Thesis, Cornell, June 1952).

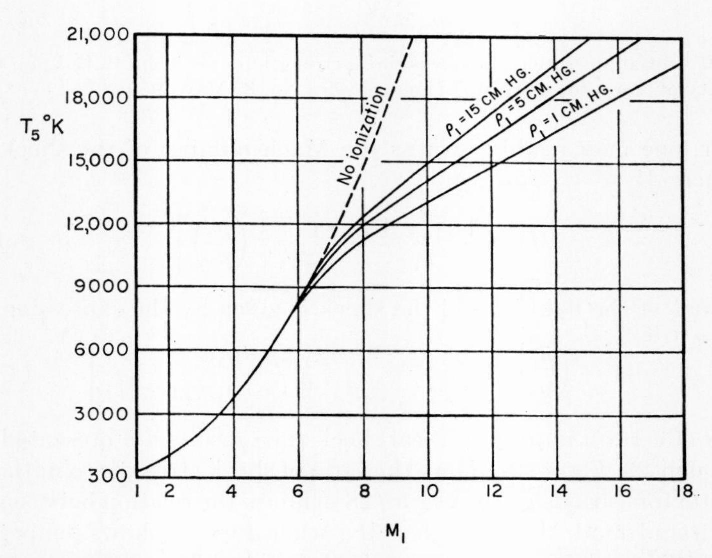

FIG. 29. Temperature behind reflected shock wave in argon (from Shao-Chi Lin, Ph.D. Thesis, Cornell, June 1952).

The ideal flow analysis predicts temperatures, shock strength, etc., somewhat in excess of those found by experiment. The theory fails because each of the assumptions fails in some degree, particularly as the shock strengths become very large. It can be seen from Fig. 29 that the energy required for the electronic excitation or ionization of the molecule, or the excitation of vibrational states or for molecular dissociation reduces substantially the translational temperature of the gas. It has been also demonstrated that the effects of viscosity and heat are after all substantial sources of attenuation of the shock wave.[114]

It should not be inferred that the utility of the shock tube is diminished in the measure that its behavior is imperfectly represented by ideal flow theory. On the contrary, phenomena of real gases are those of primary interest, and it is here as a tool in the study of the physics of real gases that the shock tube has proved of greatest worth.

[114] R. N. Hollyer, Jr., "A Study of Attenuation in the Shock Tube." University of Michigan Eng. Research Inst. Rept. Project M720-4 (1953).

5. SOUND AND VIBRATION*

5.1. General Concepts

5.1.1. Definitions

According to American Standard Acoustical Terminology,[1] "Sound is an alteration in pressure, stress, particle displacement, particle velocity, etc., which is propagated in an elastic material, or the superposition of such propagated alterations." The term vibration generally has the implication that all the particles in a finite region (particularly in a solid body) are moving in phase.

In a wave the phase changes progressively as a function of position; the distance between two successive regions having the same phase of motion is the wavelength, λ. It is related to the wave velocity c and frequency of oscillation f by $\lambda = c/f$. The frequency limits of audible sound may be taken as 20–20,000 cycles/sec. Higher frequencies are termed ultrasonic; lower, infrasonic.

In the acoustic regime, displacements are small compared to a wavelength, and particle velocities are small compared to wave velocity. These are the principal conditions which yield the linearized wave equation for a source-free space

$$\left(\nabla^2 - \frac{\partial^2}{c^2\,\partial t^2}\right) W = 0 \qquad (5.1.1)$$

where W is any mechanical parameter that characterizes the alterations occurring in the medium during the passage of the wave. A consequence of this linear wave equation is that superposition of solutions is possible. Physically this means that complicated motions may be considered as composed of the sum of simple ones. This statement is exemplified in Fourier series and integrals, in the Laplace transform, and in instruments such as wave analyzers and filters.

Waves may be characterized as longitudinal or transverse, according to whether the particle motion is parallel or normal to the direction of the wave (in an isotropic medium). In dissipationless fluids, permanent transverse waves cannot exist, but the presence of dissipation makes it possible to excite them in rapidly damped form. In solids both types exist, and usually have different velocities. The complicated combination of these two types occurring during reflection is in part responsible for

[1] Reprinted from the American Standard Acoustical Terminology, Z24.1-1951, American Standards Association, New York.

* Part 5 is by **Vincent Salmon.**

the difficulty of obtaining complete analytic solutions to many problems of wave motion in finite solids.

The magnitude of wave motion may be measured by the sound pressure in fluids, the tensor stress in solids, or by the particle velocity in all media. The sound pressure is the deviation from ambient pressure caused by the passage of the wave, and is usually measured by the rms value unless otherwise specified. Particle velocity is similarly the deviation from the undisturbed value. The lower limit measurable is set by the pressure and velocity variations due to thermal agitation in the medium.

The field of mechanical wave motion comprises communications acoustics, industrial acoustics, and medical acoustics. In the first, the main concern is with the transmission of pragmatic or esthetic intelligence between humans, often with the interposition of electroacoustic devices. The ear is the primary end organ, and its characteristics determine in great part the quantities that are measured, and the units that are used.

Industrial acoustics is concerned with the use of sonic energy for processing, testing, and control, and with the control of undesired sound and vibration. Sound at all frequencies and intensities may be employed in all media. The experimental methods utilized usually are adaptations of techniques used in communications acoustics.

At present, medical acoustics involves primarily the use of ultrasonic energy for diagnosis and therapy. Most of the applicable methods and techniques will be covered in discussing the first two fields.

5.1.2. Decibel Notation

In communications acoustics the approximately logarithmic response of the ear to sound pressure has led to the widespread use of logarithmic measures of the magnitude of all acoustic field quantities. The principal unit is the bel, or its more widely used subdivision, the decibel, abbreviated db. Originally, the decibel applied only to a power ratio; two powers P and P_0 are defined to differ by N db if

$$N = 10 \log_{10} P/P_0. \tag{5.1.2}$$

This definition has been extended to apply similarly to powerlike quantities. For example, voltage, sound pressure, and particle velocity appear squared in expressions involving electrical or acoustic power. Thus squares of ratios involving these quantities would appear in the extended definition of the decibel.

In actuality there are in use two types of decibel units, the relative and the absolute. When two quantities are being compared with no thought of one being more than a temporary reference, N is in relative decibels. When, however, the reference quantity in the denominator has a fixed and well-accepted value, the comparison is in absolute decibels, and the logarithmic measure is called the level, for which the symbol is L. In the communications industry, power level has already been standardized with respect to a reference value of 0.001 watt, and the corresponding standard decibel unit is the dbm.[2] Thus the third letter indicates the absolute nature of the logarithmic comparison, the type of quantity, and fixes the reference value. This procedure will be followed here.

The measurement of sound pressure level is of great importance in both communications and industrial acoustics. The reference value of sound pressure is commonly taken to be 0.0002 rms dyne/cm^2, corresponding to the threshold of hearing for a rather sensitive ear. For the purposes of the present text we define dbt as the unit of pressure level[3] L_p, given by

$$\begin{aligned} L_p &= 10 \log(p/p_0)^2 \\ &= 20 \log(p/p_0) \\ &= 20 \log(p/0.0002) \\ &= 74 + 20 \log p \text{ dbt} \end{aligned} \tag{5.1.3}$$

where p is in rms dyne/cm^2, and all logarithms are to the base 10. If the sinusoidal pressure amplitude is 1 atmos, the pressure level is 191.5 dbt; ordinary speech levels are about 75 dbt at 2 ft. Note that pressure squared is employed to make the ratio powerlike; this follows from the relation (for fluids) between pressure p and intensity I for sinusoidal motion and plane or spherical waves $I = p^2/\rho_0 c$, where I is the energy transmitted per unit time normal to unit area of wave front, and ρ_0 is the undisturbed density of the fluid. The combination $\rho_0 c$, is called the specific acoustic impedance of the medium, by analogy to the electrical relation between power P, voltage E, and resistance R, $P = E^2/R$. Thus pressure p, in this type of analogy, is similar to voltage. The cgs unit of wave resistance, the dyne-sec/cm^3, is also called the rayl.[4]

In underwater sound,[3] the reference for sound pressure level is commonly the microbar (1 dyne/cm^2), and the suggested unit is the dbu. As may be seen from Eq. (5.1.3), these differ by 74 db.

Intensity level in aerial acoustics is usually referred to 10^{-16} watt/cm^2, which under normal sea level conditions corresponds closely to the intensity in a plane or spherical wave at the reference pressure of 0.0002

[2] Radio-Electronics-Television Manufacturers Association (Now Electronic Industries Association), "Sound Systems." Standard RS-160, December, 1951.

[3] R. W. Young, "Scheme proposed for indicating reference quantities in level measurements." Private communication, November, 1947.

[4] L. L. Beranek, "Acoustics," p. 11. McGraw-Hill, New York, 1954.

dyne/cm². Thus for practical purposes we may take $L_I = L_p$, where we define[3] $L_I = 160 + 10 \log I$ dbe, with I in watt/cm².

Power level L_w in dbm is given by[2] $L_w = 30 + 10 \log P$ dbm, where P is in watts. However, for acoustical purposes it often turns out to be convenient to employ a reference power of 10^{-12} watt (1 picowatt). For use in the present text we define[5] $L_w = 120 + 10 \log P$ dbp, where P is in watts.

Another popular reference value for acoustical power is 10^{-13} watt,[6] for which the level abbreviation *PWL* has been used. The numbers of decibels N corresponding to each power reference level are related by N (re 10^{-13} watts) $= 10 + N$ (re 10^{-12} watts) $= 100 + N$ (re 10^{-3} watts).

In communications acoustics it is important that objective measurements provide as close correlation as possible with the evaluation of the ear. One of the most important units is the phon (pronounced phōne). The loudness expressed in these units is the sound pressure level of a 1000 cycles/sec pure tone which under specified listening conditions is subjectively judged to produce a loudness sensation equal to that of the unknown tone.[7] See Section 5.3.2.3 for discussion.

5.1.3. Radiation of Sound; Equivalent Circuits

In the process of radiating sound into a transmitting medium, a more or less well-defined surface is caused to vibrate. The surface is usually solid (as a loudspeaker diaphragm), but it may also be liquid (as the surface of the ocean), or gaseous (as in sirens). The reaction of the transmitting medium back on the radiating surface may be described in terms of two types of energy transfer: the acoustic energy which leaves the surface and does not return, and that which leaves but is stored in and returned periodically from the medium.

It is more convenient to describe radiation in equivalent terms of the complex ratio of the sound pressure produced by the reaction of the medium on a surface having an impressed velocity u normal to the surface. This ratio is termed the wave impedance z_w, defined by

$$z_w = p/u = 1/y_w.$$

Here y_w is the wave admittance, a quantity which usually has a simpler analytical form than the impedance. The wave admittance or impedance

[5] C. M. Harris, ed., "Handbook of Noise Control," pp. 2–11. McGraw-Hill, New York, 1957.

[6] Reference 4, p. 14.

[7] H. Fletcher and W. A. Munson, "Loudness, its definition, measurement and calculation." *J. Acoust. Soc. Am.* **5**, 82 (1933).

is obtained by solving the wave equation (5.1.1) for the particular radiating system involved. The value obtained will depend upon the properties of the transmitting medium, the shape of the wave front, and the shape and nature of the boundaries. Because it is often necessary for the experimenter to estimate the radiation impedance, results for several systems are summarized below.

For spherical waves in fluids, or the equivalent radiation from a radially pulsating sphere of radius a,

$$\begin{aligned} y_w &= (1/\rho_0 c)[1 + (1/jka)] \\ z_w &= \rho_0 c[1 + (j/ka)]/[1 + (1/ka)^2] \end{aligned} \tag{5.1.4}$$

where j is the imaginary unit and k is the wave number $2\pi/\lambda = \omega/c$, ω being the radian frequency $2\pi f$. For plane waves a is infinite, and

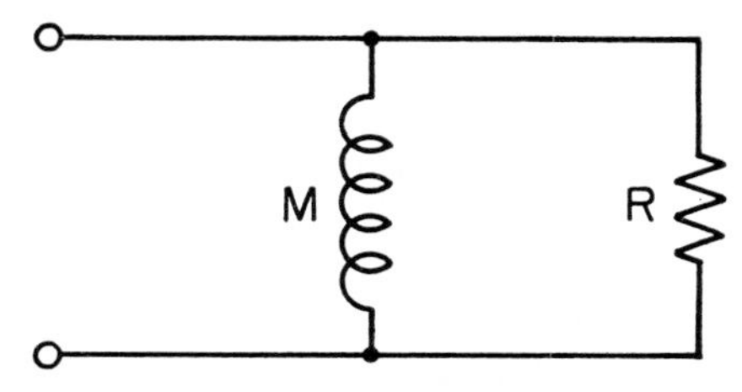

FIG. 1. Equivalent circuit for mechanical radiation impedance of radially pulsating sphere of radius a in fluid of density ρ_0 and sound velocity c.

$M = 4\pi a^3 \rho_0 \quad R = 4\pi a^2 \rho_0 c.$

$z_w = \rho_0 c$. For convenience in the interpretation of acoustical and mechanical systems the impedances may be portrayed as analogous electrical ones in circuits equivalent in behavior with respect to frequency and power. In the analogy used here, a resistor signifies energy lost to the system, an inductor is associated with kinetic energy stored in a moving mass, and a capacitor signifies energy stored elastically in potential form. Other analogs are possible, but there is no one which fits all requirements perfectly. The equivalence used herein is variously termed the classical, force-voltage, or impedance. The equivalent circuit interpretation is somewhat simpler if we employ mechanical impedance (force/velocity), where the force is the integral of the pressure over the area concerned. Bauer[8] describes a formalism for writing down an equivalent circuit. Figure 1 shows the equivalent circuit for spherical waves. Note that the shunt reactance may be conceived as due to an equivalent mass which is three times that of the fluid in a sphere of radius a.

[8] B. B. Bauer, "Transformer couplings for equivalent network synthesis." *J. Acoust. Soc. Am.* **25**, 837 (1953).

If a spherical cavity of radius a in an infinite solid with Poisson's ratio σ, density ρ, and compressional velocity c is radiating by radial pulsation, the mechanical radiation impedance is

$$z_m = 4\pi a^2 \rho c \left[\frac{1}{1 + (1/jka)} + \frac{2(1 - 2\sigma)}{(1 - \sigma)jka} \right]. \tag{5.1.5}$$

The equivalent circuit[9] is given in Fig. 2. The capacitive element corresponds to a compliance (ratio of displacement to force) which is $3(1 - \sigma)/2(1 - 2\sigma)$ times the value for a fluid sphere of radius a having the same wave velocity and density as the solid. For fluids, $\sigma = \frac{1}{2}$, the capacitance is infinite, and the circuit becomes that of Fig. 1, as it should.

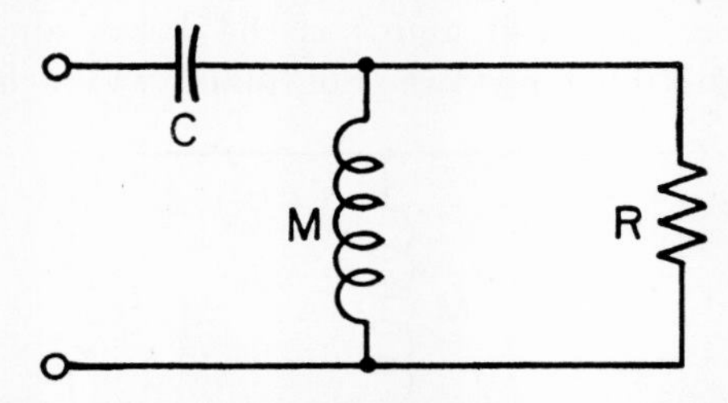

FIG. 2. Equivalent circuit for mechanical radiation impedance of radially pulsating spherical cavity of radius a in infinite solid of density ρ_0, compressional wave velocity c, and Poisson's ratio σ (after Salmon, Reference 9).

$$C = \frac{1 - \sigma}{2(1 - 2\sigma)} \frac{1}{4\pi a \rho c^2}$$

$$M = 4\pi a^3 \rho \qquad R = 4\pi a^2 \rho c.$$

If the radiator is a rigid sphere of radius a small compared to a wavelength, and is oscillating as a dipole in a fluid, the mechanical radiation impedance is

$$z_m = 4\pi a^2 \rho_0 c (ka)^4/12 + j\omega(4\pi a^3 \rho_0/3)/2. \tag{5.1.6}$$

No simple equivalent circuit exists for portraying this behavior. The equivalent mass m in the reactance $j\omega m$ is one-half that for the radially-pulsating sphere of radius a.

If a radiator can be represented by an axially vibrating rigid circular piston of radius a set flush in a closely fitting opening in a rigid infinite wall, it is called a Rayleigh piston. The radiation impedance cannot be expressed by simple functions, and is best given by a table of values[10] or a graph, as Fig. 3. The equivalent circuit for the smoothed behavior is

[9] V. Salmon, "Equivalent circuit for spherical radiation in a solid." *J. Acoust. Soc. Am.* **28,** 724 (1956).

[10] P. M. Morse, "Vibration and Sound," 2nd ed., p. 447. McGraw-Hill, New York, 1948.

closely given[11] by Fig. 4. The equivalent mass is that of the fluid medium contained in a sphere of radius $(2/\pi)^{\frac{1}{3}}a$, or in a cylinder of radius a and height $(8/3\pi)a$.

If the radiator is an infinitely long wire of radius a (as approximated by the center of a vibrating string in a musical instrument), the mechanical

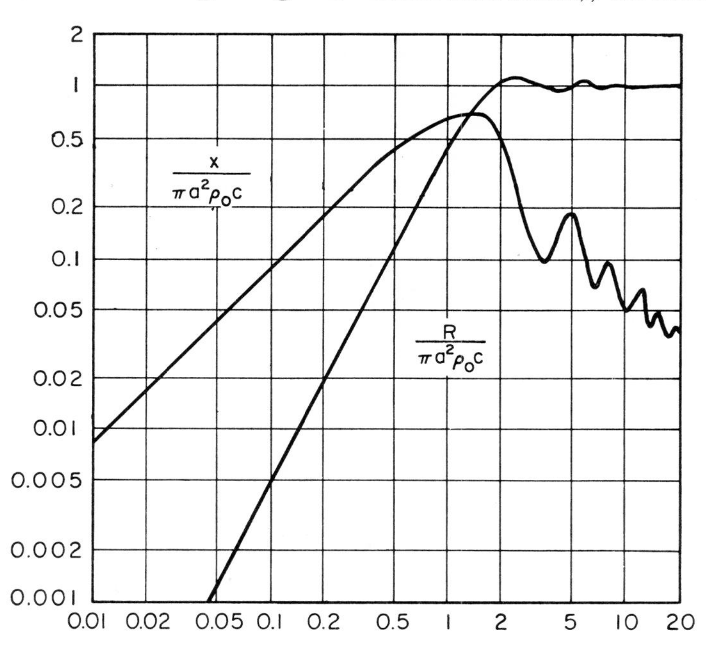

FIG. 3. Mechanical radiation impedance $R + jX$ for Rayleigh piston of radius a. Abscissa is $ka = 2\pi a/\lambda$.

radiation impedance per unit length, when the wire radius is much less than a wavelength, is given by

$$z_m = \pi a^2 \rho_0 c(\pi k(ka)^2/2) + j\omega(\pi a^2 \rho_0). \tag{5.1.7}$$

This has no simple equivalent circuit; the equivalent mass equals that contained in a cylinder of radius a.

In addition to systems in which sonic energy is directly radiated from a solid surface, the experimenter may employ horns. A horn is a rigid-walled expanding conduit in which the ratio of sound pressure to particle velocity is different at the two ends. In this fashion it acts somewhat as a transformer. It may be shown that under very general assumptions,[12]

[11] See reference 4, p. 121.

[12] See reference 10, pp. 265–271.

the radius-axial distance function $r(x)$ for a family of straight horns of circular cross section is given by $r = r_t[\cosh(x/x_0) + T \sinh(x/x_0)]$, where r_t is the radius at the throat (or small end), T is the parameter $(0 \leq T \leq \infty)$ which fixes the particular horn, and x_0 is a constant which fixes the scale of length. These are called hyperbolic exponential horns. The energy transmitted by these horns falls off markedly below a cutoff frequency f_0 given by $f_0 = c/2\pi x_0$.

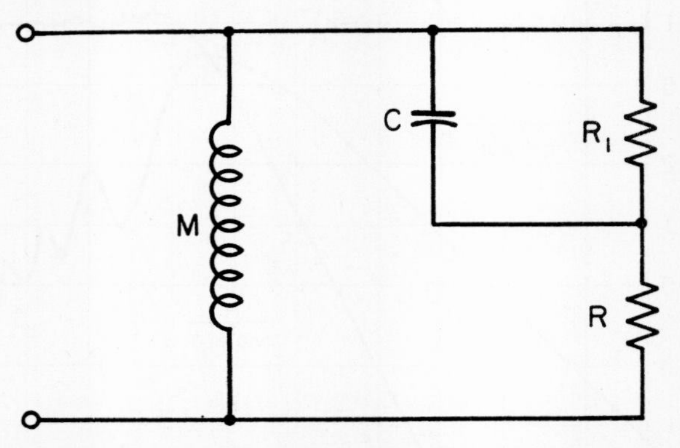

FIG. 4. Equivalent circuit for smoothed mechanical radiation impedance of Rayleigh piston of radius a (after Beranek, Reference 11).

$$M = 8a^3\rho_0/3 \qquad R = \pi a^2 \rho_0 c$$
$$R_1 = \left(\frac{128}{9\pi^2} - 1\right) \pi a^2 \rho_0 c \qquad C = \frac{0.6}{a\rho_0 c^2}.$$

The mechanical impedance at the throat of infinite horns of the above type, for frequencies above f_0, is approximated[13] by

$$z_m = \pi r_t^2 \rho_0 c \left[\frac{(1 - (1/kx_0)^2)^{\frac{1}{2}} + jT/kx_0}{1 - (1 - T^2)/(kx_0)^2}\right]. \tag{5.1.8}$$

It may be noted that in the above relations, the dimensionless combination kx_0 may be written as the frequency ratio f/f_0 where f_0 is $c/2\pi x_0$. The equivalent circuit for the above horns cannot be written completely in terms of frequency-independent elements. The representation of Fig. 5 is often used. The exponential horn corresponds to $T = 1$; when both T and x_0 increase without limit, the conical horn obtains. The horn for $T = 0.6$ is often used when the resistance must vary little over the pass band.

Many sound radiating devices are directional, and the experimenter often wishes to have a single-number measure of this characteristic. When

[13] V. Salmon, "A New Family of Horns." *J. Acoust. Soc. Am.* **17**, 212 (1946).

total power radiated is the main concern rather than details of the distribution of intensity with angle, a useful measure is the directivity index. It is based on the assumption that sound pressure is measured in the inverse square field, that is, at points sufficiently distant from the source

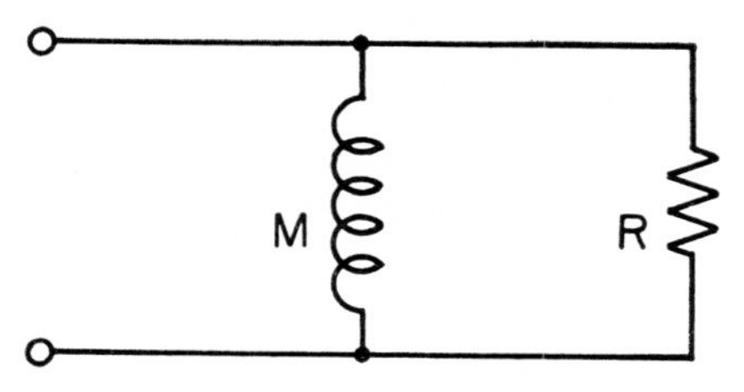

FIG. 5. Equivalent circuit for mechanical impedance above cutoff frequency f_0 at throat of infinite hyperbolic exponential horn. Throat area is S_t and family parameter is T.

$$M = S_t\rho_0 c/2\pi f_0 T \qquad R = S_t\rho_0 c/[1 - (f_0/f)^2]^{\frac{1}{2}}.$$

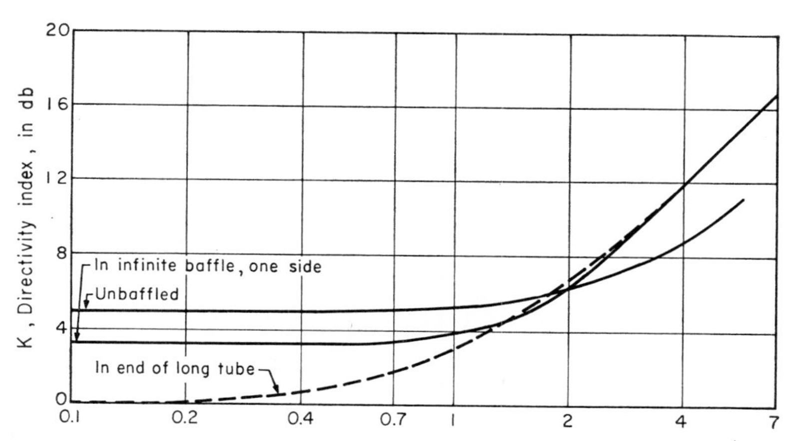

FIG. 6. Directivity indices for pistons of radius a. Abscissa is $ka = 2\pi a/\lambda$ (after Beranek, Reference 14).

so that sound pressure varies inversely with distance. The total power radiated through a sphere of radius r is, in spherical coordinates,

$$P = \frac{r^2}{\rho_0 c} \int_0^{2\pi} \int_0^{\pi} p^2(\theta,\varphi,r)\sin\theta \, d\theta \, d\varphi. \tag{5.1.9}$$

If p_0 is the sound pressure at radius r in some reference direction, usually on an axis of symmetry, then we may set $P_0 = 4\pi r^2 p_0^2/\rho_0 c$. The directivity index K in decibels is then given by $K = 10 \log(P_0/P)$. The utility of the concept lies in the fact that K may often be estimated from a knowledge of the geometry of the radiator. Values of K for typical radiators likely to be encountered by the experimenter are given[14–16] in Fig. 6.

The sound pressure level L_p (in dbt) and power level L_w (in dbm) are related to the directivity index K approximately by

$$L_p = L_w + 60 + K - 20 \log(r/10)$$

where r is the distance (in meters) from source to point of measurement. If the power level is in dbp, the numeric changes from 60 to -30; if PWL is used, the numeric becomes -40.

5.1.4. Transmission

The velocity of transmission of sound in a nondissipative elastic medium is given by $c^2 = 1/\beta\rho_0$, where ρ_0 is the rest density, and β is the dynamic compressibility $\beta = -dV/V\,dp$. In gases for which the heat transfer is poor and isentropic conditions obtain, $c^2 = \gamma p_a/\rho_0$, $\gamma = c_p/c_v$, where p_a is the ambient pressure. In air, at frequencies of the order of 0.0003 cycles/sec, heat radiation becomes important. Above about 10^9 cycles/sec heat conduction enters; outside these frequency limits the velocity changes to the isothermal value.

Since mechanical wave motion is transmitted by elastic forces, the velocity will depend on how many degrees of freedom of molecular vibration are excited. Thus velocity measurements in gases may be used to determine the details of molecular relaxation. Frequencies at which those effects occur vary widely; in nitrogen, vibrational and rotational relaxation effects have been noted at 4 cycles/sec and 240 Mc/sec respectively.[17]

Because of the quasicrystalline structure of liquids, the effects of compressibility, temperature and frequency are more complicated than in gases.[17] Again, acoustic means of analysis can be used to ascertain details of structure.

Solids differ from liquids in possessing shear rigidity. Two elastic constants are required for their description, and two primary velocities exist. For longitudinal (also called dilatational or irrotational) waves, the velocity in an infinite isotropic solid is

$$c_l = [E(1 + \sigma)/\rho(1 + \sigma)(1 - 2\sigma)]^{\frac{1}{2}} \tag{5.1.10}$$

[14] See reference 4, pp. 109–115.

[15] H. F. Hopkins and N. R. Stryker, "Proposed loudness-efficiency rating for loudspeakers and the determination of system power requirements for enclosures." *Proc. I. R. E.* (*Inst. Radio Engrs.*) **36,** 315 (1948).

[16] H. F. Olson, "Acoustical Engineering," Chapter 2. Van Nostrand, New York, 1957.

[17] T. F. Hueter and R. H. Bolt, "Sonics," pp. 401–440. Wiley, New York, 1955.

where E is Young's modulus, σ is Poisson's ratio, and ρ is the density. For many metals σ is near 0.3.

For transverse (or equivoluminal) waves,

$$c_t = [E/2\rho(1 + \sigma)]^{\frac{1}{2}}. \tag{5.1.11}$$

When waves of either type are incident upon a boundary, in general both types will be produced upon reflection and transmission.[18]

The wave velocities for finite idealized solids will depend on the geometry. In a long, thin, circular cylinder, the velocity of longitudinal waves for wavelength large compared to diameter is

$$c_l' = (E/\rho)^{\frac{1}{2}} \tag{5.1.12}$$

whereas for torsional waves the velocity is the same as for transverse waves, Eq. (5.1.11). If transverse flexural waves are excited, the wave velocity in a cylindrical bar will be dependent on frequency f and radius a, with a "bar" phase velocity of

$$c_b = (\pi a^2 E/\rho)^{\frac{1}{4}} f^{\frac{1}{2}}. \tag{5.1.13}$$

Thus the medium is dispersive for this mode of oscillation, and the waveform of a complex wave is not preserved during transmission. If the cross section of the rod is not circular, replace a^2 in the velocity by $4k^2$, where k is the radius of gyration of the actual cross section about its center line.

If the cylinder is inextensible but is perfectly flexible, and is placed under tension T, it approaches a stretched string. The phase velocity for lateral displacements in the ideal string is

$$c_s = (T/\epsilon)^{\frac{1}{2}} \tag{5.1.14}$$

where ϵ is the mass per unit length. If the string also has finite self-stiffness, as in the treble strings of a piano, then it partakes of the properties of both string and bar. In this case the stretched "wire" velocity c_w depends on the bar velocity c_b (which varies with the square root of frequency), and is given[19] by

$$c_s/c_w = [(c_s{}^4/2c_b{}^4)^2 + (c_s/c_b)^4]^{\frac{1}{2}} - (c_s{}^4/2c_b{}^4). \tag{5.1.15}$$

If the wire is very flexible and the frequency is low, c_b becomes very small, and c_w is closely equal to c_s. At higher frequencies and shorter wavelengths the self-stiffness of the wire becomes important and the velocity changes. It is this phenomenon which makes the first overtone of the short treble strings in a piano slightly higher in pitch than one octave above the fundamental frequency.

[18] J. J. Jakosky, "Exploration Geophysics," Chapter 7. Trija, Los Angeles, 1940.

[19] See reference 10, pp. 166–170.

Many electroacoustic instruments utilize a membrane or a plate as the element which couples the acoustical to the mechanical systems. A membrane is distinguished by the fact that its restoring force is due primarily to the tension T (expressed as force per unit length) with which the membrane is stretched. The velocity of flexural waves in a membrane is given by

$$c_m = (T/\sigma)^{\frac{1}{2}} \tag{5.1.16}$$

where σ is here the mass per unit area.

In a plate the self-stiffness of the material is responsible for the restoring force. The wave velocity for motion normal to the surface is given by

$$c_p = [Eh^2/12\pi^2\rho(1 - \sigma^2)]^{\frac{1}{4}}f^{\frac{1}{2}} \tag{5.1.17}$$

where h is the half thickness and σ is Poisson's ratio. As in the case of the bar, Eq. (5.1.13), the presence of the frequency term indicates that the wave motion is dispersive.

In a dispersive medium a finite wave train will consist of a group of waves having several frequencies. The velocity of the group c_g, is given

TABLE I. Density ρ_0 and Compressional Wave Velocity c in Fluids

Substance	Temperature (°C)	ρ_0 (gm/cm³)	$10^{-4}c$ (cm/sec)
Gases			
Air	0	1.29×10^{-3}	3.31
Air	20	1.20×10^{-3}	3.44
Alcohol vapor	0	—	2.31
Ammonia	0	0.771×10^{-3}	4.15
Carbon dioxide	0	1.98×10^{-3}	2.58*
Chlorine	0	3.17×10^{-3}	2.06
Ether vapor	0	—	1.79
Hydrogen	0	0.09×10^{-3}	12.70
Illuminating gas	0	—	4.90
Nitrogen	0	1.25×10^{-3}	3.37
Oxygen	0	1.43×10^{-3}	3.17
Steam	100	0.58×10^{-3}	4.05
Liquids			
Alcohol, methyl	19	0.81	11.4
Ammonia	16	0.65	16.6
Gasoline	—	0.68	13.9
Mercury	20	13.50	14.1
Turpentine	15	—	13.3
Water, fresh	13	1.00	14.4
Water, sea	15	1.03	15.1

* Below 10,000 cycles/sec.

TABLE II. Young's Modulus E, Poisson's Ratio σ, Density ρ, and Long Bar Velocity c_l' in Solids

Substance	$10^{-11}E$ (dyne/cm²)	σ	ρ (gm/cm³)	$10^{-4}\, c_l'$ (cm/sec)
Aluminum	7.3	0.33	2.7	52
Beryllium	12.7	0.33	1.8	84
Beryllium copper	12.5	0.33	8.2	39
Brass	9.5	0.33	8.4	34
Cellulose acetate sheet	0.14	—	1.3	10
Cellulose acetate butyrate	1.70	—	1.2	37
Concrete	2.5	0.30	2.6	31
Copper	11.0	0.35	8.9	35
Cork	0.0062	—	0.25	5
Dural	7.0	0.33	2.8	50
Glass, crown	7.4	0.25	2.4	55
Glass, heavy flint	7.2	0.23	5.9	35
Granite (average)	5.0	0.22	2.7	43
Ice	0.92	0.365	0.92	32
Iron, cast	9.0	0.29	7.8	34
Lead	1.7	0.43	11.3	12
Phenolic, mineral filled	1.05	—	1.8	24
Polymethylmethacrylate, cast	0.35	—	1.2	17
Magnesium	4.0	0.33	1.7	48
Nickel	21.0	0.31	8.8	49
Paraffin	0.15	—	0.9	13
Polyester	0.38	—	1.4	16.5
Porcelain	4.2	—	2.4	42
Quartz, fused	5.2	—	2.7	44
Quartz, X-cut	—	—	2.7	57.2
Rubber, hard	0.23	—	1.1	14
Steel, 0.4% carbon	20.0	0.29	7.7	51
Wood, fir, with grain	1.1	—	0.51	47
Wood, white oak, with grain	1.2	—	0.72	41
Wood, white pine, with grain	0.6	—	0.45	36

by $c_g = c - \lambda\, dc/d\lambda$, where λ is the wavelength, related to frequency by $c = \lambda f$.

Table I lists compressional velocities and densities in common fluids. In Table II data on Young's modulus, Poisson's ratio, and density are given from which the various wave velocities in solids may be calculated. The long bar, long wavelength velocity is also listed.

5.1.5. Fresnel Zone

When sound is radiated from an extended surface, interference effects will be experienced in the region close to the radiator. In this near field of Fresnel region, exact calculation of the sound pressure is possible in only a few cases. One of the most important of these is the sound pressure on the axis of a Rayleigh piston; it is given[20] by

$$p \propto \sin\{(k/2)[(r^2 + a^2)^{\frac{1}{2}} - r]\} \tag{5.1.18}$$

where r is the distance along the axis and a is the piston radius. In this situation the experimenter is usually interested in the boundary of the Fresnel region, beyond which the inverse square field (far field, Fraunhofer region) obtains. In the far field the piston may be considered substantially a point source, intensity will vary as $1/r^2$, pressure as $1/r$, and the wave impedance will be $\rho_0 c$. The limiting radius r_0 beyond which far field (inverse square) conditions obtain is given by

$$r_0/a = [(2a/\lambda) - (\lambda/2a)]/2.$$

Thus for short wavelengths (high frequencies) the position of measurement must be quite distant to be in the far field. At distance r_0 the ratio of sound pressure extrapolated from the inverse square region to the pressure actually obtaining is $(\pi/2)/[1 - (\lambda/2a)^2]$. Thus at r_0 and for very short wavelengths, the actual pressure level is at least 4 db below the extrapolated value.

5.1.6. Scattering, Reflection, Refraction

A discontinuity in the transmission properties of a medium will alter the nature of the sound field. The extent of the alteration will depend upon the nature of the discontinuity, and its size and shape. As a general matter, regions smaller than a wavelength in dimension will scatter little sound. For a rigid sphere, the total scattered long wavelength power P_s is given by[21]

$$P_s = (p^2/\rho_0 c)(256\pi^5 a^6 f^4/9c^4) \tag{5.1.19}$$

where p is the undisturbed sound pressure and a is the radius of the sphere. This is the familiar Rayleigh fourth power scattering law, for which the term in the second parenthesis is the scattering cross section area. At high frequencies and short wavelengths ($\omega a/c \gg 1$), another calculation must be made. The scattering cross section then becomes the projected area, corresponding to specular reflection. Ordinarily a surface must be at least five wavelengths across ($\omega a/c > 16$) before we may speak

[20] L. E. Kinsler and A. R. Frey, "Fundamentals of Acoustics," pp. 184–187. Wiley, New York, 1950.

[21] See reference 10, pp. 354–355.

of total reflection, for which the sound pressure at normal incidence is doubled. Under the nonlinear conditions obtaining in shock waves, however, the overpressure (in air, normal incidence) may rise to eight times the incident value.

Sound, as light, is refracted at the interface between two media. Snell's law still holds, giving $\cos \phi/\cos \phi_0 = c/c_0 = 1/n$, where the grazing angles ϕ are measured from the plane interface, and n is the index of refraction. When n is a function primarily of a single coordinate z, as in a stratified medium such as is approximated by the earth, atmosphere, or ocean, the path of a ray is not a single straight line. The intrinsic equation of the ray, expressed in terms of the angle θ from the tangent to the path as a function of distance s along the path is given by

$$d\theta/ds = -\cos \theta \, d(\ln n)/dz.$$

When the velocity gradient has a constant value a, $c = c_0 + az$, the ray takes a path of radius $r = c_0/a \cos \theta_0$, where θ_0 is the angle of entry at the interface at which the gradient starts. Since many velocity–distance functions may be approximated as a series of linear segments, this permits simple graphical means for plotting rays.

Both the atmosphere and deep ocean show minima in the velocity–depth function. Thus portions of wavefronts removed from the position of the minimum tend to curve so as to direct energy toward the position of the minimum. Acoustic energy emitted at or near such a minimum hence tends to become trapped in a layer termed a duct or channel. Since the spreading is then cylindrical rather than spherical the decrease in sound pressure with distance is considerably less, and long distance transmission is possible. This effect is the basis of the sofar[22] (SOund Fixing And Ranging) system employed for location of explosions in the ocean in rescue work. A small sofar bomb, fused for the appropriate depth, provides a signal which has been picked up over 3000 km away.[22]

5.1.7. Absorption

Viscosity (both shear and compressional) and molecular relaxation account for most of the absorption of acoustic energy in fluids. In liquids, viscosity is most important, whereas in gases viscosity and heat conduction play substantially equal roles. When the wave period equals the time for excitation of an internal mode of motion of the molecule, absorption will be anomalously large. In CO_2, the absorption coefficient (per wavelength) at 20 kc/sec is about ten times that at frequencies far removed, owing to relaxation of vibrational modes of motion.[17]

[22] See reference 20, p. 444.

Dimensional analysis of absorption in gases indicates that the appropriate frequency parameter is $\omega(\eta' + 2\eta)/\rho_0 c^2$, where η' is the dilational and η the shear viscosity.[23] Since $\rho_0 c^2 = \gamma p_a$ for nearly perfect gases, and the effective viscosity is largely independent of frequency and pressure, the frequency parameter is essentially f/p_a. Hence decrease in ambient pressure may be used as a means of increasing the effective frequency range. Data are often reported with f/p_a as the independent variable. For a variety of useful situations where the Stokes–Kirchhoff analysis applies, absorption in gases varies with frequency squared,[17] and the ratio (absorption coefficient/frequency^2) is often used in compact reporting of results.

Absorption in solids is similar to that in liquids, with complications added by the granular structure many solids have.* Intergrain heat flow[24] and scattering[25] are of importance. It is essential that the experimenter recognize that the absorption for shear and dilatation waves may be quite different. It is convenient to express this energy loss as the imaginary part of a complex elastic coefficient.[26] If the shear and bulk (dilatation) moduli are respectively G and B, then we may set

$$G = G_0(1 + jg)$$

and

$$B = B_0(1 + jb).$$

Young's modulus E is given in terms of these by

$$E = 9GB/(G + 3B) = E_0(1 + je). \tag{5.1.20}$$

The quantity e is the tangent of the angle that E makes with E_0 on the complex plane, and thus e is a complicated combination of the loss tangents g and b. It is important in comparing measured absorptions to be explicit about what elastic constant (or combination) is involved. Experimental arrangements should be chosen to minimize the possibility of measuring other than the intended constant.

If plane sound waves are being radiated into a semi-infinite rod the

* See also Vol. 6, Chapter 4.2.

[23] F. V. Hunt, "Notes on the exact equations governing the propagation of sound in fluids." *J. Acoust. Soc. Am.* **27**, 1019 (1955).

[24] C. Zener, "Elasticity and anelasticity of metals," p. 4 ff. University of Chicago Press, Chicago, 1948.

[25] W. P. Mason, "Piezoelectric Crystals and Their Application to Ultrasonics," p. 422. Van Nostrand, New York, 1950.

[26] E. Skudrzyk, "Die Grundlagen der Akustik," pp. 766–778. Springer, Vienna, 1954.

mechanical radiation impedance per unit area, $R + jX$, related to the appropriate elastic constant[27] by

$$\begin{aligned} Y &= Y_0(1 + jy) \\ R &= (Y_0\rho/2)^{\frac{1}{2}}((1 + y^2)^{\frac{1}{2}} + 1)^{\frac{1}{2}} \\ X &= (Y_0\rho/2)^{\frac{1}{2}}((1 + y^2)^{\frac{1}{2}} - 1)^{\frac{1}{2}}. \end{aligned} \tag{5.1.21}$$

In such a medium the wave amplitude will decay in the direction of propagation x as $\exp(-x/x_0)$. The components of the appropriate modulus Y are then given[27] by

$$\begin{aligned} Y_0 &= \frac{\omega^2\lambda^2[4\pi^2 - (\lambda/x_0)^2]}{[4\pi^2 + (\lambda/x_0)^2]^2} \\ y &= \frac{4\pi\lambda/x_0}{4\pi^2 - (\lambda/x_0)^2}. \end{aligned} \tag{5.1.22}$$

5.1.8. Doppler Effect

If observer, source, and medium have component velocities (along the same line) of v_0, v_s, and v_m respectively, then the observed frequency f_0 is related to f_s emitted by the source by

$$f_0 = f_e(c + v_m - v_0)/(c + v_m - v_s). \tag{5.1.23}$$

This effect may be employed to measure any of the material velocities concerned, if the others are known.

5.1.9. Cavitation

The tension in the negative phase of an intense sound wave in a liquid may exceed the dynamic tensile strength of the medium. Rupture of the liquid occurs at nuclei which may be dissolved gas, dust, or even weaknesses in the instantaneous structure of the liquid. The bubbles thus formed (of the order of a few microns diameter in water) contain liquid vapor and perhaps dissolved gases. During the positive pressure phase, these bubbles collapse, and the surrounding liquid swiftly flows radially inward. Near the center the motion is stopped violently, and the kinetic energy of the hydrodynamic mass is largely converted to an intense but short-range shock wave. Local shock pressures may reach 500 atmos. It is this high shock pressure which is largely responsible for the disruptive effects of cavitation.

Provided the medium is several wavelengths in extent, the energy required for cavitation increases with frequency. For ordinary water about 0.2 watt/cm^2 is required up to about 10 kc/sec, whereas at 1 Mc/sec,

[27] O. R. Abolafia, "Survey of the methods used to determine the dynamic mechanical properties of polymers." Picatinny Arsenal Tech. Rept. 2060, September, 1954.

over 100 watt/cm^2 is needed; the practical limit is reached near 10 Mc/sec.[28] The presence of cavitation is easily detected by the characteristic sizzling noise produced.

5.2. Instruments

5.2.1. Sources of Sound and Vibration*

Sound is generated by the motion of a material surface. The surface may be well defined as a loudspeaker cone, or it may be tenuous, as in the shear of a jet discharging into relatively still air. The experimenter must ever be aware of sources of unwanted sound which may interfere with the source intended to be measured.

The most common sources are electroacoustic transducers. These may be classified according to the motor employed to change electrical into mechanical vibratory energy. The most important motors are moving coil, variable reluctance, magnetostriction, piezoelectric, ferroelectric, and variable capacity.

Analysis of linear transducer action is best undertaken by relating the current, voltage, force, and velocity by a pair of linear equations. Of the several forms possible, one of the most useful is that due to Poincaré and discussed at length by Hunt.[29]

$$\begin{aligned} E &= Z_e I + T_{em} v \\ F &= T_{me} I + z_m v. \end{aligned} \tag{5.2.1}$$

Solution for impedances yields for the total electrical and total mechanical impedances

$$\begin{aligned} Z_{ee} &= Z_e + (-T_{em} T_{me}/z_m) \\ z_{mm} &= z_m + (-T_{me} T_{em}/Z_e). \end{aligned} \tag{5.2.2}$$

Here the added terms on the right arise because of the motion, and are called motional impedances; ordinarily this designation is reserved for last term in the first relation. Electroacoustic transducer theory is essentially the delineation of the transduction coefficient T. One of the most interesting physical facts is that when magnetic effects are involved,

* See also Vol. 2, Section 2.42.

[28] R. Esche, "Untersuchungen der Schwingungskavitation in Flüssigkeiten." *Acustica* **2** (Beih. 4), AB208 (1952).

[29] F. V. Hunt, "Electroacoustics," Chapter 2. Wiley, New York, 1954.

$T_{em} = -T_{me}$ whereas for electrical effects $T_{em} = T_{me}$. This affects the nature of the equivalent circuits realizable.[29]

For obtaining the maximum possible acoustic output, modulator type sources may be employed. In these the control signal modulates the release of energy from another source. The most common examples are the human voice and its electronic analog, the modulated airstream loudspeaker. Such a loudspeaker can reproduce sinusoidal signals over the audio range with a total acoustic power some 10 to 15 db above that from a conventional loudspeaker of the same size and weight.[30] A horn unit weighing less than 10 lb, and controlled by a 10-watt amplifier, can have an acoustic output equal to that from a conventional 500-watt loudspeaker.

When a wide band source of extremely high power is required, explosions are often used. Power levels over 60 dbm have been obtained[31] with a 10-gage shotgun blank.

For many special purposes stretched strings and membranes, rods, and plates may be set into vibration and used as sound sources. The applicable theory is covered in most standard texts.[10]

5.2.2. Acoustical Detectors

Many detectors are electroacoustic, and use generators similar to the motors in sound or vibration-generating transducers. Detectors for use in gases are called microphones; in water, hydrophones; and in solids, vibration pickups. They may be classed according to whether they are sensitive to force or motion. Most microphones are pressure-difference-sensitive. When one of the pressures is more or less constant and steady, the detector is called a pressure microphone. If both sides of the active surface are exposed, the device is a pressure-difference microphone. Since particle velocity is proportional to pressure gradient, such instruments are often termed velocity microphones. In actuality, only the hot-wire anemometer and the Rayleigh disk are true velocity microphones.

In water, hydrophones ordinarily employ piezoelectric, ferroelectric, or magnetostrictive generators.[32] Directivity is important, and in sonar work is obtained by arrays of hydrophones whose outputs are combined in the proper amplitude and phase. For small scale work, hydrophones encapsulated in rubber are available commercially.

[30] V. Salmon and J. C. Burgess, "Modulated airstream loudspeaker." *J. Acoust. Soc. Am.* **29**, 767A (1957).

[31] W. J. Galloway, B. G. Watters, and J. J. Baruch, "An explosive noise source." *J. Acoust. Soc. Am.* **27**, 220 (1955).

[32] J. W. Horton, "Fundamentals of Sonar." U.S. Naval Institute, Annapolis, Maryland, 1957.

For measurement of sound in air, a sound level meter[33] is the standard instrument. It was originally intended to measure loudness, but this complex factor still defies any simple objective measurement. The sound level meter provides three frequency weightings—*A*, *B*, and *C*—in which the *A* and *B* correspond roughly to the Fletcher–Munson[7] contours of constant loudness at 40 and 70 db loudness level respectively. On the *C* scale, the response is intended to be flat within about ±2 db from 25 to 8000 cycles/sec. Thus on the *C* scale the instrument reads sound pressure level. However, it is more correct to speak of the instrument as

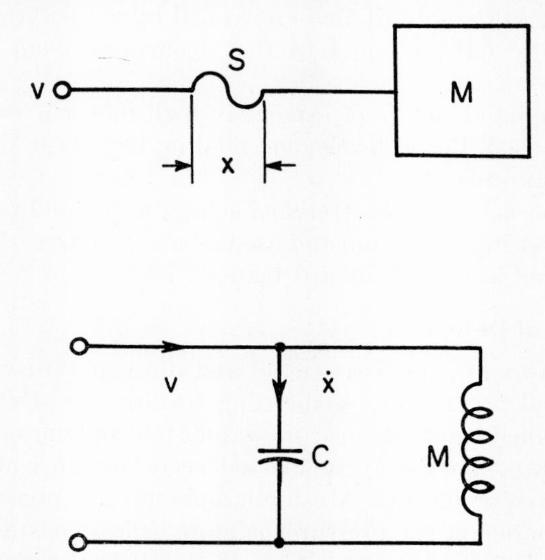

FIG. 7. Mechanical and equivalent electrical representations of undamped accelerometer comprising stiffness s and mass M, with electrical output proportional to spring displacement x. Compliance $C = 1/s$.

reading sound level, a quantity circularly defined as the reading of this standard meter. For this reason there is growing recognition of the fact that the readings should not be stated in decibels, especially for complex sounds. The term sound unit has been informally suggested; the SL (sound level) readings from the three scales would then be stated in SUA, SUB, and SUC, respectively.

Vibration pickups may be sensitive to displacement, velocity, or acceleration. The experimenter will find that the design of the mechanical elements is the most critical part of devising the instrument. It is particularly important that there be no spurious modes of motion near the

[33] American Standards Association, "Sound Level Meters for Measurement of Noise and Other Sounds." Standard Z24.3-1944.

intended one, and that the damping be linear and independent of temperature and age. Unusual means of converting to an electrical signal are valuable only insofar as they contribute to these ends.

Accelerometers are the most common type of vibration pickup. Fig. 7 shows the mechanical and equivalent electrical circuits of an undamped accelerometer in a single degree of freedom lumped system with input velocity v. Mass m is free, and is driven by the input through spring of stiffness s.

The displacement x of the spring is then given by $x = a_0/(\omega^2 - \omega_0^2)$, where a_0 is the input acceleration and ω_0 is the resonant frequency of the spring–mass combination. When $\omega_0 \gg \omega$, x is proportional to a_0. Then if the transducer element is displacement-sensitive, the instrument is acceleration-responsive. Piezoelectric accelerometers are of this type.

Damping is added to control the response at resonance, to avoid ringing. For following transient inputs closely, a damping 0.6 to 0.7 of critical is common. If most rapid rise time is required at the expense of some ringing, the damping may be much less. If the coupling between electrical and mechanical systems is high, much of the damping may be placed in the electrical system. Ordinarily most of the damping must be in the mechanical system. If grease-like or liquid damping material is required between surfaces approaching each other along a common normal, care must be taken that impulsive inputs do not drive the material completely out of the gap.

5.2.3. Level Recorders

Since much acoustical information may inhere in the amount of sound or vibrational energy expressed as a function of frequency, automatic means are often used to plot this function. In communications acoustics the information output is ordinarily written with an abscissa logarithmic in frequency and an ordinate linear in decibels. Although any graph format is suitable for research purposes, the existing standard for external presentation of such information requires that a length of a logarithmic frequency decade (10:1 ratio) be equal to that of 30 db on the ordinate.[34] This is available in standard graph papers.

To obtain the decibel scale, linear data from the detector may be processed by a logarithmic amplifier. Usually the output is proportional to the logarithm of the peak input amplitude, so the input waveform must be monitored if reliable results are to be obtained for nonsinusoidal inputs.

[34] Radio-Electronics-Television Manufacturers Association (Now Electronic Industries Association), "Speakers for Sound Equipment." Standard SE-103, April, 1949.

The recording device may itself provide the decibel scale by a servomechanism and an input voltage-dividing resistor (potentiometer) for which the output voltage is a logarithmic function of slider position.* [35] This output voltage is amplified, rectified, and subtracted from a constant reference voltage. The servo acts to reduce to zero the difference voltage by moving the arm of the logarithmic input potentiometer. In commercially available high-speed level recorders, input changes in excess of 1000 db/sec have been recorded. In some instruments interchangeable potentiometers permit rapid selection of range. Linear potentiometers are also available for recording such quantities as impedance versus frequency and voltage versus time.

5.2.4. Waveform and Frequency†

In much acoustical work the interpretation of results often depends critically on the assumption that a given waveform, usually sinusoidal, is present. Since this assumed condition does not always obtain, waveform visualization by a cathode ray oscilloscope should always be employed. A useful adjunct for the audio range is a sweep circuit which automatically provides two cycles of the input signal on the screen, regardless of the frequency.[36]

Waveform analysis may be effected by a variety of filters and analyzers. Wide-band filters with bandwidths of $\frac{1}{3}$, $\frac{1}{2}$, or 1 octave are available. In using any analyzer, consideration must be given to whether the bandwidth is constant or is proportional to center frequency. By convention, sound spectra are referred to a bandwidth of 1 cycle/sec. In vibration technology, the term spectral power density connotes the same reference. In proportional bandwidth analyzers, corrections supplied by the manufacturer permit reduction to a per-cycle/sec basis. The calculation is based on the assumption that white noise has an energy content proportional to the bandwidth. It may be noted, however, that most filters are calibrated so as to read the proper level for a constant amplitude single frequency tone, regardless of frequency in the pass band, or the bandwidth of the filter.

Most filters are passive, but some active types are available with continuously adjustable bandwidth. These provide ready means of exploring a spectrum. Another convenient spectrum-search device is the filter in

* See also Vol. 2, Chapter 8.3.

† See also Vol. 2, Chapter 9.2.

[35] W. R. Clark, "The Speedomax power level recorder." *Trans. Am. Inst. Elec. Engrs.* **59**, 957 (1940).

[36] A. Block, "An external automatic sweep generator for use with cathode-ray oscilloscopes." *J. Audio Eng. Soc.* **2**, 259 (1954).

which the search signal continually sweeps the region of interest, and the spectrum is portrayed on an oscilloscope. Where much analysis must be done, this provides a rapid means of selecting regions for more precise measurement. Automatic recording wave analyzers are also available to avoid the work of point-by-point plotting.

Frequency may be determined to 2 or 3% by obtaining the beat note with an adjustable oscillator. For more precise work an electronic counter may be employed. Some workers have used the power line frequency as a standard, with Lissajous figures and an oscilloscope employed to widen the range. If the line frequency is kept continually adjusted, it will be known to better than 0.1%. However, in some power systems the frequency is readjusted only near midnight to correct to the proper total number of cycles per day. Knowledge of current local power company practice is recommended before using the power line frequency as a precise standard.

5.2.5. Recording and Reproduction

In noise and vibration studies it is particularly important for the experimenter to retain acoustic samples for analysis, comparison, and reference. This has been made possible in a simple manner by the tape recorder. For noise studies in the audio range, machines operating at $7\frac{1}{2}$ ips can be quite satisfactory. Low distortion and good signal-to-noise ratio are more important than low wow and flutter, or extreme frequency range.

5.3. Experimental Techniques

Since the same technique may be employed in several fields, it is convenient to classify experimental methods according to function. We are interested in measuring properties of the medium, the sound field, and acoustic devices, as regards sound in both solids and fluids. Although steady signals will receive the greatest emphasis, measurement of transient and impulsive sound will be treated where particularly applicable.

5.3.1. Measurement of Properties of Medium

The linear properties of a transmitting medium may be described by the velocity c and attenuation coefficient α appropriate to the wave type. Sometimes both are combined in the propagation constant

$$\gamma = \alpha - j\omega/c = \alpha - jk.$$

Velocity may be measured by time-of-flight or by phase methods. In time-of-flight a wave packet with an identifiable beginning or end is emitted, and the time of travel to a distant point or the total travel time for a reflection is noted. Electrical or optical means are used as the "instantaneous" signal, since in air, the ratio of optical and sonic velocities is about 870,000. Care must be taken if the electrical system includes long lines or other equipment with extreme delay or low phase velocity.

5.3.1.1. Fluids. For the precision measurement of velocity and attenuation in fluids, the acoustic interferometer[37] is employed. This instrument, usually constructed by the experimenter, typically employs a piezoelectric disk as a source of plane longitudinal waves. Separated from the emitting surface is a reflector moved by a micrometer screw. For proper operation the source and reflector surfaces must be optically plane and parallel, and a precision screw must be used. More highly refined constructions are available for use in gases[38] and in liquids.[39] The transducer is driven by a source of approximately constant voltage, and the transducer current delivered is plotted as a function of reflector position. Figure 8 shows the general features of the resulting plot. The sharp maxima are a half-wavelength apart. With the wavelength thus determined and the frequency known, the velocity follows from $c = \lambda f$.

To obtain the attenuation constant, first extrapolate the current maxima to i_m and the minima to i_n, the values they would have at the source (see Fig. 8). When the reflector is a distance x from the source, the resonant and antiresonant currents (determined from the envelope) will be i_r and i_a. The attenuation constant α is then given[40] by

$$\tanh \alpha x = \left[\frac{\left(\frac{i_m}{i_r} - 1\right)\left(\frac{i_m}{i_n} - \frac{i_m}{i_a}\right)}{\left(\frac{i_m}{i_n} - 1\right)\left(\frac{i_m}{i_a} - 1\right)} \right]^{\frac{1}{2}}. \tag{5.3.1}$$

The velocity of sound in solids may be measured by exciting resonances in a suitably cut specimen, which thus becomes a fixed-path, variable frequency interferometer. If resonances in the exciting transducer are sufficiently removed from those of the specimen, extremes in the driving point mechanical impedance of the specimen will recur for specimen

[37] L. Bergmann, "Der Ultraschall," 6th ed., p. 238, 346. Hirzel, Stuttgart, 1954.

[38] A. J. Zmuda, "Dispersion of velocity and anomalous absorption of ultrasonics in nitrogen." *J. Acoust. Soc. Am.* **23**, 472 (1951).

[39] D. R. McMillan and R. I. Lagemann, "A precision ultrasonic interferometer for liquids, and some velocities in heavy water." *J. Acoust. Soc. Am.* **19**, 956 (1947).

[40] See reference 17, pp. 338–343.

thicknesses of integral half-wavelengths. This assumes that the end of the specimen remote from the source is not loaded. If f_0 is the lowest half-wave resonance frequency, then $c = 2lf_0$, where l is the specimen thickness. A more precise method, that of phase comparison,[41] is available for small specimens. However, the equipment required is more complex.

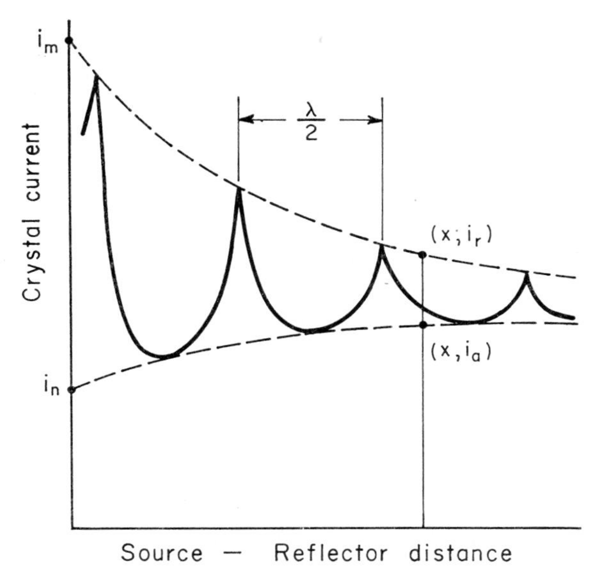

FIG. 8. Interferometer current versus source–reflector separation.

5.3.1.2. Solids. Measurement of absorption in solids is complicated by the presence of a loss factor for each of the two elastic constants. Perhaps the simplest of the measurement schemes for low frequencies and small dissipations uses a vibrating reed, and provides the real and imaginary (loss) parts of Young's modulus in the form $E = E_1 + jE_2$. The reed is clamped as a cantilever and is driven at resonance. The ratio of amplitude of the free end to that of the clamped driving point is measured by a ferrous pin cemented on the end. The pin reluctance-modulates the magnetic flux from a polarizing direct current in a coil to provide an amplitude sensitive output. A similar scheme is used to indicate the driving point motion.[42]

[41] H. J. McSkimin, "Use of high frequency ultrasound for determining the elastic moduli of small specimens." IRE Trans. Profess. Group on Ultrasonic Engineering PGUE-5, p. 25, August, 1957.

[42] S. Strella, "Vibrating reed test for plastics." *ASTM Bull.* **105,** 47 (1956).

Then

$$\left.\begin{aligned} E_1 &= \frac{3072\rho\omega^2[l^4 + (4m_p/\rho S)l^3]}{d^2[16a_0^2 - F^2]^2} \\ E_2 &= \frac{3072\rho\omega^2 F[l^4 + (4m_p/\rho S)l^3]}{a_0^3 d^2[16a_0^2 - F^2]} \\ F &= \frac{-5.478 + 2[7.502 + 6.15M^2]^{\frac{1}{2}}}{1.689M^2} \end{aligned}\right\} \tag{5.3.2}$$

where M is the ratio of free end to driver amplitude at resonance, ω is the resonant angular frequency, ρ is the reed density, l is the reed length, d is the reed thickness, S is the reed cross section area, m_p is the pin mass, and $a_0 = 1.875$, the first approximation to the root of the secular equation for the first resonant frequency. If the dissipation is small ($E_2/E_1 < 0.1$), this value is satisfactory. For large dissipation, the value can be calculated by successive approximation.[43]

The real and imaginary parts of the bulk modulus may be obtained at frequencies for which the sample is small compared to a wavelength. The sample is enclosed in a cavity with two expander-type piezoelectric or ferroelectric elements and a coupling liquid. From the transducer equations (5.2.1) the complex bulk modulus may be expressed in terms of the coupling liquid.[44]

Many other experimental techniques have been employed; reference 27 contains a useful bibliography (up to 1954) as well as a succinct statement of the main feature of each method.

5.3.2. Measurement of Sound Field

As in optics, the field in the Fresnel (near) zone is quite complicated. To make it simpler to calculate such a quantity as total energy from simple pressure measurements, every effort should be made to work in the inverse square (also called the Fraunhofer or far field) zone. Under reasonably reflection-free conditions, the boundary of this zone is determined by plotting pressure level as a function of distance from the source for a single-frequency signal. Between the limiting distance discussed in Section 5.1.5 and that at which the ratio of reflected to direct sound becomes important the level change should be 6 db/distance doubled (in three dimensions). Outside these limits the level will fluctuate to an increasing extent. The boundaries of the useful inverse square field are often taken where the level deviation exceeds ± 1 db each side of the average slope line. For many measurements the inner boundary may be

[43] D. R. Bland and E. H. Lee, "Calculation of the complex modulus of linear viscoelastic materials from vibrating reed measurements." *J. Appl. Phys.* **26,** 1497 (1955).

[44] J. S. McKinney, S. Edelman, and R. S. Marvin, "Apparatus for the direct determination of the dynamic bulk modulus." *J. Appl. Phys.* **27,** 425 (1956).

assumed to extend to a distance equal to at least three times the largest transverse dimension of the radiator.[34]

5.3.2.1. Sound Pressure. Sound pressure measurements are the easiest to make in fluids. If the sound field is not to be disturbed, the maximum dimension of the detector should be less than a quarter wavelength. For discrimination against unwanted sound, use a cardioid-pattern microphone or directional hydrophone array. If the utilization point is distant, reasonably low detector electrical impedance will be important so that long lines may be run. For lowest self-noise, microphones should ordinarily be operated into substantially an open circuit. Distortion should be low for the amplitudes expected. Temperature and moisture stability is often important, as is sensitivity to magnetic and electric fields, wind noise and support vibration disturbances. Finally, the response (open circuit voltage versus frequency for constant sound pressure or vibratory input) should be smooth and adequate in range for the measurement task.

Most acoustic detectors have low outputs and require careful shielding. Special attention must be given to grounding[45,46] so as to avoid having two separated conductors grounding the same device. This condition is known as a ground loop, and permits magnetic induction of undesired signals. In general, no ground wire should be used as a common for carrying a signal.

Pickup of sound in the presence of ambient noise can be improved by use of gradient detectors.[47] In these the output is approximately proportional to the first or higher order space derivative of the quantity measured. If placed close to the intended source, the gradient will be large, whereas ambient disturbances arising in all directions from more distant sources will have a lower gradient. If the microphone output is proportional to the nth space derivative of the sound pressure, and the intensity of desired and undesired sound is the same, the ratio of desired to undesired signal powers is $(2n + 1)$.[47]

5.3.2.2. Vibration. In the vibration of solids it would be useful to measure both the normal and transverse motion of a surface. For this purpose a special biaxial pickup has been devised.[48] It employs a wire bent at 90°, with its free ends bearing small barium titanate transducers held in contact by springs; see Fig. 9. Each portion of the drive wire lies in a plane normal to the vibrating surface, and each is 45° from the sur-

[45] W. E. Stewart, "Broadcast audio wiring practice." *Audio* **36** (8), 16 (1952).

[46] E. F. Coriell, "Planning and building a radio studio; Part 5." *Audio* **36** (10), 30 (1952).

[47] H. F. Olson, "Gradient Microphones." *J. Acoust. Soc. Am.* **17**, 192 (1946).

[48] J. S. Arnold, "Development of nondestructive tests for structural adhesive bonds." Wright Air Development Center Technical Report 54-231, Part 6 (1957).

face. When the lowest resonant frequency of the transducer or the transducer mass and drive wire stiffness system is above the region of interest, the system is acceleration responsive.

In use the bend in the wire is firmly pressed on a vibrating surface. For sinusoidal excitation the path of the contact point is an ellipse, components of which appear as electrical signals from the transducers. These signals feed, through suitable amplifiers, the deflection plates of a cathode-ray oscilloscope. If the tube is rotated 45°, it becomes a resolver such that the pattern on the screen is an enlarged version of the actual path

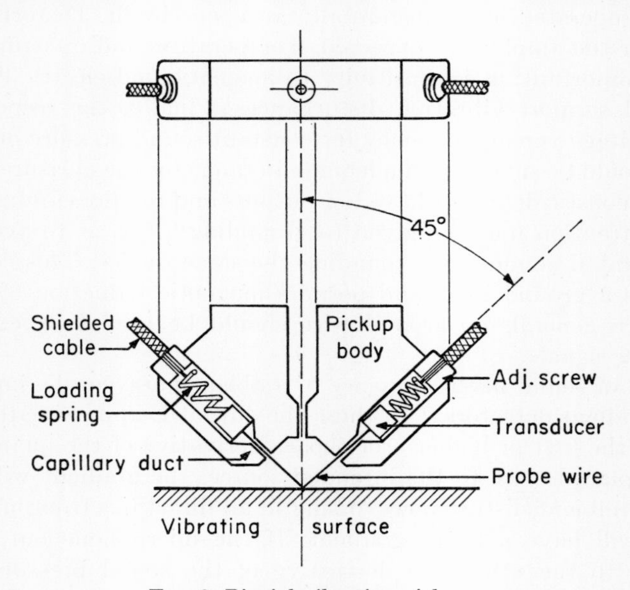

FIG. 9. Biaxial vibration pickup.

of the contact point, with vertical on the screen selected to be normal to the surface. If the motion is random the pattern will be a smear, but one direction of motion will usually be more prominent than another. A map of these directions of maximum motion over a vibrating surface will often help in locating the effective source of the vibration. If signals corresponding in phase and amplitude to the normal and transverse components of motion are desired, these may be obtained by summing and differencing the two electrical outputs.

There are many more vibration measurement means available to the experimenter.[49] Care must be taken that the presence of the instrument

[49] P. M. Pflier, "Elektrische Messung mechanischer Grössen." Springer, Berlin, 1943.

does not change the vibration being measured, and that the signal is due to the vibration, and not to one electrically introduced.

5.3.2.3. Loudness. Often the loudness of a complex tone is required more precisely than can be obtained with a sound level meter. When the sound spectrum is substantially continuous and smooth, a procedure devised by Stevens[50] has proved useful. A sound level meter[33] standardized by a calibrator[51] or similar means, feeds an octave band analyzer.[52] The sound signal, picked up in a standard manner,[53] is fed in with the sound level meter set on *C*-scale. With the analyzer, octave band sound pressure levels of the noise may then be tabulated for each octave band. Using Fig. 18 in reference 50, convert the sound pressure levels for each band to the loudness numbers called sones. In one band there will be a maximum number of sones, S_m. If ΣS is the arithmetic sum of the number of sones in all bands, then the number of sones in the total noise is given by Stevens as

$$S_t = S_m + 0.3(\Sigma S - S_m). \tag{5.3.3}$$

This loudness calculation will be in error if the spectrum contains strong single-frequency components. In this case the calculated loudness will usually be too low, and more complicated methods must be used.

5.3.2.4. Intensity. Measurement of sound field intensity is not possible directly. In the inverse square field in fluids, pressure measurement and the relation $I = p^2/\rho_0 c$ are commonly used. For high intensity ultrasonic sound in liquids, the worker in industrial acoustics may employ a radiation pressure balance.[54] If a traveling plane wave is incident normally on a plane surface several wavelengths in diameter, the time average radiation pressure p_r is given by $p_r = I(1 + r)/c$ where r is the power reflection coefficient of the reflector. In water, a force of m gm measured on a reflector area S cm^2 corresponds to an intensity

$$I = 14.65m/(1 + r)S \text{ watt/cm}^2.$$

An alternative scheme for measurement of ultrasonic intensity in liquids is to measure the temperature rise in an absorber which has a

[50] S. S. Stevens, "Calculation of the loudness of a complex noise." *J. Acoust. Soc. Am.* **28,** 807 (1956).

[51] A. P. G. Peterson and L. L. Beranek, "Handbook of Noise Measurement," p. 22. General Radio Co., Cambridge, Massachusetts, 1953.

[52] American Standards Association, "Specification for octave band filter set for the analysis of noise and other sounds." Standard Z24.10-1953.

[53] American Standards Association, "Test code for apparatus noise measurement." Standard Z24.7-1950.

[54] G. E. Henry, "Ultrasonic output power measurements in liquids." IRE Trans. Profess. Group on Ultrasonic Engineering PGUE-6, p. 17, December, 1957.

density and sound velocity close to that of the liquid medium.[55] The length of the absorber along the wave direction is to be such that the incident sonic energy is almost completely absorbed. For measurements in water, the absorber may be a thin capsule of castor oil encased in 0.075 mm polyethylene. A thermocouple of wire 0.125 mm in diameter senses the temperature rise from absorption of a one-second square wave gated pulse of energy from the source. By calibration with known sound pressures in a free field, the device becomes a quite useful intensity probe.

5.3.2.5. Taped Samples for Analysis. In most noise measurements it is common to find the character of the noise fluctuating during the waveform analysis. This is especially true if the source is factory noise, or if a long transmission path is involved. Under such conditions it is convenient to gather acoustic samples with a tape recorder, and to splice tape loops for repetitive playback and analysis of representative samples. For studies of audible noise a most useful instrument combination is a specially modified tape machine[56] or a sound level meter feeding a portable tape recorder. The sound level meter (set for *C*-scale) is connected to the radio or line input of the tape machine, and a test tone at about 400 cycles/sec is introduced acoustically into the microphone of the sound level meter by a calibrator.[51] Then with the sound level meter reading zero db on its meter scale, the recording gain control on the tape machine is set so that a 10–15 db increase in the recording level may be tolerated before harmonic distortion in the reproduced tape would reach 3–5%. This margin is needed to handle peaks in the spectrum, as well as to account for noise with marked high frequency components. If the noise is very atypical, as explosions or impact sounds, a detailed knowledge of the recording and playback equalization will be required to set the machine properly for that spectrum.

With this procedure, any on-scale reading of the sound level meter will not overload the tape machine. Moreover, the *C*-scale may be used to read sound pressure level; the reading observed visually may be transferred to the tape by speaking it into the microphone. In fact, in the hands of an experienced operator the tape also becomes a notebook for recording data on the noise source, microphone position, condition of transmitting medium, and special conditions of the test. On a stereo machine the noise sample and documentary data may be put on separate tracks.

In playback a tape loop is spliced of such length that the sample is

[55] W. J. Fry and R. B. Fry, "Determination of absolute sound levels and acoustic absorption coefficients by thermocouple probes." *J. Acoust. Soc. Am.* **26,** 294, 311 (1954).

[56] G. W. Kamperman, "Portable magnetic tape recorder for acoustical measurements." *Noise Control* **4** (1), 23 (1958).

adequate and representative. It is best to make the loop from a tape copy, so that the raw data may be preserved untouched. It will be necessary to supply holdback tension on the loop during its playback. This is easily supplied by a pressure pad at a tape idler or guide.

An advantage of a taped sample is that correlation techniques may be employed in analyzing noise to locate point of origin, and for ignoring unwanted signals.[57] In this technique two signals, recorded simultaneously on two-channel tape, are played back with an adjustable delay which is usually supplied by a fast-moving tape loop or drum with adjustable spacing between record and play heads. One of the original signals is then multiplied by the delayed second signal, integrated, and plotted as a function of added delay.[58] The correlation technique is a powerful one which is rapidly growing in importance in noise and vibration studies.

5.3.2.6. Subjective Evaluation. In noise studies for which a subjective evaluation is necessary, the use of a sound "jury" to evaluate noise on the site can become expensive. It has been established that a useful substitute is binaural recording, usually by tape. When such samples are played back binaurally, the auditor is effectively transported to the site. The tape provides means for copying, editing, and intermixing under repetitive and controlled conditions. This makes it possible to obtain data of greater statistical significance from taped than from direct samples.

In binaural recording the microphones should ideally be sunk in a dummy head at the end of simulated ear canals. An acceptable substitute is a pair of pressure dynamic "stick" microphones mounted on a common stand. Their hinge joints should permit bending in a common plane so that the microphone separation may be adjusted easily. With no barrier between the microphones, a spacing of 20–30 cm is usually satisfactory. On playback, high quality calibrated earphones should be employed; these are usually available from firms selling audiometers. With a laboratory standard microphone[59] and a coupler,[60] earphones may be calibrated directly.

5.3.2.7. Reverberation Time. In room acoustics an important attribute of the room is its reverberation time, defined as the time for the sound energy density to decay by 60 db. Usually the average slope of the important first 20 db of the decay is extrapolated to obtain the value ordinarily

[57] K. W. Goff, "Application of correlation techniques to some acoustic measurements." *J. Acoust. Soc. Am.* **27,** 236 (1955).

[58] K. W. Goff, "Analog electronic correlator for acoustic measurements." *J. Acoust. Soc. Am.* **27,** 223 (1955).

[59] American Standards Association, "Specification for laboratory standard microphones." Standard Z24.8-1949.

[60] American Standards Association, "Coupler calibration of earphones." Standard Z24.9-1949.

given. Because a single value of slope often cannot characterize the decay, the preferred procedure is to employ a high-speed level recorder to plot sound pressure level against time. Here again, a tape machine and sound level meter may be used to collect acoustic samples for later analysis in the laboratory, using an octave band analyzer or similar wide band filter. Warble tones or wide band noise should be used as the noise source to obtain appreciable bandwidth, covering many room resonances. A convenient source is a gunshot;[31] in a church, a dissonant chord on an organ flute stop may also be used. The source should be at the position where the programs usually originate; several microphone positions should be investigated.

5.2.3.8. Isolation. In building acoustics it is important to keep out extraneous sound and to keep the desired sound from leaking to neighboring rooms. A measure of the isolation (which depends mainly on the mass per unit area of the wall) is transmission loss. Suppose that in a plane wave the intensity I_0 exists. If an isolating panel is introduced so that sound cannot leak around it, the transmitted intensity will be less, say I_1. Then the transmission loss is $10 \log_{10}(I_0/I_1)$. There are various measurement schemes using normal[61] or random incident[62] waves. The mounting of a panel influences the transmission, and care must be taken to simulate conditions of use, and to avoid flanking transmission.

Thin panels will become transonant at frequencies for which the flexural velocity in the panel equals the wavefront contact velocity of an obliquely incident sound wave. The importance of this coincidence effect is becoming appreciated, and the experimenter in building acoustics will find such measurements among the most useful, but difficult.[63]

5.3.3. Measurements on Acoustic Devices

5.3.3.1. Response. The experimenter must often know the capabilities of his instruments. The response-frequency characteristic is one of the most important in determining range of applicability. A useful measure of response is the generalized gain, defined as output level minus the input level, in appropriate absolute decibel units. For example, the gain of a microphone is the output power level in dbm, minus the input sound pressure level, in dbt.[64]

The use of the gain concept is advantageous when several different

[61] See reference 4, pp. 324–331.

[62] L. L. Beranek, "Acoustic Measurements," p. 880. Wiley, New York, 1949.

[63] W. Westphal, "Zur Schallabstrahlung einer zu Biegeschwingungen angeregten Wand." *Acustica* **4** (Beih. 4), 603 (1954).

[64] Radio-Electronics-Television Manufacturers Association, "Microphones for sound equipment." Standard SE-105, August, 1949.

transducers are connected in cascade. The over-all system gain is then the sum of the separate gains of each transducer. Relative gain, taken with respect to a reference value at a frequency in the middle of the useful range, is also used as a synonym for response. Loudspeaker response is often so stated.

The experimenter in communications acoustics is often called upon to give a measure of electrical power when the utilization device does not present an ideal termination impedance to the source. The measure of power which has been demonstrated to give the most consistent results and fairest comparisons is the power available to the load.[65] Suppose the ideal load for the source would be a resistor of value R_L. To find the power available to the actual load, replace it by this resistor, and measure E_L, the voltage across it. In this process no change is to be made in the source. The power available is then E_L^2/R_L. For microphones the available power output is taken as $E_0^2/4R_m$, where E_0 is the open circuit voltage and R_m is the nominal or rating impedance assigned to the microphone by the manufacturer. If the microphone is intended to operate directly into the grid of a vacuum tube, R_m is taken as 100,000 ohm.[64]

5.3.3.2. Calibration. Absolute calibration of electroacoustic transducers may be accomplished by reciprocity techniques.[66] These are based on the presence of at least one reversible transducer. When such an instrument is employed in a free field (no reflections), the ratio of response as a microphone (or vibration pickup) to the response as a loudspeaker (or vibration generator) is independent of transducer constants. This calibration technique may be employed in a free field or with a lumped coupling element.[67] By using suitable reflector and pulse techniques, a single reversible transducer may be employed as its own pickup device.[68]

5.3.3.3. Transient Measurements. When a dynamic system has the Bode minimum-phase property, and the magnitude of its response-frequency function is known for all frequencies, the phase-frequency function is uniquely determined. Unfortunately, most radiating transducers do not have this property, and the phase (and hence transient) response cannot be inferred from steady state response magnitude measurements. A useful technique for locating regions of poor transient response in a source of vibration or sound is the delayed response method.[69]

[65] F. F. Romanow and M. S. Hawley, "Proposed method of rating microphones and loudspeakers for systems use." *Proc. I. R. E.* (*Inst. Radio Engrs.*) **35,** 953 (1947).

[66] See reference 62, p. 116 ff; reference 4, p. 377 ff.

[67] M. Harrison, A. O. Sykes, and P. G. Marcotte, "Reciprocity calibration of piezoelectric accelerometers." *J. Acoust. Soc. Am.* **24,** 384 (1952).

[68] R. M. White, "Self-reciprocity transducer calibration in a solid medium." *J. Acoust. Soc. Am.* **29,** 834 (1957).

[69] D. E. L. Shorter, "Loudspeaker transient response," *BBC Quart.* **1,** 121 (1946).

A gated signal, preferably with an integral number of whole cycles, is supplied to the transducer to be tested. The resulting signal is received by a high quality pickup device, and is examined for steady state amplitude and for amplitude at adjustable delayed times after the signal should have ceased. Response-frequency curves are plotted with the delay as the parameter. Poor transient response is usually associated with those frequency regions for which the delayed response curves show strong and persistent peaks as delay is increased.

5.3.3.4. Acoustical Impedance. The impedance of an acoustic element (horn, absorbent, isolator) which can be coupled to a tube is most conveniently measured by the tube or transmission line method.[70] A rigid-walled tube of constant cross section is closed at one end by the acoustic

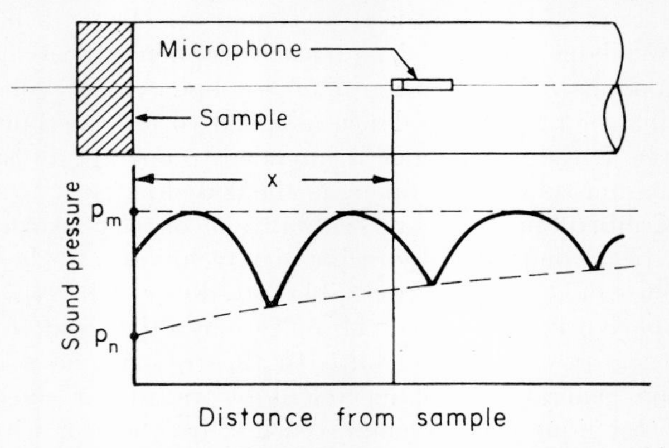

FIG. 10. Tube method for acoustic impedance.

element whose impedance is desired. At the other end is a source of sound, and a small probe microphone scans the sound field along the tube. The resulting sound pressure is shown in Fig. 10, expressed as a function of distance from the unknown impedance. The primary data are the distance x_1 to the first minimum (expressed in wavelengths) and the ratio of maximum pressure to the minimum, extrapolated to the value it would have at $x = 0$. The wavelength is twice the distance between minima. By use of charts,[70] the impedance equations may be solved graphically. Precautions include control of temperature, use of rigid-walled tubing of constant cross section (usually circular), avoiding frequency ranges with transverse modes of transmission, and keeping normal to the axis the plane of the device of unknown impedance.

5.3.3.5. Absorption. Absorption coefficients are important parameters in room acoustics. Acoustic impedance obtained by the tube method

[70] Reference 62, p. 302 ff.

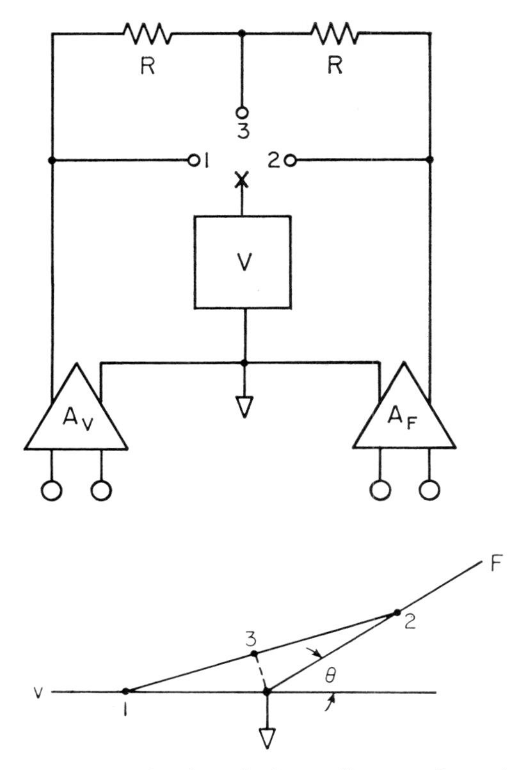

FIG. 11. Phase comparison circuit and phasor diagram. Input impedance of voltmeter V is high.

(normal incidence) gives data for a useful first approximation to the absorption coefficient[71] at normal incidence. If the normalized impedance is $r + jx$, then the normal absorption coefficient is

$$\alpha = 4r/[(r + 1)^2 + x^2]. \tag{5.3.4}$$

However, there is no substitute for the conditions of random incidence obtaining in a reverberation chamber.[72] The absorption coefficient may also be approximated from the values of the pressure maximum p_m and minimum p_n extrapolated to the sample; see Fig. 10. Then

$$\alpha = 4(p_m/p_n)/((p_m/p_n) + 1)^2. \tag{5.3.5}$$

[71] H. J. Sabine, "Sound absorption and impedance of acoustical materials." *J. Soc. Motion Picture Engrs.* **49,** 262 (1947).

[72] See reference 62, p. 860 ff.

For evaluating relative absorption of a series of samples, the box method may be employed.[73] In this a short rigid tube of adjustable length is tuned to resonance at a test frequency, and samples are inserted. The drop in level is a measure of absorption. The method is not absolute, and requires calibration with samples of known coefficient.

5.3.3.6. Mechanical Impedance. The worker in the field of vibration will find that the impedance concept furnishes a unifying point of view and is supplanting other concepts such as complex modulus of elasticity. Measurement techniques are usually based on simultaneous comparison in amplitude and phase of electric signals generated by separate force-and-motion-sensitive transducers.[74] Force may be supplied and measured by a drive rod interrupted by a pressure sensitive transducer, such as a barium titanate disk. Alternatively, a moving coil system may supply the force through a drive system of mechanical impedance much less than that measured. Thus the current in the coil is a measure of the force. The velocity signal is often obtained by electrical integration of the signal from an accelerometer. The signals may be compared with a phase voltmeter or similar device. For rapid phase comparison the circuit shown in Fig. 11 may be employed. The velocity (v) and force (F) amplifiers are polarized to deliver out-of-phase signals when the test impedance is a pure mechanical resistance. Equal resistors R are of sufficiently high resistance so as not to load the amplifiers, the gains of which are adjusted for equal output voltages, $E_1 = E_2$. Then θ, the phase angle of the impedance, is given by $\theta = 2 \sin^{-1} E_3/E_1$. The voltmeter used should not load the circuit.

[73] C. B. Boenning, S. F. Huber, and T. Mariner, "Modification of the box method of determining relative sound absorption coefficients." *J. Acoust. Soc. Am.* **23,** 114 (1951).

[74] K. Schuster, "Die Messung mechanischer und akustischer Widerstände." *Ergeb. exakt. Naturw.* **21,** 323 (1945).

6. HEAT AND THERMODYNAMICS

6.1. Thermometry*[1–15]

6.1.1. Principles of Temperature Measurement[16–21]

Most of the variables needed to specify the macroscopically observable state of a simple system are quantities whose definition and measurement are parts of classical mechanics and electromagnetism, e.g., pressure, volume, mass, electric and magnetic fields, and polarizations, etc. To complete the specification of such a state it is found necessary to introduce an additional variable of quite another type, namely, the temperature. As with some other fundamental observables the qualitative aspects of temperature are directly accessible to certain sensory receptors. However, in order to define temperature quantitatively and to measure it one must carry out, at least in principle, certain experiments which do not rely on the fallible sensations of "hotness" or "coldness."

[1] M. W. Zemansky, "Heat and Thermodynamics," 4th ed. McGraw-Hill, New York, 1957.

[2] J. K. Roberts and E. R. Miller, "Heat and Thermodynamics," 4th ed. Blackie, London, 1951.

[3] M. N. Saha and B. N. Srivastava, "Treatise on Heat," 3rd ed. Indian Press, Allahabad, India, 1952.

[4] A. G. Worthing and D. Halliday, "Heat." Wiley, New York, 1948.

[5] J. M. Cork, "Heat," 2nd ed. Wiley, New York, 1942.

[6] R. Glazebrook, "Dictionary of Applied Physics," Vol. 1. Macmillan, London, 1922.

[7] F. Kohlrausch, "Praktische Physik," 20th ed., Vol. 1. Teubner, Stuttgart, 1955.

[8] F. X. Eder, "Moderne Messmethoden der Physik," Part 2. Deutscher Verlag der Wissenschaften, Berlin, 1956.

[9] American Institute of Physics, "Temperature," Vol. 1. Reinhold, New York, 1941.

[10] American Institute of Physics, "Temperature," Vol. 2. Reinhold, New York, 1955.

[11] F. Henning, "Temperaturmessung." Barth, Leipzig, 1951.

[12] E. Griffiths, "Methods of Measuring Temperature," 3rd ed. Griffin, London, 1947.

[13] R. L. Weber, "Heat and Temperature Measurement." Prentice-Hall, New York, 1950.

[14] J. A. Hall, "Fundamentals of Thermometry." Institute of Physics, London, 1953.

[15] J. A. Hall, "Practical Thermometry." Institute of Physics, London, 1953.

[16] M. Born, "Natural Philosophy of Cause and Chance," p. 35. Oxford Univ. Press, London and New York, 1949.

[17] C. Caratheodory, *Math. Ann.* **67**, 355 (1909).

[18] See reference 1, Chapter 1.

[19] See reference 2, Chapter 1.

[20] See reference 9, Chapter 1.

[21] See reference 10, Chapters 1–4.

* Chapters 6.1 through 6.3 are by **Simeon A. Friedberg.**

Consider two systems, 1 and 2, characterized by the variables (X_1, Y_1) and (X_2, Y_2), respectively. Let them be brought into contact with one another separated by a so-called "diathermal wall" to prevent only the interchange of matter between the systems. Such contact is generally found to be accompanied by changes in X_1, Y_1, X_2, Y_2 until constant values of these variables are attained which satisfy a definite functional relation

$$F(X_1, Y_1, X_2, Y_2) = 0.$$

Thermal equilibrium is said to have been achieved, the diathermal wall having permitted thermal contact between the systems. Further experiments of this type lead us to the important conclusion that two systems in thermal equilibrium with a third are in thermal equilibrium with one another. This fact may be expressed by saying that validity of any two of the three functional relations

$$F(X_1, Y_1, X_2, Y_2) = 0, \qquad F(X_2, Y_2, X_3, Y_3) = 0, \qquad F(X_3, Y_3, X_1, Y_1) = 0$$

implies the validity of the third. This is only possible if each of these relations can be expressed as an equality of two functions each depending only upon the variables characterizing one system.[17] Thus thermal equilibrium between systems 1 and 2 is described also by

$$f_1(X_1, Y_1) = f_2(X_2, Y_2).$$

In this way the two systems are seen to have a property which, upon the establishment of thermal equilibrium, takes on the same numerical value for each. This property is the temperature.

In order to measure temperature we can designate, for example, system 1 as our thermometer. Values of the quantity $f_1(X_1, Y_1)$ might then be called empirical temperatures. Clearly any function of f_1 could serve equally well as an empirical temperature. The most convenient choice in practice turns out, fortunately, to be also the simplest. All independent variables are held constant during a measurement except one, say X, which is called the thermometric parameter. The empirical temperature $\theta(X)$ is then taken to be a particular linear function of X, $\theta(X) = aX$; (const Y) where a is a constant. The arbitrary nature of the temperature scale established by this choice should be borne in mind. An absolute scale does exist and we shall see later its relation to the empirical scales.

Two methods are available for determining the constant a and so of calibrating the thermometer. Prior to 1954 it was common practice to allow the thermometer to come to thermal equilibrium with each of two arbitrary standard systems in readily reproduced states. Corresponding to these two "fixed points" on the temperature scale two values of X

were found such that $\theta(X_1) = aX_1$ and $\theta(X_2) = aX_2$. An arbitrary number of degrees was assigned to the interval $\theta(X_1) - \theta(X_2)$ so that the constant slope could be written

$$a = \frac{\theta(X_2) - \theta(X_1)}{X_2 - X_1}$$

without specifying a numerical value for $\theta(X_1)$ or $\theta(X_2)$. In this method the fixed points were commonly taken to be (1) the temperature at which pure ice and air-saturated water coexist under a pressure of 1 atmos, and (2) the equilibrium temperature of pure water and pure steam under a pressure of 1 atmos, known respectively as the ice point and the steam point. The interval $\theta(X_2) - \theta(X_1)$ was given the value of 100°.

A serious weakness, inherent in the calibration procedure just outlined, has caused it to be abandoned. Small errors in the establishment of the ice or steam point can lead to large uncertainties in temperatures measured with the thermometer far below the ice point, particularly in and below the liquid helium region. In 1954 international agreement was finally achieved on a method of fixing the temperature scale which is free of this objectionable feature. The idea is not new, having been advocated originally by Lord Kelvin in 1854 and more recently by Giauque[22] in 1939. The thermometer is now calibrated by allowing it to come to equilibrium with only one standard system so that the scale has only one fixed point. The standard system is pure water at the particular temperature and pressure at which solid, liquid, and vapor phases can coexist in equilibrium. This unique state is called the "triple point of water" and can be readily reproduced in a simple cell developed at the U.S. National Bureau of Standards.[23] The temperature at the triple point is arbitrarily designated as 273.16 degrees Kelvin (°K). Thus, if X_3 is the value of the thermometric parameter with the thermometer at this temperature, we have $\theta(X_3) = aX_3 = 273.16°\text{K}$ and $a = 273.16°\text{K}/X_3$. The accepted empirical temperature is, thus, $\theta(X) = 273.16°(X/X_3)$ (const Y). The number chosen for $\theta(X_3)$ is, of course, one often obtained in measurements of the triple point with thermometers calibrated in the traditional way. Its adoption means that the readings of thermometers calibrated by the old and new methods will not differ significantly at high temperatures.

Of the many conceivable thermometers, only a few types are commonly employed. These will be discussed in some detail in Sections 6.1.2–6.1.6. Let us merely list here some of the important thermometric parameters. These are (1) the pressure P of a gas at constant volume, (2) the volume

[22] W. F. Giauque, *Nature* **143,** 623 (1939).

[23] H. F. Stimson, *J. Wash. Acad. Sci.* **35,** 201 (1945).

V of a gas at constant pressure, (3) the resistance R of an electrical resistance at constant pressure and tension, (4) the thermal emf E of a thermocouple at constant pressure and tension, (5) the length L of a liquid column in a glass or quartz capillary. A temperature scale for each of these thermometers is defined by the last equation of the preceding paragraph, with the appropriate parameter substituted for X. Should thermometers of the five types be brought simultaneously into equilibrium with a given system it is found that the readings will generally be different. Furthermore, thermometers of a given type but employing different materials are found to yield different results in such a comparison. Closest agreement, however, is found among gas thermometers. For this reason a gas thermometer has been chosen as the empirical standard.

6.1.1.1. Ideal Gas Temperature Scale. It is known that in the limit of infinite dilution or zero pressure, all gases exhibit the same, so-called ideal gas, behavior. This is demonstrated, for example, in a constant volume gas thermometer by making repeated measurements of the temperature of a system in a given state each time reducing the amount of gas in the thermometer and thus the value of P_3 and P. In the limit $P_3 \to 0$ a temperature will be obtained which proves to be the same no matter what gas is used in the thermometer. Similar results are obtained for a constant pressure thermometer if the pressure employed is reduced indefinitely. We may therefore define an ideal gas temperature θ either as

$$\theta = \lim_{P_3 \to 0} \theta(P) = \lim_{P_3 \to 0} 273.16° \frac{P}{P_3}; \qquad \text{(const } V)$$

or

$$\theta = \lim_{P \to 0} \theta(V) = \lim_{P \to 0} 273.16° \frac{V}{V_3}; \qquad \text{(const } P).$$

This scale is the most fundamental empirical one because it may be shown to be identical with the absolute or Kelvin scale over the wide temperature range in which gas thermometers may be employed. This range, it should be noted, does not include 0° which therefore has no meaning on the ideal gas scale although it is a well-defined point on the absolute scale. We designate ideal gas temperatures in degrees Kelvin, keeping in mind the fact that the ideal gas thermometer measures only a portion of the absolute scale.

6.1.1.2. Celsius Temperature Scale. Many practical thermometers are calibrated to indicate temperatures on the Celsius (centigrade) scale. One Celsius degree is equal in magnitude to one degree of the ideal gas scale. The zero of the Celsius scale is shifted, however, to coincide with the temperature of the ice point which on the ideal gas scale is 273.15°K,

0.01° below the triple point. The Celsius temperature t and the ideal gas temperature θ are thus related by the expression $t = \theta - 273.15°$.

6.1.1.3. Absolute Temperature Scale.* In the modern axiomatic formulation of thermodynamics the absolute temperature T (see Chapter 6.4), appears quite naturally as an integrating denominator making the quantity $đQ/T = dS$ a perfect differential. In contrast with the empirical temperature, the absolute temperature is uniquely defined to within an arbitrary multiplicative constant which is determined when the size of a degree is fixed. Furthermore, the absolute scale is defined without reference to any particular thermometer or thermometric substance as is, for example, the ideal gas scale.

This feature of the scale was pointed out by its originator, Lord Kelvin, who noted that any reversible device capable of being carried through a Carnot cycle could serve as an absolute thermometer. Such a cycle operating as an engine includes one isothermal process in which an amount of heat Q_1 is absorbed by the system from a high temperature reservoir at T_1, and another during which an amount Q_2 is given up to a reservoir at a lower temperature T_2. Two adiabatic steps complete the cycle. Since for a reversible cyclic process $\oint dS = 0$, we have in this case $Q_1/T_1 = Q_2/T_2$, or $Q_1/Q_2 = T_1/T_2$. Thus the ratio of two absolute temperatures is equal to the ratio of the heats absorbed or given up by a Carnot device operating between two reservoirs at these temperatures.

In keeping with currently accepted procedure we may calibrate the absolute thermometer by letting one of these reservoirs consist of water at its triple point, a state to which is assigned the temperature $T_3 = 273.16°K$. The absolute temperature of any other reservoir is just $T = 273.16°K(Q/Q_3)$. Should the Carnot device have as its lower temperature reservoir one for which $Q = 0$ that reservoir would then be at a temperature $T = 0°K$, i.e., the absolute zero. On the absolute scale, therefore, zero degrees is a well-defined temperature regardless of whether or not it may be achieved practically.

In order to establish the numerical identity of the ideal gas and absolute scales it is merely necessary to calculate the ratio Q_1/Q_2 for a quantity of ideal gas carried through a Carnot cycle between two reservoirs at the temperatures θ_1 and θ_2. One finds $Q_1/Q_2 = \theta_1/\theta_2$, and, since $\theta_3 = T_3 = 273.16°K$, it follows that $\theta \equiv T$, at least over the range accessible to a gas thermometer. In practice, it is possible to correct the readings of real gas thermometers to the ideal gas scale and thus to obtain absolute temperatures between $\sim$1°K and 1000°K. It is unnecessary in this interval to attempt to approximate reversible Carnot devices in order to determine absolute temperatures. Interestingly enough, in the

* See Section 6.4.3.1.

region below 1°K where gas thermometers fail, it has been found possible to carry certain paramagnetic salts through cycles which are very nearly reversible and to measure absolute temperatures directly according to Kelvin's prescription.[24]

6.1.2. Gas and Vapor Pressure Thermometry

6.1.2.1. Gas Thermometry.[21] The importance of gas thermometers is due to the fact that their readings, once reduced to the ideal gas scale, provide the most direct measure of absolute temperature between $\sim 1°$ and 1000°K. Instruments of both the constant volume and the constant pressure types have been used. Constant volume thermometers are, generally speaking, not only the most accurate but also the simplest to operate. We shall therefore consider only this type. The less common constant pressure instruments are discussed in numerous standard references.[2,3]

Basically, the constant volume gas thermometer consists of a gas-filled bulb of metal or glass connected by a capillary tube to a manometer which may be adjusted to keep the gas approximately at constant volume (see Fig. 1). The bulb and a portion of the capillary come into thermal equilibrium with the system under study. The maximum useful temperature for the thermometer is that at which these parts become significantly permeable to the thermometric gas. Liquefaction of the gas determines roughly the lowest temperature at which the thermometer is used. With helium gas at reduced pressure and thermocouple or Pirani manometers,* temperatures down to about 1.5°K have been accurately measured.

For accurate temperature determination it is necessary, generally, to apply several corrections to the data in addition to those characteristic of the manometer itself. Among the more important of these are: (1) allowance for the fact that some of the gas occupies the so-called "dead volume," i.e., capillary and manometer, and is not at the temperature being measured; (2) correction for changes in the volume of the bulb with temperature and with the pressure of the contained gas. Additional corrections of importance particularly in the determination of low temperatures are: (3) allowance for the adsorption of gas on the walls of the thermometer bulb; and (4) the thermomolecular pressure difference set up between the cold bulb and the warm manometer when the mean free path of the gas molecules is comparable with the diameter of the capillary.*

In principle it should be possible to obtain ideal gas temperatures with

* See Section 4.2.6.

[24] See for example F. E. Simon, N. Kurti, K. Mendelssohn, and J. F. Allen, "Low Temperature Physics," p. 46. Pergamon, London, 1952.

the constant volume thermometer merely by making repeated determinations in which the gas pressure P_3, at the standard fixed point, 273.16°K, is systematically reduced, and extrapolating the results to $P_3 = 0$. This is seldom done in practice. Instead, P_3 is kept fixed, usually at a value of 1000 mm Hg, and corrections to ideal gas conditions are computed with

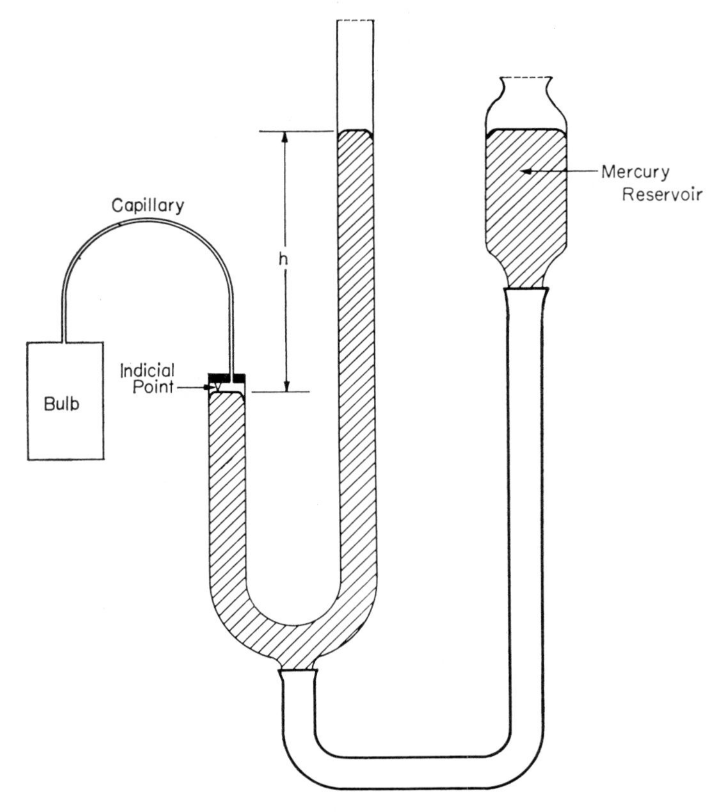

FIG. 1. Simplified constant volume gas thermometer. Mercury reservoir is raised or lowered so that meniscus at left always touches indicial point. Bulb pressure equals h plus atmospheric pressure.

the help of information about the nonideality of the particular real gas being used. Complete corrections for both constant volume and constant pressure thermometers are possible.[2,3,25,26]

6.1.2.2. Vapor Pressure Thermometry. The fact that a pure liquid or solid at a given temperature can coexist in thermodynamic equilibrium

[25] See reference 10, Chapters 5 and 7.

[26] See reference 9, p. 45.

with its saturated vapor at only one pressure serves as the basis for the so-called vapor pressure thermometer. The unique pressure–temperature relationships describing these two-phase equilibrium states are called the sublimation curve between 0°K and the triple point, and the vaporization curve from the triple point up to the critical point. Both curves have been established for many pure substances, the temperature variable having been determined experimentally either with a gas thermometer or some reliable secondary device. The data are usually available either in tabular or analytical form.[27]

A vapor pressure thermometer consists merely of a vessel, in thermal contact with the system whose temperature is being measured, in which the pure thermometric substance may be condensed and allowed to come to equilibrium with its saturated vapor. The equilibrium pressure in the vessel is measured with an appropriate manometer and converted into a temperature value by means of the known vapor pressure–temperature relation. Usually the thermometric substance is so chosen that all pressure measurements may be easily carried out with a mercury or oil-filled U-tube manometer.

Vapor pressure thermometry is particularly convenient in the region below room temperature. Here the thermometric substance is usually the same low-boiling point material used as the refrigerant. Quite often the cryostat itself will form the thermometer vessel and experimental apparatus will be immersed in the liquid (or solid) part of the two-phase system serving as both thermometer and refrigerant. Due to poor heat conduction in the liquid phase, it is very easy for significant temperature differences to exist between the position of the apparatus and the free surface of the bath, particularly in the presence of a large hydrostatic pressure head. For most liquids, stirring will insure that the vapor pressure over the bath provides an accurate measure of the temperature at the immersed apparatus. If this is not possible, a hydrostatic head correction is sometimes applied to the measured vapor pressure before the temperature is computed, although dynamical instability of the bath in the presence of a temperature gradient may make this an uncertain procedure. These problems become particularly acute for a liquid He^4 bath above its lambda point.

A list of a few of the substances often used in vapor pressure ther-

[27] See tables of physico-chemical data, e.g., "Landolt-Börnstein, Physikalisch-Chemische Tabellen," 5th ed. Springer, Berlin, 1936; J. D'Ans and E. Lax, eds., "Taschenbuch für Chemiker und Physiker." Springer, Berlin, 1949; "Handbook of Chemistry and Physics." Chemical Rubber, Cleveland, Ohio. A summary of many useful low temperature data has been prepared by C. T. Linder, Research Rept. R-94433-2-A, Westinghouse Research Laboratories, Pittsburgh, Pennsylvania, 1950.

mometry is given in Table I. The normal boiling points and triple points are shown as a rough indication of the range in which each may be most easily employed. The accuracy attainable with vapor pressure thermometers varies depending upon the reliability of the basic data. Recently, for example, the accuracy of the helium vapor pressure scale has been improved to the point where measurements accurate to within one or two millidegrees are now possible.[28]

TABLE I. Some Substances Used in Vapor Pressure Thermometry

Substance	Normal boiling point	Triple point	Triple point pressure	Critical point	Critical point pressure
He^3	3.195°K	—	—	3.35°K	1.15 atmos
He^4	4.216	—	—	5.3	2.26
H_2 (normal)	20.37	13.96°K	54.1 mm Hg	33.3	12.80
Ne	27.7	24.57	324	44.5	26.9
N_2	77.32	63.14	94	126.26	33.54
O_2	90.2	54.36	1.14	154.4	49.7
CH_4	92	90.7	700	190.7	45.7
C_2H_4	170	104	?	282.7	50.7
NH_3	239.8	195.40	45.57	405.5	111.3
SO_2	263.2	197.68	1.256	430.7	77.8
CS_2	319.4	161.1	?	552	78

6.1.3. Liquid Expansion Thermometers

The most common thermometers are of the liquid-in-glass type. These consist essentially of a glass capillary of uniform bore fitted at one end with a reservoir bulb and closed at the other. The thermometer liquid, having a volume expansivity much larger than that of the glass, fills the reservoir and part of the capillary forming a liquid column whose length becomes the thermometric parameter L. The temperature, strictly speaking, is $\theta(L) = 273.16°(L/L_3)$ where L_3 is the liquid column length at 273.16°K. In fact, most thermometers of this type now in use were calibrated by the old procedure at the ice and steam points. In addition, their scales are marked to read directly in the Celsius scale (°C).

The useful range of a particular liquid expansion thermometer is fixed at its lower limit by the freezing point of the liquid and at its upper end by either the boiling point of the liquid or the softening point of the glass. Thus a mercury-in-glass thermometer containing compressed nitrogen gas to elevate the mercury boiling point may be used between −40°C and 550°C. The upper useful limit might be extended to about 750°C by

[28] H. van Dijk and M. Durieux, *in* "Progress in Low Temperature Physics" (C. J. Gorter, ed.), Vol. 2, Chapter 14. North Holland, Amsterdam, 1957.

using quartz instead of glass or even to 1000°C if gallium were to replace the mercury. In practice this is seldom done, electrical thermometers being preferred above 500°C because of their greater accuracy. Liquid-in-glass thermometers for use below −40°C employ instead of mercury such liquids as toluene (down to −95°C), alcohol (−112°C), *n*-pentane (−132°C), and isopentane (−160°C).

The readings of liquid expansion thermometers must be corrected for several types of systematic error in any but the roughest measurements. Corrections for such defects as nonuniformity of the capillary bore and nonuniformity in or misplacement of scale markings may be determined by comparison with standard thermometers or by checking the calibration at the standard fixed point and several secondary ones. Frequent repetition of such checks is advisable since the glass thermometer bulb may completely recover its original volume only very slowly after being heated and is therefore usually varying in volume with time.

Thermometers are usually calibrated either for total or partial immersion of the mercury-filled parts in the medium whose temperature is being measured. When it is not possible to reproduce the prescribed conditions of immersion in using the instrument an estimated correction, often quite large, must be applied to each reading. This "stem correction" is just the change in length of that portion of the liquid thread at the wrong temperature produced by raising or lowering its temperature to the value it would have when properly immersed. Tables giving the stem correction for certain idealized cases are available.[29]

Reproducible changes in the volume of the thermometer bulb and thus in the calibration may be produced by varying either the internal pressure exerted by the contained liquid or the external pressure exerted by the medium in which the thermometer is immersed. A variation of the first kind results, for example, from changing the stem from a vertical to a horizontal position. Corrections for effects such as these may usually be calculated when needed.

The absolute readings of a conventional mercury-in-glass thermometer corrected in the above ways should be accurate to within ±0.05° from −10° to 150°, ±0.1° from 150° to 200°, ±0.5° from 200° to 400°, and ±1° from 400° to 500°C. Over a limited range, greater relative accuracy may be achieved with a thermometer the whole stem of which covers just this interval. The most common instrument in this category is the Beckmann thermometer, having a small range which may be adjusted by actually changing the amount of mercury in its bulb. This is accomplished by means of a reservoir at the top of the stem in which varying amounts of mercury may be trapped.

[29] See reference 9, p. 228.

6.1.4. Electrical Thermometers

6.1.4.1. Resistance Thermometers.*[30,31] The electrical resistivity of a metal may be represented ideally as $\rho = \rho_0 + \rho(T)$ where ρ_0, the residual resistivity, depends on the amount of chemical and physical impurity present in the specimen and not on the temperature, and $\rho(T)$ is that part associated with scattering of the current carriers by the thermal vibrations of the crystal lattice. At temperatures above the Debye temperature of the metal ($T > \theta_D$) it is found that $\rho(T) \propto T$, while at low temperatures ($T < \theta_D$) theory predicts $\rho(T) \propto T^5$ although smaller exponents are usually found experimentally. The resistivities of metals at room temperature are typically about 10^{-5} or 10^{-6} ohm cm and their temperature coefficients, $\alpha = (1/\rho)(d\rho/dT)$, are not large, the value for platinum at 0°C being 3.5×10^{-3} $(\text{C}°)^{-1}$, for example. The sensitivity of electrical measuring devices, however, is such that the resistance of a thin metal wire is perhaps the easiest thermometric parameter to measure with high accuracy over a wide temperature range.

Platinum wire is commonly used in resistance thermometry because of its chemical inertness, its high melting point, and the rather large interval over which its temperature coefficient varies slowly. In one commercially produced thermometer[32] widely used at room temperature and below, a spiral of the well annealed wire (diam 0.05–0.20 mm) is wound noninductively in a strain-free manner on a mica cross enclosed in a platinum thimble about 5 mm in diameter and 40 mm long. The open end of the thimble is joined to a glass bead through which pass four platinum leads (two current and two potential leads) and a hole which is sealed after the capsule thus formed has been filled with helium gas to facilitate heat transfer throughout the thermometer. In other thermometers the platinum wire may be wound on unglazed porcelain or quartz (for use above 700°C), the outer wall may be of glass, quartz, porcelain, or high melting point alloy, and the interior either filled with dried air or connected to the atmosphere. The reproducibility of the resistance–temperature relation of a given instrument depends critically upon the purity of the platinum, its heat treatment and the absence of strain. It is well to adhere to the established standards in selecting and preparing wire for thermometric use.[30,31] An important consideration in the choice of a particular thermometer design for a given application is the time required for the establishment of thermal equilibrium throughout the

* See also Vol. 6, Chapter 7.2.

[30] See reference 9, p. 162.

[31] See reference 10, Chapter 9.

[32] J. C. Southard and F. G. Brickwedde, *J. Research Natl. Bur. Standards* **9**, 807 (1932), RP 508.

device. This thermal relaxation time is governed by the heat capacity and thermal conductivity of the thermometer components. It is often rather long for most conventional platinum thermometers, making them inappropriate for the measurement of rapid temperature variations.

A typical precision platinum thermometer has a resistance of about 25 ohms at 0°C. To achieve an accuracy of 0.001°C at 0°C it is necessary to measure the resistance to within four parts in 10^6. Assuming the thermometer to be reproducible, a fact which can be determined by frequent checks at the standard and secondary fixed points, the principal obstacles to be overcome in such a measurement are the effects of lead resistance and thermal emf's. These may be effectively eliminated if the thermometer is provided with separate current and potential leads and is connected in series with a standard resistor and a source of constant current. The potential drops across the thermometer and the standard resistor are measured with current flowing in both directions and averaged separately to eliminate contributions due to thermal emf's. The resistance of the thermometer winding alone is then easily computed. This procedure requires four measurements with the system under steady-state conditions. It is, nevertheless, the method of preference in many types of precise work. The results are, of course, only as good as the potentiometer used in the measurements. It is desirable to use an instrument in which thermal emf's are negligible or easily accounted for such as the White or Diesselhorst potentiometers. In order that the thermometer be in thermal equilibrium with the system whose temperature it is measuring it is also important that the measuring current be kept small so that the dissipated Joule heat be negligible.

Numerous alternative methods for measuring the resistance of the thermometer are available many of which are more rapid than that just described although generally less precise. Most of these are bridge circuits in which compensation for lead resistance is often only partial. In addition to the four-wire thermometer connection, some circuits utilize a two-wire connection plus a set of inactive compensating leads, others a three-wire connection with two leads attached to only one end of the resistance winding. Precision comparable with that obtained with the better potentiometric circuits may be realized with modified Wheatstone bridges due to Mueller and to Smith and with a form of the Kelvin double bridge designed by Smith.[30,31]

In principle, a resistance temperature scale may be constructed as have been the other empirical scales already mentioned. Taking as the standard fixed point the triple point of water for which $T_3 = 273.16°K$ we would have $\theta(R) = 273.16°(R/R_3)$. This scale, unfortunately, diverges rather markedly from the absolute scale and so is of little practical interest. It is most useful to calibrate the platinum thermometer so that at several fixed points its readings agree exactly with the Celsius scale and that at intermediate points temperatures differing by no more than $\sim$0.1°C from the Celsius temperature may be obtained by means of a simple interpolation formula. Between 0° and 660°C the resistance varies closely with the Celsius temperature t, according to the relation $R = R_0(1 + At + Bt^2)$ where the constants R_0, A, and B are determined by measurements at the ice point (or triple point of H_2O), the steam point, and the sulfur point respectively. From -190°C to 0°C a modified relation, $R = R_0[1 + At + Bt^2 + C(t - 100)t^3]$ is used where R_0, A, and B are the same as before, and C is determined by means of a measurement at the oxygen point. The range of platinum resistance thermometry has been extended to temperatures as high as about 1300°C and as low as about -263°C (10°K) by direct or indirect calibration against a gas thermometer, the calibration relation being expressed in tabular form.[33] The readings of a properly prepared and calibrated platinum resistance thermometer are reliable to within 0.01°C up to 500°C and to within 0.1°C up to about 1300°C.

Various other metals and alloys have been used in resistance thermometry when, for reasons of economy or insensitivity, platinum was considered inappropriate. Copper and nickel are common substitutes at ordinary temperatures. Lead and indium still show large temperature coefficients below 20°K, i.e., in the region in which the platinum thermometer has become rather insensitive. These metals are used down to 7° and 3.4°K respectively at which temperatures they become superconducting. The alloy Constantan is sensitive in the liquid hydrogen region while phosphor-bronze or brass with small amounts of added lead have proved extremely useful below 7°K. Unfortunately, thermometers of most of these materials are usually much less reproducible in their characteristics than platinum and thus require frequent recalibration.

Resistance thermometers of high sensitivity over comparatively limited temperature ranges often employ electronic semiconductors rather than metals.[34] The resistivity of a typical semiconductor is much larger than that of a metal at a given temperature and generally increases rapidly with falling temperature, i.e., α is large and negative. This striking difference between metals and semiconductors stems from the fact that in the latter the charge carriers are free to transport current only after

[33] H. J. Hoge and F. G. Brickwedde, *J. Research Natl. Bur. Standards* **22,** 351 (1939), RP 1188; see also reference 10, Chapter 11.

[34] Reference 10, Chapter 20 gives original references through 1954. For recent work on germanium see J. E. Kunzler, T. H. Geballe, and G. W. Hull, *Rev. Sci. Instr.* **28,** 96 (1957).

they have been thermally excited into conducting energy states. The number of these free carriers is therefore usually small and drops rapidly as T falls below that temperature for which kT is approximately equal to the carrier activation energy. The resistance of a typical semiconducting thermometer is rather high ($\sim 10^2$ or 10^3 ohms) reducing the importance of lead resistance and making possible the use of comparatively simple measuring circuits.

Resistances fabricated of semiconducting metallic oxides, e.g., NiO, Mn_2O_3, known as "thermistors" are produced commercially and are quite widely used in thermometry and temperature control between about $-60°$ and 100°C. A typical thermistor of $NiO + Mn_2O_3$ might exhibit a temperature coefficient α, of $-44 \times 10^{-3}(K°)^{-1}$ at 0°C, opposite in sign and more than ten times as large as that for platinum. The resistance of such a thermometer varies with the absolute temperature in its useful range in the manner typical of semiconductors namely $R = AT^{\beta}e^{\Delta\epsilon/2kT}$ where A is a constant, β is a small number positive or negative, k is Boltzmann's constant, and $\Delta\epsilon$, the carrier activation energy, is of the order of several tenths of an electronvolt. Calibration is most conveniently carried out by comparison with a reliable secondary thermometer, e.g., platinum, in the particular restricted interval in which the resistor is to be used. Frequent recalibration is probably advisable in most cases where the thermometer is to be used for absolute measurements. The sensitivity of the thermistor makes it extremely useful in the detection of small temperature changes, even when these are no greater than 10^{-4}°C at 0°C. Thermistors are available in a variety of shapes with or without protective envelopes. A typical "bead" type may be merely a small piece of semiconductor less than a millimeter in diameter with two metal leads attached. Such a device has an extremely small thermal relaxation time and is easily placed in intimate contact with the system whose temperature is being measured so that rapid temperature variations are readily followed.

Semiconducting compounds have been employed in resistance thermometry to temperatures as high as 1373°K (Al_2O_3) and as low as 1°K (ZnO, SnO_2) while the semiconducting element germanium has been used between 1° and 25°K. Graphite, which exhibits many of the properties of a semiconductor including large negative temperature coefficients of resistance, has been applied in various forms in thermometry from 0.1°K to about 80°K. The most common carbon thermometers consist either of ordinary radio resistors or of thin films of colloidal graphite prepared from alcohol or water suspensions. While most carbon thermometers show lack of reproducibility upon repeated warming and cooling, one type, made by Allen-Bradley, has been found to have remarkably stable characteristics. These resistors, in the $\frac{1}{2}$ watt and smaller sizes with room temperature resistances of approximately 50 ohms have been very widely used in precision calorimetry at hydrogen and helium temperatures. Clement[35] has shown that the resistance–temperature relation for these resistors in this region is approximately represented by

$$\left[\frac{\log R}{T}\right]^{\frac{1}{2}} = a + b \log R$$

where a and b are adjustable quantities. Calibration is accomplished by bringing the thermometer into indirect contact with either the liquid hydrogen or helium bath and using the known vapor pressure–temperature relations. A reasonable value of b is chosen and a is determined for each of several calibration points. The quantity a is plotted against log R and a smooth curve drawn through the points. This curve then provides a value of a to be used in the equation for the precise calculation of the absolute temperature T corresponding to a given value of R. Temperatures in the region between 4.2° and 10° may also be determined with reasonable accuracy in this way in spite of the absence of calibration data in this interval.

6.1.4.2. Thermocouples.[*][36] If wires of two unlike metals are joined at their ends to form a closed circuit and the two junctions maintained at different temperatures, an electromotive force is set up in the circuit. The magnitude of this thermal emf depends only on the nature of the two metals constituting the thermocouple and the temperature difference between the junctions if the wires are homogeneous. If one junction is kept at a fixed temperature the other junction may serve as the sensitive element of a thermometer, the thermal emf being the thermometric parameter. The fixed temperature chosen for the reference junction is usually 0°C. Here again the empirical thermoelectric temperature scale established as in Section 6.1.1 is of little practical interest and the thermometer is commonly calibrated directly in terms of the Celsius scale.

It should be noted that the introduction into the thermoelectric circuit of additional metal parts does not affect the emf in that circuit so long as all added junctions are isothermal. The thermal emf is thus measured by inserting in the circuit a potentiometer or a millivoltmeter, the connections being arranged as shown in Fig. 2. While a millivoltmeter is often used as an indicator in rough measurements, one must remember that because of its comparatively small internal resistance the reading of such an instrument is not the true thermal emf of the particular combina-

* See also Vol. 6, Chapter 7.6, and this volume, Chapter 8.7.

[35] J. R. Clement, cited in reference 10, p. 382.

[36] See reference 9, p. 180.

tion of metals comprising the thermocouple. A potentiometer, on the other hand, measures this quantity directly since it is read with no current flowing in the thermoelectric circuit. Potentiometers of either the hand operated or automatic recording variety are always used in precise thermoelectric thermometry not only because they are capable of great sensitivity but also because measurement of the true thermal emf simplifies the calibration of many thermocouples, as we shall see.

The four most commonly used thermocouples permit coverage of the temperature range −200° to 1500°C with an accuracy second only to that obtained with the platinum resistance thermometer. Copper–constantan (40% Ni, 60% Cu) is used between about −200° and 400°C; iron–constantan between 0°C and about 800°C; Chromel (90% Ni, 10% Cr)–Alumel (98% Ni, 2% Al) up to about 1300°C; and platinum–platinum–10% rhodium up to about 1750°C. Tables of thermal emf versus

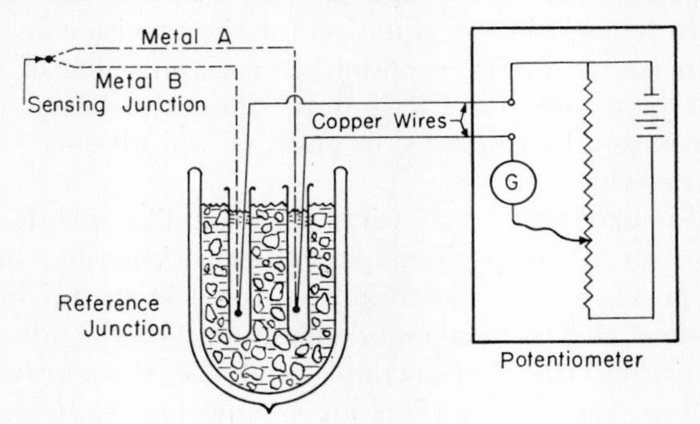

FIG. 2. Thermoelectric thermometer circuit.

Celsius temperature have been prepared for standard specimens of these and other frequently used thermocouples with the reference junctions at 0°C.[36,37] Properly prepared thermocouples (well annealed homogeneous wire, welded junctions) of these compositions used with a potentiometer are assumed for many routine applications to obey the same calibration relations. This assumption is quite good for the platinum–platinum–rhodium couple, the errors being no greater than 2 or 3°C at the highest temperatures. Other thermocouples often exhibit calibration deviations from the tabulated values of 5% or more. For accurate work it is necessary then that the thermocouple be calibrated at several temperatures either against a calibrated secondary thermometer or by means of appropriate fixed points. A curve of deviations of the calibration data from the tabulated thermal emf's against temperature is then prepared. The

[37] "Handbook of Chemistry and Physics," 40th ed. Chemical Rubber, Cleveland, Ohio, 1958.

table together with this correction curve provide a thermal emf–temperature relation whose accuracy is limited by the number and reliability of the calibration points chosen. Temperature measurement at 600°C, for example, accurate to within a degree or two is commonly achieved in this way. With care, the uncertainty can be reduced to ~0.1°C.

Thermoelectric thermometry outside the range −200°C to 1500°C has not yet been standardized to the same extent as that within these limits. Typical of the thermocouples used at very high temperatures are those of tungsten and molybdenum; 99% Mo/1% Fe–W is used up to 2200°C while 75% W/25% Mo–W can be used up to 3000°C. The characteristics of these couples vary strongly from specimen to specimen, however, making calibration difficult. In addition, they must be carefully protected from oxygen while in use. At temperatures from −200°C to about −263°C the copper–constantan couple has been used with considerable success and rather complete calibration tables have been prepared.[38] This couple while quite insensitive at liquid helium temperatures has, nevertheless, been used also for rough temperature indication in this range. A thermocouple of copper–gold/cobalt has been found to be fairly sensitive at hydrogen and helium temperatures and to have rather reproducible characteristics.[39]

Over an extended range, the thermal emf–temperature relation of a given thermocouple is apt to be rather complicated. Over a restricted interval it is often possible, however, to approximate it quite closely with the formula $E = a + b(t - t_0) + c(t - t_0)^2$ or if t_0 the temperature of the reference junction is taken to be 0°C, $E = a + bt + ct^2$. The constants a, b, and c are determined by calibration at fixed points in the chosen interval. For example, the freezing points of gold, silver, and antimony are used to calibrate the platinum–platinum–10% rhodium thermocouple in the range 630°–1063°C. In this case temperatures calculated with the three constant formula agree with the Celsius scale to within 0.1°C over the whole range.

Several of the common error-producing difficulties in thermoelectric thermometry can, fortunately, often be completely avoided. The simplest of these is variation of the temperature of the reference junction. A small Dewar flask containing melting ice is usually completely adequate for maintaining a reference temperature of 0°C and is generally used in laboratory work. The triple point cell used to establish the standard fixed point should also provide a practical means of fixing the reference junction temperature in the most precise work. Many industrial thermoelectric thermometers, however, are designed to operate with the reference junc-

[38] See reference 9, p. 219.

[39] N. Fuschillo, *J. Phys. Chem.* **61**, 644 (1957).

tion at the temperature of the ambient air and often have compensating devices to correct automatically for small changes in that temperature. Perhaps a more difficult problem is that posed by inhomogeneities in the wires of the thermocouple resulting from contamination or strain. Should such a portion of the circuit be placed in a temperature gradient an additional emf will be produced, changing the characteristic of the couple and often making standard calibration data inapplicable. Proper annealing, careful handling, protection from chemical contamination, and strain-free mounting of the thermocouple wire will reduce such effects. It is often helpful to test a wire for inhomogeneity by connecting its ends to a galvanometer, passing it slowly through a strong temperature gradient, and watching for significant deflections. Assuming that precautions such as these have been taken to insure homogeneity it is, nevertheless, still advisable to reproduce as closely as possible during measurements with a thermocouple the temperature gradient existing during the calibration.

The thermocouple has several rather unique properties which are often exploited in practical application. For example, it may be made of extremely fine wires joined to produce a sensing junction of very small dimensions and heat capacity. Thermocouples are, therefore, often used to measure the temperatures of small objects or of systems in which the temperature varies rapidly with time. Another important class of applications arises from the fact that the thermoelectric circuit is sensitive essentially to temperature differences. Differential thermocouple arrangements are quite useful, for example, as the sensing elements in thermostats of many types or in the measurement of temperature gradients particularly where the absolute value of the temperature need not be accurately known or is measured separately. As a last example of the unique flexibility of the thermocouple we may note that several of them operating with their junctions at the same temperatures may be connected in series electrically to yield a much larger total emf. Such a device, called a thermopile, can be made sufficiently sensitive and of small enough heat capacity to serve as a detector for thermal radiation.

6.1.5. Radiation Thermometers[40,41]

It is often impossible or impractical to bring a thermometer into direct mechanical contact with the object whose temperature we wish to know, particularly if that temperature is a very high one. In this case we avail ourselves of the fact that the object is continuously emitting electromagnetic radiation whose spectral distribution and total intensity are func-

[40] W. P. Wood and J. M. Cork, "Pyrometry," 2nd ed. McGraw-Hill, New York, 1941.

[41] See reference 9, Chapter 12.

tions of its absolute temperature. These functional relations are known precisely only if the source is a perfect emitter and absorber, a so-called blackbody. Let us assume this to be the case initially since sources of nearly blackbody character are often encountered, e.g., a furnace with uniformly heated interior walls viewed through a small aperture. Planck's distribution law gives the radiant energy of wavelength λ emitted by the blackbody

$$E(\lambda) = \frac{2\pi hc^2}{\lambda^5} \frac{1}{e^{ch/\lambda kT} - 1} \text{ erg sec}^{-1} \text{ cm}^{-3}. \quad (6.1.1)$$

The total energy of all wavelengths emitted by the blackbody is

$$E = \int_0^\infty E(\lambda)\, d\lambda = \frac{2\pi^5 k^4}{15c^2h^3} T^4 = \sigma T^4 \quad (6.1.2)$$

the familiar Stefan–Boltzmann law, where

$$\sigma = 5.669 \times 10^{-5} \text{ erg/cm}^2 \text{ deg}^4 \text{ sec}.$$

Practical radiation thermometers or pyrometers may be divided into two categories depending upon whether they are sensitive essentially to E or to $E(\lambda)$. Those responding to E are called total radiation pyrometers, the oldest and still typical one being that due to Féry. In this instrument, some of the radiation from the source, e.g., furnace opening, is concentrated by a concave mirror on one junction of a thermocouple or a thermopile. So long as the image just fills the aperture in a diaphragm in front of the receiving element the reading of the instrument may be shown to be independent of its distance from the source, neglecting atmospheric absorption. This condition is met by visual adjustment. The emf developed in the thermocouple circuit is then proportional to $(T^b - T_0{}^b)$ where T_0, the temperature of the receiver, is usually negligible in comparison with the source temperature T, and b may range from 3.8 to 4.2 depending upon the instrument. That b is not exactly 4 results from (1) the thermal emf not being a linear function of the temperature difference between the thermocouple junctions, (2) strong reflection within the instrument, (3) heating, by conduction, of the cold junction of the thermocouple which is built into the device, and (4) the rate of heat loss by the receiver not being proportional to the temperature difference between it and its surroundings. Since the simple fourth power law is seldom directly applicable, it is necessary to calibrate the total radiation pyrometer against a standard blackbody source at numerous temperatures determined by another thermometer. Up to the gold point (1063°C), the platinum–platinum–10% rhodium thermocouple is used. Above that temperature, the reference thermometer is a standard optical pyrometer of the type which we shall now discuss.

The term optical pyrometer is used to describe those devices responding to radiation of a narrow band of wavelengths, usually in the visible range, whose intensity is given by Planck's formula. The disappearing filament pyrometer is the simplest and most commonly used instrument of this type. It is essentially a telescope focused on the source. The filament of a standard lamp is placed in the focal plane of the objective and is seen through the eyepiece superposed on the real image of the source. A red filter in the eyepiece restricts the light reaching the eye to a relatively limited range of wavelengths. The current through the lamp filament is adjusted until the filament and source are indistinguishable to the viewer. At this point $E(\lambda)$ for the wavelengths passed by the filter is the same for the source and filament, i.e., they are at the same temperature. An ammeter in the lamp circuit serves as the indicator and is usually calibrated to read temperatures directly. Again the distance between source and pyrometer is unimportant if atmospheric absorption can be neglected.

The optical pyrometer is calibrated up to 1063°C by sighting it on a blackbody at various temperatures determined with the platinum–platinum–10% rhodium thermocouple. One gets then $E(\lambda)$ or T as a function of filament current. To measure the temperature of a blackbody above 1063°C, a rotating sector is placed between the source and the optical pyrometer. By this means $E(\lambda)$ for the source can effectively be cut down in a known way to a value corresponding to a source temperature in the calibrated range, usually 1063°C. If θ is the angular aperture of the sector, the energy reaching the pyrometer is just $\theta/2\pi$ what it would be in the absence of the sector. Thus

$$\frac{E_T(\lambda)}{E_{\mathrm{Au}}(\lambda)} = \frac{e^{c_2/\lambda T_{\mathrm{Au}}} - 1}{e^{c_2/\lambda T} - 1} = \frac{2\pi}{\theta} \tag{6.1.3}$$

if we assume the radiation to be strictly monochromatic, and we may calculate the unknown temperature T in terms of that of the gold point, $T_{\mathrm{Au}} = 1336.15°\mathrm{K}$. Since the radiation is not monochromatic we must use an effective wavelength in the calculation. A correction for this fact is described in treatises on the subject.

The lower limit of usefulness of an optical pyrometer of the type just described is determined by the color of the light employed. For visible red light this limit may be as low as 600°C. Total radiation pyrometers of special design have been used successfully at room temperature and even below. There is, in principle, no upper limit to the temperatures measurable with either kind of device. Extension of the range of the optical pyrometer beyond the gold point by means of a rotating sector or, alternatively, a calibrated filter has been described. Such extension for the total radiation pyrometer is achieved usually by reducing the aperture.

The most important errors in radiation thermometry result from deviations of the properties of the emitting surface under examination from those of a blackbody, i.e., its total emissivity and its spectral emissivities for particular wavelengths may be less than one. Since both total radiation and optical pyrometers are calibrated against a blackbody they will indicate temperatures less than the true one for such a surface. These temperatures are usually designated as the "radiation temperature" or "brightness temperature" depending upon whether they have been measured with the total radiation or the optical pyrometer. The emissivities of various surfaces prepared in different ways have been measured and tabulated[42] so that an estimate of the discrepancy is often possible. Determination of the emissivities can, of course, be carried out directly in the range where the true surface temperature may be measured independently with a thermocouple.

6.1.6. Fixed Points and the International Scale[43,44]

The role of the standard fixed point in the calibration of thermometers including those measuring absolute and ideal gas or Celsius temperatures has already been discussed. It has also been noted that the empirical temperature scales of some of the most convenient thermometers differ sufficiently from the absolute scale to make them of little practical interest. As a result it has become common practice to calibrate such devices as thermocouples and resistance thermometers to read approximate ideal gas or Celsius temperatures directly. This is accomplished by means of a number of easily reproduced auxiliary fixed points, usually the normal melting or boiling points of readily obtained pure substances, for which the ideal gas or Celsius temperatures have been accurately determined with a gas thermometer. Table II lists the currently accepted fixed points and their temperatures. The temperature of the standard fixed point is, of course, arbitrarily set. Those of the so-called basic and secondary fixed points are measured quantities. Prescriptions for the accurate reproduction of the fixed points have been published by the various national standards laboratories.

Any thermometer may, in principle, be calibrated at several fixed points and an interpolation procedure developed which permits Celsius temperatures to be determined with it to the desired accuracy. These operations are more easily and reliably carried out with some thermom-

[42] See reference 9, p. 1184.

[43] See reference 10, Chapters 8 and 9.

[44] H. F. Stimson, *J. Research Natl. Bur. Standards* **42,** 209 (1949), RP 1962.

eters than with others. Over their respective useful ranges, the platinum resistance thermometer and the platinum–platinum–rhodium thermocouple have been found to be both reproducible and very simply calibrated to measure Celsius temperatures with sufficient accuracy for most practical purposes. On the basis of international agreement reached originally in 1927 and revised in 1948 a so-called International Temperature Scale has been set up in which these two thermometers, calibrated

Table II. Temperatures of Fixed Points[a]

	Fixed points	Temperature (°C)	Temperature (°K)
Standard	Triple point of water	0.01	273.16
Basic	nbp of oxygen (oxygen point)	− 182.97	90.18
	Equilibrium of ice and air-saturated water (ice point)	0.00	273.15
	nbp of water (steam point)	100.00	373.15
	nbp of sulfur (sulfur point)	444.60	717.75
	nmp of antimony (antimony point)	630.50	903.65
	nmp of silver (silver point)	960.80	1233.95
	nmp of gold (gold point)	1063.00	1336.15
Secondary	nbp of helium	− 268.93	4.22
	nbp of hydrogen	− 252.78	20.37
	nbp of neon	− 246.09	27.06
	nbp of nitrogen	− 195.81	77.34
	nmp of mercury	− 38.86	234.29
	Transition point of sodium sulfate	32.38	305.53
	nbp of naphthalene	217.96	491.11
	nmp of tin	231.85	505.00
	nbp of benzophenone	305.90	579.05
	nmp of cadmium	320.90	594.05
	nmp of lead	327.30	600.45
	nmp of zinc	419.50	692.65

[a] After M. W. Zemansky, "Heat and Thermodynamics," 4th ed. McGraw-Hill, New York, 1957.

at the basic fixed points, serve as convenient interpolation devices over the range −182. 97°C to 1063.0°C. A temperature, t, on the International Temperature Scale is designated °C or °C (Int. 1948) since it differs from the corresponding Celsius temperature at most by about 0.1°C. The International Temperature Scale may be conveniently divided into four ranges depending upon the interpolation method used. We shall merely outline here the rules adopted for each of these ranges referring the reader for details to the papers of Stimson or Hall.[43,44]

(1) Between 0° and 630.5°C a standard platinum resistance thermometer is used, the platinum being of such purity and state of anneal that

$R_{100}/R_0 \geq 1.391$. The temperature is given by $R_t = R_0(1 + At + Bt^2)$, the constants R_0, A, and B having been determined at the ice point, steam point, and sulfur point. Rather than determine R_0 directly at the ice point, it is now recommended that a triple point cell be used and an ice point value be calculated taking the latter point to be 0.0100°C below the triple point of water.

(2) Between −182.97° and 0°C the same platinum resistance thermometer is employed, the temperature being given now by the formula $R_t = R_0[1 + At + Bt^2 + C(t - 100)t^3]$. The constants, R_0, A, and B are the same as in the higher range, and C is fixed by calibration at the oxygen point.

(3) Between 630.5° and 1063°C the scale is defined in terms of the emf of the platinum–platinum–10% rhodium thermocouple, one junction of which is kept at 0°C. The temperature is given by $\epsilon = a + bt + ct^2$ where a, b, and c are determined at the antimony, silver, and gold point respectively. Actually no reliance is placed on the antimony point. The normal melting point of the particular antimony specimen employed is determined with the platinum resistance thermometer calibrated as described above. Alternatively, the thermocouple may be calibrated directly against the platinum resistance thermometer at a temperature near 630.5°C.

(4) Above 1063°C a standard optical pyrometer is used to extrapolate the scale. The intensity of radiation of a particular wavelength λ from a blackbody at the unknown temperature E_t is compared with the intensity of radiation from a blackbody at the gold point E_{Au} having as nearly as possible the same wavelength. The unknown temperature, t, is then calculated from the formula

$$\frac{E_t(\lambda)}{E_{\text{Au}}(\lambda)} = \frac{\exp[c_2/\lambda(t_{\text{Au}} + T_0)] - 1}{\exp[c_2/\lambda(t + T_0)] - 1} \tag{6.1.4}$$

where c_2 is 1.438 cm degrees and T_0 is the temperature of the ice point, 273.15°K. It is not possible in practice to work with light of the same wavelength in the two measurements and a small correction for the difference must be included.

6.1.7. Miscellaneous Methods

Conventional thermometry utilizes only a few of the many temperature dependent parameters familiar even to the beginning student of physics. Physical constraints existing in some experimental situations make conventional thermometers inapplicable. We shall conclude our discussion of thermometry by considering briefly some illustrative examples of thermometers based on familiar temperature sensitive properties of matter which have been exploited when standard devices could not be used.

6.1.7.1. Magnetic Thermometry.[45] The magnetic susceptibilities of many salts containing well-separated paramagnetic ions can be approximately represented to very low temperatures by the relation (Curie–Weiss law)

$$\chi = \frac{C}{T - \Delta}$$

where C is the appropriate Curie constant, and Δ depends on the shape of the specimen.* Δ is zero for a sphere of an isotropic crystal in a homogeneous field. For simplicity we assume these conditions to be met in the following discussion. At low temperatures the susceptibilities of these salts are rather large and may be conveniently measured by placing the spherical specimen inside a pair of coils and observing the change in their mutual inductance by ballistic or ac methods. This arrangement, it will be noted, does not require direct mechanical contact between the measuring device and the salt. Thus the coils may be wound on the outside of the experimental chamber in which the salt is located, perhaps as part of a thermally isolated system.

Magnetic thermometers are usually calibrated directly against the vapor pressure of the refrigerant baths used to achieve the low temperatures at which they are employed. As interpolation devices they are particularly useful because of the simplicity of the calibration relation which they obey. In a recent application, a magnetic thermometer was actually used to detect significant irregularities in the vapor pressure–temperature relation for liquid He^4. It was assumed that the vapor pressure–temperature data were reliable at only a few well-determined points and that the susceptibility was a smooth function of the temperature as given above. Discrepancies between the calibration thus established and other vapor pressure–temperature points were then attributed to inaccuracy of the latter values.[46]

The full potentialities of the magnetic thermometer are realized below 1°K where it serves as the principal temperature measuring device. Quite often the same salt specimen acts both as refrigerant and thermometer. Cooling is achieved by isothermally magnetizing the salt at a temperature T_i near 1°K in a large external field H_i, thermally isolating the magnetized salt and then demagnetizing it adiabatically. The external

* See also Vol. 6, Chapter 9.1.

[45] See reference 10, Chapters 13 and 16.

[46] See reference 10, Chapter 12.

field needed for the susceptibility measurement can be kept small enough (a few oersteds) so that no appreciable reheating takes place even at the lowest temperatures. Having measured the susceptibility above 1°K and so calibrated the thermometer at known temperatures, one might hope to be able simply to extrapolate this calibration curve, $\chi T = \text{const}$, into the region below 1°K. However, this must be done with caution. If H_i/T_i is large, the final temperature reached in an adiabatic demagnetization will be of the order $\Delta\epsilon/k$ where $\Delta\epsilon$ is the splitting of the ground state of the paramagnetic ions due to their mutual interaction. This is just the region in which Curie's law breaks down. Use of an extrapolated calibration curve may, therefore, yield temperatures significantly different from the absolute temperatures of the salt. It is customary, nevertheless, in work below 1°K to make this extrapolation for convenience, being careful to label the temperatures thus defined as "magnetic" temperatures, T^*.

In some cases the relation between T^* and T for a given salt has been established empirically making possible the determination of absolute temperatures below the useful range of a gas thermometer. This has usually been done by a method first suggested by W. H. Keesom. It consists essentially of carrying the salt through a series of Carnot cycles between a known temperatures T_i and various final temperatures T determined according to Kelvin's definition. Full descriptions of this procedure and many other aspects of magnetic thermometry will be found in the reviews of de Klerk,[45,47] Ambler and Hudson,[48] van Dijk[45] and the books of Casimir[49] and Garrett.[50]

6.1.7.2. Sound Velocity Thermometry.[51] The velocity of sound in an ideal gas is given by the familiar formula $v = \sqrt{\gamma RT/M}$ where γ is the ratio of specific heats at constant pressure and constant volume, M is the molecular weight, and R is the gas constant. So long as ideal gas conditions prevail and γ is not a function of temperature, the absolute temperature of a gas may be obtained from the measured sound velocity by means of this formula. The method has been used successfully in gases at very high temperatures, e.g., in electric arcs. Some attention has been given the possibility of using the sound velocity in gaseous hydrogen or helium as a thermometric parameter at very low temperatures. Deviations

[47] D. de Klerk, *in* "Handbuch der Physik—Encyclopedia of Physics" (S. Flügge, ed.), Vol. 15, p. 38. Springer, Berlin, 1956.

[48] E. Ambler and R. P. Hudson, *Repts. Progr. in Phys.* **18,** 251 (1955).

[49] H. B. G. Casimir, "Magnetism and Very Low Temperatures." Cambridge Univ. Press, London and New York, 1940.

[50] C. G. B. Garrett, "Magnetic Cooling." Harvard Univ. Press, Cambridge, Massachusetts, 1954.

[51] See reference 10, Chapter 21.

from ideal gas behavior make the velocity–temperature relations more complicated in this case.

6.1.7.3. Interferometer Thermometry.[52] The density of an ideal gas at constant pressure varies inversely with temperature. Since the index of refraction of the gas is proportional to the density, it, too, will vary inversely with temperature. If one of the interfering beams in an interferometer passes through a gas of nonuniform temperature the observed interference fringes will be distorted. This effect has been made the basis of a method for the determination of the temperature distribution in a gas near warm surfaces. The method is attractive since the perturbing effect of most thermometers in this situation is large enough to distort the distribution severely.

6.2. Calorimetry[1–8,53,54]

By the methods of calorimetry we measure the quantities of heat absorbed or given up by physical systems experiencing (a) temperature changes, (b) phase transformations, or (c) chemical reactions. Such observations were made long before the nature of heat was properly understood, essentially by noting the temperature rise produced in a standard substance, water, when a system undergoing one of these changes was placed in contact with it. Thus we have heat traditionally measured in such units as the 15° calorie, the amount required to raise the temperature of one gram of water from 14.5° to 15.5°C.

The first law of thermodynamics is a generalization of the observation that heat, energy, and work are commensurable quantities, transformable into one another and obeying a common conservation principle. For an infinitesimal process it may be expressed as follows

$$đQ = dE + đW \tag{6.2.1}$$

where $đQ$ is the amount of heat added to a system, dE the increase in its internal energy, and $đW$ the work which the system does on its surroundings; $đQ$ and $đW$, it will be noted, are not exact differentials (see Chapter 6.4).

[52] See reference 9, p. 685.

[53] A. Eucken, *in* "Handbuch der Experimentalphysik" (W. Wien and F. Harms, eds.), Vol. 8, Part 1. Akademische Verlagsges., Leipzig, 1929.

[54] J. M. Sturtevant, *in* "Physical Methods of Organic Chemistry" (A. Weissberger, ed.), 2nd ed., pp. 731–845. Interscience, New York, 1949.

Calorimetry, as a source of data of wide physical significance, may be said to be based upon the first law of thermodynamics. First of all the law permits the magnitudes of quantities of heat and energy or work to be expressed in the same units, the accepted equivalence being 1 cal ≡ 4.1854 abs joules.* Secondly, it extends the list of methods available for the measurement of quantities of heat to include mechanical and particularly electrical techniques. And finally, the first law permits us to relate measured quantities of heat to internal energy changes of systems, making calorimetry one of the most important sources of information about the structure of matter in bulk.

Before discussing calorimetric procedures, let us summarize briefly the quantities which are commonly measured calorimetrically. For simplicity we may mention first a single component, single phase system whose state may be specified fully by means of the variables P, V, and T. The analogous quantities for systems for which other variables are relevant are given in numerous standard references.[1,2] The first law for processes involving such systems becomes $đQ = dE + P\,dV$. Consider the addition of a quantity of heat $đQ$ to a single component–single phase system at constant volume. The temperature will increase by an amount dT such that $đQ = C_v\,dT$ where $C_v = (dQ/dT)|_v$ is called the heat capacity at constant volume. From the first law we see $C_v = (\partial E/\partial T)_v$. If $đQ$ be added at constant pressure and no second phase appear, then a temperature rise is observed such that $đQ = C_P\,dT$ where $C_P = (đQ/dT)|_P$ is defined as the heat capacity at constant pressure. This process is accompanied not only by an increase in the internal energy by an amount $dE = C_v\,dT$ but also by the performance of an amount of work $P\,dV$. According to the first law

$$C_P = \left[\frac{\partial(E + PV)}{\partial T}\right]_P = \left(\frac{\partial H}{\partial T}\right)_P$$

where $H = E + PV$ is the enthalpy function. The quantities C_P and C_v, the amounts of heat needed to raise the temperature of a system by one degree under the two most common conditions of constraint, will in general vary with temperature. C_P and C_v are expressed in cal/deg or joules/deg and refer to a particular system of mass m. It is convenient to use the heat capacities per gram, the specific heats, or the heat capacities per mole or gram atom, the so-called molar or atomic heats. Typical units for these quantities would be cal/gm deg or cal/mol deg. Since C_P and C_v are not usually measured with equal ease it is often necessary to make use of the general relation connecting them, namely,

* This figure represents a weighted average of the measured results given in references 76–78.

$$C_P = C_v + \frac{TV\beta^2}{\kappa} \tag{6.2.2}$$

where β is the volume expansion coefficient and κ the compressibility of the substance.

If, upon the addition of heat to a system at constant pressure, a second phase begins to form, the internal energy of the system will be increased, but its temperature will not rise until all of the first phase has disappeared. Since the specific volumes of the two phases will be different in such a "first order" transformation,[55] work is also performed. The quantity of heat absorbed in the transformation of a unit mass of a substance from one phase to another at constant temperature and pressure is called the latent heat of transformation l. It is commonly given in cal/gm and will in general be a function of temperature. In the transformation of m grams of a substance then, according to the first law, $ml = \Delta E + P\,\Delta V$. The heats of transformation commonly studied include those of fusion, vaporization, and sublimation as well as those accompanying polymorphic transitions in the solid state and the formation of adsorbed phases.

The establishment of equilibrium in a mixture of different kinds of molecules at a given temperature and pressure is usually accompanied by the evolution or absorption of what may be termed generally a heat of reaction. These quantities of heat are often named so as to identify further the kind of reaction which has occurred. Among the more common we may list heats of formation, dissociation, polymerization, combustion, neutralization, solution, and dilution. It will be noted that not all of these reactions involve the appearance of new molecular species. Most reactions are carried out at constant pressure and are accompanied by volume changes, i.e., the performance of work. Thus the heat of reaction at constant pressure Q_P is just the change in the enthalpy of the system ΔH, i.e., the enthalpy of the products less that of the reactants. ΔH will in general be a function of both temperature and pressure although the pressure variation is often negligible. If heat is given up, ΔH is negative and the reaction is termed exothermic. A reaction in which heat is absorbed is called endothermic and is characterized by a positive ΔH. This sign convention, unfortunately, is not universally followed necessitating caution in the use of tabulated data of this type. One common reaction, namely that of combustion, is usually carried out under conditions of approximately constant volume in a bomb calorimeter. In this case the heat of reaction is essentially Q_v or ΔE, the internal energy change for the process. It follows from the definition of the enthalpy that at constant pressure $\Delta H = \Delta E + P\,\Delta V$, permitting us to work with either quantity.

[55] See reference 1, Chapter 15.

Calorimetric methods may be grouped roughly according to the manner in which the heat absorbed or given up by the system of interest is measured. The oldest class includes those methods in which this heat is transmitted to or from a standard substance of known heat capacity, e.g., water or copper, and the resultant temperature change observed. A second class includes methods in which the heat to be measured is added to or extracted from a standard two-phase system at constant temperature and pressure and the extent of the resultant transformation noted. A third class covers all methods in which known amounts of energy are dissipated in an electrical heater. Less common methods not belonging to any of these categories include, for example, techniques in which heat is supplied in known amounts by gamma ray or electron bombardment, or eddy current heating.

6.2.1. Specific Heats

6.2.1.1. Solids.* The specific heats of solids near room temperature and above have generally been measured by the method of mixtures which belongs to the first category listed in the preceding paragraph. The calorimeter consists essentially of a metal vessel of known heat capacity containing a known amount of water whose temperature is accurately measured with an electrical thermometer or perhaps a Beckmann thermometer. The specimen is heated to a measured temperature in a steam jacket or electric furnace and quickly transferred by a suitable mechanism into the calorimeter. Stirring speeds the equalization of the temperature throughout the calorimeter. A continuous record of the calorimeter temperature is kept. Ideally it will rise from one steady value to another after introduction of the specimen. Assuming all heat given up by the specimen to be absorbed by the calorimeter, an average value of the specific heat at constant pressure C_P of the specimen is easily calculated for the interval through which its temperature has changed.

The major difficulty encountered in measurements of this kind is the exchange of heat between the calorimeter and its surroundings particularly when the temperature of the latter also varies during an experiment. Various precautions are taken to minimize these losses. The metal calorimeter vessel is polished to reduce radiation loss. In the simpler devices it is mounted on pointed insulating supports inside a double-walled shield, the annular space of which is filled with water at a constant temperature chosen to reduce heat exchange. Often the calorimeter will be mounted inside a Dewar vessel.

During a typical experiment the temperature of the calorimeter will initially show a gradual linear rise with time as a result of the heat input

* See also Vol. 3, Chapter 8.1, and Vol. 6, Part 5.

due to continuous stirring (fore period). Upon insertion of the specimen the temperature rises rapidly, generally reaching a maximum after a short interval and then beginning to fall gradually (after period). Many methods are available for determining the temperature rise which the calorimeter would have experienced in the absence of unwanted heat leaks.[53,56] In rough work it is sometimes sufficient to extrapolate the initial and final linear portions of the temperature–time curve to a time midway between the end of the fore period and the maximum and to take the temperature difference at that point. A better approximation to the

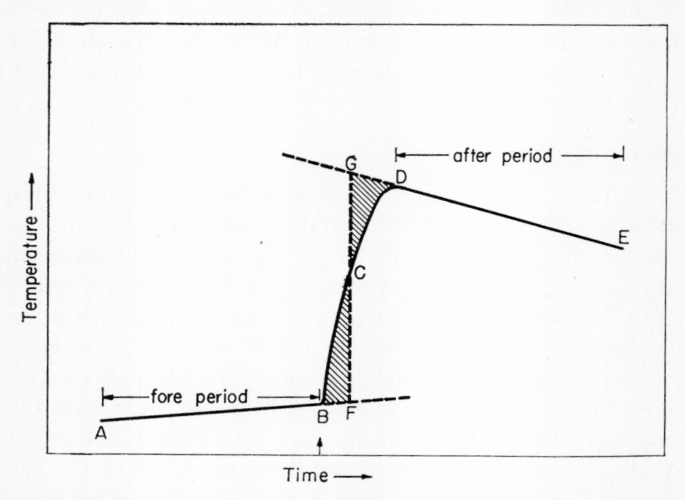

FIG. 3. Calorimeter temperature as a function of time for a typical experiment. The ideal temperature rise of the calorimeter is given by the length of the vertical line *FG*, drawn so as to make the areas *BFC* and *CDG* equal.

ideal temperature rise is obtained by the graphical construction shown in Fig. 3.

Heat exchange between the calorimeter and its environment can be effectively eliminated by surrounding the calorimeter with an adiabatic shield, a metal enclosure whose temperature is kept the same as that of the calorimeter. This is usually done with an electrical heater controlled by a differential thermocouple and servoregulator. If many measurements of high accuracy are planned the inclusion of an adiabatic screen in the apparatus is advisable.

The water calorimeter has several severe limitations: it is useful only in a restricted temperature range; water has such a large specific heat that the observed temperature changes are usually quite small; the water

[56] W. P. White, "The Modern Calorimeter." Chemical Catalog, New York, 1928.

may evaporate in the course of a set of measurements. These objections may be overcome by replacing the water-filled calorimeter with a heavy copper block drilled to receive the specimen. Measuring the block temperature with thermocouples or a resistance thermometer, one can carry out calorimetry by the method of mixtures exactly as above. This was first done at low temperatures (down to solid CO_2 temperatures) by Nernst and co-workers[57] who mounted the block inside a Dewar vessel. More recently the same method has been employed up to about 1600°C

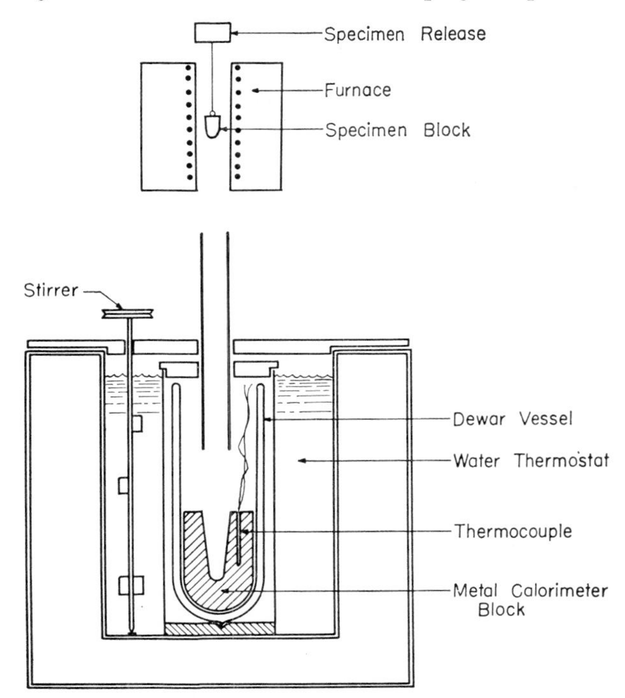

FIG. 4. High temperature metal block calorimeter (schematic).

by Jaeger and Rosenbohm.[58] In this work an aluminum block and Dewar vessel enclosure were used and very elaborate precautions taken to reduce heat exchange (see Fig. 4). The accuracy of the results was estimated to be 0.1%.

There are several calorimetric methods, in which a cooling or warming specimen is placed in contact with a standard two-phase system and the

[57] W. Nernst, F. A. Lindemann, and F. Koref, *Sitzber. deut. Akad. Wiss. Berlin* **12,** **13,** 247 (1910).

[58] F. M. Jaeger and E. Rosenbohm, *Koninkl. Ned. Akad. Wetenschap., Proc.* **33,** 457 (1930); **34,** 808 (1931).

extent of the resultant transformation used to measure the specific heat of the specimen. In Bunsen's[59] ice calorimeter the reduction in volume of an ice-water system upon the addition of heat from a cooling specimen is used to measure the amount of melted ice and thus the quantity of heat. In Joly's steam calorimeter,[60] the measured quantities are the mass of water formed by the condensation of steam on the specimen and the temperature rise of the latter. Dewar[61] has measured average specific heats of solids at low temperatures by cooling them first to 0°C and then dropping them into liquid O_2 or H_2 and measuring the volume of gas evolved. Because the latent heats of vaporization of the liquids used are quite small, these volumes are large and easily measured.

All of the methods discussed in the preceding paragraphs yield specific heat values which are averages over rather wide temperature intervals. At room temperature and above, the specific heats of most solids vary slowly with temperature and such averages may be adequate. At low temperatures, however, the specific heat is varying rapidly with temperature and it is important to reduce as much as possible the temperature interval over which the average is taken. This is readily accomplished in the vacuum calorimeter,[62–64] due originally to Nernst and now employed almost exclusively in the measurement of C_P for solids between liquid helium temperatures and room temperature, and sometimes even at temperatures several hundreds of degrees above room temperature.* In its simplest form, illustrated in Fig. 5, the vacuum calorimeter consists of a metal chamber, which may be either highly evacuated or filled with helium gas at low pressure, immersed in one of several refrigerant baths, e.g., liquid H_2, N_2, etc. The specimen is suspended inside the chamber, usually by threads of low thermal conductivity material, and may be in thermal contact with the bath or well isolated from it depending upon whether the helium "transfer" gas is present or not. An electrical heater winding and an electrical resistance thermometer or thermocouple are attached to the specimen, often as parts of a detachable assembly whose heat capacity may be separately determined. Sometimes it is convenient to use one resistance as both heater and thermometer. In the most com-

* See Vol. 6, Chapter 5.1.

[59] R. Bunsen, *Ann. Physik* [2] **141,** 1 (1870); see also reference 6, pp. 46 and 563; W. Swietoslawski, "Microcalorimetry," Chapter 5. Reinhold, New York, 1946.

[60] J. Joly, *Proc. Roy. Soc.* **41,** 352 (1856); see also reference 6, p. 52.

[61] J. Dewar, *Proc. Roy. Soc.* **A76,** 325 (1905); **A89,** 158 (1913).

[62] W. Nernst, *Sitzber. deut. Akad. Wiss. Berlin,* pp. 262, 306 (1910); *Ann. Physik* [4] **36,** 395 (1911); A. Eucken, *Physik. Z.* **10,** 586 (1910).

[63] P. H. Keesom and N. Pearlman, *in* "Handbuch der Physik—Encyclopedia of Physics" (S. Flügge, ed.), Vol. 15, p. 282. Springer, Berlin, 1956.

[64] See reference 54, p. 760; also reference 8, p. 224.

mon mode of operation of the vacuum calorimeter, the specimen is first cooled to the temperature of the refrigerant bath via "transfer" gas, then isolated and a known amount of heat added electrically to produce a small temperature rise. This rise may perhaps be about 0.1°K in the helium range or one or two degrees at higher temperatures so that the

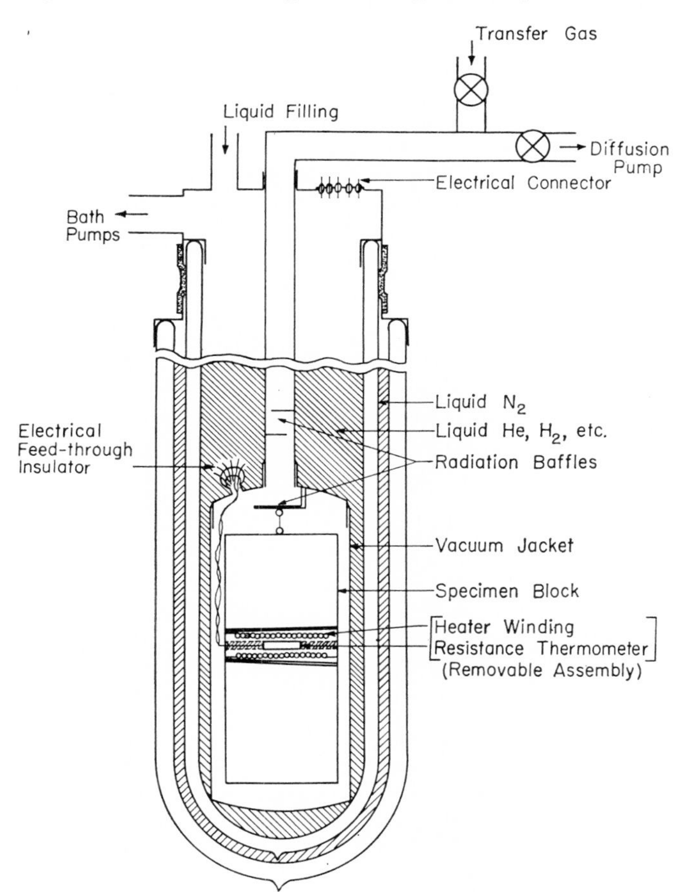

FIG. 5. Low temperature vacuum calorimeter.

measured average heat capacity will always closely approximate the true heat capacity at a given temperature. Often several successive heatings of the specimen for a fixed bath temperature may be carried out before excessive heat transfer between specimen and bath via electrical leads and residual gas makes determination of the true temperature rise difficult. If the temperature difference between specimen and vacuum chamber wall is kept small, and the pressure quite low it is not uncommon for the temperature to be essentially constant before and within a few minutes after the addition of heat. In this case the true temperature rise is easily deduced. Various graphical techniques have been proposed for obtaining this quantity even in the presence of rapid heat transfer.[65,66] The existence of such techniques, however, should not encourage the experimenter to feel that excessive heat transfer due to poor vacuum, for example, need be tolerated. Other methods for determining C_P with the vacuum calorimeter have been developed and are described in the literature.* [67]

The achievement of a high degree of thermal isolation of the specimen in the vacuum calorimeter presents numerous technical problems particularly in the liquid helium region. For example, the removal of "transfer" gas by pumping may be slow or incomplete below 4°K. This has led some workers to abandon transfer gas entirely in favor of mechanical devices[68] by means of which the specimen may be placed in direct contact with the vacuum chamber wall and so with the bath. Another problem is the proper trapping of thermal radiation coming down the pumping tube from warm portions of the system. Yet another is the problem of bringing electrical leads into the vacuum chamber and to the specimen so as to minimize heat exchange with the bath. Some workers bring these leads through vacuum tight feed-through insulators mounted in the wall of the vacuum chamber while others bring them down the pumping tube being careful to maintain thermal contact between bath and leads near the entrance to the chamber. Many of these problems can be more or less eliminated by surrounding the specimen with an electrically heated adiabatic screen to which all leads are attached thermally and which is kept automatically at the sample temperature.[69] Such devices have been included in many of the vacuum calorimeters used at hydrogen temperatures and above. The vacuum calorimeter has been refined to the point where C_P values accurate to 1 or 2% may be obtained at helium temperatures and an accuracy of perhaps 0.5 or 1% achieved at higher temperatures.

While the method of mixtures and the vacuum calorimeter are probably the most commonly used techniques in the accurate determination of

* See Vols. 3 and 6.

[65] W. H. Keesom and J. A. Kok, *Koninkl. Ned. Akad. Wetenschap., Proc.* **35,** 294 (1932); also *Leyden Rijksuniv., Kamerlingh Onnes Lab., Commun.* No. **219c.**

[66] J. S. Kouvel, *J. Appl. Phys.* **27,** 639 (1956).

[67] See for example reference 1, Chapter 4.

[68] K. G. Ramanathan and T. M. Srinivasan, *Phil. Mag.* [7] **46,** 338 (1955).

[69] F. Lange, *Z. physik. Chem.* **110,** 343 (1924); for additional references see reference 54, p. 761.

the specific heats of solids other, usually less accurate, often ingenious methods are available which may prove useful under special circumstances. Concise descriptions of several of these are given in the books of Saha[3] and Eucken.[53]

It should be borne in mind that each of the methods mentioned above yields essentially the specific heat at constant pressure C_P. For purposes of comparison with theory it is usually necessary to reduce the data to values of C_v. This is done by means of Eq. (6.2.2). Unfortunately, the thermal expansion coefficient β and the compressibility κ of most substances are not known over wide temperature ranges. There are, however, two approximate schemes which make the calculation possible without these complete data.[70,71] Equation (6.2.2) may be rewritten, using molar quantities, as

$$c_v = c_P(1 - Ac_PT) \qquad \text{where} \qquad A = \frac{\beta^2 v}{\kappa c_P{}^2}.$$

Now A has been shown theoretically and experimentally to be nearly independent of temperature so that only small errors will be introduced by using a value of A obtained from quantities measured, for example, at room temperature. When even these data are lacking A has often been estimated from the approximate relation shown by Nernst and Lindemann[72] to connect it and the melting point of the solid T_M, namely

$$A = \frac{0.0214}{T_M} \frac{\text{mol}}{\text{cal}}.$$

6.2.1.2. Liquids and the Mechanical Equivalent of Heat. Many of the techniques used in the measurement of the specific heats of solids may be adapted for use with liquids. A liquid not reacting chemically with water or some other appropriate calorimetric substance might be studied by the method of mixtures without any essential change in procedure. If, instead of being allowed to mix directly, the heated liquid is passed into a vessel immersed in the water calorimeter, the method of mixtures may be applied in the measurement of C_P for any liquid regardless of its reactivity. The observations are subject to the same errors and limitations as are those for solids.

The vacuum calorimeter has also been used extensively with liquids under pressure of their saturated vapors particularly those having very low boiling points including helium and hydrogen. The specimen is held in a capsule of known heat capacity carrying an electrical heater and a

[70] See reference 1, Chapter 13.

[71] See reference 53, p. 208.

[72] W. Nernst and F. A. Lindemann, *Z. Elektrochem.* **17,** 820 (1911).

thermometer and mounted in the vacuum chamber in the same manner as a solid sample. Often a capillary of low thermal conductivity material, e.g., glass or German silver, connects the capsule with the outside of the system to facilitate filling. As in the case of solids, the temperature changes of the liquid specimen in the vacuum calorimeter may be kept small so that the measured specific heats are very nearly the true values at the temperatures in question.

A very accurate method applicable to fluids in general and liquids in particular is that of the steady-flow electric calorimeter of Callendar and Barnes.[73,74] The apparatus, shown schematically in Fig. 6, consists essentially of a tube with an electrical heating element, usually of platinum, passing through it. The specimen substance flows through the tube at a

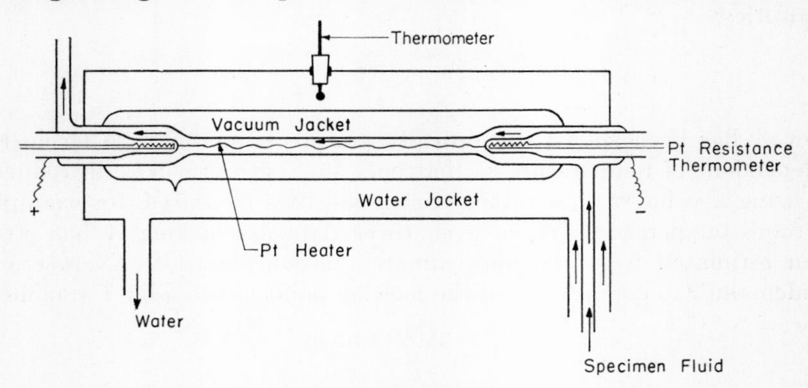

FIG. 6. Steady flow electric calorimeter.

constant rate and the steady temperature difference established between the entering and leaving liquid is measured with platinum resistance thermometers connected differentially as arms of a special bridge. Loss of heat is reduced by mounting the flow tube inside a water-thermostated vacuum jacket. As long as it is constant, this heat loss may at least in principle be accurately measured and eliminated by adjusting the heater current to produce the same temperature rise in two experiments in which the flow rates are different. The measured quantities in a given experiment are the mass M of the liquid flowing through the tube in a time t, the power dissipated in the heater, and the temperature rise ΔT. Combining the results of two experiments one then obtains a value of the average specific heat of the liquid over the interval ΔT. This interval may be kept as small as 2°C if, as in the original experiment, ΔT is meas-

[73] H. L. Callendar, *Phil. Trans. Roy. Soc. London, Ser.* **A199,** 55 (1902); H. T. Barnes, *ibid.* **199,** 149 (1902).

[74] See reference 54, p. 764.

ured accurately to 0.001°C. To achieve an accurate result care must be taken, (1) to maintain steady conditions during an experiment, (2) to mix the flowing liquid so that the temperature through any cross section of the tube is uniform. Requirement (2) may be met by using a spiral heater. The advantages of the method are first that the heat capacity of the calorimeter itself need not be known since all parts of it are at fixed temperatures throughout the measurement, and secondly, steady conditions insure adequate time for the accurate measurement of all quantities, there being no problem such as thermometric lag, etc.

In the Callendar and Barnes apparatus one measures the temperature rise accompanying the expenditure of a measured amount of electrical energy. If the liquid passing through the calorimeter is water and the temperature rises, for example, from 14.5° to 15.5°C, the specific heat being by definition 1 cal/gm, the experiment yields a value of the ratio of the calorie to the unit of energy, the joule. This ratio is the so-called mechanical equivalent of heat J. The existence of a unique value of J, or more precisely, the first law of thermodynamics, has been pre-supposed in earlier parts of this section.

It is appropriate at this point to mention some of the techniques employed in the determination of J since each, in principle, is also a method for measuring the specific heat of a liquid. Ideally, each of them involves the addition of just enough energy, measured in joules, to a given mass of water to cause its temperature to rise from 14.5° to 15.5°C, i.e., to produce the same effect as the addition of one 15° calorie per gram of water. The so-called direct methods accomplish this by the dissipation of mechanical energy. In his famous refinement of Joule's experiment, Rowland[75] rotated a set of paddles which meshed with vanes fixed to the inner wall of a water-filled cylindrical calorimeter. The mechanical work done was measured by observing the torque required to prevent rotation of the calorimeter about its axis and the number of revolutions of the paddles. Rowland found that at 15°C, $J = 4.188$ joule/cal. In addition he showed that if the specific heat of water be assumed equal to 1 cal/gm at all temperatures then J must vary, reaching a minimum value above 30°C. Accepting rather the invariance of J, he deduced for the first time the actual temperature variation of the specific heat of water between about 10° and 35°C. Direct determinations of J have been carried out by other investigators. The most accurate is that of Laby and Hercus[76] who employed an electromagnetic brake. They found $J = 4.186$ joules/cal at 15°C.

[75] H. Rowland, *Proc. Am. Acad. Arts Sci.* **7** (1880).

[76] T. H. Laby and E. O. Hercus, *Phil. Trans. Roy. Soc. London, Ser.* **A227,** 63 (1927); E. O. Hercus, *Proc. Phys. Soc.* (*London*) **48,** 282 (1936).

Techniques in which electrical energy is added to the water by dissipation in an immersed conductor are described as indirect. The continuous flow calorimeter of Callendar and Barnes described above was first used for such a measurement yielding $J = 4.182$ joules/cal. The temperature variation of the specific heat of water was also studied. An alternative indirect procedure consists of observing the temperature rise of a static mass of water accompanying the dissipation in it of a measured amount of electrical energy. Joule first used this procedure. Jaeger and Steinwehr[77] refined the method by using a large mass of water and never heating it very much above the temperature of its surroundings. In this way the calorimeter correction was kept very small and heat losses accurately accounted for. The small temperature change was, however, difficult to measure. The most recent and probably most accurate of all determinations of J is that of Osborne and co-workers.[78] By the static indirect technique they have found $J = 4.1858$ joules/cal as well as the temperature variation of the specific heat of water between 0° and 100°C. It should be noted that the energies obtained by electrical measurements in the indirect procedures are commonly expressed in international joules, units based upon the Weston standard cell of emf, and not in absolute joules, the units based upon the meter, kilogram, and second. The conversion factor relating international and absolute joules is a measured quantity, the currently accepted equivalence being

$$1 \text{ int joule} = 1.000165 \text{ abs joule.}$$

The results of indirect determinations of J quoted above have been converted to absolute joules to make them comparable with directly measured values.

6.2.1.3. Gases. Numerous methods are available for the measurement not only of C_P and C_v for gases but also of their ratio, $C_P/C_v = \gamma$. Unlike solids and liquids whose thermal expansion coefficients are usually small, gases do large amounts of work when heated at constant pressure, i.e., $C_P - C_v$ is large. The general expression for this difference may be written*

$$C_P - C_v = T\left(\frac{\partial P}{\partial T}\right)_V \left(\frac{\partial V}{\partial T}\right)_P.$$

When the equation of state for a gas is known the partial derivatives may be evaluated. Thus for one mole of an ideal gas

* See Section 6.4.3.3.4.

[77] W. Jaeger and H. von Steinwehr, *Ann. Physik* [4] **64,** 305 (1921).

[78] N. S. Osborne, N. F. Stimson, and D. C. Ginnings, *J. Research Natl. Bur. Standards* **23,** 197 (1939).

$$Pv = RT \qquad \text{and} \qquad c_p - c_v = R \simeq 2 \frac{\text{cal}}{\text{deg mole}}$$

indicating the magnitude of the difference between the quantities with which we shall be concerned. The equations of state of many gases have been studied in some detail over extended temperature ranges making possible the accurate conversion of one type of specific heat data into the other when necessary.

The following discussion is restricted to representative techniques for the measurement of C_P, C_v, and C_P/C_v. Extensive compilations of experimental methods are to be found in the treatises of Eucken[53] and Partington and Shilling[79] as well as the textbooks of Roberts[2] and Saha.[3]

6.2.1.3.1. Specific Heat at Constant Pressure. The specific heats of gases at constant pressure are measured either by a variant of the method of mixtures or by a continuous flow electrical technique. Regnault's experiments are typical of the application of the method of mixtures to gases.[80] The gas first flows at the desired pressure through a coiled tube immersed in a bath at a measured constant temperature. The gas emerges at that temperature and passes immediately into another coiled tube immersed in a water-filled calorimeter. While flowing through this coil, the gas gives up heat to the water raising its temperature and finally leaves, having come to the same temperature as the calorimeter. The quantity of gas flowing through the calorimeter during the experiment is measured by collecting it in a gasometer. Knowing the temperature drop experienced by this quantity of gas and the temperature rise of the calorimeter, C_P may then be calculated in the usual way. In addition to the correction for heat exchange with the surroundings required of most water calorimeters, a special correction is necessary in this case for heat flowing into the calorimeter via the tube bringing gas from the heating bath. An experiment in which no gas is passed through the system serves to determine the latter correction. Holborn and Henning[81] have employed a similar technique in measurements between 100° and 1400°C. Millar[82] has adapted the metal block calorimeter for measurements of this type in the range below room temperature.

A continuous flow method quite like that used with liquids by Callendar and Barnes (see Section 6.2.1) was first employed in the measurement of C_P for gases by Swann.[83] The technique is particularly well-

[79] J. R. Partington and W. G. Shilling, "The Specific Heats of Gases." Benn, London, 1924.

[80] H. Regnault, *Mém. acad. Sci. Paris* **26,** 1 (1862).

[81] L. Holborn and F. Henning, *Ann. Physik* [4] **23,** 809 (1907).

[82] R. W. Millar, *J. Am. Chem. Soc.* **45,** 874 (1923).

[83] W. F. G. Swann, *Proc. Roy. Soc.* **A82,** 147 (1909).

suited to the study of the temperature variation of C_P. Perhaps the most accurate measurements of this type are those of Scheel and Heuse[84] between $-180°$ and 0°C. Their apparatus employed platinum resistance thermometers in the measurement of the temperatures of the incoming and outgoing gas. It was so constructed that before passing through the tube containing the electrical heater, the gas stream flowed around this tube absorbing heat radiated from the heater which would otherwise have been lost. This arrangement, plus the almost complete enclosure of the system in a vacuum jacket immersed in a constant temperature bath served to reduce the heat losses considerably. As with liquids, repetition of the measurements with different flow rates permits correction for the remaining heat loss.

The continuous flow method has been used in the determination of C_P for gases at high pressures (up to 30 atmos) by Knoblauch and others.[85] A modification of the method by Blackett and associates[86] has been used in measurements accurate to about 1% up to 370°C and is presumably applicable at considerably higher temperatures. This technique actually avoids the difficulties associated with temperature measurement and heat loss encountered in the conventional procedure. Short discussions of this rather elaborate method are given by Roberts[2] and Saha.[3]

6.2.1.3.2. SPECIFIC HEAT AT CONSTANT VOLUME. The first accurate determinations of C_v for gases were those of Joly[87] who employed his steam calorimeter. In this method the temperature of a container filled with the gas is raised by condensing steam on it. The amount of added heat is determined by weighing the condensate. Among the many corrections applied to the observations are those for changes in the volume of the container due to the rise in temperature and gas pressure during an experiment. The actual apparatus employs two identical spheres, one empty, suspended from the arms of a balance. Steam is allowed to condense on both spheres so that the heat capacity of the container may be eliminated in a single differential measurement.

Eucken[88] has adapted the vacuum calorimeter for the measurement of C_v for gases. Its application is restricted to the low temperature region where the specific heats of metals are relatively small. The heat capacity of a specimen held in a metal capsule is then a reasonable fraction of the

[84] K. Scheel and W. Heuse, *Ann. Physik* [4] **37,** 79 (1912); [4] **40,** 473 (1913); [4] **95,** 86 (1919).

[85] O. Knoblauch and E. Raisch, *Z. Ver. deut. Ing.* **66,** 418 (1922); lists earlier references.

[86] P. M. S. Blackett, P. S. H. Henry, and E. K. Rideal, *Proc. Roy. Soc.* **A126,** 319 (1930).

[87] J. Joly, *Proc. Roy. Soc.* **A48,** 440 (1890).

[88] A. Eucken, *Sitzber. deut. Akad. Wiss. Berlin,* p. 141 (1912); p. 682 (1914); A. Eucken and K. Hiller, *Z. physik. Chem.* (*Leipzig*) **B4,** 142 (1929).

total heat capacity and may be accurately determined. Electrical heater windings and thermometer are attached to the capsule just as to a solid specimen, and the measurements are carried out by one of the several techniques used with such specimens. Values of C_v for helium and ortho- and parahydrogen mixtures have been determined by this method as functions of both temperature and pressure.

Values of C_v for gases at temperatures up to about 3000°C have been determined by the explosion method originally devised by Bunsen.[89] The gas under study is contained, at a temperature t_1, in a heavy-walled steel vessel together with a gaseous explosive mixture with which it does not react. All are at known partial pressures. The explosion is set off electrically and a quantity of heat Q is liberated raising the temperature of the gas and reaction products very rapidly to a maximum temperature t_2. The pressure inside the vessel is measured in one version of the apparatus[90] by observing the motion of a thin corrugated steel diaphragm closing a small aperture in the wall. Knowing the change in the number of molecules in the vessel and the ratio of maximum to initial pressure one can then calculate t_2. Using known values for the heat capacities of the reaction products and the heat of reaction, the specific heat of the specimen gas may be calculated. Corrections are necessary for heat loss to the walls of the vessel before the maximum temperature has been reached as well as for incomplete combustion and time lag in the manometric device. Measurements to check these corrections are usually made on a gas whose specific heat is known to be temperature independent in this region, e.g., argon. The accuracy of C_v values obtained by the explosion method has often been uncertain. Perhaps the most important limitation on the usefulness of some of the C_v results is the fact that they may represent averages over very large temperature intervals.

For many gases, the most reliable available C_v values are actually not those determined calorimetrically but rather those calculated from spectroscopic data. From the observed spectrum of a dilute gas it is possible to deduce at least part of the energy level scheme of an individual molecule belonging to that gas. Usually included are all those levels normally populated even at quite high temperatures. Knowing the spacing of the levels and their statistical weights one can then calculate the molecular partition function and from it the specific heat C_v, assuming that interactions between molecules can be neglected.[91] Modern recording

[89] R. Bunsen, *Ann. Physik* **131,** 161 (1867).

[90] M. Pier, *Z. physik. Chem. (Leipzig)* **66,** 795 (1909); see also R. W. Fenning and A. C. Whiffen, *Phil. Trans. Roy. Soc. London, Ser. A* **238,** 149 (1939).

[91] See for example J. Mayer and M. G. Mayer, "Statistical Mechanics." Wiley, New York, 1940.

spectrometers and high speed computing machines have combined to make the method much less laborious than most calorimetric procedures. Its results are particularly reliable at high temperatures, the region of great interest in chemical technology. Consequently, the heat capacities of a great many gaseous substances have been obtained in this way for use in thermochemical calculations of practical interest.

6.2.1.3.3. THE RATIO C_P/C_v. A useful type of thermal measurement on a gas is one in which the ratio $C_P/C_v = \gamma$ is directly measured. In these experiments the gas is allowed to undergo one or more adiabatic changes of state for whose description γ is an essential parameter. For an ideal gas experiencing an adiabatic change from state $(P_0V_0T_0)$ to a state (P,V,T) sufficiently close that C_v and C_P may be regarded as constant during the process one may show:

$$\frac{P}{P_0} = \left(\frac{V_0}{V}\right)^{\gamma} = \left(\frac{T}{T_0}\right)^{\gamma/\gamma-1}.$$

Analogous relations may be derived for adiabatic processes in real gases. These are generally rather complicated. Perhaps the simplest useful example is the relation obtained by Nernst[92] for a gas obeying Berthelot's equations of state, assuming γ constant and pressure changes small:

$$\left(\frac{P}{P_0}\right)^{(\gamma-1)/\gamma} \simeq \frac{T}{T_0}\left\{1 + \frac{27}{32}\frac{P + P_c}{2P_c}\left(\frac{2T_c}{T + T_c}\right)^3\right\}$$

where T_c and P_c are the critical temperature and pressure respectively. In principle, γ can be determined from either type of relation upon substitution of the measured values of the initial and final state variables.

The experimental methods used to obtain γ may be divided roughly into three groups depending upon whether, (1) only a single rapid adiabatic expansion or compression occurs, (2) rather slowly alternating adiabatic compressions and expansions of the specimen gas take place, or (3) rapidly alternating adiabatic expansions and compressions occur as in the passage of a sound wave through the gas. The earliest measurement of γ was that of Clément and Désormes[93] by means of a technique of the first type.

In a later single expansion procedure due to Lummer and Pringsheim[94] the temperature change accompanying an adiabatic expansion is followed by means of a sensitive, low heat capacity resistance thermometer (bolometer) suspended inside the large spherical flask from which a part

[92] W. Nernst, "Theoretical Chemistry," 5th ed., p. 259. Macmillan, London, 1923.
[93] M. Clément and C. B. Désormes, *J. phys. chim. hist. nat. et arts* **89,** 321, 428 (1819).
[94] O. Lummer and E. Pringsheim, *Ann. Physik* **64,** 555 (1898).

of the contained gas is allowed to escape. If the temperature and pressure before expansion are P_1 and T_1 and the temperature is T_0 at a pressure of P_0 then $T_1{}^\gamma/P_1^{\gamma-1} = T_0{}^\gamma/P_0^{\gamma-1}$ and $\gamma = 1\Big/\left(1 - \frac{\log T_1 - \log T_0}{\log P_1 - \log P_0}\right)$ for an ideal gas.

Partington[95] further improved the Lummer–Pringsheim method by, among other things, increasing the speed of response of the thermometer so that temperature variations during the pressure oscillations accompanying the expansion could be followed. The aperture through which the expansion occurred could then be reduced to the point where oscillations did not take place. If used near room temperature, where cooling corrections are negligible, Partington's method yields quite accurate results. He found, for example, $\gamma = 1.4034$ for air at 17°C, essentially the currently accepted value.

Most of the methods of measuring γ by carrying the gas cyclicly through relatively slow expansions and compressions are intrinsically inaccurate because adiabatic conditions cannot be maintained. Among the simplest of these is the method due to Rüchardt[96] in which a close-fitting steel ball is allowed to oscillate freely in a vertical tube attached to the top of a gas-filled flask. The period of oscillation is measured and γ calculated on the assumption that the pressure changes in the flask are adiabatic, a condition only partially fulfilled. Clark and Katz,[97] however, have succeeded in measuring γ with high accuracy in an apparatus very similar in principle. The gas is held in two vessels at either end of a cylinder in which a steel piston is forced to oscillate by magnetic fields provided by external coils. The frequency of forced oscillation is varied by changing the frequency of the emf applied to the coils, so that the resonant frequency of the system may be determined. From this frequency and constants of the apparatus γ may be calculated. The technique has been used at pressures up to 25 atmos. At 23°C, Clark and Katz find for helium, extrapolating to zero pressure, $\gamma = 1.6669$ compared with the theoretical value of 1.6667.

The last group of methods for measuring γ is that involving the measurement of the velocity of propagation of sound in a gas. It may be shown that in general the velocity of sound in an extended gas is[98] $u = \sqrt{\gamma(\partial P/\partial \rho)_T}$ where ρ is the density. The derivative may be evaluated if the equation of state of the gas is known. Since for an ideal gas

[95] J. R. Partington, *Proc. Roy. Soc.* **A100,** 27 (1921).

[96] E. Rüchardt, *Physik. Z.* **30,** 57 (1929).

[97] A. L. Clark and L. Katz, *Can. J. Research* **A18,** 23 (1940); **A19,** 111 (1941); **A21,** 1 (1943).

[98] G. Joos, "Theoretical Physics," 2nd ed., p. 210. Blackie, London, 1951.

$P = \rho RT/M$ where M is the molecular weight, we get the familiar relation $u = \sqrt{\gamma RT/M}$. Measurements of u and T in this case are sufficient to determine γ. As an example of a formula applicable to a slightly non-ideal gas we may quote the result obtained using Berthelot's equation of state[99]

$$u^2 = \frac{\gamma RT}{M}\left\{1 + \frac{9}{64}\frac{P}{P_c}\frac{T_c}{T}\left(1 - 6\frac{T_c^2}{T^2}\right)\right\}.$$

In this case we must also measure the pressure of the gas and know its critical pressure and temperature in order to determine γ.

It should be emphasized that the equations for the sound velocity given above refer to an unconfined gas. Only for air has it been possible to carry out measurements under conditions where they are strictly valid. Hebb[100] performed such a measurement using a technique in which sound of known frequency was reflected from paraboloidal mirrors 5 ft apart. His results for dry air at 0°C are $u_a = 331.41$ meter/sec, essentially the same as the best recent value 331.36 meter/sec. Most sound velocity determinations for gases other than air have been carried out in tubes. In many cases, they have been relative measurements relying ultimately on the value for air quoted above.

The observed velocity of sound in a gas confined in a tube, u' differs from that for a free gas according to the relation $u' = u(1 - kc)$ where k is a constant depending upon the properties of the tube and on the frequency of the sound, and c depends upon the properties of the gas. Kirchhoff showed that[101]

$$c = \sqrt{\frac{\eta}{\rho}}\left\{1 + \sqrt{\frac{K}{\eta c_v}}\left(\frac{\gamma - 1}{\gamma}\right)\right\}$$

where η is the viscosity of the gas and K its thermal conductivity. Using dry air for which u is known and c may be calculated, k for a particular apparatus and frequency may be determined by a measurement of u'.

The earliest relative determinations of sound velocities in tubes are those of Kundt.[102] His apparatus consists of two similar glass tubes each of which has a movable plunger in one end so that its effective length may be varied. Standing sound waves are set up in a tube by means of a piston in the other end driven by a metal rod vibrating longitudinally. One sounding rod with a piston on either end excites waves of identical frequency in the two tubes, one filled with specimen gas, the other with air. The effective lengths of the tubes are varied until each just equals an

[99] See reference 2, p. 185; also reference 3, p. 195.
[100] T. C. Hebb, *Phys. Rev.* **20,** 89 (1905); [2] **14,** 74 (1919).
[101] G. Kirchhoff, *Ann. Physik* [2] **134,** 177 (1868).
[102] E. Kundt, *Ann. Physik* [2] **127,** 497 (1866).

integral number of half wavelengths of the sound in the gas it contains. Attainment of the standing wave pattern is observed by means of a fine powder distributed along the tube which collects in small heaps at the displacement nodes. The wavelength λ is just twice the distance between successive nodes. Since the frequency of the sound is the same in both tubes, we have $u_g'/u_a' = \lambda_g/\lambda_a$, where g refers to the specimen gas and a refers to air. Correcting the observations as indicated above one obtains u_g and from it γ for the gas. Behn and Geiger[103] have improved Kundt's method and their technique has been employed by other investigators, particularly Partington and Shilling.[104]

Resonating gas columns form the basis of many newer techniques for measuring sound velocity in gases. Kundt's sounding rod and dust pattern detector have given way to more modern devices. Partington and Shilling[105] developed an apparatus in which a telephone receiver driven by an audio oscillator provides a continuous source of audible sound and resonance is detected by ear. The frequency f_g is accurately known and the wavelength λ_g is determined from the length of the resonant column. Thus $u_g' = f_g\lambda_g$. The method has been used at temperatures between 90° and 1273°K. Pierce and others[106] have developed ultrasonic techniques in which the sound is radiated from piezoelectric-quartz plates connected in electronic oscillator circuits. The sound is reflected from a movable mirror. When the gas column between source and mirror is of resonant length the amplitude of the oscillations of the crystal is a sharp maximum. This condition is observed by measuring electrically the accompanying changes in the characteristics of the driving circuit. A very accurate measure of the wavelength is obtained, permitting the velocity to be calculated. An attractive feature of the technique is that it also is capable of providing attenuation data of considerable intrinsic interest. The method has been employed extensively at low temperatures by van Itterbeek and others.[107] A low temperature acoustic interferometer is shown schematically in Fig. 7.

Probably the most refined methods of measuring the velocity of sound in fluids including gases are the ultrasonic pulse techniques.* Pulses of

* See Vol. 3, Chapter 8.2.

[103] U. Behn and H. Geiger, *Verhandl. deut. physik. Ges.* **9,** 657 (1907).

[104] J. R. Partington and W. G. Shilling, *Phil. Mag.* [6] **45,** 416 (1923).

[105] J. R. Partington and W. G. Shilling, *Trans. Faraday Soc.* **18,** 386 (1923).

[106] G. W. Pierce, *Proc. Am. Acad. Arts. Sci.* **60,** 271 (1925). See also review articles by W. T. Richards, *Revs. Modern Phys.* **11,** 36 (1939); C. Kittel, *Repts. Progr. Phys.* **11,** 205 (1946–1947); J. J. Markham, R. T. Beyer, and R. B. Lindsay, *Revs. Modern Phys.* **23,** 353 (1951).

[107] A. van Itterbeek, *in* "Progress in Low Temperature Physics" (C. J. Gorter, ed.), Vol. 1, Chapter 17. Interscience, New York, 1955.

ultrasound (frequencies as high as 10^8 cycles/sec) lasting about 1 μsec are sent through the gas from a quartz transducer. These are reflected from a stationary mirror and finally received either by the emitter or a second quartz crystal having traversed a path of known length. The

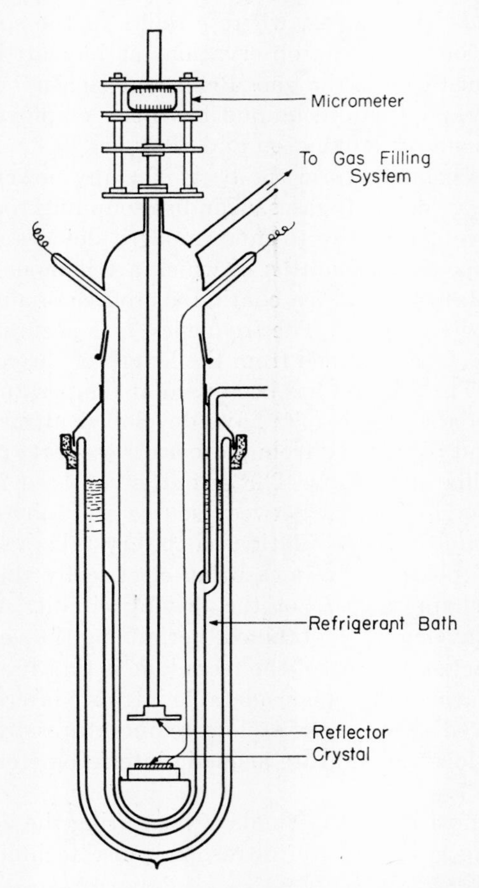

FIG. 7. Low temperature acoustic interferometer.

time delay between the initial pulse and the echo may be very accurately measured, yielding the sound velocity directly. Measurements of this type on gases have been used primarily in the study of attenuation and dispersion.[108]

[108] J. R. Pellam and J. K. Galt, *J. Chem. Phys.* **14,** 608 (1946); see also J. J. Markham, R. T. Beyer, and R. B. Lindsay, *Revs. Modern Phys.* **23,** 353 (1951).

6.2.2. Heats of Transformation and Reaction[7,53]

6.2.2.1. Latent Heats of Transformation. The common phase transformations: fusion, vaporization, sublimation, polymorphic change, all occur at constant temperature and pressure and are characterized by discontinuous changes in the specific volume of the substance and the absorption of a latent heat. Latent heats are either measured directly by calorimetric procedures or calculated from pressure–temperature data describing the states in which the two phases in question coexist in equilibrium. The latter indirect method is based upon the well-known Clausius–Clapeyron equation* relating the slope of a two-phase equilibrium curve in the P,T plane to the specific volume and entropy changes accompanying a transformation at a point on that curve:

$$\frac{dP}{dT} = \frac{\Delta s}{\Delta v} = \frac{l}{T\,\Delta v}.$$

Knowing, for example, the slope of the sublimation curve of a substance and the change in its specific volume at a particular sublimation temperature, one can easily calculate the corresponding latent heat of sublimation. The problem of determining latent heats is thus reduced to one of constructing, by pressure and temperature measurements, the relevant portions of the equilibrium phase diagram of the substance in question, and of measuring differences in density between adjacent phases. Description of such measurements are given, for example, in the books of Roberts, Worthing and Halliday, etc.[2,4]

The indirect method is used extensively in determining latent heats of vaporization and particularly latent heats of sublimation. In the case of sublimation and of vaporization well below the critical point, the specific volume of the vapor is much larger than that of the condensed phase. In addition, the vapor is often nearly an ideal gas. When these conditions exist the Clausius equation assumes a simple form. The latent heat of sublimation becomes, for example,

$$l_{\text{sub}} \simeq \frac{RT^2}{P}\frac{dP}{dT} = -R\,\frac{d\ln P}{d(1/T)}. \tag{6.2.3}$$

Thus a plot of the logarithm of the vapor pressure of a solid versus $1/T$ may be differentiated directly to give the latent heat of sublimation. For most solids the vapor pressure becomes so small at low temperatures that data exist only over rather limited high temperature intervals. Within such a range l_{sub} may not vary much, i.e., $\ln P$ versus $1/T$ will be a straight line whose slope may be determined even with rather meager data. The

* See Section 6.4, Eq. (6.4.28).

temperature variation of the latent heat of sublimation in the region in which vapor pressure data do not exist may be calculated by means of Kirchhoff's equation if the specific heats of both the solid and vapor are known as functions of the temperature.

Direct methods for determining latent heats of transformation utilize the same basic principles employed in heat capacity measurements. For example, the latent heat of vaporization (or condensation) of substances such as water or ethyl alcohol at their normal boiling points can be measured with reasonable accuracy by means of a water calorimeter and the method of mixtures. The technique was first employed by Berthelot.[109] Vapor at the boiling temperature is transferred from an external boiler into a condenser vessel immersed in the calorimeter and the temperature rise of the system measured. Knowing the mass of the condensate, its heat capacity, and the heat capacity of the calorimeter and condenser, one can easily calculate the latent heat released. Refinements of this simple method usually involve precautions to ensure that the vapor is not superheated and that it condenses only where its latent heat will be taken up by the calorimeter.

Awberry and Griffiths[110] have measured heats of vaporization by a technique in which vapor is continuously condensed in a tube cooled by water flowing in a surrounding jacket. The steady-state temperature rise and flow rate of the water, the specific heat, mass, and final temperature of the condensate are known or measured so that the latent heat is readily calculated. The method yields results of greater accuracy than does Berthelot's simpler procedure.

Techniques in which measured amounts of heat supplied electrically are used to vaporize (or sublime) measured quantities of condensed material are applicable over the widest range of temperatures and are probably the most commonly employed in accurate work. This method is well illustrated by the low temperature measurements of Simon and Lange[111] on liquid hydrogen (nbp $\sim$21°K). A metal capsule with an electrical heater winding holds the liquid and is mounted inside a vacuum chamber. A capillary of low thermal conductivity alloy connects the capsule with the external filling and measuring system. When a current is passed through the heater, liquid in the capsule vaporizes. Constant pressure during evaporation is maintained by controlling the flow of gas from the capillary and the quantity of gas evolved is measured at room temperature in auxiliary apparatus. Ideally, the latent heat of this mass of hydrogen is just the energy dissipated in the heater. In practice, corrections may

[109] D. Berthelot, *J. physique* **6,** 337 (1877).

[110] J. H. Awberry and E. Griffiths, *Proc. Phys. Soc.* (*London*) **36,** 303 (1924).

[111] F. E. Simon and F. Lange, *Z. Physik* **15,** 312 (1923).

be necessary either because heat leaks to the capsule from its surroundings, if these are at a higher temperature, or because some heat flows from the heater to the bath. Both types of heat leakage may be measured in separate experiments and can be kept small. Electrical heating methods used at room temperature and above are not usually able to exploit the almost ideal thermal isolation provided by the vacuum calorimeter at low temperatures and so tend to be rather elaborate.

Heats of fusion of substances melting at room temperature and above have been measured most accurately by the method of mixtures. The measurement consists essentially of transferring the molten specimen (at a known temperature) to a water or ice filled calorimeter and noting the temperature rise of the latter. Many refinements have been introduced particularly by Awberry and Griffiths[112] in their work on metals.

Latent heats of fusion at low temperatures are measured by the electrical heating method in the vacuum calorimeter.[113]

Latent heats of transformation between two solid phases of a given substance have been measured either by the method of mixtures or the vacuum calorimeter method depending upon the temperature at which the transformation occurs.

For full discussions of both direct and indirect procedures for latent heat determination the reader is referred to the books of Eucken[53] and Roberts[2] and the article of Sturtevant.[54]

6.2.2.2. Heats of Chemical Reaction. The heat evolved or absorbed in a given chemical reaction may be determined either directly by a calorimetric procedure or indirectly by thermodynamic calculation using either data on other reactions or the results of various noncalorimetric observations. Many methods of both types appropriate to particular kinds of reactions are described by Eucken[53] and Sturtevant.[54] We shall mention here only a few examples from each group.

The water calorimeter is a basic instrument in many direct techniques. An appropriate vessel containing the reacting substances is first allowed to come into thermal equilibrium with the calorimeter and then the reaction is initiated. The temperature change of the calorimeter accompanying the reaction is measured. Knowing the heat capacities of the apparatus and of the reacting substances, one can calculate the heat of reactions. If, for example, the reactants were in liquid solution the reaction vessel might consist of two chambers one above the other connected by a tube with a remotely operated valve which when opened would allow the solutions to mix. Such a device could also be used when one of the reactants

[112] J. H. Awberry and E. Griffiths, *Proc. Phys. Soc.* (*London*) **38,** 378 (1926).

[113] See reference 111; see also J. S. Dugdale and F. E. Simon, *Proc. Roy. Soc.* **A219,** 291 (1953).

was either liquid or in liquid solution and the other a solid or even a gas. Venting the reaction vessel to the atmosphere permits the reaction to take place at constant pressure. Any gaseous reaction products would be allowed to leave via a coiled tube also immersed in the calorimeter to insure that they came to the same final temperature as the rest of the system.

The heats of combustion of gases, liquids, and solids at nearly constant volume are determined by using as a reaction vessel a heavy walled bomb

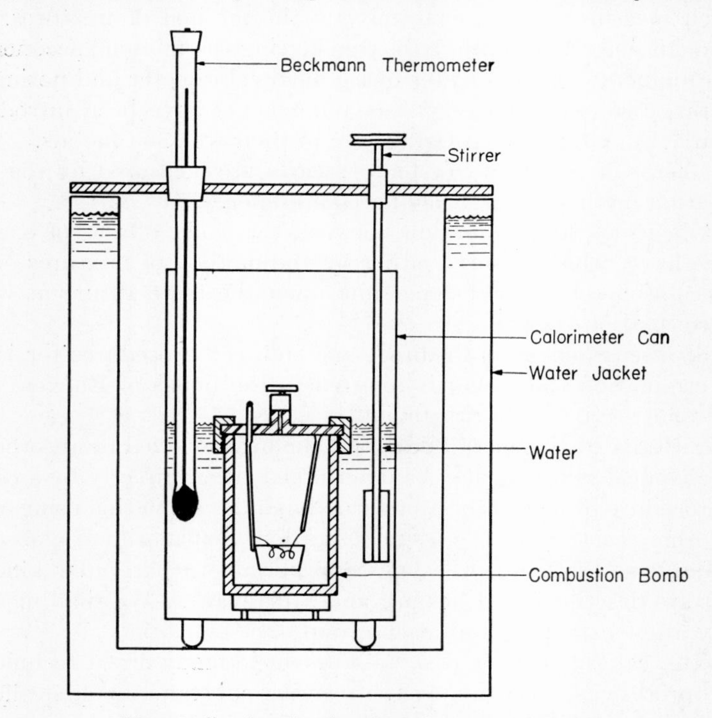

FIG. 8. Combustion bomb and water calorimeter.

of steel or a corrosion resistant alloy. The bomb, containing the specimen and more than enough oxygen for complete combustion, is immersed, as before, in a water calorimeter (see Fig. 8). Closing an electrical circuit causes a small piece of wire inside the bomb to fuse and the reaction to begin. The heat capacity of the apparatus may be determined by carrying out an experiment with a known amount of a standard specimen substance, e.g., benzoic acid[114] or by an electrical heating procedure.

[114] W. Jaeger and H. von Steinwehr, *Z. physik. Chem.* (*Leipzig*) **135,** 305 (1928); see, however, W. A. Roth and G. Becker, *ibid.* **179,** 450 (1937).

Heats of combustion of gases at constant pressure are commonly measured in a continuous-flow calorimeter.[115] The specimen gas is fed at a measured rate into a burner located inside a cooling jacket. The flowrate of the cooling water through the jacket and its steady-state temperature rise are measured and the heat of combustion easily calculated. The reaction products leave the jacket by a tortuous path which insures complete transfer of the heat of combustion to the cooling water. This is checked by measuring the temperature of the emergent products. Completeness of combustion is sometimes determined by collecting and measuring the amount of one of the reaction products. As in other types of continuous flow calorimetry the maintenance of steady state conditions is essential in obtaining accurate results.

It should be noted that the heats of reaction in many cases are quite small. This is often true particularly for heats of solution or dilution. Conventional apparatus often proves too insensitive for measuring these quantities and consequently special "microcalorimetric" techniques, many of a differential nature have been developed. Many of these are described by Swietoslawski[116] and Sturtevant.[54]

Perhaps the most common indirect technique for the determination of a heat of reaction is the algebraic addition of the heats of a series of reactions which connect the same initial and final states as does the reaction in question.[117] The possibility of such a calculation is a simple consequence of the first law of thermodynamics according to which the internal energy U and enthalpy H of a system are functions only of its state. Thus all processes connecting the same two states of a system are characterized by the same values of ΔU and ΔH, the heats of reaction at constant volume and pressure respectively. As a simple example of the usefulness of this procedure we might consider the reaction

$$\text{C (graphite)} + \tfrac{1}{2}\text{O}_2 \text{ (gas)} \rightarrow \text{CO (gas)}$$

whose heat of reaction is almost impossible to determine accurately by direct calorimetry. The heats of two related reactions are, however, well known, namely,

$$\begin{aligned} \text{C (graphite)} + \text{O}_2 \text{ (g)} &\rightarrow \text{CO}_2 \text{ (g)} \qquad \Delta H = -393500 \text{ joules} \\ \text{CO (g)} + \tfrac{1}{2}\text{O}_2 \text{ (g)} &\rightarrow \text{CO}_2 \text{ (g)} \qquad \Delta H = -283000 \text{ joules.} \end{aligned}$$

Subtracting the second from the first we get just the desired reaction and for it $\Delta H = -110500$ joules.

[115] See reference 54; see, however, G. Pilcher and L. E. Sutton, *Phil. Trans. Roy. Soc. London, Ser. A* **248,** 23 (1955).

[116] W. Swietoslawski, "Microcalorimetry." Reinhold, New York, 1946.

[117] W. J. Moore, "Physical Chemistry," 2nd ed., p. 39. Prentice-Hall, New York, 1956.

As an example of an indirect method based entirely on noncalorimetric observations we may mention the use of the reversible chemical cell. This is a device which enables the energy made available in a given reaction to be used in moving electric charge through a circuit. If the reaction is carried out reversibly at constant temperature and pressure this energy is the decrease in the Gibbs function of the system

$$-\Delta G = -\Delta H + T\,\Delta S = -\Delta H + T\left(\frac{\partial\,\Delta G}{\partial T}\right)_P.$$

Per mole of reactant consumed, this is just $nF\epsilon$ where n is the valence change, F is the Faraday (96500 coulombs), and ϵ is the reversible emf of the cell. ϵ is measured by means of a potentiometer which, when balanced, allows no current to flow between the terminals of the cell so that it operates reversibly. Substituting $-\Delta G = nF\epsilon$ we obtain for the heat of reaction

$$\Delta H = -nF\left(\epsilon - T\frac{d\epsilon}{dT}\right).$$

Thus determination of ΔH requires the measurement of ϵ and also its temperature coefficient $d\epsilon/dT$. A common laboratory cell (Clark's cell) employs the reaction $Zn + Hg_2SO_4 = ZnSO_4 + 2Hg$ for which $n = 2$. At 15°C one finds for this cell $\epsilon = 1.4324$ volts and

$$\frac{d\epsilon}{dT} = -0.00119 \text{ volts/deg}$$

so that

$$\Delta H = -2 \times 96519(1.4324 + 288 \times 0.00119) = -342710 \text{ joules.}$$

Calorimetric methods give for this reaction $\Delta H = -339500$ joules with an accuracy much less than that of the electrical method.

The detailed construction of a cell depends upon the type of reaction being studied.[118] Many varieties are in common use and constitute probably the most accurate sources of heats of reactions in a great number of cases.

[118] See reference 117, p. 473ff.

6.3. Heat Transfer[1,2,7,8,119,120]

Heat may be transferred from one body at a given temperature to another at a lower temperature by one or more of the following processes: (1) conduction, the passage of heat through and by means of intervening matter without any evident motion of that matter; (2) convection, in which bulk matter, having been heated by contact with the warm body, moves to the cooler body and gives up heat to it; (3) radiation, in which energy is transmitted through the intervening space without the necessary presence of matter.

In the three sections that follow we shall be concerned with the measurement of the properties of matter which determine the rate at which heat is transferred either through, by, from, or to various substances. Only those properties uniquely identified with heat transport will be considered in any detail.

6.3.1. Conduction

6.3.1.1. Solids. The establishment of a temperature difference between different parts of a conducting medium is opposed by the flow of heat. Quantitative measurement has shown that the amount of heat per second crossing a unit area perpendicular to the direction of flow at a point in an isotropic conductor is proportional to minus the temperature gradient at that point. Thus we may write $\dot{\mathbf{q}} = -K\nabla T$ where $\dot{\mathbf{q}}$ is the heat current density, ∇T the temperature gradient, and K a positive scalar quantity, the thermal conductivity. The quantity K is found to depend upon temperature and other parameters. In metals, for example, it is affected by applied magnetic fields.

In crystalline media, $\dot{\mathbf{q}}$ is not in general parallel to ∇T and it is necessary to write[121]

$$\dot{q}_i = -\sum_{j=1}^{3} K_{ij} \frac{\partial T}{\partial X_j} \tag{6.3.1}$$

where $i = 1, 2, 3$. The coefficients K_{ij} are components of the second rank

[119] O. Knoblauch and H. Reiher, *in* "Handbuch der Experimentalphysik" (W. Wien and F. Harms, eds.), Vol. 9, Part 1. Akademische Verlagsges., Leipzig, 1929; also C. Fabry, "Propagation de la Chaleur." Armand Colin, Paris, 1949.

[120] Engineering texts on heat transfer, for example, M. Jakob and G. A. Hawkins, "Elements of Heat Transfer and Insulation," 2nd ed. Wiley, New York (1950); or M. Jakob, "Heat Transfer," Vols. 1 and 2. Wiley, New York.

[121] J. F. Nye, "Physical Properties of Crystals," p. 195. Oxford Univ. Press, London and New York, 1957.

thermal conductivity tensor which may be shown to be symmetrical, i.e., $K_{ij} = K_{ji}$. It is always possible to choose axes in such a way that only the three diagonal components of the conductivity tensor are different from zero. In cubic crystals all three of these so-called principal conductivities will be the same so that, as with amorphous or polycrystalline conductors, one measurement suffices to fix the thermal conductivity for given values of the temperature and external parameters. For crystals of lower symmetry, as many as three independent measurements may be required to provide the same information. Scalar conductivities and tensor conductivity components alike are commonly expressed in units such as cal/cm sec °C or watts/cm °C.

The requirement of energy conservation leads to an equation of continuity for heat flow.[122] It is usually written $c\rho(\partial T/\partial t) = -\boldsymbol{\nabla} \cdot \dot{\mathbf{q}} + P$ where c is the specific heat of the conductor, ρ its density, and P the rate of heat production per unit volume. Let us consider an isotropic conductor for which $\dot{\mathbf{q}} = -K\boldsymbol{\nabla}T$. Substituting this expression for $\dot{\mathbf{q}}$ into the equation of continuity we get the so-called equation of heat flow

$$c\rho(\partial T/\partial t) = K\nabla^2 T + P. \tag{6.3.2}$$

At a point where heat is not being produced this reduces to

$$\nabla^2 T = \frac{c\rho}{K}\frac{\partial T}{\partial t}.$$

This is a partial differential equation identical in form with that describing the diffusion of matter. Heat flow is thus often referred to as a diffusion process, the ratio $K/c\rho \equiv D$ being called the thermal diffusivity. Applying appropriate boundary conditions, the heat flow equation may be solved to give the temperature distribution within a conducting medium and its time variation. General methods of solving this equation and many solutions of specific problems in heat flow are given in the treatises of Carslaw and Jaeger[123] and Ingersoll *et al.*[124]

Thermal conductivities of solids are measured either by steady state or dynamical methods. In experiments of the first type a nonuniform temperature distribution within the specimen is kept stationary by a measured heat input. Appropriately placed thermometers determine enough of the temperature distribution so that the temperature gradient

[122] J. C. Slater and N. H. Frank, "Introduction to Theoretical Physics," Chapter 18. McGraw-Hill, New York, 1933.

[123] H. S. Carslaw and J. C. Jaeger, "Conduction of Heat in Solids." Oxford Univ. Press, London and New York, 1948.

[124] L. R. Ingersoll, O. J. Zobel, and A. C. Ingersoll, "Heat Conduction." University of Wisconsin Press, Madison, Wisconsin, 1948.

may be calculated at a point for which the heat current density is also easily computed. The thermal conductivity is then, by definition, $K = -(\dot{\mathbf{q}}/\nabla T)$. In the dynamical experiments, the temperature distribution in the specimen is caused to vary with time, often periodically. The thermal diffusivity is determined from measurements of the temperature as a function of position and time according to the heat flow equation. Although heat current densities need not be measured in these techniques, it is necessary to know the specific heat of the substance and its mass density in order to determine the conductivity K.

Let us consider first what is, in principle at least, the simplest steady state method. A specimen in the form of a long solid cylinder of uniform cross section A is heated at a constant rate at one end. The other end is kept at a fixed temperature through contact with a flowing coolant or a two-phase system such as ice water. Steady state conditions are reached quickly in good conductors and the temperature distribution is given by Laplace's equation $\nabla^2 T = 0$. If losses across the surface of the rod are negligible, heat flows parallel to the specimen axis only and the problem is one dimensional, i.e., $d^2T/dX^2 = 0$ so that $T = ax + b$ where a and b are constants of integration. Thus the temperature varies linearly along the bar so that the gradient is simply determined by measuring the temperature difference ΔT between two points a measured distance ΔX apart. If $\dot{Q}$ is the total amount of heat entering or leaving the bar per second then the thermal conductivity $K = -(\dot{Q}/A)(\Delta X/\Delta T)$. A common apparatus of this type due to Searle[125] is well suited to measurements on high conductivity materials near room temperature, e.g., copper for which $K \simeq 0.9$ cal/cm sec °C at 20°C. Lateral heat flow is minimized by covering the bar with low conductivity material such as asbestos lagging ($K \simeq 4 \times 10^{-4}$ cal/cm sec °C). Heat is supplied by a steam jacket or electrical heater and is absorbed by water flowing through a cooling coil or jacket. The heat current through the bar is measured by observing the flow rate and steady state temperature rise of the cooling water. The temperature gradient is determined by means of two or more mercury-in-glass thermometers or, preferably, thermocouples inserted into small holes drilled into the bar.

The essential idea of the simple experiment just described has been preserved in the techniques used to measure thermal conductivity and its temperature variation in solids over the range <1°K to room temperature. Lees[126] pioneered this important development using small specimen rods ($\sim$0.6 cm diam and $\sim$8 cm long), electric heater windings, and

[125] G. F. C. Searle, "Experimental Physics," p. 273. Cambridge Univ. Press, London, 1934.

[126] C. H. Lees, *Phil. Trans. Roy. Soc. London, Ser. A* **208,** 381 (1908).

platinum resistance thermometers. His apparatus was housed in a Dewar vessel so that it could be cooled to liquid air temperatures (~90°K). The technique was brought essentially into its present form by Grüneisen and Goens[127] who enclosed the specimen rod in a high vacuum jacket.

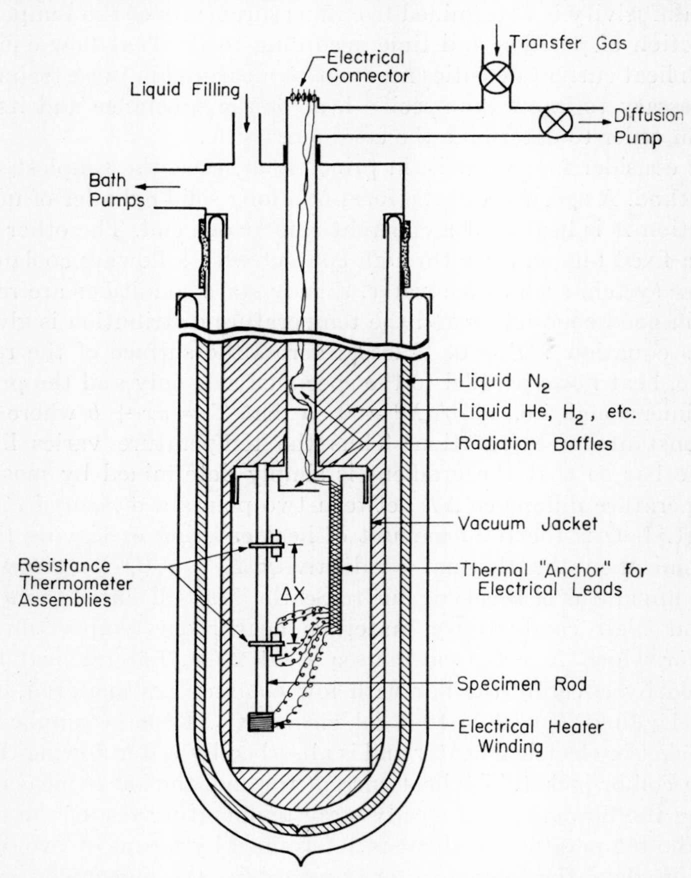

FIG. 9. Apparatus for measuring thermal conductivities of solids at low temperatures.

In one common type of apparatus[128] (see Fig. 9), one end of the specimen rod is soldered or clamped to the wall of the metal jacket and is thus kept in good thermal contact with the surrounding refrigerant (liquid helium,

[127] E. Grüneisen and E. Goens, *Z. Physik* **44,** 615 (1927).

[128] P. G. Klemens, "Handbuch der Physik—Encyclopedia of Physics" (S. Flügge, ed.), Vol. 14, pp. 189–281. Springer, Berlin, 1956. Full list of references included in this survey of theory; see also reference 8 for useful summary.

hydrogen, etc.). An electrical heating coil is attached to the other end. The temperature gradient in the rod is measured by means of thermometers fastened at two points along the rod with lacquer, solder, or small clamps. The thermometers in common use include thermocouples, platinum or carbon resistances, and miniature helium gas thermometers. The thermometer leads or capillaries and the heater leads are carried out of the vacuum can either via the pumping tube or directly through the walls. Maintenance of high vacuum ($<10^{-6}$ mm Hg) around the specimen insures negligible heat loss from its surface due to gaseous conduction or convection. The temperature gradient in the rod is usually small so that its mean temperature is very close to that of the jacket. Thus at low temperatures radiation losses can also be kept negligible. Perhaps the most important leakage path between specimen and bath is along the heater and thermometer connections. Heat transfer by this means can be minimized by using leads of low thermal conductivity alloy, care being taken to insure that electrical power dissipation in them is not significant. The thermal isolation of the specimen is generally complete enough so that the heat current through the rod is accurately calculated by dividing the power dissipated in the heater by the cross sectional area of the rod so that again $K = -(\dot{Q}/A)(\Delta X/\Delta T)$. This technique has been employed successfully not only with high conductivity specimens such as pure metals but also with poor conductors such as disordered alloys or glasses. In this sense it is much more flexible than the room temperature procedures of Searle and others.

It should be noted that oriented single crystals are often conveniently grown in the form of small rods suitable for measurement in the manner just described. The method also lends itself to measurements in a magnetic field. The specimen, vacuum can, refrigerant bath, and enclosing Dewar are readily mounted in a solenoid or in the gap of an iron-core electromagnet depending upon the field strength and orientation desired. The review articles of Olsen and Rosenberg,[129] Berman,[130] and Klemens[128] should be consulted for further information about low temperature heat conduction measurements and their interpretation.

Steady state measurements on specimens having other than rod-like geometry are often made at ordinary temperatures, particularly for low conductivity substances. One common technique employs a slab of sample material having plane parallel faces. One face is heated uniformly and the other maintained at a uniform lower temperature. Between a limited portion (area A) of the hot surface near its center and the corre-

[129] J. L. Olsen and H. M. Rosenberg, *Advances in Phys.* (*Phil. Mag. Quart. Suppl.*) **2,** 28 (1953).

[130] R. Berman, *Advances in Phys.* (*Phil. Mag. Quart. Suppl.*) **2,** 103 (1953).

sponding portion of the cooled surface heat flow will be one dimensional, the surrounding material providing a so-called "guard ring." The heat current density $\dot{Q}/A$ in the useful cylindrical part of the slab is determined much as in Searle's experiment. Two thermometers imbedded well within this cylindrical zone measure the temperature gradient. The conductivity, as before, is $K = -(\dot{Q}/A)(\Delta X/\Delta T)$.

The guard ring principle is also employed in an apparatus consisting of a long cylindrical heater covered by a cylindrical shell of specimen material.[131] Under steady state conditions, the inner surface of the shell (radius r_1) is kept at a uniform temperature T_1 and the outer surface (radius r_2) assumes the temperature T_2. In a section of the shell remote from the ends heat flow is purely radial. The temperature at a point in the shell is then found to be inversely proportional to the logarithm of the radial distance outward from the cylindrical axis. Differential thermocouples on the inner and outer surfaces of the shell measure $T_1 - T_2$. Temperature measurements are made at several places along the central section of the cylinder as a check for any possible longitudinal heat flow. If the power per unit length dissipated by the heater core is $\dot{Q}/l$ then one may show that

$$K = \frac{\dot{Q}}{2\pi l} \frac{\ln(r_2/r_1)}{(T_1 - T_2)}. \tag{6.3.3}$$

Kohlrausch[132] has described a very simple experiment in which the ratio of thermal to electrical conductivities K/σ (Wiedemann–Franz ratio) in electrical conductors may be measured. A potential difference V is maintained across the ends of a specimen wire which are kept at the same temperature T_1. Joule heating causes a parabolic temperature distribution to be established along the wire with the maximum temperature T_2 at the center. It is easy to show that under steady state conditions,

$$\frac{K}{\sigma} = \frac{1}{8} \frac{V^2}{(T_2 - T_1)}. \tag{6.3.4}$$

Jaeger and Diesselhorst[133] used this technique extensively at ordinary temperatures, Meissner[134] made measurements of this type on copper down to 20°K. Angell,[135] Worthing and Langmuir[136] used a modified version of the method at very high temperatures. It should be noted that

[131] See reference 4, Chapter 7; reference 5, Chapter 4.

[132] F. Kohlrausch, *Ann. Physik* [4] **1,** 132 (1900).

[133] W. Jaeger and H. Diesselhorst, *Sitzber. deut. Akad. Wiss. Berlin* **38,** 719 (1899); *Physik.-tech. Reichsanstalt, Wiss. Abhandl., Charlottenburg* **3,** 269 (1900).

[134] W. Meissner, *Ber. deut. physik. Ges.* **12,** 262 (1914).

[135] M. F. Angell, *Phys. Rev.* **33,** 421 (1911).

[136] A. G. Worthing, *Phys. Rev.* **4,** 535 (1914); I. Langmuir, *ibid.* **7,** 151 (1916).

the quantity $K/\sigma T$ is expected to be a simple constant independent of temperature above the Debye temperature of a metal if the flow of heat current in metals, like that of electric current, is a transport effect in a gas of electrons.[128] Experiments on pure metals bear out this expectation quite well, indicating heat conduction in them to be predominantly electronic. In many alloys, however, deviations are observed which suggest that vibrations of the crystal lattice are contributing noticeably to the heat conduction. Lattice vibrations, of course, account for all heat transport in electrical insulators.

Let us consider now one rather elegant dynamical method for thermal conductivity determination due originally to Ångström.[137] In more recent versions[138] of this technique a rod specimen is used, one end of which is equipped with an electrical heating coil while the other is connected to some heat sink. Insulation or a vacuum jacket prevents heat loss from the surface of the rod so that the only significant flow occurs parallel to the axis (x-direction). The current in the heater at $x = 0$ is varied in such a way that the temperature at this point varies sinusoidally about a mean value T_0 with an angular frequency ω and an amplitude T_m. The heat flow equation, which in this case is simply

$$\frac{\partial^2 T}{\partial X^2} = \frac{1}{D}\frac{\partial T}{\partial t} \tag{6.3.5}$$

may then be solved, subject to the condition that at $x = 0$,

$$T = T_0 + T_m \sin \omega t$$

to give the temperature distribution in the rod. A satisfactory solution apart from transient effects is found to be[139]

$$T = T_0 - gx + T_m e^{-x\sqrt{\omega/2D}} \sin(\omega t - x\sqrt{\omega/2D}).$$

The first two terms represent a uniform fall of temperature from a value T_0 at $x = 0$ with a gradient $-g$. Inclusion of the third term superposes on this stationary gradient a periodic wave of temperature variation whose amplitude falls exponentially with increasing x from a value T_m at $x = 0$. The wavelength of this "temperature wave" is just $\lambda = 2\pi\sqrt{2D/\omega}$ while its velocity is $v = \sqrt{2\omega D}$. The attenuation coefficient is seen to be $\sqrt{\omega/2D}$. Of these quantities, each of which would yield a value for D, it proves simplest to measure the wave velocity. Two thermocouples or resistance thermometers, not necessarily calibrated, are attached to the

[137] A. J. Ångström, *Phil. Mag.* [4] **25,** 130 (1863).
[138] R. W. King, *Phys. Rev.* **6,** 407 (1915).
[139] See reference 5, p. 123; also reference 123.

rod at points a known distance apart. The temperatures at these points are measured as functions of the time. From the plotted data it is easy to read off the time taken for a given maximum to travel between the two thermometers and thus to measure v. Since

$$D = \frac{K}{c\rho} = \frac{v^2}{2\omega}$$

we have

$$K = \frac{v^2 c\rho}{4\pi f}$$

where the specific heat c, and density ρ of the specimen as well as the frequency of the temperature wave f are known.

Recently Mendoza[140] has used essentially the method just described to measure the thermal diffusivity of metals below 4.2°K. The object of these experiments, however, has not been the measurement of thermal conductivities but rather of specific heats which are the more elusive quantities at these temperatures. A steady state measurement of K is performed in the same apparatus used to determine D and the specific heat computed. Zavaritsky[141] has combined this method with adiabatic demagnetization techniques to measure specific heats down to 0.15°K.

6.3.1.2. Liquids. Measurement of the thermal conductivity of a liquid is complicated by the fact that convection currents are easily set up when a temperature difference is established in a specimen. The thermal conductivities of most liquids, except molten metals, are quite low ($\sim 10^{-4}$ cal/cm sec °C) so that heat transfer via convection, if present, may well be completely dominant. The most common way of avoiding this difficulty is to heat the liquid at the top so that the densest (coldest) material is at the bottom of the specimen and there is no tendency for natural convection to occur. A cylindrical column of liquid in a container of known low thermal conductivity may be treated in much the same way as a bar or slab of a poor solid conductor if this precaution is observed. Weber,[142] for example, carried out measurements of this type, surrounding the liquid column with a guard ring to eliminate lateral heat losses.

It usually proves easier to determine the heat current density through the liquid if the column height is reduced until the specimen is merely a film perhaps 1–3 mm thick. In the method of Lees[143] the specimen film,

[140] D. H. Howling, E. Mendoza, and J. E. Zimmerman, *Proc. Roy. Soc.* **A229,** 86 (1955).

[141] N. V. Zavaritsky, *Zhur. Eksptl. i Teoret. Fiz.* **33,** 1085 (1957).

[142] R. L. Weber, *Ann. Physik* [4] **11,** 1047 (1903).

[143] C. H. Lees, *Phil. Trans. Roy. Soc. London, Ser. A* **191,** 418 (1898).

contained by a thin ebonite ring, is sandwiched between copper disks. Above this combination are stacked a glass disk, another of copper, a heater, and an uppermost copper disk. Heat flowing from the heater through the specimen must first traverse the glass disk. Thermocouples in the alternate copper disks measure the temperature drop across both the glass and the specimen. The thermal conductivity of the glass is assumed known so that the heat flux is determined. Applying an appropriate correction for the ebonite ring and for lateral losses one easily calculates the thermal conductivity of the liquid specimen. This technique, it should be noted, is adapted from one used extensively by Lees with poorly conducting solids. Jakob,[144] in his work on water, and subsequent investigators[145] have also made measurements on film specimens held, often merely by capillary action, between two metal plates. As in the earlier work, heat is supplied from above. The heat current, however, has generally been determined by direct calorimetry rather than by the somewhat uncertain comparative procedure. The papers of Bowers[146] and Grenier[147] describe the extension of measurements of this type to liquid HeI and are particularly instructive.

Thin specimen films make possible the use of many other measuring techniques which do not involve heating from above. With decreasing thickness the contribution of conduction to heat flow through a liquid layer increases while the convective contribution, initially constant, eventually drops. Thus, with sufficiently thin liquid layers conduction will be the dominant process for any orientation of the thermal gradient. This condition is achieved in a widely used technique in which the film ($\sim$1 mm thick) fills the annular space between a cylindrical metal heater and a coaxial cooled metal shell. The small temperature difference maintained between the two cylinders is measured with resistance thermometers or thermocouples. This experimental arrangement is easily adapted to high pressure and/or low temperature observations.[148] In a common variant of this method, the outer shell is merely a metal capillary and the heater a coaxial platinum wire.[149] Difficulty in centering the wire, however, makes absolute conductivity measurements with such a device rather unreliable. It is used primarily as a comparative scheme.

All of the techniques mentioned above apply to static liquid specimens

[144] M. Jakob, *Ann. Physik* [4] **63,** 537 (1920).

[145] F. Meyer and M. Eigen, *Z. Naturforsch.* **8a,** 500 (1953); W. Hamann, *Ann. Physik* [5] **32,** 593 (1938). Thick Film.

[146] R. Bowers, *Proc. Phys. Soc.* (*London*) **A65,** 511 (1952).

[147] C. Grenier, *Phys. Rev.* **83,** 598 (1951).

[148] P. W. Bridgman, *Proc. Am. Acad. Arts Sci.* **59,** 141 (1923).

[149] R. Goldschmidt, *Physik. Z.* **12,** 417 (1911).

under steady state heat flow conditions. In his work on continuous flow calorimetry Callendar[150] described a steady state method for measuring the thermal conductivity of a liquid flowing through a metal tube along which a constant temperature gradient is maintained. This technique does not appear to have been extensively exploited. Various nonsteady state procedures with static liquids have also been described. E. F. M. van der Held *et al.*[151] have perfected such a technique and demonstrated its utility in measurements on a number of liquids.

6.3.1.3. Gases.* The same problems, somewhat accentuated, are encountered in the measurement of the thermal conductivities of gases as occur with liquids. Precautions against convection become particularly important because of the greater ease with which such currents are established. In magnitude, the conductivity of a typical gas may be as much as ten times smaller than that of a liquid or insulating solid, e.g., $K \simeq 6 \times 10^{-5}$ cal/cm sec °C for N_2 at 20°C.

Several of the basic techniques employed in measurements on liquids have been used successfully with gases. Hercus and associates[152] have worked with thin films of specimen gas between flat horizontal metal plates, the upper one heated and the lower one cooled. A third plate above the heater and at the same temperature and a guard ring insure that heat flows only downward through the specimen. A correction for heat transfer between the plates via radiation was determined in auxiliary experiments. Convection was effectively eliminated. Ubbink and de Haas[153] have adapted this method for use at very low temperatures in their study of heat conduction in helium gas down to 1.6°K.

Most of the recent measurements on gases have been made by the hot wire method due originally to Schleiermacher.[154] The apparatus consists of a metal or glass tube (i.d. 1–10 mm) with an axial platinum wire (0.05–1 mm diam) sealed to its ends (see Fig. 10). The specimen gas forms a cylindrical shell between heater and tube of inner and outer radii r_1 and r_2. In the steady state, the heater dissipates a measured power per unit length $\dot{Q}/l$ and stays at a temperature T_1 while the tube is kept at a temperature T_2. Temperature T_1 is usually obtained by measuring the resistance of the heater. The conductivity of the gas is

* See also Section 4.2.1.5.

[150] H. L. Callendar, *Phil. Trans. Roy. Soc. London, Ser. A* **199,** 110 (1902).

[151] E. F. M. van der Held and F. G. van Drunen, *Physica* **15,** 865 (1949); E. F. M. van der Held, J. Hardebol, and J. Kalshoven, *ibid.* **19,** 208 (1953).

[152] G. R. Hercus and T. H. Laby, *Proc. Roy. Soc.* **A95,** 190 (1919); G. R. Hercus and G. Southerland, *ibid.* **A145,** 599 (1934).

[153] J. Ubbink and W. J. de Haas, *Physica* **10,** 451 (1943); *Leyden Rijksuniv., Kamerlingh Onnes Lab., Communs.* **No. 266c, d.**

[154] A. A. Schleiermacher, *Ann. Physik* [3] **34,** 623 (1888).

$$K = \frac{\dot{Q}}{2\pi l} \frac{\ln(r_2/r_1)}{T_1 - T_2}.$$

Several corrections must be applied if the results are to be meaningful. Heat loss from the ends of the tube may be determined by repeated measurements with similar tubes of quite different lengths. Heat transfer by radiation is measured in experiments in which the apparatus is highly

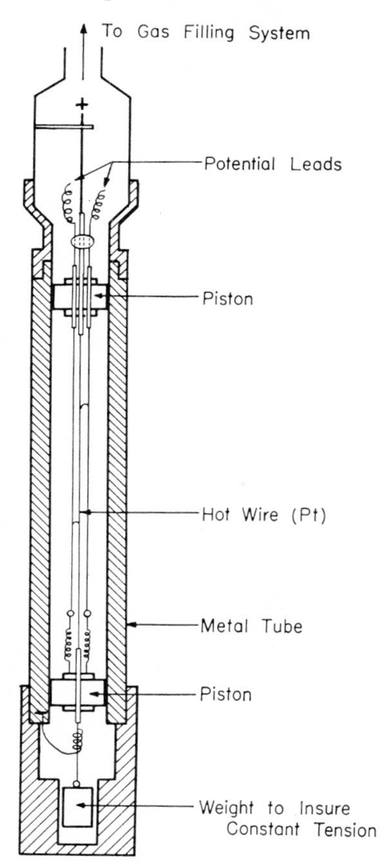

FIG. 10. Cell for measuring thermal conductivities of gases by hot wire method.

evacuated. Convection may be detected by operating the apparatus in vertical and horizontal positions and by varying the pressure of the gas. At pressures below about 150 mm Hg convective transfer becomes quite small. The conductivity, on the other hand, remains constant until the pressure is reduced to the point where the mean free path is comparable with the dimensions of the container. Thus by working at sufficiently low pressures convection can be effectively eliminated. However, it is in this region that a discontinuous change in temperature between the gas and the heater or tube wall becomes apparent, producing an error in the measured gradient. This "temperature-jump" effect reflects the failure of a gas to behave as a viscous fluid over distances of the order of a few mean free path lengths. The theory of the effect[155] is complicated by the fact that gas molecules colliding with a solid surface may not achieve equilibrium with that surface, i.e., have an accommodation coefficient <1. Kannuluik[156] has minimized this correction by using a thick heater wire and has developed a satisfactory method of calculating the end correction which is increased thereby. In what is probably the most refined application of the hot wire method, Taylor *et al.*[157] however, have chosen a thin wire ($\sim$0.01 in diam) and avoided all end corrections by using only a portion of the heater extending between two potential leads. In this work the conductivity was measured as a function of pressure and the temperature jump correction obtained by an extrapolation procedure. The conductivities of several gases were determined in the temperature range 80° to 380°K with an estimated accuracy of 0.5%.

The theory of gaseous heat conduction is comparatively well developed[158] in contrast with the theory of heat conduction in liquids. It provides relations between the conductivity and molecular parameters such as interaction cross sections as well as with other measurable macroscopic quantities. For example one finds $K = f\eta c_v$ where η is the viscosity, c_v the specific heat at constant volume, and f a number characteristic of the gas. While a general theory of the factor f has not been developed, Eucken[159] has obtained by nonrigorous arguments a relation between f and $\gamma = c_p/c_v$ which successfully correlates many observations, namely, $f = \frac{1}{4}(9\gamma - 5)$. The fact that K is independent of pressure at ordinary pressures is consistent with the theory which in addition predicts accurately the effects of high pressure on K. Enskog[160] has shown that

[155] See for example I. Estermann, *in* "Thermodynamics and Properties of Matter," (F. D. Rossini, ed.), Section I, pp. 759–771. Princeton Univ. Press, Princeton, New Jersey, 1955.

[156] W. G. Kannuluik and L. H. Martin, *Proc. Roy. Soc.* **A144,** 496 (1934); W. G. Kannuluik and P. G. Law, *Proc. Roy. Soc. Victoria* **58,** 142 (1947).

[157] W. J. Taylor, H. L. Johnston, and E. R. Grilly, *J. Chem. Phys.* **14,** 219, 233, 435 (1946).

[158] S. Chapman and T. G. Cowling, "The Mathematical Theory of Non-Uniform Gases," 2nd ed. Cambridge Univ. Press, London and New York, 1953; see also reference 155.

[159] A. Eucken, *Physik. Z.* **14,** 324 (1913).

[160] D. Enskog, *Kgl. Svenska Vetenskapsakad. Handl.* **63,** No. 4 (1921).

$$K = K^0 b\rho \left(\frac{1}{b\rho x} + 1.2 + 0.7575 b\rho x \right) \tag{6.3.6}$$

where K^0 is the conductivity at unit pressure, b the van der Waals constant per gram, ρ the density, and x a numerical factor. At pressures low enough so that the mean free path of molecules in the gas is comparable with the dimensions of the container, intermolecular collisions become much less important than collisions with the walls of the vessel. The notion of a conductivity as used in the preceding discussions loses its precise meaning. Under these conditions the heat carried away per second from a unit area of a surface at T_2 by a dilute gas whose molecules have an average kinetic energy corresponding to a temperature T_1 is given theoretically[161] as

$$\dot{q} = \frac{1}{4} \frac{\gamma + 1}{\gamma - 1} \sqrt{\frac{2R}{\pi M}}\, pa \frac{(T_2 - T_1)}{\sqrt{T_1}} \tag{6.3.7}$$

where R is the gas constant, M the molecular weight, p the pressure, and a the accommodation coefficient. Experiment generally confirms this relation, which has many practical implications. It has been used for example in the determination of accommodation coefficients. The operation of the Pirani vacuum gage* depends on the indicated proportionality of $\dot{q}$ and p (see Section 4.2.3.4).

6.3.2. Convection[119,120,162] †

In convective heat transfer a portion of a fluid is warmed at the surface of a hot solid object and then moves, giving up heat eventually to another cooler solid surface or, after mixing, to the rest of the fluid. Two types of convection may be distinguished: (1) natural convection in which the fluid moves as a result merely of differences in density of portions at different temperatures; (2) forced convection in which the motion is maintained by a pump or a fan.

Before being convected away from the solid surface, heat must be transferred by conduction to the fluid stream. The details of this process are obscure and render the analysis of convection problems quite difficult. It is generally believed that the initial step is one of conduction through a thin boundary layer or film next to the solid surface in which fluid flow is laminar even when the main stream is in turbulent motion. It is assumed that due to turbulent mixing the main stream is at a uniform temperature

* See Section 4.2.6.

† See also Section 4.2.1.

[161] See reference 2, p. 274; also reference 155.

[162] See reference 2, p. 294; reference 4, Chapter 12.

different from that of the solid surface by an amount ΔT, this drop occurring across the effective thickness Δx of the boundary layer. If K is the thermal conductivity of the film then the rate of heat loss from a surface of area A due to convection becomes, in the steady state

$$\dot{Q} = KA \frac{\Delta T}{\Delta x} = hA \, \Delta T. \tag{6.3.8}$$

The quantity $h = K/\Delta x$ is called the film coefficient. It depends not only upon the intrinsic properties of the fluid such as its thermal conductivity, viscosity, density, and specific heat, but also upon the magnitude and direction (horizontal or vertical) of the fluid velocity, the shape of the solid surface, etc.

A detailed theoretical calculation of the film coefficient h is usually not feasible. Dimensional analysis, however, can indicate the general way in which h depends upon the pertinent variables leaving to experiment the evaluation of the detailed functional relations. Forced convection was first treated in this way by Rayleigh,[163] natural convection by Davis.[164] In the case of forced convection one finds for geometrically similar bodies

$$h = \frac{K}{l} F\left(\frac{lv\rho}{\eta}\right) f\left(\frac{c_p\eta}{K}\right) \tag{6.3.9}$$

where l is a linear dimension of the body, v the velocity of the fluid past its surface, and ρ, η, c_p, K are the density, viscosity, specific heat, and thermal conductivity respectively of the fluid at the mean boundary layer temperature.

The dimensionless ratios $lv\rho/\eta$, $c_p\eta/K$, and hl/K are known respectively as the Reynolds, Prandtl, and Nusselt numbers. The functions F and f must be determined by actual observations of the rate of heat loss from bodies of a given geometry under forced convection. The measurements are essentially calorimetric, utilizing basic techniques previously discussed (see Chapter 6.2). As an example of the kind of semiempirical formula obtained in this way we may quote the result for circular pipes of diameter D cooled by fluid forced through them[120]

$$h = 0.023 \frac{K}{D} \left(\frac{Dv\rho}{\eta}\right)^{0.8} \left(\frac{c_p\eta}{K}\right)^{0.4}. \tag{6.3.10}$$

This formula is found to hold for all but the more viscous fluids when the Reynolds number exceeds 2100.

[163] Lord Rayleigh, *Nature* **95,** 66 (1915).
[164] A. H. Davis, *Phil. Mag.* [6] **40,** 692 (1920).

For geometrically similar bodies cooled by natural convection dimensional arguments lead to the expression

$$h = \frac{K}{l} F\left(\frac{L^3\rho^2 g\beta\,\Delta T}{\eta^2}\right) f\left(\frac{c_p\eta}{K}\right) \tag{6.3.11}$$

where g is the gravitational acceleration, β the volume expansion coefficient of the fluid, and L a characteristic length corresponding to the distance along the warm surface traversed by the passing fluid and in general different from l. The dimensionless ratio $L^3\rho^2 g\beta\,\Delta T/\eta^2$, the Grashof number, may be written $\frac{1}{2}(LV\rho/\eta)^2$ to emphasize its close analogy with the Reynolds number. Here $V = \sqrt{2g\beta\,\Delta TL}$ is the velocity of a unit mass of fluid raised through a distance L by buoyant forces as a result of a temperature increase ΔT. Again the functions F and f must be determined experimentally. A general feature of the results obtained with gases is that $h \propto \Delta T^{\frac{1}{4}}$. Thus we can say that for a body cooled by natural gaseous convection

$$\dot{Q} = hA\,\Delta T \propto \Delta T^{\frac{5}{4}}.$$

It is interesting to note that for forced convection, h is not a function of ΔT since the flow velocity does not depend on the temperature difference between the bulk fluid and the warm surface. Thus, if a body be cooled by forced convection $\dot{Q} \propto \Delta T$ which of course is just Newton's law of cooling. The fact that this relation describes only forced convective cooling was clearly stated by its discoverer but ignored by many later authors.

6.3.3. Radiation* [120,165–167]

A solid or liquid body emits a continuous spectrum of electromagnetic radiation whose spectral distribution is a function of its absolute temperature. Below about 500°C most of this radiation occurs in the range of infrared or longer wavelengths, visible light becoming evident as the temperature is raised. If the emitter is a perfect one, a blackbody, the energy $E(\lambda)$ which it loses per second per unit surface area in the form of radiation of wavelength λ is given by Planck's distribution law (Section 6.1.5). The energy which a blackbody loses per second per unit area as

* See also Part 7.

[165] W. Wien and C. Müller, *in* "Handbuch der Experimentalphysik" (W. Wien and F. Harms, eds.), Vol. 9, Part 1, pp. 347–475. Akademische Verlagsges., Leipzig, 1929.

[166] See reference 4, Chapter 13.

[167] W. E. Forsythe, "Measurements of Radiant Energy." McGraw-Hill, New York. 1937.

radiation of all wavelengths is given by the Stefan–Boltzmann formula $E = \sigma T^4$ where $\sigma = 5.669 \times 10^{-5}$ erg/cm² deg⁴ sec.

The effectiveness of a nonblackbody as an emitter of radiation is measured by its total emissivity e, the ratio of the rate of energy loss per unit surface area of the body to the corresponding quantity for a blackbody at the same temperature. The value for a blackbody is unity. The analogous quantity for radiation of only one wavelength is referred to as a spectral emissivity $e(\lambda)$. For a nonblackbody, the Stefan–Boltzmann formula becomes $E = e\sigma T^4$.

The emissivities have values between 0 and 1 and are characteristic of the substance. They depend upon the condition of the surface as well as upon the temperature. The emissivities of nonmetals are usually rather high, a typical value being that for smooth glass at room temperature, $e \sim 0.9$. Clean, polished metals on the other hand have low emissivities which generally increase with temperature. Typical values for polished copper are $\sim$0.04 at 50°C and $\sim$0.08 at 500°C. Oxidation greatly increases the value. Heavily tarnished copper at 50°C for example might have $e \simeq 0.5$. Several summaries of emissivity data are available.[168]

A given body is not only continuously losing radiant energy but also receiving it. It may be easily proved that the fraction of incident radiation absorbed by a body, its absorptivity, is just equal to its emissivity. Knowing the emissivities, absolute temperatures, separation, and relative orientation of two surfaces the net rate of energy transfer between them can be calculated, usually under the assumed validity of Lambert's law. The resulting formulas for many particular cases are to be found in treatises on heat transfer.[120] It will suffice to mention here only the simplest result of this type namely the net energy gain of a small convex body at a temperature T_1 inside a cavity whose interior walls are maintained at a temperature T_2. If the surface area of the body is A and its emissivity at T_1 is e, then $\dot{Q} = Ae\sigma(T_2^4 - T_1^4)$.

The formula just given serves as the basis for several methods of determining the constant σ. In one of the simplest of these[169] a small hollow blackened metal sphere ($e = 1$), containing an electrical heater and a thermocouple, is suspended inside an evacuated enclosure whose walls are kept at a constant temperature T_2. Electrical energy is dissipated in the heater at a measured rate P. When the temperature of the sphere reaches a constant value T_1 it is losing energy by radiation at just this rate. Thus

$$\sigma = \frac{P}{4\pi r^2(T_1^4 - T_2^4)} \tag{6.3.12}$$

[168] See reference 4, Appendix II, Table V; also reference 9, p. 1184.

[169] See reference 5, p. 158.

where r is the radius of the sphere. If the sphere were not blackened and the known value for σ assumed, such an experiment would measure the total emissivity of the surface of the sphere. Various calorimetric procedures and radiator or absorber geometries have been used in the measurement of emissivities. In one particularly interesting investigation Wexler[170] has measured the emissivity of polished copper at 4.2°K exposed to 77° and 297°K blackbody radiation. Heat transfer rates were measured by observing the rate at which liquid helium evaporated from a Dewar vessel made essentially of two concentric copper spheres. The observed emissivities were found to be very small, a fact important to the operation of vessels for storing liquid helium and also of considerable theoretical interest.

Emissivities at high temperatures have often been studied by pyrometric methods.[171] Worthing[172] has used for such work small tubes of the specimen metal heated to incandescence. The surface of the tube emits normal radiation while through a small hole in the tube blackbody radiation of the same temperature emerges. Comparison of the external surface brightness with that of the adjacent hole yields the emissivity.

Emissivities are, of course, directly related to the optical constants of a substance, i.e., its index of refraction and absorption coefficient. In some instances determination of these constants by optical procedures proves an appropriate way of measuring the emissivity. For some substances, particularly metals, the theory of the optical properties[173] is sufficiently reliable to make possible the approximate calculation of emissivities in certain spectral regions from the known electromagnetic properties.

6.4. Thermodynamics*

The great value of thermodynamics as a working tool for the physicist lies partly in the following features. Thermodynamics provides a precise way of characterizing real physical systems in an important class of phenomena, namely, the phenomena associated with equilibrium. It makes general and precise statements about the properties and behavior

[170] A. Wexler, *Natl. Bur. Standards (U.S.) Cir.* **519,** 195 (1952).
[171] See reference 9, Chapter 12.
[172] A. G. Worthing, *Phys. Rev.* **10,** 377 (1917).
[173] M. P. Givens, *Solid State Physics* **6,** 313–352 (1958).

* Chapter 6.4 is by **Paul M. Marcus.**

of such systems, even though it is not a fundamental theory which contains a complete description of the systems being considered. It specifies the number of independent variables which fix the state, and the number of equations of state which contain the equilibrium properties of any given system. It provides relations among the equations of state which reduce the number of measurements required to determine those equations.

In view of its close relationship to experimental measurements on real physical systems, it seems worthwhile to develop the subject in a form which emphasizes these qualities. Accordingly, the development to follow is made in a postulational form which proceeds from known and basic concepts and operations. Definitions are introduced based on known quantities, and postulates are introduced which generalize the results of observation and lead to new concepts and operations. The development contains interspersed sequences of numbered definitions (up to eleven), numbered postulates (up to five), and theorems (up to five), plus two special observations which establish a choice between theoretical possibilities. Proofs are in general omitted for brevity, but references are given to fuller discussions.

The primitive concepts, undefined here and taken over from mechanics and electromagnetism, are the concepts of generalized force and displacement and their combination in the form of work. These are assumed measurable as required, by known techniques that need not be explicitly stated, and (a sufficient number of them) furnish a complete characterization of the state of any system. An appropriate operational definition of equilibrium is introduced, and thermodynamics then concerns itself primarily with properties of equilibrium states. In particular it considers the influence on equilibrium of the presence of microscopic transfer processes, beneath the level of macroscopic observation, but indicating their presence indirectly by the behavior of the internal energy, temperature, entropy, and chemical potential with and without such transfer processes.

The formulation of the theory achieves a neatness of statement by the use of theoretical or idealized operations and measurements, which can be carried out *in principle* as the limiting form of actual operations or measurements. In practice, of course, no practicable way may be known to provide these, and awkward questions of sensitivity and accuracy occur, but these are bypassed in the present discussion.

6.4.1. Thermodynamic Definitions, Operations, and Variables

The contents of mechanics and electromagnetism, including concepts, laws, and measurement procedures are assumed, and the values of mechanical and electrical quantities* are therefore available at any time or position.

The first basic concept in the development of thermodynamics is that of a physical system.

Definition 1. A *physical system* (system for short) is a limited definite portion of matter. It is prescribed as the contents of a closed surface at a given instant of time.

Note that a given system may be recognized by following continuously the motion of the matter at its boundary surface; this requires a mechanical measurement which has been assumed possible. In addition measurements of the mechanical or electrical forces and displacements at every point of a system at any time are possible, hence the work done on any part of the system may be measured. Such measurements can be made without interference with the system.

The first basic operation on physical systems is total isolation.

Definition 2. A system is *totally isolated* if all alterations of physical conditions outside the system do not alter the mechanical or electrical quantities of the system (or any part) and their behavior with time.

An especially important and simple condition for a physical system can now be defined, called the equilibrium state.

Definition 3. A system is in an *equilibrium* state if the mechanical and electrical quantities of the system are constant in time, and if this constancy persists when the system is totally isolated.†

The implication of this definition is that sufficient time has elapsed for all macroscopic motions or changes to have died out. It is necessary to make the test of total isolation to distinguish the equilibrium state from steady states in which no change is perceptible in the system, but microscopic currents of energy or matter may be flowing through the system. Steady states also have certain simple properties, but they are not the states with which ordinary thermodynamics is concerned.

Definition 3.1. An important practical supplement to true equilibrium states concerns states which when totally isolated show no change with time during the period of measurement, but show changes over much longer periods of time. Such states may be designated as *pseudoequilib-*

* Examples of such quantities are given later. They are the various generalized forces and displacements.

† The observations may be made with respect to coordinates fixed in the system, so that systems in motion may still be in equilibrium in some coordinate frame. The motion then has to be rigid body motion, with at most uniform translational and rotational acceleration. Such accelerations will then introduce steady reaction forces in the system-fixed coordinates.

rium states, and all thermodynamic postulates and theorems are assumed to hold for such states as well.*

Henceforth the word *state*, without modifiers, will mean equilibrium state (true or pseudo), and nonequilibrium states will be specifically named when intended.

A second basic operation on physical systems required in the development is a partial or controlled isolation of a system, designated as adiabatic isolation.

Definition 4. A system in an equilibrium state is *adiabatically isolated* if any change in any mechanical or electrical quantity (produced by alteration of external conditions) is necessarily accompanied by work done on some part of the system.

Thus, an adiabatically isolated system starting in equilibrium cannot undergo a change in which no work is done on any part.†

Thus far the mechanical and electrical quantities used to identify a system and follow its behavior have not been specifically illustrated. We now give some examples. We shall refer to the *basic thermodynamic variables* as those which occur in conjugate pairs constituting a generalized displacement variable and a generalized force variable. The products then give a work term. As basic mechanical variables, we have the position of each part of a system, and the stress or various force components per unit area acting at that part; this includes the volume and pressure of a fluid system. These may in general vary with time and position. In addition there are the mass density and the gravitational potential, and the electrical variables such as the electric charge density and electrical potential, the electric and magnetic polarization densities, and electric and magnetic fields, etc. In any actual physical process to which thermodynamic analysis is being applied, only a few of these variables are likely to change, and attention may be focused on them. Thus in the absence of long range interactions we may expect homogeneity in equilibrium states, at least throughout macroscopic-sized regions (i.e., phases), which greatly reduces the number of variables, and of course in equilibrium states no changes with time, or velocities, need be specified.‡

* Pseudoequilibrium states may be regarded as containing an effective constraint which retards attainment of equilibrium for certain degrees of freedom. A true constraint, as used later, permits no change in the degrees of freedom it affects, hence gives true equilibrium.

† The physical idea embodied here is that ordinary or nonadiabatic enclosure of a system permits it to be changed without work being done on any part, namely by heating it, and appropriately constraining it so no displacements take place. Heat will be given a precise meaning later, but is not used in this definition.

‡ Solids may be inhomogeneous without long range forces, but in pseudoequilibrium because of the slowness of diffusion processes.

We may regard the equilibrium states as fixed by the values of an appropriate number of the basic variables, namely the variables entering the expression for work in the changes of interest. As the equilibrium state changes continuously, the values of (at least) one or more of the basic variables change continuously.

A third basic operation on a physical system greatly extends the possible equilibrium states, and can change the state even of an isolated system. This operation is the alteration or removal of a *constraint,* which refers generally to a restriction on the freedom of a displacement to assume arbitrary values. We introduce

Definition 5. A constraint freezes some displacements at their initial values.

Examples of constraints are rigid walls, or sets of rigid walls. We shall assume constraints may be imposed or released as desired on any physical system.

6.4.2. The First Law of Thermodynamics, Internal Energy, Heat, and Empirical Temperature

Physical information is now introduced into the development by the postulate which extends conservation of energy to include all forms of work for any system.

Postulate 1. The net work done when a system changes adiabatically between two states is independent of the path.

Note that alteration of an adiabatically enclosed system by definition comes only through work. Thus, all energy changes can be accounted for in a system so enclosed, regardless of the path, by work in some form.

A useful function of the state of the system related to the work done on a system may now be defined. First, the change in such a function between two states may be defined as some appropriate function of the unique and measurable work done in an adiabatic process between the two states. If W is the *work done by the system* in a process, and we introduce the explicit relation

Definition 6.

$$\Delta E = -W \qquad \text{(adiabatic changes)} \tag{6.4.1}$$

then we have defined the internal energy function E up to an additive constant for all states (assuming an adiabatic path may be found between them). Actually, as will be shown later, such a path exists always in one direction. Then E is fixed by specifying its value in a reference state, which is henceforth assumed done. The function of state E is a measure of the energy stored in the system, since it takes proper account of additions to that energy.

The content of thermodynamics is enriched by the following observation:

Observation 1. The internal energy of a system may be altered without doing net work on it.

This states that not all envelopes are adiabatic, and that processes exist which transfer energy in a form not directly observable by macroscopic measurement.

Definition 6 and Observation 1 now permit the definition of *heat.*

Definition 7. The heat Q absorbed by a system in a change from one state to another with internal energy greater by ΔE, and during which change the system does work W, is defined by

$$Q = \Delta E + W. \tag{6.4.2}$$

Equation (6.4.2) is commonly referred to as the First Law of Thermodynamics.

Since ΔE may be measured by use of appropriate adiabatic paths between the states, and W may be measured directly by procedures assumed known (Section 6.4.1), Q is then a measurable quantity by Definition 7.*

A general contact between physical systems may now be defined by

Definition 8. Two systems are in *diathermal contact* if they may exchange heat. We assume none of the possible states (equilibrium or nonequilibrium) of either system is altered by the establishment of the heat exchange, so that the systems retain their separate identities.†

The next important concept, the empirical temperature, may now be introduced by

Postulate 2. The possible equilibrium states of a system A, in diathermal contact with system B in a definite state, are limited by a functional relation among the variables of state.

The states of a system may therefore be classified according to whether or not they are in equilibrium while in diathermal contact with a given reference state. Equilibrium while in diathermal contact will be referred to as *thermal equilibrium* of two or more systems. Two systems in thermal equilibrium have a common property that might be qualitatively characterized as an equal readiness to give up heat. For compactness and ease of reference it is convenient to speak of an isothermal "surface" as containing all the states which satisfy a functional relation in a "thermo-

* The meaning of E and Q can presumably be extended to nonequilibrium states which are sufficiently simple to be described by measurable force and displacement coordinates.

† The initial state of each system may of course be altered when the heat exchange begins.

dynamic space" whose axes are a sufficient number of forces and displacements to fix the state of the system. (The second law will later determine how many of those form such a complete set.)

The usefulness of Postulate 2 is extended by the theorem of the associativity of equilibrium in diathermal contact

Theorem 1. If systems A and C are in equilibrium when separately in diathermal contact with system B in a definite state, then A and C will be in equilibrium when in diathermal contact with each other.

The proof rests on the observation that putting A and C simultaneously in diathermal contact with B leaves them all unaltered and still in equilibrium—since by Definition 8 diathermal contact with either A or C does not disturb B—but constitutes diathermal contact between A and C via B.

Then clearly we may regard the states of an isothermal surface of system A as being in equilibrium with each other, and all states of the system may be classified into unique, nonintersecting, isothermal surfaces. If we introduce any function of state, constant on the isothermal surfaces, and varying continuously and monotonically from surface to surface, we may use such a function as an *empirical temperature*, uniquely defined for each state. An example of such a function would be any continuous monotonic function of one variable, varying along a continuous line intersecting the isothermal surfaces (such as a coordinate axis, a particular force or displacement). The empirical temperature of a reference system may then be used to characterize the states of any other system in thermal equilibrium with the reference system, which serves therefore as an empirical thermometer.*

Theorem 1 then leads to the associativity of the empirical temperature, and is frequently referred to as the Zeroth Law of Thermodynamics.†

6.4.3. The Second Law of Thermodynamics

It is useful at this stage in the development to make a distinction between reversible and irreversible processes and to consider the simpler reversible processes first. Reversible processes permit a detailed analytical description at every stage, so that differential analysis may be applied.

* Familiar examples of empirical thermometers are gas thermometers at constant pressure or volume, mercury or alcohol expansion thermometers. In practice empirical thermometers also use properties of the steady state such as resistance or thermoelectric voltage. These are of great practical importance, as illustrated in Section 6.1, but are conceptually more difficult to introduce.

† In the development given here Postulate 2 has been preferred to Theorem 1 as a more direct and powerful statement of the significance of equilibrium under thermal contact, which leads to the empirical temperature and to Theorem 1.

Accordingly we now introduce

Definition 9. A reversible process is one in which the system passes entirely through equilibrium states.

Such a process is, of course, only attainable as an idealization or limit of real processes, although it may be closely approached and reasonable extrapolations made. The course of the process is then specified by the variables of state for each intermediate equilibrium state, and requires no time derivatives. No distinction is implied between forward and reverse sequence, so that either may be considered as constituting the process, the only difference being a reversal of the signs of heat and work exchanged with the surroundings.

We now proceed to formulate the second law in the inaccessibility form given by Caratheodory in 1909;[1] recent discussions have been given by Buchdahl[2] and Wilson.[3] This has a natural separation into a reversible and an irreversible form, and resembles the previous postulates in making a simple statement about restrictions on equilibrium states.

6.4.3.1. The Restricted Form of the Second Law. The restricted form of the inaccessibility postulate may be stated as

Postulate 3a. In the vicinity of any state of a system, there are states not accessible by reversible adiabatic processes.

Vicinity means here any neighborhood of the given state, hence indefinitely close to the given state. Note that E can change in a reversible adiabatic process, since work can be done, and also the empirical temperature can change. Nevertheless, restriction to reversible adiabatics imposes a restriction on the variables of state.

By considering the differential expression for the possible reversible adiabatic paths, obtained by setting $dQ = 0$, it may be shown that the condition expressed by Postulate 3a requires the existence of new functions of state. These functions are constant for adiabatic reversible processes (they are integrals of $dQ = 0$), and their differentials are proportional to the heat absorbed in a small reversible change.* Any of these functions may be designated the *empirical entropy* function. However, by considering the properties of a composite system made up of systems in diathermal contact, a natural way to define an absolute entropy and an absolute temperature appears. Namely, by requiring that the entropy for systems in diathermal contact be the sum of the entropies of the

* See, for example, Wilson,[3] Chapter 4, pp. 74–80.

[1] C. Caratheodory, Investigations on the foundations of thermodynamics. *Math. Ann.* **67,** 355 (1909).

[2] H. A. Buchdahl, On the unrestricted theorem of Caratheodory and its application in the treatment of the second law of thermodynamics. *Am. J. Phys.* **17,** 212 (1949).

[3] A. H. Wilson, "Thermodynamics and Statistical Mechanics." Cambridge Univ. Press, London and New York, 1957.

separate systems, a unique one of the empirical functions is chosen (up to a constant factor and an additive constant) and may be defined as the *absolute entropy* S. Correspondingly, an *absolute temperature* T may be defined by using the proportionality of dQ and dS, and putting $dQ = T\,dS$.*

The constant factor in T and dS may be fixed by specifying the size of the degree, as by specifying the values of T at fixed points, and this is done for the Kelvin scale, as described in Section 6.1. It is noteworthy that the constant factor could be chosen negative, with no inconsistency, making all ordinary temperatures negative. The absolute entropies are then fixed by specifying the entropy in a reference state. Subsequent use of T and S will assume the Kelvin scale and a reference state for S.

6.4.3.2. Consequences of the Restricted Form of the Second Law.

6.4.3.2.1. The Number of Independent Variables and of Equations of State. The combined first and second laws may be written for infinitesimal reversible processes

$$T\,dS = dE + dW = dE - \sum_{i=2}^{n} f_i\,dX_i \qquad (6.4.3)$$

where dW is the work done by the system and is expressed as a differential form in a set of $(n - 1)$ generalized forces f_i and corresponding conjugate displacements X_i. For a simple fluid system, with pressure the only force variable, Eq. (6.4.3) becomes

$$T\,dS = dE + p\,dV. \qquad (6.4.4)$$

Only displacements that may change for the system under observation are included in the X_i in Eq. (6.4.3). All other displacements are assumed constrained to initial values for purposes of a particular discussion. This includes systems in pseudoequilibrium (Definition 3.1) with variables effectively constrained during the time of a measurement.

Some simple general conclusions about the number of variables in a complete set, the number of equations of state, and the relations among thermodynamic derivatives, follow from Eq. (6.4.3), and will now be obtained.

Take Eq. (6.4.3) in the form

$$dE = \sum_{i=1}^{n} f_i\,dX_i; \qquad f_1, X_1 = T, S. \qquad (6.4.5)$$

Then E may be regarded as a function of n independent variables, the X_i, $i = 1$ to n. [No additional variables can appear in E since they do not

* Wilson,[3] p. 80.

appear in dE in Eq. (6.4.5).] All the f_i are then determined as functions of the X_i by

$$f_i = \left(\frac{\partial E}{\partial X_i}\right)_{X_j, j=1 \text{ to } n, \neq i} \equiv \left(\frac{\partial E}{\partial X_i}\right)_{X_1 X_2 \cdots X_{i-1} X_{i+1} \cdots X_n}; \qquad i = 1 \text{ to } n.* \tag{6.4.6}$$

Thus the state of the system with $2(n - 1)$ force and displacement variables (forming $(n - 1)$ conjugate pairs) is fixed by specifying n variables, X_i, $i = 1$ to n. Such a system will be referred to as an n-variable system. Alternatively, the n relations, Eq. (6.4.6), may generally be solved for any n of the $2n$ variables f_i, X_i, $i = 1$ to n, in terms of the other n. Hence any n variables can form a complete set or *basis* whose values fix the state of the system.†

The physical description of the equilibrium states of the general n-variable system is therefore contained in n *equations of state* which relate the n other variables to the n chosen as the basis. For example Eq. (6.4.6) gives the n equations of state for the displacement variable basis. The function E is then determined by integration of Eq. (6.4.5). Of course the differential form for dE in any basis must be an exact differential, which leads to thermodynamic relations among the different equations of state. Section 6.4.3.3 will study these relations in more detail and discuss the minimum amount of information required to specify the system completely.

For the simple fluid system (one force variable, p), $n = 2$; hence six bases of two variables can be chosen from the set of four T, S, p, V; for each of these six there are two equations of state, e.g., $S(T,V)$, $p(T,V)$ or $T(S,V)$, $p(S,V)$, etc. For $n = 3$, many more bases are possible, each with three equations of state, e.g., $S(T,X_2,X_3)$, $f_2(T,X_2,X_3)$, $f_3(T,X_2,X_3)$. An explicit example is provided by a simple magnetizable fluid in a uniform external magnetic field H, for which

$$dE = T\,dS - p\,dV + H\,dM \tag{6.4.7}$$

where M is the magnetic moment of the fluid.

It is possible now to complete the formulation in (6.4.1), which made use of all the forces and displacements to characterize a physical system,

* The notation for partial derivatives is the conventional thermodynamic one in which the remaining independent variables, in addition to the one being varied, are stated explicitly as subscripts, since change of basis is common. For short the set without X_i is designated X_j, $j = 1$ to n, $\neq i$.

† Exceptions occur for restricted systems, such as multiphase systems, in which fewer than n variables form the independent set in one of the relations like Eq. (6.4.6); then certain sets of n variables are not independent. (See Section 6.4.3.3.3.)

but no further quantities could be used or had even been defined at that point. It was assumed then that a sufficient number of the f_i, X_i could serve as a set of basic variables whose values would fix the equilibrium state of the system, and this set was described as a complete set which could form the basis of the thermodynamic space. Now that the thermodynamic postulates have been stated, and led to the functions of state E, T, S and the relation Eq. (6.4.3), the above analysis verifies and completes the initial assumption. The number of basic variables required to fix the state of a system is one-half the number of forces and displacements plus one (e.g., in the discussion above $n = \frac{1}{2}2(n - 1) + 1$). Hence for the simplest system p and V are just enough, for all more complex systems the $2(n - 1)$ variables f_i, X_i are redundant and satisfy $n - 2$ relations among themselves, e.g., for $n = 3$, eliminating T between $f_2(T,X_2,X_3)$ and $f_3(T,X_2,X_3)$ gives a relation among f_2, f_3, X_2, X_3. Finally we note that the $2n$ basic variables could contain E instead of S, the discussion being based on Eq. (6.4.3) divided by T, instead of Eq. (6.4.5); thus $E(T,X_2 \ldots X_n)$ would be an equation of state for the basis T, X_j, $j = 2$ to n.

6.4.3.2.2. EXAMPLES OF EQUATIONS OF STATE. To illustrate the general discussion, we give a few examples of equations of state for simple fluid systems, obtained either from measurement or from statistical mechanics and a microscopic model, or frequently both. They are limited in range and in accuracy.

Gases are described by the simple two variable system, and have two equations of state. In the limiting case of the ideal gas, possible forms of the equations of state are (for N moles)

$$pV = NRT \tag{6.4.8}$$

$$S(T,V) = S_0 + \int_{T_0}^{T} \frac{C_v(T)}{T}\, dT + NR \ln\left(\frac{V}{V_0}\right) \tag{6.4.9}$$

or

$$S(T,p) = S_0' + \int_{T_0}^{T} \frac{C_p(T)}{T}\, dT - NR \ln\left(\frac{p}{p_0}\right). \tag{6.4.10}$$

In place of Eq. (6.4.9) or Eq. (6.4.10) we can give

$$E(T) = E_0 + \int_{T_0}^{T} C_v(T)\, dT. \tag{6.4.11}$$

In Eqs. (6.4.9) to (6.4.11) S_0 and E_0 are values at the reference temperature T_0 and volume V_0 (S_0' is the reference value at T_0 and p_0). For ideal gases $C_v(T)$, the heat capacity at constant volume, may be a complex function of T, but not of V, and may combine various temperature-dependent contributions from the translational, the rotational, vibrational, and electronic degrees of freedom of the molecules constituting the gas. Similar dependence holds for the heat capacity at constant pressure, $C_p(T) = C_v(T) + NR$. Real gases approach the ideal at low p and high T.

Small deviations from ideality may be accurately allowed for by a virial expansion. This may be given in the forms

$$\frac{pV}{NRT} = 1 + \frac{NB(T)}{V} + \frac{N^2C(T)}{V^2} + \cdots$$
$$= 1 + \frac{B(T)p}{RT} + \frac{C'(T)p^2}{RT} + \cdots \quad (6.4.12)$$

where B and C (or C') are the second and third virial coefficients, which may be found by measurement or by statistical calculation and an intermolecular force law. Correspondingly, to the approximation of linear terms in p in Eq. (6.4.12),

$$S(T,p) = S_0 + \int_{T_0}^{T} \frac{C_p(T)}{T}\, dT - NR \ln \frac{p}{p_0} - N \frac{dB}{dT}(p - p_0) \quad (6.4.13)$$

where $C_p(T)$ is the same as for the ideal gas, i.e., C_p in the low pressure limit.*

Equations of state for liquids retain the simplicity of a description by the single force-displacement pair, p and V. However, it is more difficult to develop simple, general functional forms than for gases or solids, because a good simple initial approximation is not available, as in those cases. In fact one proceeds by trying to modify either a gaseous or a crystalline description, but this involves quite rough approximations.

Useful empirical formulas exist, such as the equation for the p dependence of $V(p,S)$

$$\frac{V(p,S)}{V(0,S)} = \left(\frac{A(S)}{p + A(S)}\right)^{1/n}.\dagger \quad (6.4.14)$$

The principle of corresponding states is also useful, and implies a common functional form (whatever it may be) for equations of state of large classes of liquids (see reference 4, p. 235).

A systematic approach from a plausible theoretical model necessarily falls back on the general expressions of statistical mechanics for equilib-

* A recent comprehensive discussion of equations of state and thermodynamic functions for gases may be found in Hirschfelder *et al.*[4]

† A modified form of the Tait equation; see Hirschfelder *et al.*,[4] p. 261. This is also a general reference for liquid equations of state, particularly Chapter 4.

[4] J. O. Hirschfelder, C. F. Curtiss, and R. B. Bird, "Molecular Theory of Gases and Liquids." Wiley, New York, 1954.

rium properties of systems with many interacting and randomly moving particles. These expressions are unfortunately not simple, and the final formulas for equations of state are crude because of uncontrolled mathematical approximations in evaluation of the general expressions.

6.4.3.3. Thermodynamic Relations. 6.4.3.3.1. THE GENERAL MAXWELL RELATIONS FOR AN n-VARIABLE SYSTEM. The general differential relation for dE in Eq. (6.4.5) serves not only to establish the number of independent and dependent variables for a given system (defined with certain constraints), but provides also a number of relations among those dependent variables. Thus the $n(n-1)/2$ Maxwell relations,

$$\begin{aligned}\left(\frac{\partial f_i}{\partial X_j}\right)_{X_h, h=1 \text{ to } n, \neq j} &= \left(\frac{\partial f_j}{\partial X_i}\right)_{X_l, l=1 \text{ to } n, \neq i} \\ &= \left(\frac{\partial^2 E}{\partial X_i\, \partial X_j}\right)_{X_k, k=1 \text{ to } n, \neq i,j}; \qquad i, j = 1 \text{ to } n,\ i \neq j\end{aligned} \tag{6.4.15}$$

follow and are, in fact, the necessary and sufficient conditions that

$$\sum_{i=1}^{n} f_i\, dX_i$$

be an exact differential. They reduce the n^2 derivatives of f_i, $i = 1$ to n, with respect to X_j, $j = 1$ to n, to $n(n+1)/2$ independent derivatives.

Since any n of the $2n$ variables f_i, X_i may be chosen as independent, the Maxwell relations have many equivalent forms, depending on the basis. These may all be simply and compactly expressed by the use of Jacobian notation. We introduce the Jacobian determinant in n variables, and the compact generalized partial derivative notation:

$$\frac{\partial(x_1, x_2, \cdots x_n)}{\partial(y_1, y_2, \cdots y_n)} \equiv \begin{vmatrix} \frac{\partial x_1}{\partial y_1} & \frac{\partial x_2}{\partial y_1} & \cdots & \frac{\partial x_n}{\partial y_1} \\ \frac{\partial x_1}{\partial y_2} & \frac{\partial x_2}{\partial y_2} & \cdots & \vdots \\ \vdots & & & \vdots \\ \frac{\partial x_1}{\partial y_n} & \cdots & \cdots & \frac{\partial x_n}{\partial y_n} \end{vmatrix}$$

$$\frac{\partial x_i}{\partial y_j} \equiv \left(\frac{\partial x_i}{\partial y_j}\right)_{y_k, k=1 \text{ to } n, \neq j}. \tag{6.4.16}$$

Three useful rules for manipulation of the Jacobians will be needed.

(a) The sign rule: an odd permutation on either x's or y's changes the sign of the Jacobian in Eq. (6.4.16); an even permutation does not.

(b) The reduction rule: if any x's or y's are identical, these common variables may be cancelled (after bringing the x's and y's into a vertical line, with proper sign changes) and the Jacobian reduced to one of lower order in which all derivatives have the common variables held constant.

(c) The substitution rule:

$$\frac{\partial(x_1,x_2, \cdots x_n)}{\partial(y_1 \cdots y_n)} = \frac{\partial(x_1 \cdots x_n)}{\partial(z_1 \cdots z_n)} \Big/ \frac{\partial(y_1 \cdots y_n)}{\partial(z_1 \cdots z_n)}.^{*}$$

Then rules (a) and (b) put Eq. (6.4.15) in the form:

$$\left(\frac{\partial(f_i,X_i)}{\partial(X_i,X_j)}\right)_{X_k} = -\left(\frac{\partial(f_j,X_j)}{\partial(X_i,X_j)}\right)_{X_k, k=1 \text{ to } n, \neq i,j}. \tag{6.4.17}$$

Equation (6.4.17) may be generalized to any one of a large class of bases, the *nonconjugate bases.*

Definition 10. A nonconjugate basis for an n-variable system is a set of n independent variables chosen one, and only one, from each conjugate pair.

Transforming to a general nonconjugate basis $Y_1, \cdots Y_n$, with $y_1, \cdots y_n$ the conjugate set of dependent variables, and using the rules for Jacobians, gives

$$\left(\frac{\partial(f_i,X_i)}{\partial(Y_i,Y_j)}\right)_{Y_k} = -\left(\frac{\partial(f_j,X_j)}{\partial(Y_i,Y_j)}\right)_{Y_k, k=1 \text{ to } n, \neq i,j}. \tag{6.4.18}$$

Equation (6.4.18) generalizes Eq. (6.4.17) to any of 2^n nonconjugate bases. In addition the nonconjugate condition may be relaxed to include the bases (f_i,X_i,Y_k) and (f_j,X_j,Y_k) for any set of Y_k, since multiplying Eq. (6.4.18) by

$$\left(\frac{\partial(Y_i,Y_j)}{\partial(f_i,X_i)}\right)_{Y_k}$$

and using rule (c) substitutes the basis (f_i,X_i,Y_k). Thus each of the $n(n-1)/2$ Maxwell relations has $(2^n + 2)$ forms.

These variations may be illustrated by the two-variable fluid case with f_1, $X_1 = T, S$; f_2, $X_2 = -p, V$. The six choices of Y_i, Y_j in Eq. (6.4.18) give

$$\frac{\partial(T,S)}{\partial(T,-p)} = -\frac{\partial(-p,V)}{\partial(T,-p)} = -\left(\frac{\partial S}{\partial p}\right)_T = \left(\frac{\partial V}{\partial T}\right)_p \tag{6.4.19}$$

* These rules follow readily from the properties of determinants; they are discussed by Crawford.[5]

[5] F. H. Crawford, Thermodynamic variables in n-variable systems in Jacobian form, Part I. *Proc. Am. Acad. Arts Sci.* **78,** 165 (1950).

similarly

$$\left(\frac{\partial S}{\partial V}\right)_T = \left(\frac{\partial p}{\partial T}\right)_V; \qquad \left(\frac{\partial T}{\partial p}\right)_S = \left(\frac{\partial V}{\partial S}\right)_p; \qquad -\left(\frac{\partial T}{\partial V}\right)_S = \left(\frac{\partial p}{\partial S}\right)_V$$

$$\frac{\partial(T,S)}{\partial(T,S)} = -\frac{\partial(-p,V)}{\partial(T,S)} = 1 = \left(\frac{\partial p}{\partial T}\right)_S \left(\frac{\partial V}{\partial S}\right)_T - \left(\frac{\partial V}{\partial T}\right)_S \left(\frac{\partial p}{\partial S}\right)_T$$

similarly

$$1 = \left(\frac{\partial T}{\partial p}\right)_V \left(\frac{\partial S}{\partial V}\right)_p - \left(\frac{\partial S}{\partial p}\right)_V \left(\frac{\partial T}{\partial V}\right)_p.$$

The relations Eq. (6.4.19) apply, of course, to any two-variable system. A useful example is a simple magnetic system in which pressure-volume work can be ignored, hence Eq. (6.4.7) simplifies to $dE = T\,dS + H\,dM$, and applies to solid magnetic systems as well. Then replacing $-p$ by H and V by M in Eq. (6.4.19), the first and third Maxwell relations become

$$\left(\frac{\partial S}{\partial H}\right)_T = \left(\frac{\partial M}{\partial T}\right)_H \tag{6.4.20}$$

$$\left(\frac{\partial T}{\partial H}\right)_S = -\left(\frac{\partial M}{\partial S}\right)_H = -\left(\frac{\partial M}{\partial T}\right)_H \Big/ \left(\frac{\partial S}{\partial T}\right)_H. \tag{6.4.21}$$

Equations (6.4.20) and (6.4.21) give a quantitative description of the method of magnetic cooling for attainment of low temperatures.* Thus since $(\partial M/\partial T)_H < 0$ holds usually, Eq. (6.4.20) gives the decrease in S for isothermal magnetization. Then since $(\partial S/\partial T)_H > 0$ is always true, Eq. (6.4.21) gives the decrease in T in a subsequent adiabatic demagnetization or decrease of H.

6.4.3.3.2. Independent Measurements Determining the Equations of State. The Maxwell relations permit a very useful reduction in the number of independent measurements required to establish the equations of state. In Section 6.4.3.2 it was pointed out that the equilibrium properties of the n-variable system are completely described by n equations of state, functions of n independent variables. If the n independent variables Y_i are a nonconjugate set, it is not necessary to measure the functional dependence of the equations of state y_i on all of the Y_i. By using Eq. (6.4.18) the first $(i - 1)$ variables may be held constant in measuring y_i, and the variation of y_i with these variables found from other measurements. Thus a complete set of measurements can be: $y_1(Y_1 \ldots Y_n)$, $y_2(Y_1^{(0)}, Y_2 \ldots Y_n)$, $\ldots$ $y_n(Y_1^{(0)}, Y_2^{(0)} \ldots Y_{n-1} Y_n)$, where the $Y_i^{(0)}$ are held constant. Complete information on the equilibrium properties may be obtained by measuring $n(n + 1)/2$ functional variations, rather than n^2.

* See, for example, Wilson,[3] p. 308.

For the two-variable fluid system, the four possible bases are $V(T,p)$, $S(T,p)$; $p(T,V)$, $S(T,V)$; $T(S,V)$, $p(S,V)$; $S(p,V)$, $T(p,V)$; in each of these, only three functional variations need be measured, and then Eq. (6.4.19) employed for the fourth. However, if the bases p, V or T, S are used, which violate the nonconjugate condition, four functional variations must be measured; thus from $T(V,p)$, $S(V,p^{(0)})$ we cannot find $S(V,p)$. Although Eq. (6.4.19) gives

$$\left(\frac{\partial S}{\partial p}\right)_V = \left[\left(\frac{\partial T}{\partial p}\right)_V \left(\frac{\partial S}{\partial V}\right)_p - 1\right] \Big/ \left(\frac{\partial T}{\partial V}\right)_p$$

$\left(\frac{\partial S}{\partial V}\right)_p$ is known only at $p^{(0)}$, and cannot be integrated over p.

6.4.3.3.3. RESTRICTED SYSTEMS. Important special systems exist in which the preceding general analysis of thermodynamic relations in an arbitrary basis is subject to certain restrictions. These are the so-called *restricted systems* in which some equations of state relate fewer than $(n + 1)$ variables, i.e., the dependent variable depends on only *part* of the complete set of independent variables and some variables are missing. Two consequences are: (1) the set of n or fewer variables in those equations of state with missing variables may not be used as part of a complete basis, since they are functionally related, (2) certain derivatives vanish, namely those with respect to the missing variables. If the missing variable is not conjugate to the dependent variable (as in the first example below), a pair of derivatives vanish by a Maxwell relation and a second equation of state has a missing variable, otherwise—the conjugate case—a single derivative vanishes (as in the second example below). In either case the number of independent derivatives in the set of n^2 derivatives with respect to n independent variables is reduced to fewer than $(n^2 + n)/2$, hence the general relations among derivatives simplify.*

We consider some examples:

Example 1. The simple magnetic system in Eqs. (6.4.20) and (6.4.21) satisfies $dE = T\,dS + H\,dM$. If the magnetic equation of state (basis T, H) is $M = \chi H$, and the susceptibility χ is independent of T (e.g., diamagnetism) then M is independent of T, and S is independent of H. There are only two independent derivatives and the thermal and magnetic behaviors are not coupled.

Example 2. A homogeneous additive fluid phase of one component, considered as an open system, so that the mole number N may vary.

* Restricted systems and the Jacobian method have been discussed by Crawford.[6]

[6] F. H. Crawford, Polyphase polycomponent chemical systems. *Proc. Am. Acad. Arts Sci.* **83**, 191 (1955).

Anticipating the results of Section 6.4.5, the system becomes a three-variable system in which

$$dE = T\,dS = p\,dV + \mu\,dN. \qquad (6.4.22)$$

In consequence of the assumptions of homogeneity and additivity, (hence E, S, V are proportional to N at constant T, p), the chemical potential μ is a function of the intensive variables T, p alone, and Eq. (6.4.22) gives directly, since

$$\left(\frac{\partial E}{\partial N}\right)_{T,p} = \frac{E}{N}; \qquad \left(\frac{\partial S}{\partial N}\right)_{T,p} = \frac{S}{N}, \qquad \text{etc.}$$

$$\mu = \frac{(E - TS + pV)}{N}, \qquad d\mu = -\frac{S}{N}\,dT + \frac{V}{N}\,dp. \qquad (6.4.23)$$

The set (T,p,μ) is then not a possible basis, because $\mu(p,T)$, but we may add an extensive variable and use, say, (T,p,N). The set of six independent derivatives of (S,V,μ) with respect to (T,p,N) has one zero,

$$\left(\frac{\partial \mu}{\partial N}\right)_{T,p} = 0$$

and two trivial reductions

$$\left(\frac{\partial S}{\partial N}\right)_{T,p} = -\left(\frac{\partial \mu}{\partial T}\right)_{p,N} = \frac{S}{N}; \qquad \left(\frac{\partial V}{\partial N}\right)_{T,p} = +\left(\frac{\partial \mu}{\partial p}\right)_{T,N} = \frac{V}{N}.$$

This leaves only three independent derivatives,

$$\left(\frac{\partial S}{\partial T}\right)_{p,N}, \left(\frac{\partial V}{\partial T}\right)_{p,N} = -\left(\frac{\partial S}{\partial p}\right)_{T,N}, \left(\frac{\partial V}{\partial p}\right)_{T,N}$$

just as for the closed system, as might be expected, since the simple character of the dependence on N should not lead to any new physical phenomena for the open system compared to the closed system.

Example 3. A one-component, two-phase fluid system has two homogeneous phases in material contact and the system is additive in the two phases. The system as a whole, if closed, is a simple two-variable system. However, (p,T) is no longer a possible basis and the system is restricted. Anticipating Theorem 4 (Section 6.4.5), the chemical potentials of the two phases are equal, as are T and p, hence

$$\mu_1(T,p) = \mu_2(T,p). \qquad (6.4.24)$$

Also by the additivity,

$$V = N(xv_1 + (1 - x)v_2), \qquad x = N_1/N \qquad (6.4.25)$$

$$S = N(xs_1 + (1 - x)s_2) \qquad (6.4.26)$$

where v_1, v_2 are the molar volumes, s_1, s_2 the molar entropies of phases 1 and 2, all functions of T and p, N the total number of moles (a constant), and x the fraction in phase 1. Thus the equations of state for the basis T, V are $S(T,V)$, $p(T)$.

There are just two independent derivatives:

$$\left(\frac{\partial S}{\partial T}\right)_V, \left(\frac{\partial p}{\partial T}\right)_V = \frac{dp}{dT} = \left(\frac{\partial S}{\partial V}\right)_T \qquad \text{whereas} \qquad \left(\frac{\partial p}{\partial V}\right)_T = 0.$$

Various relations are simpler:

$$\left(\frac{\partial S}{\partial T}\right)_p = C_p = \left(\frac{\partial E}{\partial T}\right)_p = \left(\frac{\partial V}{\partial T}\right)_p = \left(\frac{\partial V}{\partial p}\right)_T = \infty \tag{6.4.27}$$
$$\left(\frac{\partial V}{\partial T}\right)_S = -\frac{C_v}{T(dp/dT)}; \qquad \left(\frac{\partial V}{\partial p}\right)_S = -\frac{C_v}{T(dp/dT)^2}$$

where $C_p = T\left(\frac{\partial S}{\partial T}\right)_p$, $C_v = T\left(\frac{\partial S}{\partial T}\right)_V$ are the usual heat capacities at constant p and V respectively. For this system it is of considerable interest to express the thermodynamic derivatives in terms of quantities referring to the separate phases. A natural description of the system which achieves this introduces the internal coordinate x and the basis (T,x). Then using Eqs. (6.4.25), (6.4.26), we can transform the Maxwell relation to depend on properties of the two phases,

$$\frac{dp}{dT} = \left(\frac{\partial S}{\partial V}\right)_T = \frac{\partial(S,T)}{\partial(T,x)} \bigg/ \frac{\partial(V,T)}{\partial(T,x)} = \frac{s_1 - s_2}{v_1 - v_2} \tag{6.4.28}$$

which is the Clausius–Clapeyron equation.

6.4.3.3.4. TABLE OF THERMODYNAMIC RELATIONS. By changing the basis, a great variety of thermodynamic relations may be obtained which are mixtures of Maxwell relations and mathematical identities. Systematic procedures have been given to find all relations among first derivatives,[5] based on the fact that any derivative may be expressed in terms of $(n^2 + n)/2$ derivatives from a given nonconjugate basis (see Eq. (6.4.15) and the following text), hence a relation exists among any $1 + (n^2 + n)/2$ derivatives. The two-variable case has been described by Bridgman,[7] who tabulates all 45 Jacobians of T, S, p, V, E, H, F, G, Q, W with respect to T and p in terms of three standard derivatives

$$\left(\frac{\partial S}{\partial T}\right)_p, \left(\frac{\partial V}{\partial T}\right)_p \text{ and } \left(\frac{\partial V}{\partial p}\right)_T.$$

[7] P. W. Bridgman, "A Condensed Collection of Thermodynamic Formulas." Harvard Univ. Press, Cambridge, Massachusetts, 1925.

Any derivative is then expressed in terms of these three by the relation

$$\left(\frac{\partial x}{\partial y}\right)_z = \frac{\partial(x,z)}{\partial(T,p)} \Big/ \frac{\partial(y,z)}{\partial(T,p)}.$$

Methods for second derivative relations and for the two-phase, one-component system are also discussed.

The following short table of common and useful relations (in addition to Maxwell relations) illustrates some of the results. For compactness, only relations involving T, S, p, V, E are given; Eqs. (6.4.29) introduce only first derivatives of T, S, p, V; Eqs. (6.4.30) bring in E as a variable; Eqs. (6.4.31) note a few relations involving second derivatives.

$$C_p - C_v = T\left(\frac{\partial T}{\partial p}\right)_V\left(\frac{\partial V}{\partial T}\right)_p, \qquad \left(\frac{\partial V}{\partial p}\right)_S = \left(\frac{\partial V}{\partial p}\right)_T + \frac{T}{C_p}\left(\frac{\partial V}{\partial T}\right)_p^2 \tag{6.4.29}$$

$$\left(\frac{\partial V}{\partial p}\right)_S \Big/ \left(\frac{\partial V}{\partial p}\right)_T = C_v/C_p, \qquad \left(\frac{\partial V}{\partial T}\right)_S \Big/ \left(\frac{\partial V}{\partial T}\right)_p = \left(\frac{\partial S}{\partial p}\right)_V \Big/ \left(\frac{\partial S}{\partial p}\right)_T$$

$$\left(\frac{\partial T}{\partial p}\right)_S = \frac{T}{C_p}\left(\frac{\partial V}{\partial T}\right)_p, \qquad \left(\frac{\partial p}{\partial T}\right)_S = \frac{C_p}{T(\partial V/\partial T)_p}$$

$$\left(\frac{\partial T}{\partial V}\right)_S = -\frac{T}{C_v}\left(\frac{\partial p}{\partial T}\right)_S$$

$$\left(\frac{\partial E}{\partial V}\right)_T = T\left(\frac{\partial p}{\partial T}\right)_V - p, \qquad \left(\frac{\partial E}{\partial T}\right)_V = C_v \tag{6.4.30}$$

$$\left(\frac{\partial E}{\partial p}\right)_T = -T\left(\frac{\partial V}{\partial T}\right)_p - p\left(\frac{\partial V}{\partial p}\right)_T, \qquad \left(\frac{\partial E}{\partial T}\right)_p = C_p - p\left(\frac{\partial V}{\partial T}\right)_p$$

$$\left(\frac{\partial T}{\partial V}\right)_E = \frac{1}{C_v}\left[p - T\left(\frac{\partial p}{\partial T}\right)_V\right]$$

$$\left(\frac{\partial C_p}{\partial p}\right)_T = -T\left(\frac{\partial^2 V}{\partial T^2}\right)_p, \qquad \left(\frac{\partial C_p}{\partial p}\right)_V = -T\left(\frac{\partial^2 V}{\partial T^2}\right)_p + \left(\frac{\partial C_p}{\partial T}\right)_p\left(\frac{\partial T}{\partial p}\right)_V \tag{6.4.31}$$

$$\left(\frac{\partial C_v}{\partial V}\right)_T = T\frac{\partial^2 p}{\partial T^2}, \qquad \left[\frac{\partial}{\partial p}\left(\frac{\partial V}{\partial p}\right)_T\right]_V = \left(\frac{\partial^2 V}{\partial p^2}\right)_T + \frac{\partial^2 V}{\partial p \partial T}\left(\frac{\partial T}{\partial p}\right)_V.$$

6.4.3.4. General Form of the Second Law, Stability Theory. The restricted form of the second law, Postulate 3a, has led to the existence of entropy as a function of state, to the definitions of absolute entropy and temperature, and to relations among the variables describing equilibrium states. A generalization of the postulate now leads to conclusions about the direction of change in irreversible processes, and about the general conditions for stability of thermodynamic systems. This generalization is simply stated in the inaccessibility form as

Postulate 3b. In the vicinity of any state of a system there are states that are not accessible by any adiabatic processes.

The distinction from Postulate 3a is the reference to *any* adiabatic process. Since this includes the reversible adiabatics, the conclusions from Postulate 3a about reversible processes, the existence of S and T and of equations of state and thermodynamic relations are all valid. To discuss irreversible adiabatic processes which produce changes in entropy, we consider the accessibility of neighboring states with different entropies.

Application of Postulate 3b leads easily to the conclusion that all physical systems either increase their entropy or decrease their entropy in an irreversible adiabatic process (to a new equilibrium state) but not both. A single experimental observation is all that is necessary to fix the direction, hence we introduce

Observation 2. A system undergoing an irreversible adiabatic process increases its entropy (absolute temperature being measured on the Kelvin scale).

Postulate 3b and Observation 2 thus lead to the theorem of entropy increase,

Theorem 2. The entropy of a system always increases or remains constant in adiabatic changes between states.

As a special application of Theorem 2 note that the relaxation of a constraint never decreases the entropy of a system. Since no work or heat is involved, a constraint may be relaxed in a totally isolated system, hence we are led to the fundamental theorem of thermodynamic stability.

Theorem 3. A totally isolated system in equilibrium has a maximum entropy with respect to all states constrained with respect to the given one, and linked to it by a process in which neither heat nor work is exchanged.

Theorem 3 may be applied to determine whether a system is stable, since, if changes in the variables can be found which increase S, but involve no heat or work, such changes will occur irreversibly, and the system is initially unstable.

A useful corollary to Theorem 3 considers the stability of a system which can exchange heat and work with a reservoir, where we use

Definition 11. A *reservoir* is a system so large and with such rapid transport processes that exchange of heat or work with it has a negligible effect on its variables of state which remain at their initial values throughout. Its exchanges of heat or work can be measured, however.

Starting with the system and reservoir in equilibrium at temperature T, consider an exchange of heat and work with the reservoir, which satisfies Eq. (6.4.3) for the reservoir changes (which are considered reversible), and takes the system to some possible constrained state, with defined

S and E. Applying Theorem 3 to the isolated composite system formed by them both, requires for all such processes

$$(\Delta S)_s + (\Delta S)_r = (\Delta S)_s - \frac{(\Delta E + W)_s}{T} = \frac{(T\,\Delta S - \Delta E - W)_s}{T} \leq 0 \tag{6.4.32}$$

(where subscripts s and r refer to system and reservoir respectively), otherwise the change would occur irreversibly. Thus for $T > 0$, the general stability condition for a system exchanging work and heat with a reservoir requires

$$\Delta E - T\,\Delta S + W \geq 0 \tag{6.4.33}$$

where ΔE, ΔS refer to the system, and T to the reservoir temperature (or initial system temperature), and the changes are of arbitrary size.

The general stability condition, Eq. (6.4.33), for a system in contact with a reservoir leads naturally to the introduction of the free energy to obtain a simple formulation of the stability condition in certain frequently occurring situations. We define a new function of state, the *Helmholtz free energy*, by:

$$F \equiv E - TS. \tag{6.4.34}$$

Then for a system which exchanges heat with a reservoir at temperature T, but not work, the stability condition for states at temperature T may be written

$$(\Delta F)_T \geq 0, \ (W = 0). \tag{6.4.35}$$

Thus F is a minimum in the stable state at T compared with all the constrained states at T which are linked by a process in which heat is exchanged with the reservoir at T, but not work.

Similarly for a system with $-p$ and V as one set of conjugate variables, such as a fluid system, it is useful to introduce the *Gibbs free energy*, G, defined by:

$$G \equiv E - TS + pV. \tag{6.4.36}$$

Now consider a system which exchanges heat and pressure-volume work with a reservoir at temperature T and pressure p, but no other forms of work. Then the stability condition for equilibrium states of such a system at definite T and p with respect to constrained states at the same p and T, and which are linked by a process involving exchanges of heat and pressure–volume work, may be written

$$(\Delta G)_{T,p} \geq 0 \qquad (W' = 0) \tag{6.4.37}$$

where $W' = W - p\,\Delta V$.

Equation (6.4.37) is useful in discussing multiphase systems, in which transfer between phases may be considered at constant T and p.

By consideration of infinitesimal changes, Eq. (6.4.33) leads to stability conditions in terms of properties of the initial state.* For the simple fluid system, variables T, S, $-p$, V, the necessary and sufficient differential conditions for stability are either

$$\left(\frac{\partial T}{\partial S}\right)_V > 0, \qquad -\left(\frac{\partial p}{\partial V}\right)_T > 0 \tag{6.4.38}$$

equivalent to

$$C_v = T\left(\frac{\partial S}{\partial T}\right)_V > 0, \qquad K_T \equiv -\frac{1}{V}\left(\frac{\partial V}{\partial p}\right)_T > 0$$

or

$$-\left(\frac{\partial p}{\partial V}\right)_S > 0, \qquad \left(\frac{\partial T}{\partial S}\right)_p > 0 \tag{6.4.39}$$

equivalent to

$$C_p = T\left(\frac{\partial S}{\partial T}\right)_p > 0, \qquad K_S \equiv -\frac{1}{V}\left(\frac{\partial V}{\partial p}\right)_S > 0.$$

Of course all four inequalities are necessary conditions for stability. However, the combination C_p and $K_T > 0$ (or C_v and K_S) would not be sufficient alone, since

$$K_S = K_T - \frac{T}{VC_p}\left(\frac{\partial V}{\partial T}\right)_p^2, \qquad C_v = C_p - \frac{T}{VK_T}\left(\frac{\partial V}{\partial T}\right)_p^2$$

and K_S and C_v could still be <0.

6.4.4. The Third Law of Thermodynamics

The standard postulational content of thermodynamics is now completed by a postulate about the special behavior of systems as they approach 0°K, usually referred to as the Third Law of Thermodynamics. This will be assumed in the following form:

Postulate 4. The entropies of all states of a system approach the same finite value as T approaches 0°K.

Of course the entropies of all states are referred to the same standard state. Note that the postulate applies, of course, to constrained states, since they are equilibrium states, and also to pseudoequilibrium states which are effectively constrained on the time scale of measurements.

The third law provides a useful alternative way of measuring entropy differences between states of a system, namely by referring the entropy

* See Landau and Lifshitz,[8] Chapter 6.

[8] L. Landau and E. Lifshitz, "Statistical Physics." Oxford Univ. Press, London and New York, 1938.

of these states to their values at 0°K, say by specific heat measurements down to 0°K, and then using the equality of the entropies at 0°K. This procedure is particularly valuable if the reversible path between these states is not experimentally possible or convenient.*

The third law has other interesting consequences: All heat capacities, $T(\partial S/\partial T)_{Y_i}$, (with any $(n - 1)$ variables Y_i held constant, except of course, S and T) must approach zero as $T \to 0°K$, otherwise S would have a logarithmic infinity at 0°K.

All thermal expansions of the form

$$\left(\frac{\partial V}{\partial T}\right)_p, \left(\frac{\partial X_i}{\partial T}\right)_{fi, Y_k, k=2 \text{ to } n, \neq i}$$

approach zero as $T \to 0°K$ (the $(n - 2)$ variables Y_k are any nonconjugate set with $k \neq 1, i$). This follows from the Maxwell relation

$$\left(\frac{\partial X_i}{\partial T}\right)_{f_i, Y_k} = \left(\frac{\partial S}{\partial f_i}\right)_{T, Y_k}$$

(which is Eq. (6.4.18) with Y_i, $Y_j = f_i$, T), whose right hand side vanishes as $T \to 0°K$.

Phase transition lines on the p-T plane will generally become horizontal as $T \to 0°K$, since by the Clausius–Clapeyron equation (6.4.28), $(dp/dT) = (\Delta s/\Delta v)$, and $\Delta s \to 0$ as $T \to 0°K$ whereas Δv will not vanish in general.

The third law is sometimes given in another form known as the unattainability of the absolute zero. This may be stated as

Theorem 4. The temperature of a system cannot be reduced to 0°K in a finite number of steps, i.e., a finite sequence of adiabatic and isothermal processes.†

6.4.5. Quantity of Matter as a Thermodynamic Variable; Chemical Potential

Physical systems may be influenced in another important way besides the transfer of work and heat, namely by the transfer of matter. Systems in which the quantity of a component can be changed are commonly called open systems. Inclusion of such changes in the description of systems containing chemical reactions is essential, and for the description

* However, if a reversible path is not possible in principle, as appears to be true for certain complex metastable states, the assignment of entropy by this procedure cannot be checked by measurement and is accordingly of no physical significance.

† For a discussion see Wilson,[3] p. 191.

of systems of several phases it is rather convenient. (See Section 6.4.3.3.3.) Accordingly, a necessary generalization of the formulation considers such changes and greatly enlarges the scope of thermodynamics.

By proper choice of a reference state (the vaporized state of separated components is suitable) the ΔS and ΔE of a process in which ΔN moles of a component are added to a system may be measured. We may then define a new function of state, the chemical potential* of the qth component μ_q by

$$\mu_q \equiv \left(\frac{\partial E}{\partial N_q}\right)_{S,X_i,N_k}, \qquad i = 2 \text{ to } n, \qquad k = 1 \text{ to } r, \neq q \tag{6.4.40}$$

which goes with the differential relation

$$dE = T\,dS + \sum_{i=2}^{n} f_i\,dX_i + \sum_{q=1}^{r} \mu_q\,dN_q. \tag{6.4.41}$$

To show the role played by the chemical potential in the equilibrium of systems which can exchange matter (material contact for short), a process may be carried out on a composite system made up of two subsystems in material contact and in equilibrium. The additivity of E, S, and W for the two subsystems then leads easily to

Theorem 5. Two systems in material contact with respect to component q, and in equilibrium, have equal values of μ_q.

Thus the chemical potential plays a role with respect to matter exchange similar to that played by the temperature with respect to heat exchange. In both cases the system is being influenced by microscopic transfer processes beneath the level of macroscopic observation, and the effect of this influence is equality of T or μ for equilibrium between two coupled systems.

When two systems are brought into material contact with different μ_q's, we may conclude from Eq. (6.4.41) that component q will flow from the system with greater μ_q to the other system. Consider the two systems as a totally isolated composite system, exchanging heat but not matter, hence at the same T. When the constraint on matter exchange is removed, Eq. (6.4.41) gives for ΔS of the composite system, on cancelling the energy and work terms, and using $dN_q^{(2)} = -dN_q^{(1)}$,

$$T\,\Delta S = T(\Delta S^{(1)} + \Delta S^{(2)}) = -(\mu_q^{(1)} - \mu_q^{(2)})\,dN_q^{(1)} > 0$$

since by Theorem 3, $\Delta S > 0$. Hence $dN_q^{(1)} > 0$ if $\mu_q^{(2)} > \mu_q^{(1)}$.

Thus far no assumption of homogeneity or additivity has been made.

*This is also called the partial molar free energy for simple fluid systems.

For the usual fluid chemical systems, these assumptions hold either for the whole system, or for the separate phases of which it is made.

The chemical potential of a general system can be measured by using the definition, Eq. (6.4.40). However, it is more convenient if it is measured by using the value of μ_q for a simple system in matter contact with the given system. For example, a reference system of pure component q in contact with the system through a semipermeable membrane, can be separately calibrated, (i.e., μ_q determined as a function of state). If q happens to be in the gaseous state and ideal or nearly ideal, then simple theory and measurement will serve to calibrate it.

6.4.6. Fluctuations in Equilibrium States

The idea of an equilibrium state, introduced among the earliest definitions, and basic to almost all the statements made since, rests on the observation of time-independent behavior. The facts of observation show that this is never exactly satisfied, although it is usually very closely satisfied. Hence for precision of formulation and to include certain additional physical phenomena in the thermodynamic description, it is necessary to modify the notion of equilibrium state and certain other notions based upon it.

What is in fact observed (although usually of such a small magnitude that it requires special systems and arrangements to make the observation) is that all thermodynamic variables are continually fluctuating around their mean values. These fluctuations are inherent, having nothing to do with variations in external conditions, and persist indefinitely. Accordingly it is necessary to re-define equilibrium state as a situation in which the *time-average values* of basic variables of an isolated physical system do not change with time. Strictly speaking, the times involved must be indefinitely long to obtain complete accuracy. If we settle for finite times, we settle for inherent errors in our observations of the basic variables and an inherent uncertainty in whether we are dealing with an equilibrium state. All previous statements about equilibrium states must then be interpreted in this time-average sense (but are otherwise unmodified) to include these new phenomena. In addition certain statements on accessibility must be modified.

A quantitative description of these fluctuations is provided by the following postulate:

Postulate 5. An isolated system will occupy states of entropy less than the equilibrium entropy with a probability proportional to $e^{\Delta S/k}$, where k is the gas constant per molecule and ΔS is the (negative) entropy change from the equilibrium entropy.

Application of Postulate 5 to obtain explicit formulas for various mean square fluctuations around equilibrium may be made.* For example, we have in general for an n-variable system in equilibrium with and exchanging heat and work with a reservoir

$$\begin{aligned} \overline{\Delta X_i^2} &= kT \left(\frac{\partial X_i}{\partial f_i}\right)_{f_j, j=1 \text{ to } n, \neq i}, & i &= 1 \text{ to } n \\ \overline{\Delta f_i^2} &= kT \left(\frac{\partial f_i}{\partial X_i}\right)_{X_j, j=1 \text{ to } n, \neq i}, & i &= 1 \text{ to } n \\ \overline{\Delta f_i \, \Delta X_i} &= kT, & i &= 1 \text{ to } n. \end{aligned} \tag{6.4.42}$$

Fluctuations in the simple fluid system are obtained from Eq. (6.4.42) by putting $f_1, X_1, f_2, X_2 = T, S, (-p), V$.

* See Landau and Lifshitz,[8] Chapter 6.

7. OPTICS

7.1. Geometrical Optics*[1]

7.1.1. Definitions: Optical Properties of Matter

Optical effects in which wave phenomena (Section 7.2.1) may be ignored, or, alternatively, in which light propagates along rays, are classed under geometrical optics.† The directions of such rays may be changed by reflection or refraction at an optical surface, i.e., one bounding two media of different indices of refraction (see below). The angle between the incident ray and the normal to the surface is designated the angle of incidence, ϕ, the angle between the reflected ray and the normal the angle of reflection, r, and the angle between the refracted ray and the normal the angle of refraction, ϕ'. If both media are isotropic, all rays and the normal are coplanar and the angle of reflection equals the angle of incidence; if either or both media are anisotropic, however, this is not necessarily true (Chapter 7.4).

The ratio of the phase velocity of light in vacuum to that in the medium is called the refractive index of the medium, and for isotropic materials it is independent of direction.‡ In anisotropic media the refractive index may depend both on the direction in the medium and the polarization (Chapter 7.4) of the light. For isotropic materials the angles of refraction and incidence are related by Snell's law (Fig. 1a):

$$n \sin \phi = n' \sin \phi' \qquad (7.1.1)$$

where n and n' are refractive indices. One defines the critical angle between two media, ϕ_c, as the angle of incidence at which the angle of refraction becomes 90° (this occurs only in going from a medium of larger n to one of smaller, see Fig. 1b). Hence,

$$\sin \phi_c = n'/n. \qquad (7.1.2)$$

At any angle of incidence greater than the critical angle light is totally reflected.

† Fermat's principle, which is the basis of all geometrical optics, states that the actual path of a ray is such that all paths which are infinitely close to the actual path differ from it in optical path lengths (see below) only by terms of the second or higher order.

‡ For the measurement of refractive indices in solids, see Vol. 6, Part 9.

[1] A. Maréchal, *in* "Handbuch der Physik—Encyclopedia of Physics" (S. Flügge, ed.), Vol. 24, p. 51. Springer, Berlin, 1955.

* Chapters 7.1 through 7.9 are by **John R. Holmes** and **G. L. Weissler.**

For any ray passing from one point to another through various media, the refractive index for each medium times the distance traversed in that medium and summed over the entire path ($\Sigma_i n_i s_i$) is called the optical path length between the two points. For a medium with variable n, the optical path is defined as $\int_1^2 n\, ds$. In a perfect optical instrument one of the necessary conditions is that all optical paths between an object point and the corresponding image point are equal, as illustrated in Fig. 2a and 2b by paths I and II. Two surfaces have been chosen each of such a

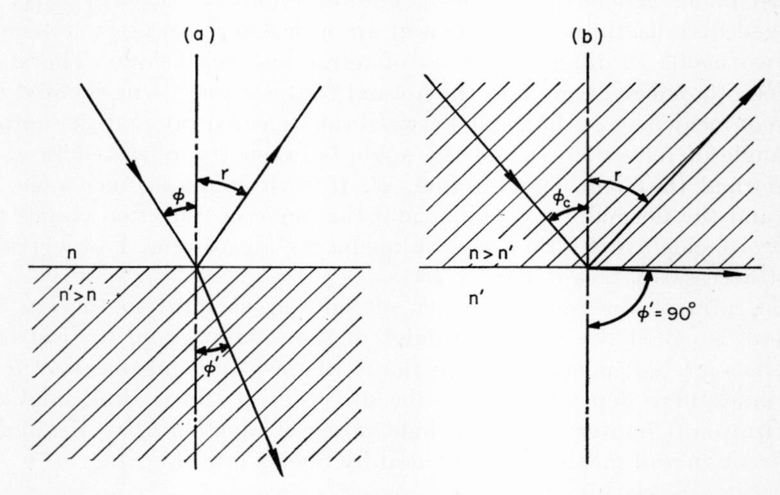

FIG. 1. Reflection and refraction at an optical surface; (a) rare to dense, (b) dense to rare, showing the critical angle, ϕ_c.

curvature as to bring together at Q' (image point) all rays originating from Q (object point). In this connection an aplanatic surface may be defined as one which contains all points for which the sum of the optical path lengths from two fixed object and image points is constant; a bundle of rays from the object point will therefore converge on or diverge from the image point after leaving the aplanatic surface. A surface may be nonaplanatic but approximate the aplanatic condition for paraxial rays, i.e., rays which are restricted to the vicinity of the axis of the surface.

The first focus of a surface is a point on the axis in the object space such that all rays from this point which reach the surface are made parallel to the axis (Fig. 3), while the second focus is a point in the image space such that all rays parallel to the axis which reach the surface are made to converge to or diverge from this point. The same definitions apply to combinations of such surfaces or lens systems.

Dispersion refers to the fact that the refractive index of any medium is a function of the wavelength, λ, of the light. Usually the index at shorter wavelengths is larger and varies more rapidly with λ. Hence there may be two chromatic aberrations, lateral and longitudinal (Section 7.1.2), caused by dispersion when light of more than one wavelength

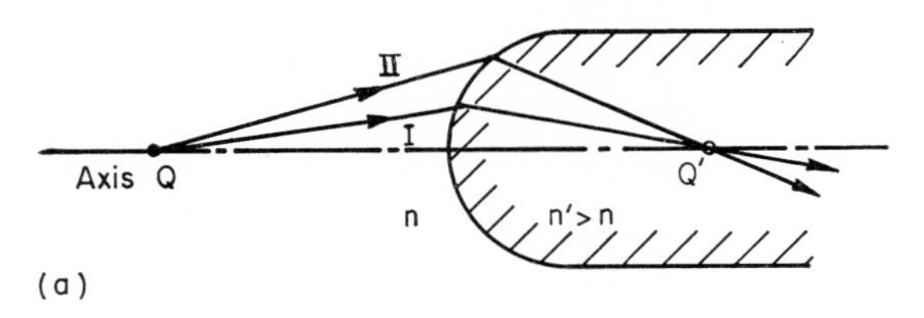

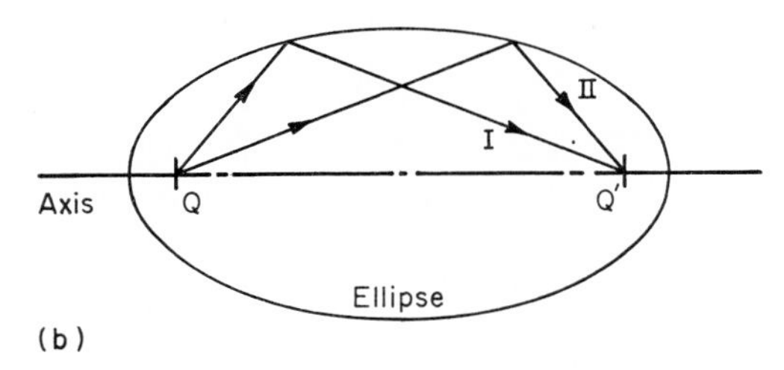

FIG. 2. Aplanatic surfaces; (a) for refraction, (b) for reflection.

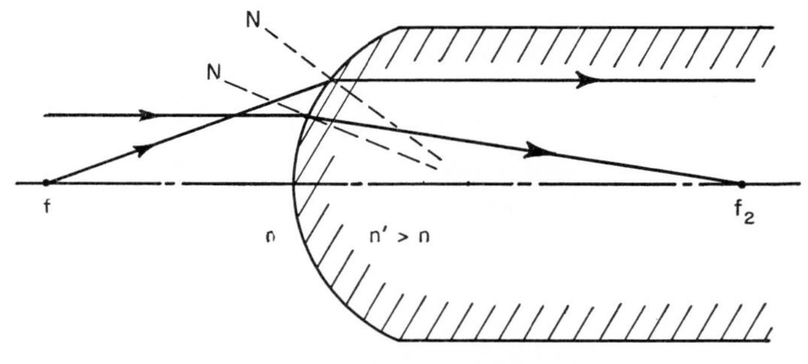

FIG. 3. Focal points for a refracting optical surface; N = normal.

is sent through a refracting optical system. Longitudinal chromatic aberration refers to the displacement of an image along the axis as the wavelength is changed, while the variation in size of the image with wavelength is called lateral chromatic aberration.

There are also five monochromatic image aberrations which may occur with light of one wavelength: spherical aberration; coma; astigmatism; curvature; and distortion.

Spherical aberration occurs when rays from a point on the axis which strike the outer zones of the surface are deflected, so that their direction does not intersect the axis at the same point as for the paraxial rays. Thus, the image of an object point is the smallest circle (circle of least confusion) containing all rays from the point. When the marginal rays intersect the axis closer to the surface than do the paraxial ones, the spherical aberration is called positive and the surface is said to be undercorrected (Fig. 4). In the opposite case the aberration is negative and the surface overcorrected.

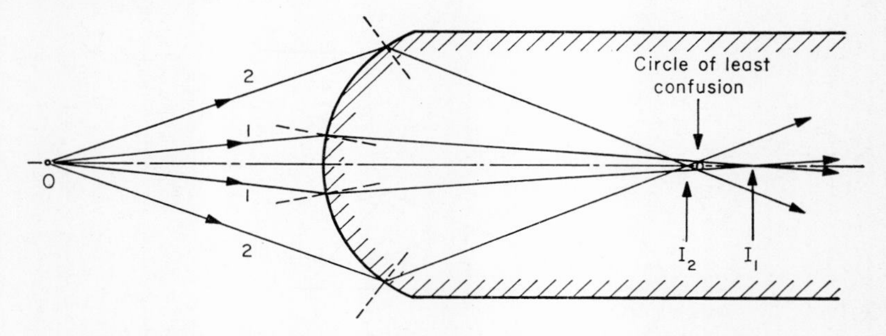

FIG. 4. Spherical aberration of a refracting spherical surface.

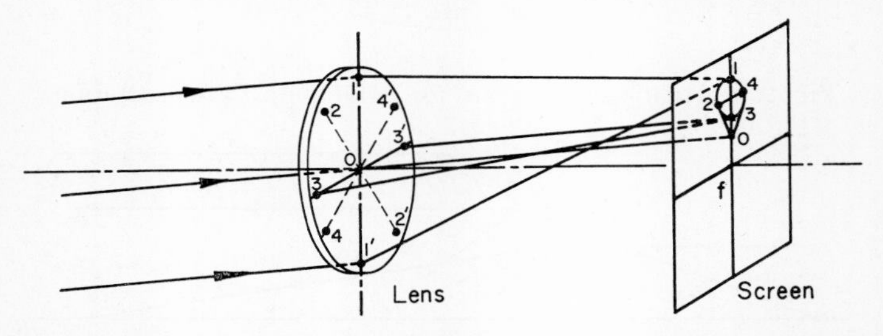

FIG. 5. Coma of a lens; ring image (1,2,3,4) formed by rays in cylindrical shell, e.g., 3.3′; inner shells give rise to smaller ring images, each progressively closer to point 0.

Coma is an aberration in which rays from a slightly off-axis object point are deflected by the surface in such a way that the magnification (Section 7.1.2) is different for rays which strike different zones of the surface (Fig. 5). The resulting image is, therefore, not a point but is asymmetrically blurred so as to have a comet-like appearance.

Astigmatism means that an object point is imaged in two lines which lie in planes perpendicular to each other and to the ray through the center of the system (Fig. 6). In general, this occurs for off-axis object points and is present even if spherical aberration and coma are completely eliminated for object points near the axis. At positions between the astigmatic line images, the image consists of an ellipse which degenerates at one place to a circle, called the circle of least confusion, where the best image is obtained. Astigmatism may also occur for on-axis object points if the surface is nonaplanatic.

If in the case of a plane object the image points (or the circles of least confusion) do not lie also in a plane, then there exists the aberration called curvature of the field. Even with astigmatism removed there still may be such curvature, which can be minimized by means of a stop (or limiting aperture) on the object side of the lens.

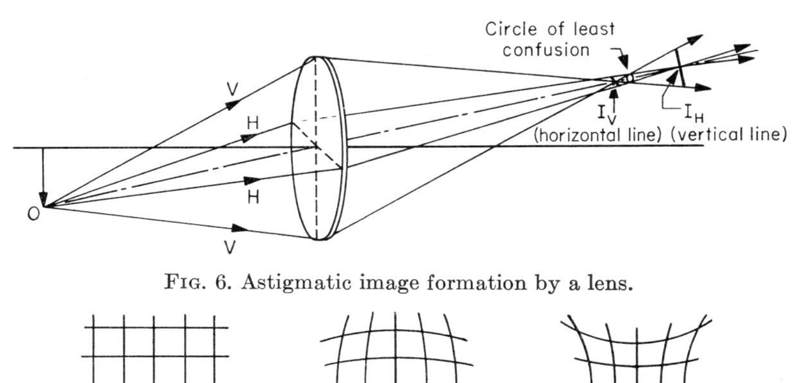

FIG. 6. Astigmatic image formation by a lens.

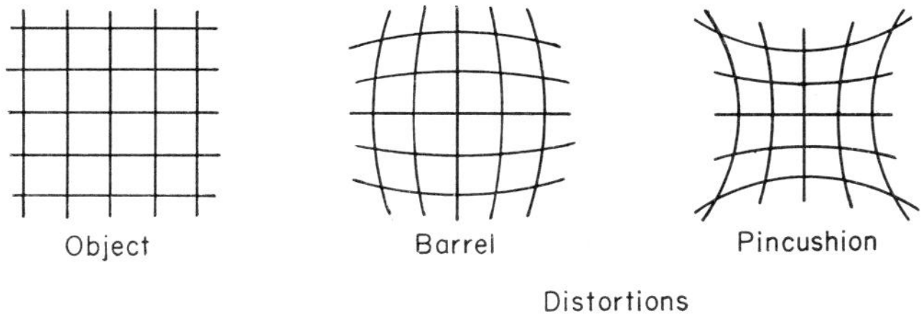

Object
Barrel
Pincushion
Distortions

FIG. 7. Geometric image distortions.

After corrections have been made such as to generate a sharp image point for each object point and to eliminate curvature, there may still be distortion if the geometric relation between image points is not the same as that between object points (Fig. 7). Such distortion is caused by the fact that the magnification (Section 7.1.2) is not constant for all rays traversing the lens, and distortion can thus be strongly influenced by the location of various stops in an optical system.

An aperture stop limits the quantity of light transmitted. In the example given in Fig. 8 the image of the aperture stop formed by the left lens is called the entrance pupil, and similarly the image of the aperture stop formed by the right lens is called the exit pupil.

The optical elements to the object side of the stop form an image of each stop. The image whose edge subtends the smallest angle at the center of the entrance pupil limits the field to the greatest extent and is called the entrance window, and the physical stop of which the entrance window is the image is called the field stop (Fig. 12). The image of the field stop formed by the optical elements to the image side is called the exit window.

The stops described may be used to control the aberrations in an optical system. Since one cannot correct the optical surfaces for all aberrations simultaneously, however, the practical problem becomes that of designing the stops to select a tolerable range of each aberration appropriate to the particular purpose. In many cases the periphery of a lens constitutes a stop.

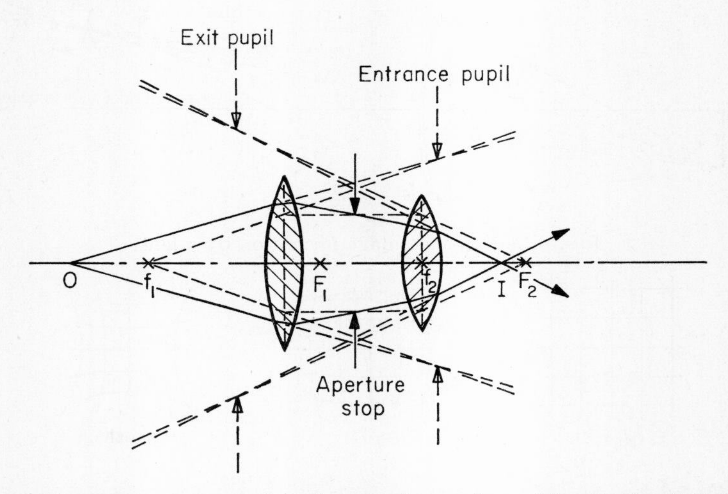

FIG. 8. The aperture stop and its images; image formed by the front (left) lens: entrance pupil, and that formed by the back (right) lens: exit pupil.

Absorptance of a surface is defined as the fraction of incident radiation which is not reflected or transmitted by the surface. On the other hand, the absorption coefficient, μ, of a medium specifies the fraction of the *entering* light absorbed per unit thickness when the thickness is very small, or $dI/I = -\mu\, dx$. The extinction coefficient* of a material is a number, K_0, such that the light intensity falls to $e^{-4\pi K_0}$ of its entering value in going a distance in the medium equal to one wavelength in vacuum. Hence, it is related to the absorption coefficient:

$$K_0 = \frac{\mu\lambda_{\text{vac}}}{4\pi}. \tag{7.1.3}$$

* This name is applied in the literature to various other quantities. Before using a numerical value be sure it is defined by Eq. (7.1.3).

Bouguer's law, rediscovered by Lambert, describes the decrease in intensity relative to the entering radiation, I_o, with x in the medium:

$$I = I_o e^{-\mu x} = I_o e^{-4\pi K_0 x/\lambda_{vac}}. \quad (7.1.4)$$

The emittance of a surface measures the total energy radiated per square centimeter of surface per second, while the reflecting power (or reflectance) is the ratio of the intensities of the reflected and incident beams. The ratio of the emittance to the absorptance for any surface is equal to the emittance of a blackbody at the same temperature (Kirchhoff's law). This is true at each wavelength. The transmittance corresponding to any thickness of a material is the ratio of the transmitted to the *entering* intensity. (See Chapter 7.4 for the effect of polarization on the above quantities.)

7.1.2. Simple Optical Systems

A system will form an image if the rays of a homocentric bundle (i.e., rays intersecting in a common point) which enter the system are deflected so that they are also homocentric on leaving it. Thus, the center of the exit bundle, usually constructed from the intersection of rays, will be the image of the center of the entering bundle. If it is possible to project the image onto a screen, it is real; if not, it is virtual and is obtained from the intersection of backward ray extensions.

7.1.2.1. Lenses. For a thick lens it is possible to compute the image position by applying the equation for a single refracting surface,

$$\frac{n'}{q} - \frac{n}{p} = \frac{n' - n}{R} \quad (7.1.5)$$

to each surface successively. This method is cumbersome and it is preferable to use a procedure based on the cardinal points of the lens: the focal points (Section 7.1.1), and the principal points. The primary principal plane (Fig. 9) of a lens is the locus of points of intersection of the direction of incident rays from the primary focus with the extension of their paths as they leave. Similarly, the secondary principal plane passes through the intersection of the extension of an incident ray parallel to the axis, with its direction as it leaves towards the secondary focus. Thus, the principal points are defined as the intersections of the principal planes with the axis. When the principal points and foci have been located the image may be computed from:

$$\frac{1}{q} - \frac{1}{p} = \frac{1}{f} \quad (7.1.6)$$

where f is the focal distance and all distances, positive to the right and negative to the left, are measured from the corresponding principal points. The first and second focal distances are the same if the medium is the same on both sides of the lens. A thin lens, clearly, is one in which the principal points coincide at the geometric center of the lens.*

For a thin lens immersed in a homogeneous isotropic medium the image distance, q, may be computed from:

$$\frac{1}{q} - \frac{1}{p} = \frac{(n' - n)}{n}\left(\frac{1}{R_1} - \frac{1}{R_2}\right) = \frac{1}{f} \tag{7.1.7a}$$

where n' is the refractive index of the lens, n that of the medium, p is the object, and q the image distance, R_1 the radius of the first and R_2 that of the second lens surface, f is the image distance when the object is at

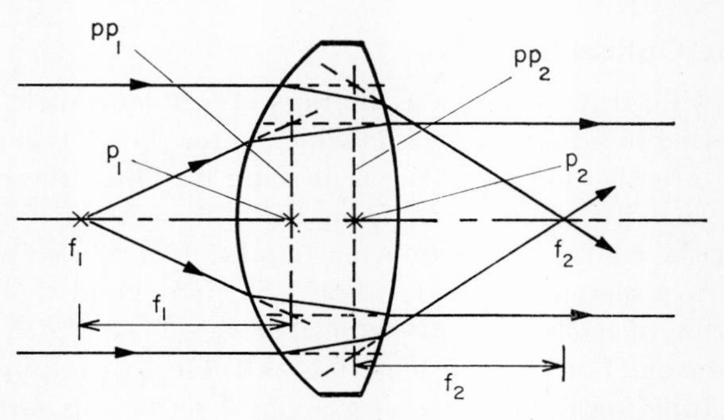

FIG. 9. Principal planes (pp), principal points (p), and focal distances (f) of a thick lens.

infinity. Distances are measured from the center of the lens: those to the left are negative, and those to the right positive. Magnification produced by the lens is the ratio of a transverse dimension in the image to the same dimension in the object and equals q/p.†

In order to make a lens system entirely achromatic, both the focal length and the positions of the principal planes must be made independent of wavelength. In the case of a cemented combination of a thin biconvex and a thin biconcave lens, the system can be corrected for both chromatic aberrations by making the focal length the same for two wavelengths.

* The focus and quality of the lens can be precisely determined with the Foucault test: J. Strong, "Procedures in Experimental Physics," p. 69. Prentice-Hall, New York, 1938. If the focal length is specified by the manufacturer the principal points are determined by measuring from the focus. The nodal and principal points can be located in the laboratory by the method described in: J. Morgan, "Introduction to Geometrical and Physical Optics," p. 68. McGraw-Hill, New York, 1953.

† The power of a lens in diopters is the reciprocal of the focal length in meters.

Spherical aberration may be corrected in a thin lens for a given object point by polishing the lens surfaces to the required nonspherical shape. As this is laborious, it is common to use spherical surfaces and reduce spherical aberration by a judicious choice of the radii of curvature ("bending" the lens). The spherical aberration is a minimum when the deviation of a ray is most equally divided between the two surfaces of the lens. Spherical aberration can be eliminated from a system of lenses by spacing them so that the spherical aberration introduced by the positive elements is equal and opposite to that introduced by the negative elements. Astigmatism and curvature of the field are intimately related to each other and neither of them can be eliminated by bending a thin lens or by placing a positive and negative element in contact. Astigmatism and curvature can, however, be simultaneously eliminated if a positive and negative element are separated. In general, a prime factor in minimizing these aberrations is a properly positioned stop.

7.1.2.2. Mirrors. The image-forming properties of mirrors in a homogeneous medium are independent of the refractive index of the medium. They do not produce chromatic aberration but are subject to monochromatic aberrations. Paraboloidal mirrors are aplanatic for objects on the axis at infinity. They do, however, show astigmatism and coma for off-axis objects. Ellipsoidal mirrors are aplanatic for the two foci of the ellipse (Fig. 2b). Spherical mirrors are only approximately aplanatic but are less expensive and, if their diameter is small compared with the focal length, give sharp images of objects on the axis. In that case, the image distance may be computed from the equation:

$$\frac{1}{q} + \frac{1}{p} = \frac{2}{R} = \frac{1}{f} \tag{7.1.7b}$$

where R is the radius of curvature of the mirror. Spherical aberration is inherent in spherical mirrors, and for off-axis object points astigmatism also exists.

7.1.2.3. Prisms. Prisms are used for deviation, dispersion, and reflection. A prism of refracting angle A, i.e. the angle between the sides of the prism through which the light enters and leaves, and index n (Fig. 10) produces a deviation, which has a minimum, D_m, if the rays traverse the prism symmetrically with respect to A such that

$$n = \frac{\sin[(D_m + A)/2]}{\sin(A/2)} \tag{7.1.8}$$

where D_m is the minimum angle between the entering and exit rays. The wavelength separation produced by the prism depends not only on the

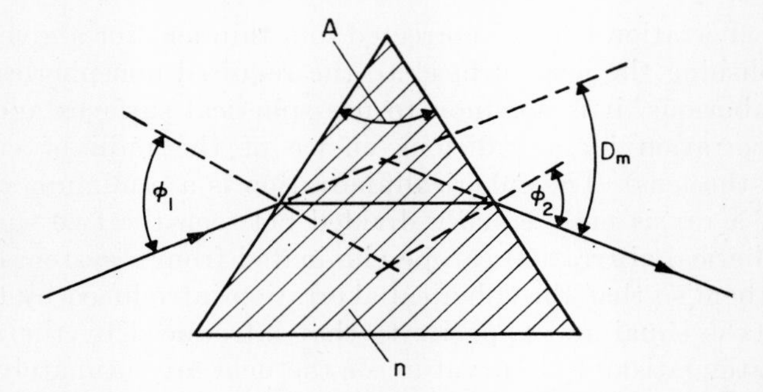

FIG. 10. A prism used at minimum deviation D_m, where $\phi_1 = \phi_2$.

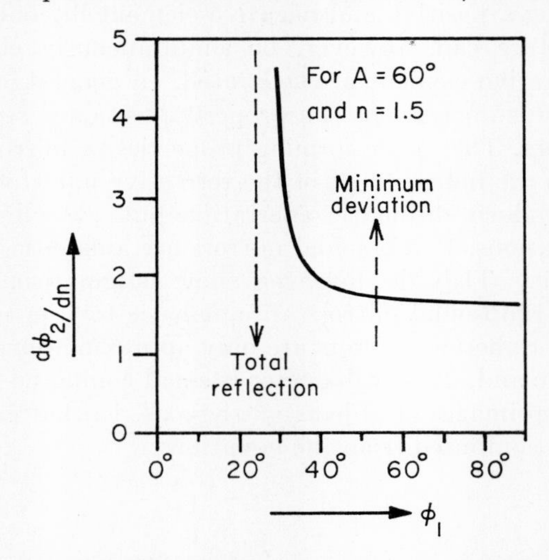

FIG. 11. Variation of prism dispersion, $d\phi_2/dn$, with angle of incidence, ϕ_1.

dispersion of the medium, $dn/d\lambda$, and on A, but also on the angle of incidence: At minimum deviation

$$\frac{d\phi_2}{d\lambda} = \frac{2\,\sin(A/2)}{\cos\,\phi_1} \cdot \frac{dn}{d\lambda}. \tag{7.1.9}$$

Dispersion increases as the angle of incidence approaches the normal (Fig. 11) but, at the same time, loss of light by internal reflection increases until finally the beam is completely lost at the critical angle.

7.1.3. Telescopes and Microscopes

7.1.3.1. Telescopes. A refracting telescope consists of an objective lens of long focal length and an eyepiece. Its magnification, defined as the ratio of the size of the retinal image when an object is viewed through the instrument to the size of the retinal image when the object is viewed with the unaided eye, is equal to the ratio of the focal lengths of the objective and eyepiece.

Most eyepieces[2] (oculars) are designed with two thin lenses of the same glass separated by a distance approximately equal to half the sum of their focal lengths. This spacing minimizes lateral chromatic aberration since then the effect of the aberration in the first lens is most nearly cancelled in the second, a most important consideration. In the Huygens ocular (Fig. 12a) the first or field lens has a focal length several times that of the second, or eye lens. This ocular is positioned so that the image, formed by the objective of the total system, falls in the first focal plane of the eye lens. Thus, the rays from each point emerge as a parallel beam

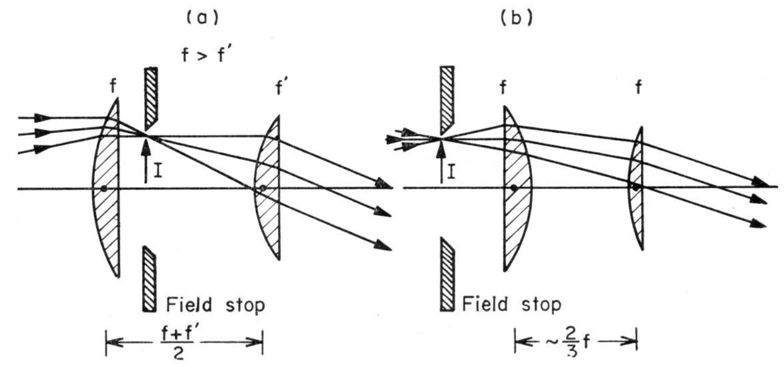

FIG. 12. Eye pieces; (a) Huygens', and (b) Ramsden's.

through the exit pupil into the eye. The Ramsden ocular (Fig. 12b) has lenses of equal focal length separated by a distance slightly less than the focal length, since separating them by the focal length would bring into focus in the eye the scratches and dust on the field lens. The Ramsden ocular has a much flatter field than the Huygens.

In a telescope objective[3] correction for chromatic aberration, spherical aberration, and coma are of primary importance, but astigmatism, curvature, and distortion are less serious because of the small field. The oblique aberrations, coma and astigmatism, are less severe in a refracting than in a reflecting telescope (see below) and, therefore, reflecting objectives are used only in large astronomical telescopes in which their freedom from spherical and chromatic aberrations is of great advantage.

[2] F. A. Jenkins and H. E. White, "Fundamentals of Optics," 2nd ed., p. 170. McGraw-Hill, New York, 1950.

[3] A. Elliott and J. H. Dickson, "Laboratory Instruments," p. 261. Chemical Publishing, New York, 1953.

Since paraboloidal reflecting objectives produce considerable coma in off-axis images it is sometimes desirable to use a spherical reflector, which is free from coma but has a great deal of spherical aberration which may be corrected by the Schmidt system of placing a diverging refraction plate at the center of curvature of the mirror (Fig. 13). The correction plate is so shaped that it corrects for spherical aberration for each zone of the mirror and thus also eliminates astigmatism. The aspherical surfaces of the Schmidt corrector are difficult to make and are avoided in the Maksutov corrector which is a meniscus refracting plate placed

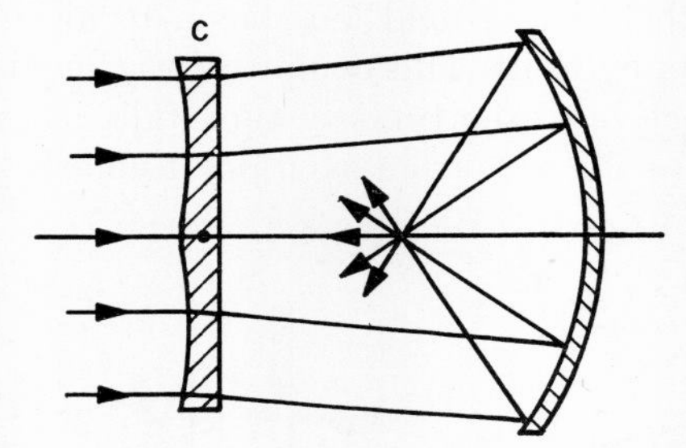

FIG. 13. Schmidt plate, *C*, to correct spherical aberration.

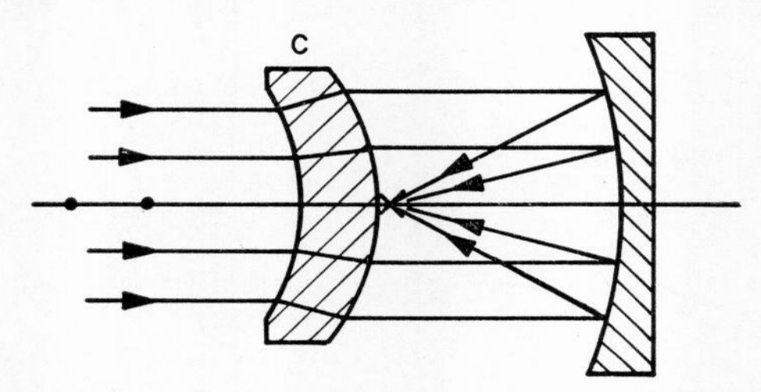

FIG. 14. Maksutov meniscus plate *C*, to correct spherical aberration.

between the center of curvature and the focus, and has spherical surfaces of approximately equal radii of curvature shaped so that the spherical aberration introduced by the corrector just cancels that of the mirror (Fig. 14). Any correcting plate introduces some chromatic aberration but this is usually not serious.

7.1.3.2. Microscopes. The microscope consists of a short focal length objective which forms a real image, and an ocular which forms a final virtual magnified image of the image formed by the objective (Fig. 15).[4] Its magnifying power is defined as the ratio of the size of the retinal image when the object is viewed with the instrument to the size of the retinal image when the object is viewed with the eye alone. For a simple magni-

[4] A. Elliott and J. H. Dickson, "Laboratory Instruments," pp. 264, 339. Chemical Publishing, New York, 1953.

fying lens the magnifying power is $25/f$, where f is the focal length measured in centimeters and 25 cm is the distance of most distinct vision. For a compound microscope with the final image at infinity the magnifying power, M, is given by

$$M = \frac{25L}{f_e f_o} \tag{7.1.10}$$

where f_e is the eyepiece focal length, f_o the objective focal length, and L the distance from the second focal point of the objective to the first focal point of the eyepiece (optical tube length, $L \leqq f_e + f_o$). In commercial microscopes L is standardized at 18 cm.

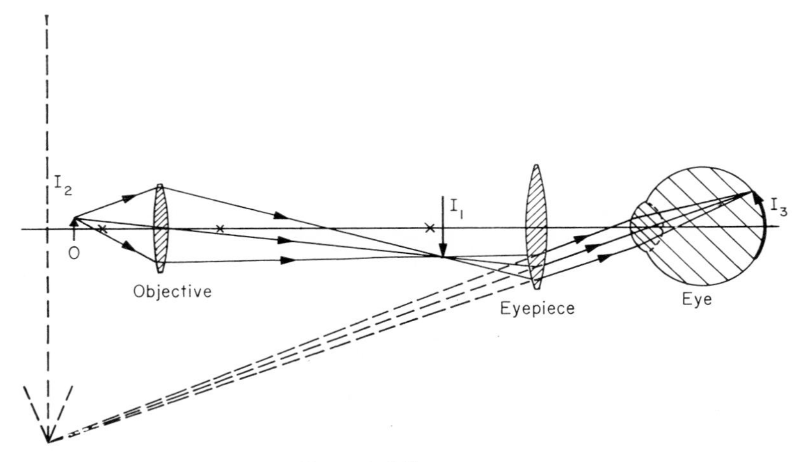

FIG. 15. Microscope.

The resolving limit, s, of a microscope refers to the smallest separation of two object points giving image points which just can be distinguished as separate (Section 7.2.1). A working rule (obtained from diffraction theory) for calculating this is

$$s = \frac{\lambda}{2n \sin i} \tag{7.1.11}$$

where λ is the vacuum wavelength of light used, n the refractive index of the medium between object and objective, and i the half-angle subtended at the object by the objective. The product ($n \sin i$) is called the numerical aperture. The oil immersion microscope resolves smaller objects since it produces a larger numerical aperture by having the object and first surface of the objective immersed in oil of large n. Furthermore, the oil also decreases the amount of light lost by reflection at the first lens surface. The largest numerical aperture thus obtainable is about 1.6 in contrast to 0.9 for dry objectives.

Another application of Eq. (7.1.11) is the ultraviolet microscope[5] which attains a smaller resolving limit by using light of shorter wavelength. This may necessitate the use of photography in examining the image, and is further complicated by the fact that the microscope and slides must be made of materials transparent to the ultraviolet. In addition the optical system must be specially corrected for the wavelength used. These needs have been satisfied in the color translating ultraviolet microscope invented by Land.[6]

In the dark field microscope[7] the slide is illuminated so obliquely that with no object present the field of vision is completely dark, since no direct light can enter the objective. As a consequence of this method of illumination, transparent objects with n different from the surrounding medium are made to appear as bright spots on a dark background in the image, since they will scatter some light into the objective. (This is true also of opaque objects smaller than the resolving limit, but their shape remains indefinite.) Thus at ransparent microorganism, such as *Treponema pallidum,* produces an image with recognizable evidence of the characteristic structure and may be identified with dark-field illumination, whereas it would not be with parallel illumination.

The phase-contrast microscope[8] is a modification of the ordinary microscope designed to make transparent objects visible by a method which converts variations of phase on the wavefront leaving the object into variations of intensity in the image (see Section 7.2.1) (Fig. 16a). This is done with two essential additions to an ordinary microscope: a phase plate P and an annular diaphragm D. D is placed in the front focal plane of the substage condenser and an image of the light source is focused on it by the concave mirror. The object on the slide is therefore illuminated by a hollow cone of parallel light. If there were no objects on the slide, this light would be focused again by the first three lenses of the objective to form an image of D on the phase plate P. This consists of a glass plate upon which is evaporated an annular layer of transparent material to such a thickness that it increases the optical path by one quarter of a wavelength of green light. The size of the retarding ring matches the image of D. The ring usually has also a thin metallic film to reduce its transmission. If there are any microscopic objects on the slide they will

[5] F. Twyman, *J. Roy. Microscop. Soc.* **56,** 365 (1936).

[6] E. Land, E. R. Blout, D. S. Grey, M. S. Flower, H. Husek, R. G. Jones, C. H. Matz, and D. P. Merrill, *Science* **109,** 371 (1949).

[7] L. C. Martin, "Technical Optics," Vol. 2, p. 128. Pitman, New York, 1950.

[8] *Ibid.,* p. 124.

cause diffraction of the light even though they are transparent (Fig. 16b). The light of the central maximum will experience in P a phase retardation of $\pi/2$ with respect to the diffracted light, which is already $\pi/2$ behind the central maximum, so that the two are brought into phase by the phase plate and constructive interference occurs at the corresponding point in the final image, whereas without the annular diaphragm and phase plate it would have been almost invisible.

An alternative way to make transparent objects visible, if they are double refracting (Section 7.4.2), is with the polarizing microscope in which

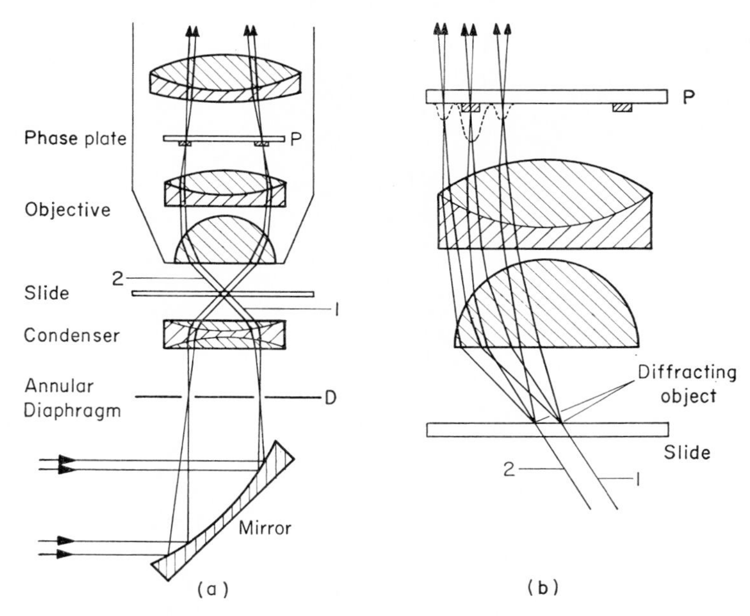

FIG. 16. Phase contrast microscope; (a) arrangement of components, and (b) diffraction pattern due to transparent object on slide.

the sample is illuminated by plane polarized light (Section 7.4.1) and its image is viewed through an analyzer (Section 7.4.2) between objective and eyepiece.

Interference microscopes[9] are also designed to convert small phase differences introduced by the object into intensity differences in the image. The sample to be examined is placed between two parallel semi-reflecting metallic films that are at so small a separation that the system acts like an interference filter (Section 7.2.2). Even though the sample may be transparent, differences in thickness or in refractive index between

[9] T. Merton, *Proc. Roy. Soc.* **A189,** 309 (1947); **A191,** 1 (1947).

various regions introduce path differences which, by interference, are converted into intensity differences in the microscope image. By illuminating with the green and violet lines of mercury it is possible, for example, to make a transparent microorganism appear as a green object on a violet background. In addition to making transparent microscopic objects visible, the interference microscope can be used to measure the thickness of objects as thin as 10^{-4} mm.[9a]

7.1.4. Simple Refractometers

Instruments designed to measure the refractive index are called refractometers and may make use of any property of the material which

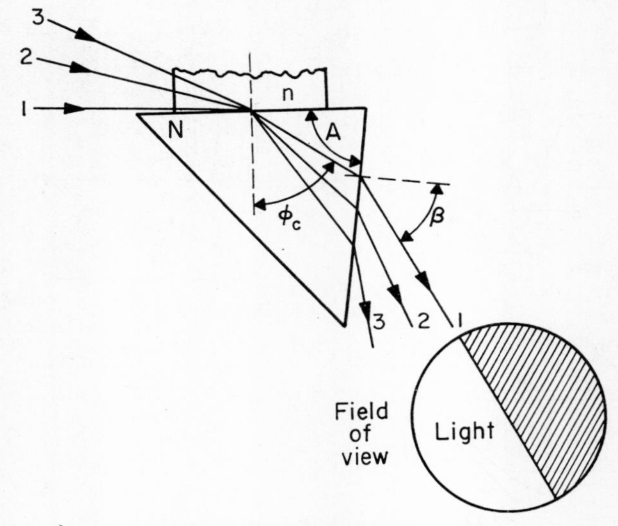

FIG. 17. Pulfrich refractometer.

depends upon the refractive index, for example, refracting power, critical angle, polarizing angle (see Section 7.4.2), and change of phase on transmission through a known thickness (interferometer methods, see Section 7.2.4).

The absolute refractive index of a liquid is usually measured by a deviation method, while interferometers are commonly used to measure small changes of refractive index, or to compare the indices of two liquids. In order to obtain the absolute values of the refractive index of a liquid, the Pulfrich refractometer (Fig. 17) may be used. This method consists of measuring the critical angle (Section 7.1.1) between the liquid and a block of glass of larger known index. If the index of the sample is n and

[9a] J. Dyson, *Physica* **24,** 532 (June, 1958).

that of the glass block is N, there will be a sharp boundary between light and dark in the field of view at an angle β such that:

$$n = \sin A \sqrt{N^2 - \sin^2 \beta} + \cos A \sin \beta.$$

The angle A is usually constructed to be 90°, within 10″, so that

$$n = \sqrt{N^2 - \sin^2 \beta}. \tag{7.1.12}$$

The Pulfrich refractometer may, of course, be used for solids by placing the solid sample in optical contact with the glass block by means of a drop of oil of higher index. The index can be measured with an accuracy of five to ten units in the fifth decimal place and the dispersion to an accuracy of one to two units in the fifth decimal place. A more rapid critical-angle instrument is the Abbe refractometer in which a liquid sample is placed between the hypotenuses of two 45° prisms about 0.1 mm

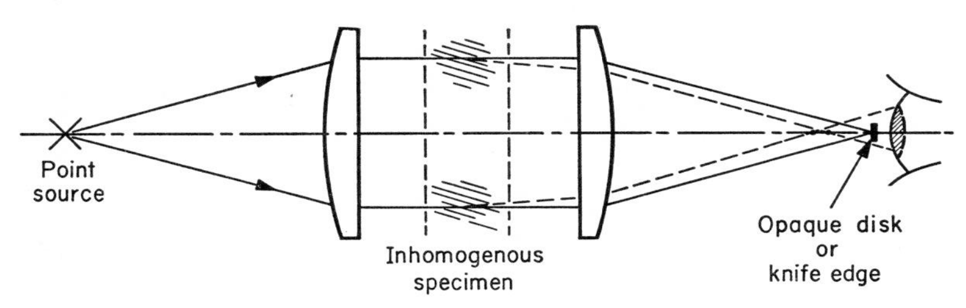

FIG. 18. "Schlieren" method for observing optical inhomogeneities.

apart and the critical angle is measured. The precision is two units in the fourth decimal place.

For solids which are not properly shaped to be measured on the refractometer, for example geological crystals, it is possible to prepare a liquid of the same index as the solid by an immersion method, adjusting the liquid index until the solid becomes least visible, and then measuring the index of the liquid with the Pulfrich or Abbe refractometer. An accuracy of a few units in the fourth decimal place is then obtainable.

An absolute measurement of the index of a liquid also may be obtained by placing it inside a hollow glass prism and measuring the angle of minimum deviation (see Section 7.1.2).

A sensitive way of observing local inhomogeneities in a sample is the "Schlieren" method developed by Toepler. In one application of this method (Fig. 18) a point source is imaged by two high quality achromats of long focal length. The image is masked by a small disk or knife edge behind which the eye is placed, and the eye is focused on the lens nearer the image. If the space between the lenses is optically homogeneous the lens will be dimly visible and appear uniformly illuminated. If, on the other hand, the space between the lenses contains a sample having optical inhomogeneities the light will be refracted by these local regions so that it is not focused on the disk and hence enters the eye. Thus the inhomogeneities appear as bright regions on a dark background, even though they would be invisible by ordinary observation. In such a fashion sound waves, for example, may be photographed (Section 7.2.2).

7.2. Diffraction and Interference

7.2.1. Diffraction

Diffraction[10] refers to the deviations from rectilinear propagation of light in a homogeneous medium, and apertures are essential for its observation. It can be satisfactorily explained only by assuming a wave character for light; and then rectilinear propagation, or geometrical optics in general, represents a limiting situation wherein the wavelength approaches zero. Thus geometrical optics gives a good approximation when the wavelength is very small compared with the dimensions of the apparatus and the different rays are independent or noncoherent. Phenomena in which wave effects become discernible are classified under physical optics. The reason for using geometrical optics in designing optical systems for which it is appropriate is simply ease of visualization and computation.

The basic principle of physical optics is Huygens' principle, which states that the new wavefront can be obtained by treating each point on the old wavefront as a small source of secondary waves, the secondary waves then combining amplitudes to form the new wavefront.

Fresnel diffraction[11] refers to cases in which either or both the light source and the place at which the diffraction effect is observed are at finite distances from the aperture causing the diffraction. This class of phenomena is easy to observe but requires a much more difficult mathematical treatment than Fraunhofer diffraction, which occurs when both light source and observation point are effectively at infinity (parallel rays). The solution in both classes may always be obtained by applying Huygens' principle to the wavefront as it goes through the aperture, and then summing the disturbances produced at the observation point by

[10] M. Françon, *in* "Handbuch der Physik—Encyclopedia of Physics" (S. Flügge, ed.), Vol. 24, p. 268. Springer, Berlin, 1955.

[11] *Ibid.*, p. 357.

each point in the wavefront; the phase relations of such disturbances must, of course, be considered in the summation. For Fresnel diffraction the mathematical integration over the aperture is in general difficult, because the optical path distance from different points in the aperture to a point of observation varies in a way which is not simple, and thus differences in phase and amplitude which develop by the time the disturbances reach the observation point complicate the form of the equations. Consequently, various geometrical and graphical artifices have been developed for obtaining approximate solutions to problems in Fresnel diffraction.

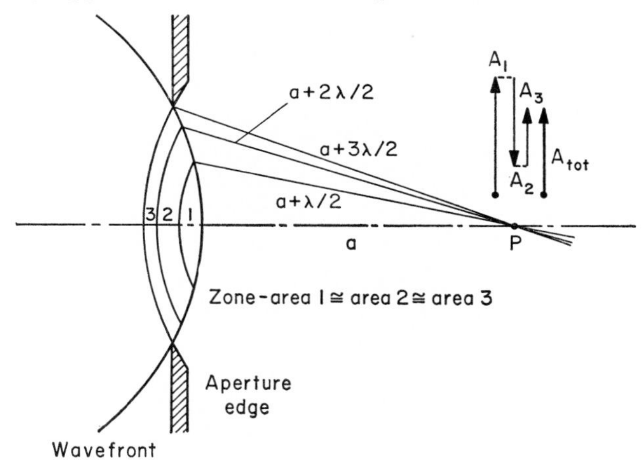

FIG. 19. Fresnel half-period zones and corresponding amplitude vectors, A_1.

These include dividing a spherical wavefront into concentric circular half-period zones, each on the average $\lambda/2$ farther from the point of observation than the previous one, and taking the vector sum of the disturbances from these zones (Fig. 19). If the light source and screen distance are allowed to become very large, the problem becomes approximately one of Fraunhofer diffraction.

Fraunhofer diffraction,[12] in which the wave is made plane, e.g., by a lens, before it reaches the aperture and is brought to a focus after leaving the aperture, is easier to deal with mathematically since the optical paths from a point of observation vary in a simple fashion, and thus the integration over the aperture can frequently be performed in a simple analytical or vectorial manner as, for instance, in the case of a single slit (Fig. 20). Here the aperture of width D is subdivided into small equal

[12] M. Françon, *in* "Encyclopedia of Physics" (S. Flügge, ed.), Vol. 24, p. 269. Springer, Berlin, 1955. For graphical treatment of amplitudes, as in Fig. 20 see F. A. Jenkins and H. E. White, "Fundamentals of Optics," 2nd ed., p. 286. McGraw-Hill, New York, 1950.

strips $a, b, c, d \ldots k$, each of which makes an equal contribution to the amplitude received at each point in the focal plane of the lens. These contributions must be summed vectorially, since at a given point there is a constant phase difference between the contributions from successive strips. Thus, the phase difference, δ, between the resultant displacement, A, at this point and that of the central maximum is half the sum of the phase differences of the contributions from the individual strips. For most optical instruments the methods of Fraunhofer diffraction are adequate.

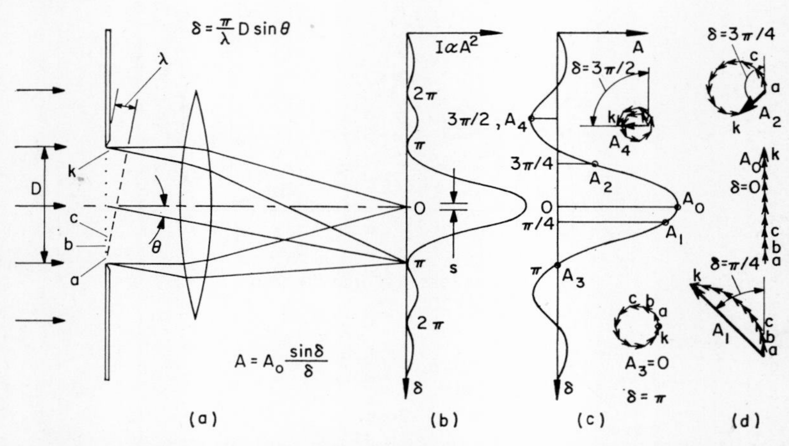

FIG. 20. Fraunhofer single slit diffraction; (a) geometry showing path differences, (b) intensity (I) distribution with the geometric image, S of the source slit (not shown), (c) amplitude-(A) distribution, (d) small amplitude vectors corresponding to elements a, b, c, d, etc. in the slit giving resultant amplitudes and their phases, $\delta = (\pi/\lambda)D \sin\theta$, at indicated points.

The presence of diffraction provides an inherent limitation on the resolving power and accuracy of even an ideal optical system. This limitation does not appear in the description of geometric optics (Chapter 7.1) but is a consequence of the finite wavelength of light and hence becomes smaller if shorter wavelengths are used. If one describes image formation from the viewpoint of physical optics, one sees that for each point in an object the optical system forms a diffraction pattern of finite size, rather than a point, in the image (Fig. 20). Thus, for example, if two object points are so close together that their diffraction patterns in the image overlap seriously it may be impossible to ascertain the existence of the separate points by an examination of the image. The net effect is the inability of even the most perfect optical system to reproduce with complete fidelity the fine details of such an object. The Rayleigh criterion

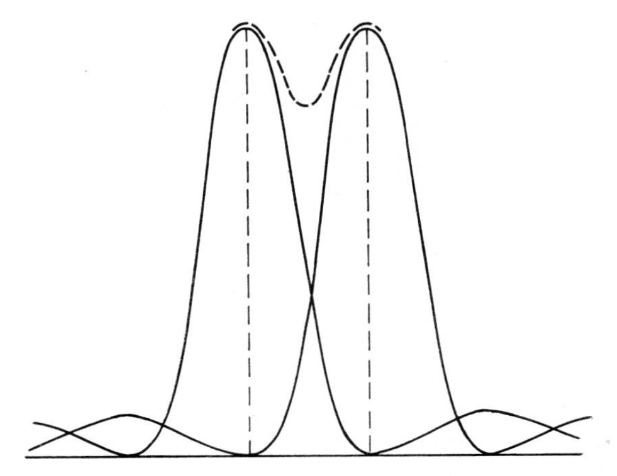

FIG. 21. Rayleigh criterion for resolution of two points.

that two such image points be resolved is that the central maximum of one point fall on the first minimum of the other (Fig. 21), with an intensity of 81% in the center.

7.2.2. Interference

When light waves cross, the resultant displacement vector at any point is merely the vector sum of the displacement vectors at that instant due to each wave separately, thus producing the phenomena of interference. This principle of superposition predicts exactly the intensity distribution in all optical experiments. Its validity is demonstrated in the common observation that two intersecting beams of light are unaltered after passage through each other. In a region common to the two beams there will be interference effects because of the fact that the intensity, i.e., the energy flux, at any point is proportional to the square of the amplitude (maximum displacement) at that point and consequently may be larger or smaller than the sum of the intensities of the individual beams. The principle may be extended to any number of beams and written as a vector sum

$$A = A_1 + A_2 \cdot \cdot \cdot + A_n \tag{7.2.1}$$

where the A's are wave displacements, and the intensity is then proportional to the square of the resultant amplitudes.

In practice interference effects may be produced either by wavefront division or by amplitude division. The first method is exemplified by a system of regularly spaced slits, a diffraction grating or a double slit (see stellar interferometer below), and the second by such instruments as the

Michelson interferometer in which an inclined plane of glass in the path of a beam acts as a beam-splitter and divides the light into two parts following separate paths, which are subsequently recombined to produce interference.

In the case of Haidinger fringes, monochromatic light from an extended source is incident on a thick film bounded by plane-parallel semireflecting surfaces, and sharp concentric interference rings appear in the transmitted beam. A single ray from a point in the source is partly transmitted and partly reflected at the second surface of the film and the reflected part is reflected again at the first surface, so it travels an extra path equal to $2nd \cos \phi'$ before reaching the second surface and being again partly reflected and partly transmitted (Fig. 22). Consequently each incident ray produces a group of coherent and parallel transmitted rays with a

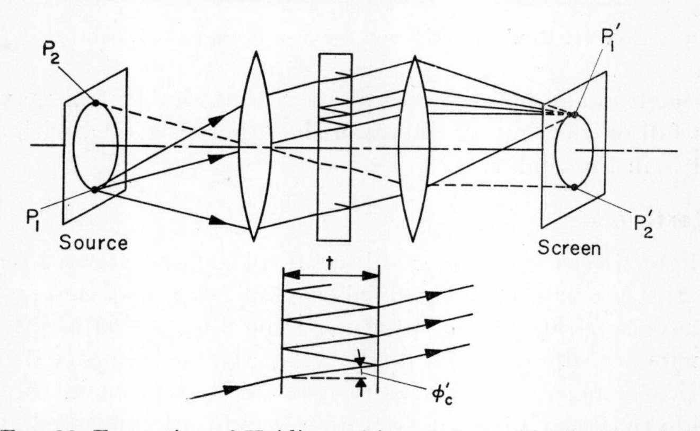

FIG. 22. Formation of Haidinger fringes of equal inclination (ϕ').

constant phase difference between successive members so that if the rays are collected by a lens they will interfere in the focal plane. The locus of points in the source giving a constant ϕ' is a circle and therefore, with an extended source, the fringe system consists of concentric bright and dark rings depending on whether $2nd \cos \phi'$ is an integral or half-integral number of wavelengths. Note that if the wavelength is decreased ϕ' increases and each ring expands.

By the use of two Fabry–Perot interferometers in series (Figs. 23 and 37) it is possible to obtain interference in white light (Brewster fringes). The two interferometers are adjusted to exactly the same optical thickness, or else one to some exact integral multiple of the other, and they are inclined to each other at an angle of 1° or 2°. Thus, a ray that bisects the angle between the normals to the two sets of plates can then be split into two, each of which after two or more reflections emerges, having traversed the same path. Since the two interfering rays are derived from the same ray and have traversed the same path, constructive interference is obtained for all wavelengths. A ray incident at any other angle will give a path difference between the two emerging ones which increases with the angle, so that a system of straight white-light fringes with colored borders is produced. The usefulness of Brewster's fringes lies chiefly in the fact that when they appear, the ratio of the interferometer optical spacings is very exactly an integer.

The results of Faraday, Maxwell, and others have established that light consists of an electromagnetic wave propagated through free space with a velocity c which is completely independent of frequency, amplitude, or the velocity of the source. Another series of experiments, initiated by Wiener,[13] has established that the electric rather than the magnetic part of the wave is the disturbance responsible for the generally

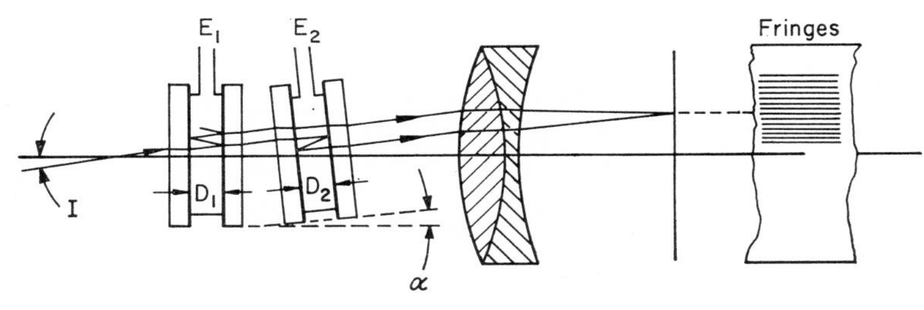

FIG. 23. Refractometer for the absolute refractive index of a gas, using Brewster's fringes from two Fabry–Perot etalons E_1 and E_2.

observed effects, visual perception, photographic detection, fluorescence, absorption, and the photoelectric effect. The classical boundary conditions on electric fields at the surface of a conductor state that the electrical field must be continuous and must be zero inside the conductor and therefore the electric field has a node at the surface of a mirror. One application of this result is that a film of an absorbing material can be deposited as a protective coating on a mirror surface without causing appreciable absorption if the film is so thin that its entire thickness is close to the node at the mirror surface.

The boundary conditions at the surface of a transparent medium show that the reflectance at normal incidence equals $(n' - n)^2/(n' + n)^2$, where n' and n are the indices of the two media bounding the surface. Clearly, in an optical system with many air–glass surfaces the loss of light by reflection will be substantial. Also the reflected light produces a background haze which reduces the clearness of the image. This loss of light

[13] O. Wiener, *Ann. Physik* [3] **40,** 203 (1890).

by reflection may be substantially reduced by coating[14] (blooming) the surface with a thin film of transparent material that has a refractive index equal to $\sqrt{n'n}$ and an optical thickness of $\lambda_{vac}/4$. With such an arrangement the wave reflected from the outer surface of the film is exactly half-wavelength out of phase with that reflected from the inner surface and has the same amplitude so that complete destructive interference occurs, and consequently there is no reflected beam. For a glass of index 1.5 such a nonreflecting film would require an index of 1.22. Even though the refractive indices of suitable materials are all higher than this, magnesium fluoride with a refractive index of about 1.39 is widely used for this purpose because of its good mechanical properties and the fact that it substantially reduces reflection, especially on a dense flint glass.

Even if the film index is correct, blooming is perfect at only one wavelength. Fortunately, the change with wavelength is small so the effect is sufficient for many purposes. More satisfactory measures, however, are possible by the use of multiple films. With two layers,[15] one of low index and one of high, there are four independent variables (two thicknesses and two n's) and therefore available materials may provide zero reflectance at two wavelengths plus low reflectance over a broad band.

If it is desired to increase[16] rather than decrease the reflectance of a dielectric surface, this may be done by depositing a thin film of high refractive index of optical thickness $\lambda/4$. The higher the refractive index the higher is the reflectance, since the reflectance of a surface of index n_2 coated with a $\lambda/4$ film of index n_1 and in a medium of index n_0 is given by

$$R = \left\{\frac{n_0 n_2 - n_1^2}{n_0 n_2 + n_1^2}\right\}^2. \tag{7.2.2}$$

Values of reflectance obtained from single layers of common materials on a surface of index 1.5 are listed in Table I. For the wavelengths given the absorptions are very small.

In a stack of an odd number of quarter-wavelength films, alternately of high and low index, the reflected beams from all the interfaces are in phase on leaving the uppermost boundary and thus interfere constructively. With such layers of dielectric films reflectances very close to 100% and with negligible light loss through absorption are obtained. Table II shows this variation with number of layers of ZnS plus cryolite on glass and Sb_2S_3 plus CaF_2 on quartz.

[14] O. S. Heavens, "Optical Properties of Thin Solid Films," p. 209. Academic Press, New York, 1955.

[15] *Ibid.*, p. 210.

[16] *Ibid.*, p. 215.

TABLE I.[17] Reflectance of Surface of Index 1.5 When Coated with Various $\lambda/4$ Single Films

Film material	Refractive index	Wavelength	Reflectance with single film
ZnS	2.3	5461 Å	0.31
TiO	2.6	5461 Å	0.40
Sb_2S_3	2.7	1 μ	0.43
Ge	4.0	2 μ	0.69

TABLE II.[18] Reflectance at Normal Incidence

Number of layers	ZnS + cryolite on glass, $n = 1.5$ ($\lambda = 5893$ Å)	$Sb_2S_3 + CaF_2$ on quartz, $n = 1.45$ ($\lambda = 1\ \mu$)
3	0.695	0.835
5	0.891	0.900
7	0.964	0.977
9	0.988	0.995

If the reflecting surface must be extremely flat it is undesirable to use a large number of layers, since the films do not have exactly the shape of the substrate and the deviations from flatness become severe with more layers. Usually materials of high refractive index in the visible are opaque in the near ultraviolet and therefore unsuitable for the preparation of high reflectance films for the ultraviolet. One of the few materials which has a high index and is also transparent in the near ultraviolet is $PbCl_2$, but it is also hygroscopic. This difficulty has been circumvented by applying protective coatings to the $PbCl_2$ producing successful ultraviolet reflectors of high reflectance and low absorption.[19] Such films may be especially useful since the reflectance of most metals is low in this region. Sb_2O_3 is also useful as a high-index, low-absorption film for the near ultraviolet and is nonhygroscopic.[19]

7.2.3. Interferometers

The stellar interferometer[20] is a device for measuring the diameter of a star or the angular separation in a double star system. The instrument can be understood by considering a conventional Young's double slit

[17] O. S. Heavens, "Optical Properties of Thin Solid Films," p. 217. Academic Press, New York, 1955.

[18] *Ibid.*, p. 220.

[19] W. L. Bair and F. A. Jenkins, *J. Opt. Soc. Am.* **46**, 141 (1956).

[20] A. A. Michelson, *Phil. Mag.* [5] **30**, 1 (1890); A. A. Michelson, "Studies in Optics." University of Chicago Press, Chicago, 1927.

system in front of the objective of a telescope (Fig. 24). If the light source is a narrow single slit, S_1, thereby providing coherent illumination for the double slits, a system of evenly spaced interference bands is seen in the focal plane. With a separation, d, of the double slit centers, the angular distance, α, between bright fringes is λ/d. Any transverse motion of the primary slit will produce an equal angular shift of the interference bands, and, therefore, two parallel primary slits, S_1 and S_2, with an angular separation, θ, produce two fringe systems separated by an angle, θ. Thus the angular separation, θ, of two narrow light sources could be measured by adjusting the separation, d, of the double slits just in front of the telescope objective until the double slit fringes become least distinct (Fig. 24). When this situation is reached, $\theta = \alpha/2$ or $\theta = \lambda/2d$. Similarly, the angular width of a single slit can be measured, except that

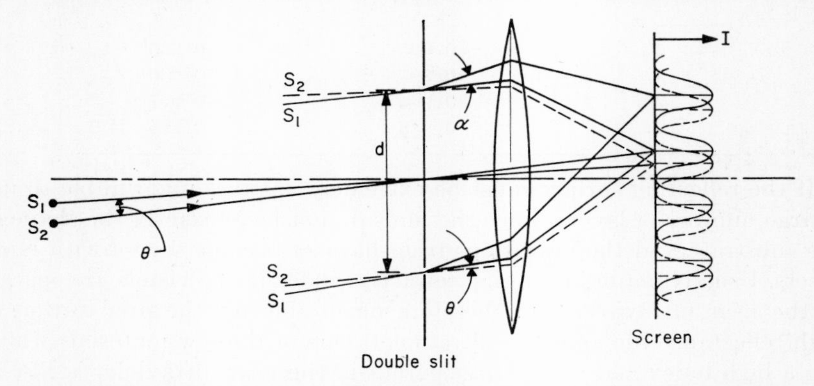

FIG. 24. Young's double slit interference with two point sources, S_1 and S_2.

it is λ/d. In case the single source is a disk of uniform brightness the fringes disappear when the angular diameter of the source is $1.22\lambda/d$, where 1.22 is a geometric factor which derives from the difference in intensity distribution between circular and rectangular openings. As the angular sizes of stars are of the order of 0.01 sec, the slit separations required are many feet, which is greater than the diameter of most astronomical objectives. In addition to this difficulty, another serious problem arises from the fact that the angular separation of the fringes equals λ/d, and therefore a large magnification is required to examine them if d is more than a few feet. Michelson circumvented both difficulties by an ingenious periscopic arrangement of mirrors such that the light from the star was first reflected from two mirrors capable of very large separation, and then reflected into the telescope from two mirrors of moderate separation (Fig. 25). The angular size of the fringes is then λ/d, where d is the separation of the mirrors in front of the telescope, and the angular size of

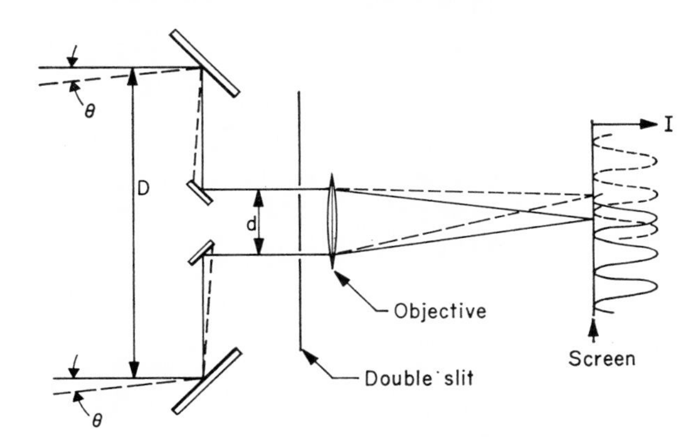

FIG. 25. Michelson stellar interferometer.

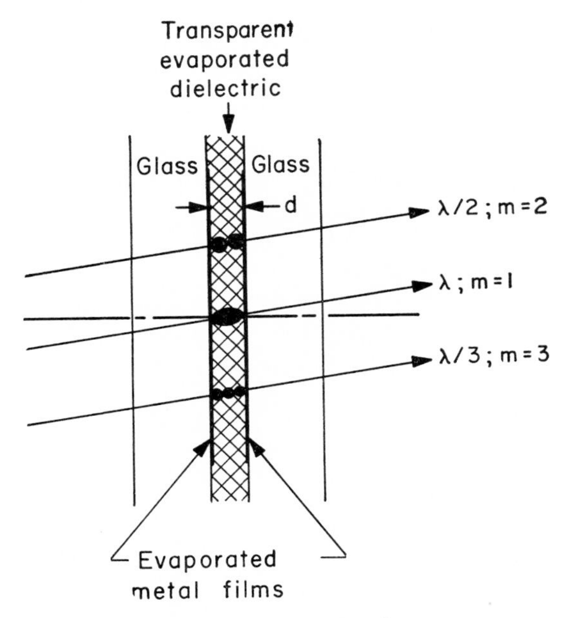

FIG. 26. Interference filter.

the star is $1.22\lambda/D$, where D is the separation of the outer mirrors and can be much greater than the size of the telescope objective. In a reverse form this sort of device is used in the Williams modification of the Rayleigh refractometer (Section 7.2.4).

The interference filter consists of two reflecting surfaces separated by a spacer (Fig. 26), as in a Fabry–Perot etalon (Section 7.3.3) of low order. When white light is passed through such a system and subsequently dispersed by a spectroscope, a banded spectrum appears since only certain wavelengths satisfy the conditions for constructive interference. If the reflecting surfaces have a high reflectance, the transmission bands in the spectrum are very narrow. The optical thickness $t(= nd)$ of the spacer layer determines the width of the transmission bands, and it is possible to make this thickness such that only one transmission band occurs in the visible spectrum. Transmission bands will, therefore, occur at wavelengths given by

$$2t \cos \theta' = m\lambda \tag{7.2.3}$$

where $m = 1, 2, 3, \ldots$ Thus, at normal incidence the system possesses transmission peaks at wavelengths $2t$, t, $2t/3$, $t/2$, etc. If t is equal to 5460 Å, transmission peaks are centered at 10,920 Å, 5460 Å, 3460 Å, 2730 Å, etc. Such a filter would serve to isolate the green line of mercury. The width W of the transmission band at wavelength λ is given by

$$W = \frac{(1 - R)\lambda}{\pi m R^{\frac{1}{2}}}. \tag{7.2.4}$$

For the filter referred to above using silver films with $R = 0.90$ the bandwidth at 5460 Å is 92 Å. Assuming an absorption of 0.04, the transmission at the peak is 0.36. If stacks of dielectric films are used in place of the silver films, vastly improved transmission values may be attained, e.g., a bandwidth of 22 Å at 4600 Å with a transmission of 70%. The practical lower limit for the attainable bandwidth with increasing number of dielectric films is set by lack of homogeneity in the spacer layer. With evaporated layers the lower limit seems to be of the order of 3–4 Å, given by a 39-layer filter.

The Michelson interferometer is a device in which each ray from an extended source is split by one surface of a glass plate placed in the beam at a 45° angle. The two parts of the ray then travel widely separated paths, 90° to each other, and each is reflected directly back to the beam splitter by means of a mirror (Fig. 27). At the beam splitter the two rays are then recombined and enter the eye of the observer in a direction at 90° to the original ray. Since the two rays were formed from a single one from the source they are coherent and able to interfere. This interference may be understood in terms of what the observer sees, namely the two reflecting mirrors (one real, the other imaged by the beam-splitter) in tandem and separated by a distance d. In effect, then, there is interference similar to the Haidinger fringes formed by a parallel-sided film of thickness d, except that there are only two interfering rays rather than a large number as in the ordinary case and, therefore, the fringes are broad. The Michelson interferometer has an advantage in that the two paths are widely separated physically so that the various samples can be placed in each path with space for separate temperature regulation, etc.

A device similar to the Michelson interferometer with important applications in wind-tunnel studies is the Mach–Zehnder interferometer.[21] In this arrangement mirror M_1 of the Michelson interferometer (Fig. 27) is rotated clockwise 45° and M_2 counterclockwise 45° so that the interfering beams are not reflected back to the first half-silvered plate but meet at a second half-silvered plate diagonally opposite the first one. Thus, if a sample is placed in one side of the rectangle formed only one of the beams

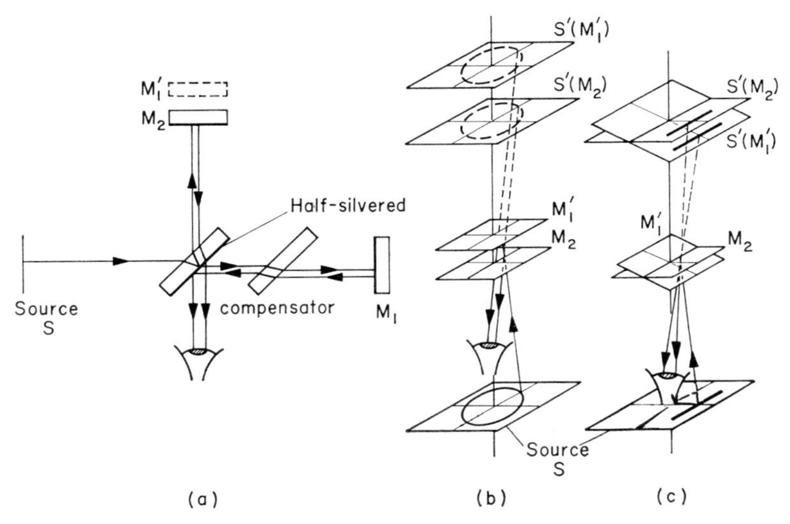

FIG. 27. Michelson interferometer; (a) arrangement, (b) fringes of equal inclination when mirror images (S') are parallel, (c) fringes of equal thickness if images are slightly inclined.

passes through it. In this way, differences in the optical density at different regions of the sample can be observed and measured and the interpretation is simpler than in the Michelson interferometer.

Twyman and Green modified the Michelson interferometer for the specific purpose of testing optical systems.[22] They introduced a high quality lens to parallelize the light before it enters the interferometer and another high quality lens to focus the light as it leaves, placing the eye at the focus of the second lens. With the ordinary Michelson interferometer it is necessary to use an extended source, since a point source would give only a point in the field of view and no fringes would be visible. With the

[21] J. Winckler, *Rev. Sci. Instr.* **19,** 307 (1948).

[22] F. Twyman, *J. Roy. Microscop. Soc.* **56,** 422 (1936).

Twyman and Green modification, on the other hand, a point source is necessary and even then the field of view is filled. Thus, if one mirror is accurately parallel to the other, the Michelson interferometer with its extended source produces concentric rings, fringes due to rays of equal inclination, whereas the Twyman and Green instrument in such a case, with its point source, gives a field of uniform tint. Often in practice, a

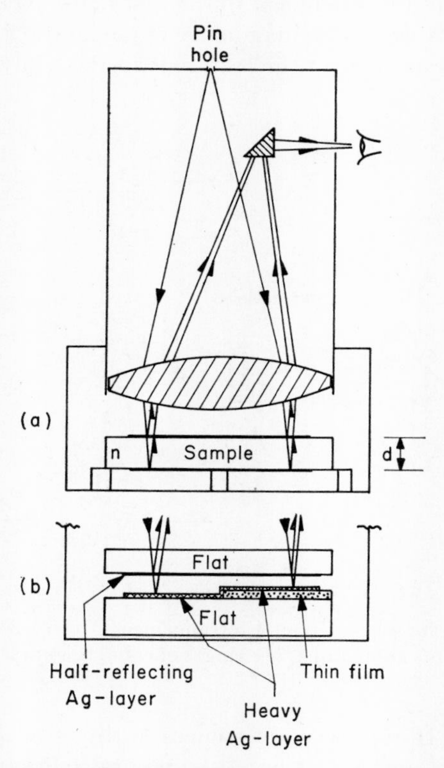

FIG. 28. Fizeau interferometer; (a) for testing transparent sample, (b) for measuring thickness of thin film.

small angle between one mirror and the image of the other is permitted, and therefore a parallel fringe system corresponding to points of equal thickness is seen, rather than a uniform field. An optical system to be tested is placed in the path of one of the beams and the interference lines examined for changes. In a perfect optical system all optical paths are equal for light of one color which leaves an object point and reaches an image point, and therefore the Twyman and Green fringes should be parallel everywhere. Lord Rayleigh showed that in practice the system will be only slightly inferior so long as the difference between the shortest and the longest optical paths does not exceed $\lambda/4$. Any flaw exceeding this is easily detectable in the Twyman and Green interferometer by adjusting the mirrors to give a uniform field and looking for the appearance of fringes after the test object is inserted. If such fringes appear they form a contour map of imperfections in the object, which then may be polished locally until the fringes disappear. To test different types of optical elements it is necessary to make appropriate modifications of the interferometer, e.g., in the mirrors.

The Fizeau interferometer is an alternative and simpler method for testing the optical homogeneity of a sample or for measuring the thickness of a thin film (Fig. 28). With this arrangement fringes of equal thickness are obtained by interference between rays reflected from the top and bottom surfaces of the sample. Thus the method tests the constancy of nd, since the wavefront which goes through the sample and is reflected back is superposed with the wavefront reflected from the top surface. On the other hand, the Twyman and Green interferometer tests the constancy of $(n - 1)d$, since the wave which goes through the sample is superposed with a wave which has taken an equal air-path through the other arm of the instrument and the extra optical path introduced by the sample is proportional to the difference, $n - 1$, between the refractive index of the sample and air. Therefore, if the same sample is examined both in a Fizeau and in a Twyman and Green interferometer, the variations in n and d may be calculated separately.

7.2.4. Interference Refractometers

For the absolute measurement of the refractive index of a gas, two Fabry–Perot (Section 7.3.3) etalons, E_1 and E_2 (Fig. 23) of roughly equal spacing, are used in tandem (Section 7.2.2). Suppose the etalons have lengths D_1 and D_2 and the second one is tilted at a slight angle α to the optic axis. Monochromatic light from an extended source to the left passes through the etalon and to the objective of a telescope focused for parallel light. The objective produces in its focal plane parallel and equally spaced Brewster's fringes perpendicular to the plane of the paper. A ray inclined at angle I to the optic axis is divided into two parts at the second semireflecting surface of E_1, and each part subsequently experiences a relative retardation by being twice internally reflected, the one in E_1 and the other in E_2. If both etalons are evacuated the relative retardations are $2D_1 \cos I$ and $2D_2 \cos(\alpha + I)$ in E_1 and E_2, respectively. The condition for the formation of a bright fringe in λ is that p be integral in the equation:

$$2D_1 \cos I - 2D_2 \cos(\alpha + I) = p\lambda \tag{7.2.5}$$

where λ is the wavelength of the light used. A convenient angular separation of the fringes can be obtained by adjusting the angle α, since the angular separation of two adjacent fringes is $\lambda/2\alpha D$.

When a gas is slowly admitted to E_1, the fringes move across the focal plane of the telescope in a direction perpendicular to their lengths, because the optical length of E_1 is $2nD_1 \cos I$ and the refractivity of the gas, $n - 1$, is proportional to the pressure.

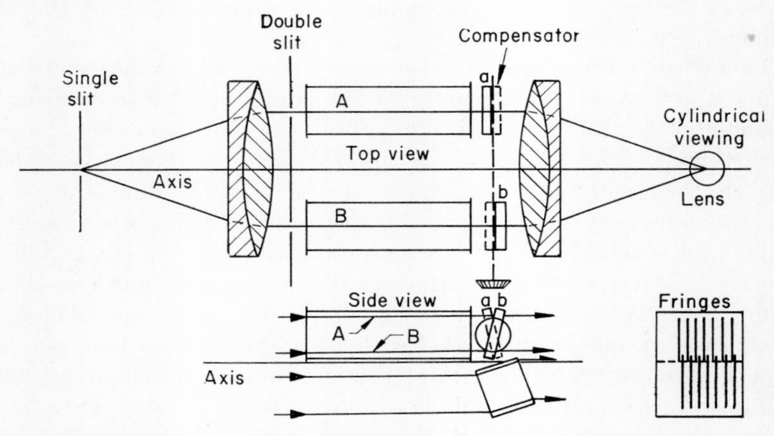

FIG. 29. Rayleigh refractometer for comparing indices in A and B.

The fringe displacement is observed at a convenient reference point such as the optic axis. Here, $I = 0$, so that when both etalons are evacuated,

$$2D_1 - 2D_2 \cos \alpha = p_1\lambda.$$

When the gas is admitted to E_1 while E_2 remains evacuated, this becomes

$$2nD_1 - 2D_2 \cos \alpha = p_2\lambda.$$

Subtracting and setting D_1 and D_2 equal, one obtains

$$n - 1 = (p_2 - p_1)\lambda/2D. \tag{7.2.6}$$

For comparing refractive indices, the Rayleigh refractometer is most widely used. Light from a narrow slit is collimated by a lens and passes through two parallel slits, each followed by a tube; then the emergent beams are recombined by a telescope and form interference fringes in its focal plane (Fig. 29). When the materials in the two tubes differ in refractive index, the fringe system is displaced sideways; this displacement may be measured, preferably by comparison with a second fringe system formed by two beams which have traveled identical paths, and, hence, the refractive indices of the materials in the tubes compared. As a result of the necessary narrowness of the initial slit, low intensity is a difficulty with this instrument. This slit must be so narrow that the light reaching the subsequent two slits is coherent; thus the greater the separation of the double slits the narrower must be the single slit. The double slits are about 12 mm apart in order to provide space for the following tubes, which means also that the double slit fringes formed are very close together and must be examined under high magnification, thereby again lowering the intensity.

The intensity problem has been solved in an ingenious modification,[23] which in effect keeps the double slits close together but separates the two beams after they pass the slits, thus obtaining at least a twentyfold increase in intensity (Fig. 30).

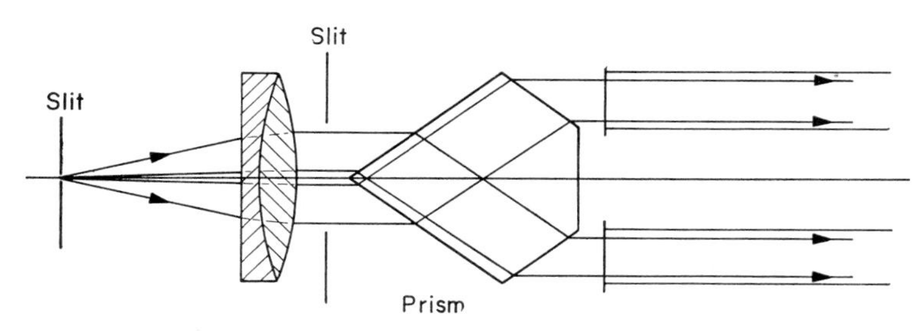

FIG. 30. Williams' modification of Rayleigh refractometer.

When the Rayleigh refractometer is used to compare the composition of two solutions, it is capable of measuring differences of two parts in a million of solvent. For example, it has been used to measure the concentration of deuterium oxide in ordinary water, with a precision of 0.01%. This is as well as can be done with a pycnometer and in addition, the method is rapid, 10 to 15 min, and can be used with samples as small as 0.05 cm^3. In general, the Rayleigh refractometer will measure a small change in refractive index more accurately than any other instrument, surpassing the Jamin[24] refractometer because the comparison is made with an undisplaced set of fringes rather than a cross-hair, thus ensuring that distortion of the apparatus does not affect the readings. With gas chambers 1 cm long and $\lambda = 5461$ Å, one unit in the sixth decimal place is the least change observable.

[23] W. E. Williams, *Proc. Phys. Soc. London* **44,** 451 (1932).

[24] J. Valasek, "Introduction to Theoretical and Experimental Optics," p. 146. Wiley, New York, 1949.

7.3. Spectroscopy in the Visible Region

7.3.1. Prism Spectrographs

The prism spectrograph consists of an entrance slit at the focus of a high quality collimator lens which makes the rays from the slit parallel and sends the beam through a refracting prism, with refracting edge parallel to the entrance slit, and then onto a camera lens which forms in its focal plane an image of the entrance slit, the spectral line. Since the deviation by the prism is a function of its refractive index and hence a function of wavelength (Section 7.1.2), each wavelength in the beam forms a separate image of the slit, which is recorded on a photographic plate. Different wavelengths are dispersed along the plate in a direction at right angles to the line images (Fig. 31). Usually the prism is oriented

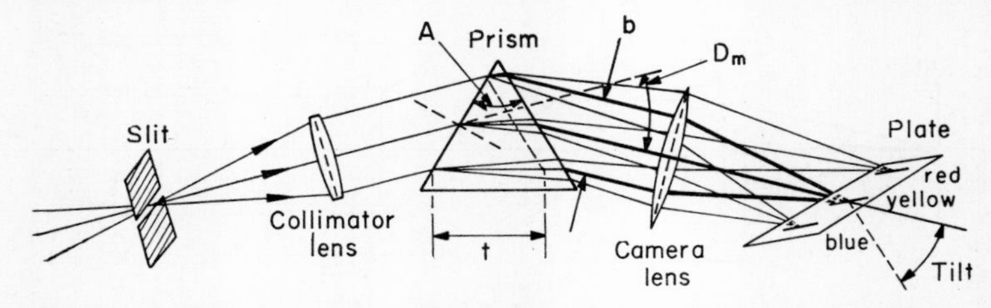

FIG. 31. Prism spectrograph showing plate tilt to compensate for chromatic aberration.

so that the refracted light for one wavelength leaves the second surface at the same angle the incident light enters the first surface (position of minimum deviation).

The linear separation, $dl/d\lambda$, of adjacent wavelengths on the plate is proportional to the angular dispersion of the prism (set at minimum deviation: $\phi_1 = \phi_2$), as given in Eq. (7.1.9). In addition $dl/d\lambda$ is proportional to the focal length, f_c, of the camera lens, and depends also on the tilt the plate may have to correct for chromatic aberration in the camera lens. Thus, exclusive of tilt,

$$\frac{dl}{d\lambda_{\text{vac}}} \sim f_c \frac{d\phi_2}{d\lambda_{\text{vac}}} \sim f_c \frac{\tan \phi_1}{n} \frac{dn}{d\lambda_{\text{vac}}} \sim f_c \frac{\sin(A/2)}{\cos \phi_1} \frac{dn}{d\lambda_{\text{vac}}} \tag{7.3.1}$$

where one may substitute the slope of the n versus λ_{vac}-curve for $dn/d\lambda$ or make use of the Cauchy relation

$$n = \text{constant} + \frac{\text{constant}}{\lambda_{\text{vac}}^2} \tag{7.3.2}$$

to obtain

$$\frac{dn}{d\lambda_{\text{vac}}} = -\frac{2\ \text{constant}}{\lambda_{\text{vac}}^3}. \tag{7.3.3}$$

The angular separation of two adjacent emergent wavelengths can also be shown to be

$$d\phi_2 = \frac{t}{b}\, dn \tag{7.3.4}$$

where t is the difference in prism thickness traversed by the two outside rays and b is the breadth of the emergent beam (Fig. 31). Then the resolving power of a prism is

$$\frac{\lambda}{\Delta\lambda} = t\,\frac{dn}{d\lambda} \tag{7.3.5}$$

provided that the incident beam is of rectangular cross section. For dense flint glass $dn/d\lambda$ is about 1200 cm^{-1}.

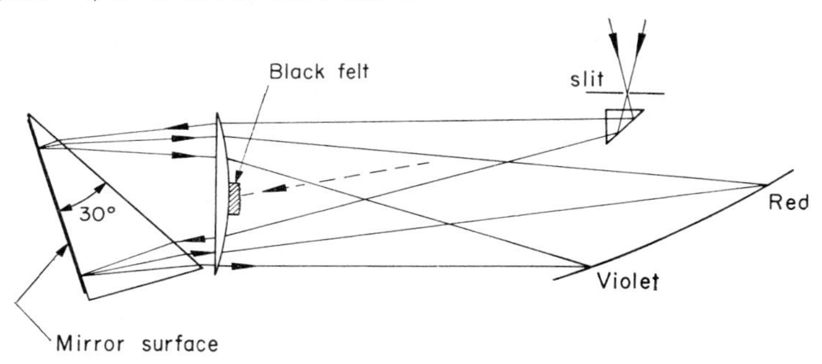

FIG. 32. Littrow mounting; felt mask reduces background.

Large angular dispersions can be obtained by passing the light through several prisms in series, since dispersions are additive, or by passing the light twice through one prism. In this latter method the effect of flaws in the prism tends to be doubled. Both methods are subject to considerable light loss by reflection at the refracting faces and by absorption in the prism. Large dispersion can also be obtained by using an angle of incidence less than the angle for minimum deviation, but this also introduces an increased light loss at the second refracting surface and, since both t and b are reduced, decreases the resolving power. In practice the prism angles are set at 60° as a compromise, because smaller angles would yield a smaller t and larger angles cause an excessive light loss by reflection at the refracting surfaces.

Various arrangements of the prism and lenses are used. To cite one example, the Littrow mounting (Fig. 32) has the advantage of using the same lens as both collimator and camera lens and of requiring only a back-silvered or -aluminized 30° prism through which the light goes twice. Its big advantage lies in economy of space since the light beam doubles back on its path. Thus, it is especially favored for spectrographs of long focal length. One serious disadvantage is the presence of considerable background from the light which is reflected at the first face of the lens directly onto the photographic plate. This can be reduced by placing a horizontal strip mask of black velvet across the middle of the first surface of the lens, since the light reflected from the lens above and below the mask would miss the plate. The Littrow mounting has another disadvantage if it is used in series with a Fabry–Perot interferometer (Section 7.3.3) in that the interferometer must then be mounted external to the spectrograph, whereas with other spectrographs the interferometer may be mounted internally with some advantages such as economy of lenses.

The photographic speed, i.e., power per unit area in the spectral image, is determined mainly by the size of the aperture and the focal length of the camera lens. If the instrument is used at its maximum resolution the focal length of the collimator has less influence on the speed since a short focal length here necessitates a narrower spectrograph slit and thus less light is admitted to the instrument. For maximum resolution the slit width should be no more than one quarter-wavelength times the ratio of focal length to useful diameter of the collimator. In practice, wider slits than the above are commonly used because, until excessive widths are reached, the intensity of the central part of the line image increases almost linearly with slit width whereas the resolving power does not diminish very rapidly. For example, with a slit width of four times the optimum, namely of one wavelength times the ratio of collimator focal length to diameter, the intensity is three-fourths of the maximum value for a very wide slit, while there is a loss of only about 20% from maximum resolving power obtained with the narrowest slit.

Illumination of the slit is important in the consideration of spectrograph speed since, of the light which enters the slit, only that part can be effective which falls on the collimator lens. The slit, therefore, should be illuminated so that the cone of light which goes through it just fills the collimator lens. Any larger cone contains rays which not only contribute nothing to the image but may also increase background haze, while any smaller cone fails to utilize the full aperture of the instrument. Usually a condenser lens is used to form an image of the light source on the slit, an image which, incidentally, is never brighter than the source. If the condenser is arranged in such a fashion that the collimator lens is just filled with the cone of light maximum illumination of the spectral lines is achieved. A shorter focal length condenser may give a brighter image on the slit but will not brighten the dispersed spectrum. If the light source has considerable depth it is generally better to arrange the condenser to image the back surface of the source onto the slit.

For some applications liquid prisms have advantages over solid prisms in terms of cost, dispersion, size, and variety of materials available. Their principal disadvantage is that the refractive index is a steep function of temperature so that temperature control is a problem during long exposures. Also temperature gradients introduce optical inhomogeneities. In spite of this they are used in special circumstances with such liquids as water, carbon disulfide, monobromonaphthalene, ethyl cinnamate, and aqueous solution of barium mercuric bromide. An inexpensive ultraviolet monochromator, for example, can be made by immersing a mirror in water at an angle so that the horizontal surface of the water is the front face of the prism and the arrangement is a Littrow mounting.

The ideal prism material should be transparent over a large wavelength range, isotropic, available in large homogeneous pieces, capable of taking an optical polish, resistant to atmospheric exposure and laboratory fumes, and have a large dispersion ($dn/d\lambda$) and a small temperature coefficient of refractive index (dn/dT). No material satisfies all these requirements, and compromises must be accepted. Table III gives properties of some useful materials.

The prism introduces some astigmatism into an optical system, but this is at a minimum if (1) the prism is traversed by parallel light; (2) the slit is parallel to the prism edge; (3) the light rays are parallel to a "principal section" of the prism, i.e., a section perpendicular to the edge; and (4) the rays pass through at minimum deviation. Astigmatism is then generally much less than in a grating spectrograph of conventional mounting. Although the astigmatism is a minimum and the definition and resolution a maximum when the prism is used at minimum deviation, the resolving power declines only very slowly as the prism is turned from minimum deviation while the angular dispersion increases rapidly (see Fig. 11), which may be useful in some instances.

If birefringent materials (Section 7.4.2) are used in prisms it is necessary that the optic axis of the material be in a principal plane and parallel to the base of the prism in order to avoid doubling of the spectrum. This means, furthermore, that the prism must be used at minimum deviation. If the material is also optically active (Section 7.4.2), e.g., crystal quartz, the rotation of the plane of polarization must be balanced out, again to avoid doubling the spectrum. The Littrow mounting automatically provides the necessary compensation since the light goes through such a prism once in each direction, thus cancelling the rotation. With a conventional mounting a Cornu prism, made of a right-handed and a left-handed crystal cemented together so that the optical activity in the two halves cancels, must be used. Dispersing polarized light in such spectrographs may still present difficulties because of polarization effects in the instrument, such as partial polarization on refraction (Chapter 7.4).

TABLE III

Material	Useful wave-length range	n_D*	$\left(\frac{d\theta}{d\lambda}\right)_D$* in radians/angstrom (60° prism)	λ of residual rays
Uviol crown glass	3000 Å–4000	1.5035	0.616×10^{-5}	
Dense flint	4000 Å–2.5 μ	1.6555	1.703×10^{-5}	
Water	1800 Å–2200 Å	1.3330	0.416×10^{-5}	
Carbon bisulfide	3500 Å–30,000 Å	1.6276	2.885×10^{-5}	
Monobromonaphthalene	Visible	1.6576	2.501×10^{-5}	
Ethyl cinnamate	Visible	1.5604	2.212×10^{-5}	
Potassium iodide	2500 Å–31 μ	1.6634	2.881×10^{-5}	
Potassium bromide	15 μ–28 μ	1.5581	1.449×10^{-5}	82 μ
Lithium fluoride	1100 Å–6 μ	1.39177	0.286×10^{-5}	20 μ
Fused quartz	2000 Å–3.5 μ	1.45848	0.517×10^{-5}	
Crystal quartz	1850 A–3.5 μ	1.54426	0.628×10^{-5}	9 μ; 21 μ
Calcium fluoride (fluorite)	1200 Å–9 μ	1.43385	0.333×10^{-5}	33 μ
Sodium chloride (rock salt)	1750 Å–16 μ	1.54431	0.938×10^{-5}	52 μ
Potassium chloride (Sylvine; much more hygroscopic than KBr)	1800 Å–21 μ	1.49038	0.729×10^{-5}	63 μ
42% Tl Br + 58% TlI (KRS − 5)	24 μ–40 μ	2.6316		120 μ
Silver chloride (windows only)	1 μ–28 μ			
Sapphire	1500 Å–6 μ			

* The n_D and $(d\theta/d\lambda)_D$ are measured for the yellow lines of sodium.

7.3.2. Grating Spectrographs

In a grating spectrograph the dispersing element is a surface ruled with a total number N, of equidistant lines, a diffraction grating (Fig. 33). The dispersion produced by the grating is due to the fact that there exists a path difference between two wavelets leaving from adjacent rulings in a given direction, θ, and if constructive interference in this direction occurs for one wavelength it will not occur for an adjacent one.

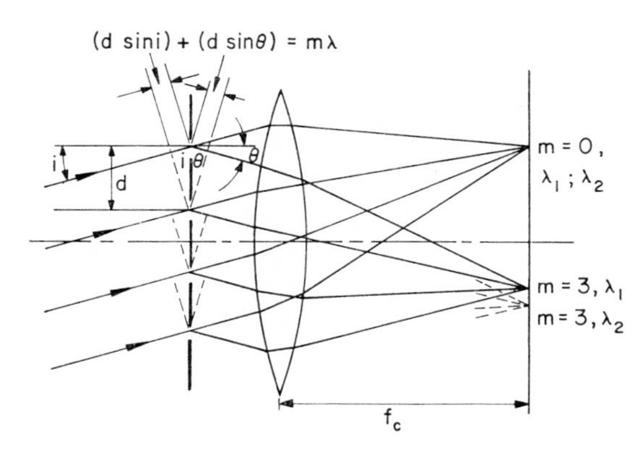

FIG. 33. Transmission grating.

For the case of a plane wave incident at an angle i upon a plane grating with a ruling spacing d, the path difference between waves from adjacent rulings will be $(d \sin i) \pm (d \sin \theta)$, and these waves will interfere constructively along θ (when collected for example by a lens) if

$$d(\sin i \pm \sin \theta) = m\lambda \tag{7.3.6}$$

where m is an integer called the order number.

The angular dispersion, $d\theta/d\lambda$, is obtained by differentiating the grating equation (7.3.6),

$$\frac{d\theta}{d\lambda} = \frac{m}{d \cos \theta} \tag{7.3.7}$$

and can be made large by using a large m or a small d. Frequently the linear separation of two wavelength, $dl/d\lambda$, is of interest and may be obtained by multiplying Eq. (7.3.7) with the focal length, f_c, of the camera lens, neglecting tilt of the plate.

The resolving power of a grating is given by

$$\frac{\lambda}{\Delta\lambda} = mN \tag{7.3.8}$$

or, in a given direction θ, by

$$\frac{\lambda}{\Delta\lambda} = \frac{Nd(\sin i \pm \sin \theta)}{\lambda} \tag{7.3.9}$$

which shows a dependence on only the grating width (Nd), a fact used in the echelon grating (see Section 7.3.3).

Gratings are commonly ruled on a metallic surface, and aluminum is preferred since it is malleable enough to minimize wear on the ruling diamond, has a high reflectivity, and resists tarnishing. It may be cleaned with $\frac{1}{10}$ normal KOH or with precipitated chalk, or a detergent, rubbing gently in the groove direction. Such a ruled metallic surface is a reflection grating and, if concave, will focus the light without a collimator or camera lens. A disadvantage of the concave grating is that there exists considerable astigmatism in the image if the incident light comes from an off-axis direction.[25]

There is a tendency for the speed of a grating to be less than that of a prism of the same size because the grating distributes the incident energy into several orders. In modern gratings, however, this disadvantage has been mostly overcome by shaping the reflecting surface of the ruled groove so that most of the energy is thrown into one order for a particular wavelength range. This is accomplished by sloping the reflecting side of the groove in such a fashion that the direction in which the desired diffracted spectrum appears is also the direction in which a ray of light would be reflected from the groove face. Such a grating is said to be blazed for the wavelength for which this condition is satisfied. For a plane grating, the width of the blazed region of the spectrum is approximately equal to the blazed wavelength divided by the order. For example, the blazed region for a grating blazed in the first order at 7500 Å would be 7500 Å wide and would extend from 3750 to 11,250 Å in the first order. In the second order the same grating would have a blaze from 1875 Å to 5625 Å. The stipulated wavelength of the blaze is on the assumption that the angle of incidence equals the angle at which the light leaves the grating and is on the same side of the grating normal, i.e., the light enters and leaves in nearly the same direction. The wavelength of the blaze will be different for any other arrangement of the grating and can be computed from $\lambda_2 = \lambda_1 \cos(\gamma/2)$, where λ_1 is the blazed wavelength when the incident and reflected light are undirectional and λ_2 is the blazed wavelength when γ is the angle between the incident and reflected rays.

If the wavelength of the incident light is much larger than the groove separation the grating acts chiefly as a mirror and puts most of the light into the zero order. Consequently, coarse gratings must be used in the

[25] H. G. Beutler, *J. Opt. Soc. Am.* **35**, 311 (1945).

infrared and the coarseness makes blazing easier.[26] R. W. Wood gave the name echelette to such coarse blazed gratings which are of special importance in the infrared because of the low intensity of infrared light sources.

Harrison[27] has developed for the visible region extremely coarse blazed gratings called echelles in which the energy is thrown into a very high order and, consequently, a high resolution is obtained. For example, a 25-cm grating with 1000 grooves has in the 1000th order a theoretical resolving power of 1,000,000 and, at 5000 Å, a free spectral range of 5 Å without overlapping orders. The echelle must, therefore, be used in series with some other dispersing system and must be carefully ruled to maintain groove shape identical to a fraction of a wavelength across its surface.

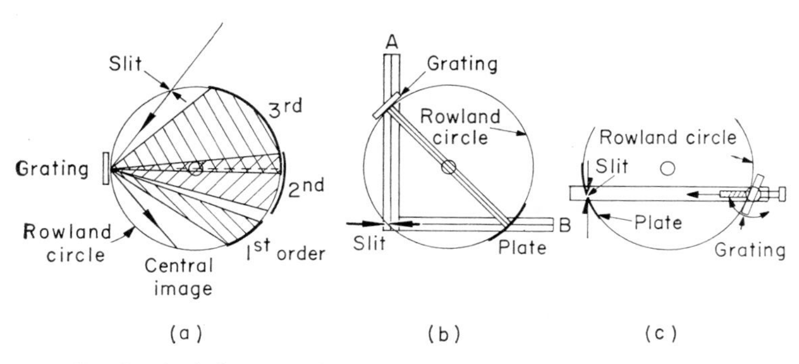

FIG. 34. Rowland circle mountings; (a) Paschen–Runge, (b) Rowland, and (c) Eagle mounting.

All the mountings of the concave grating, except that of Wadsworth, are adaptations of Rowland's principle that if the slit lies on a circle, the Rowland circle, to which the grating is tangent and which has as its diameter the radius of curvature of the grating blank the diffracted spectrum lies on the same circle (Fig. 34). The most popular mounting for large gratings is the Paschen–Runge mount in which one or more fixed slits are placed along the Rowland circle to give appropriate angles of incidence for various problems, and where photographic plates can be placed nearly anywhere along the circle to obtain a large part of the spectrum with one exposure, Fig. 34a. The dispersion is nonlinear except for the region near the grating normal. In the Rowland mounting, Fig. 34b, the dispersion is linear. With the Paschen–Runge mounting, reflecting mirrors on the Rowland circle can be used to reflect the light

[26] R. W. Wood, *Phil. Mag.* [6] **20,** 770 (1910).

[27] G. R. Harrison, J. E. Archer, and J. Camus, *J. Opt. Soc. Am.* **42,** 706 (1952).

back to the grating for multiple-passes and thus multiply the dispersion and resolution obtainable.[28]

A more compact mounting similar to the Littrow, is the Eagle mounting in which the slit and plate holder are close together on one end of a rigid bar while the other end carries the concave grating, Fig. 34c. As in the Littrow mounting it is necessary to rotate the grating around its vertical axis and simultaneously to move it along the bar in order to adjust for different spectral range. In addition the plate holder must be rotated around some axis in the plate surface.

In the Wadsworth mounting (Fig. 35) the grating is illuminated with parallel light, usually from a concave mirror. Since the slit must be placed beside the grating the mounting is quite compact. The instrument is usually arranged to work near the normal because of the rapid increase of

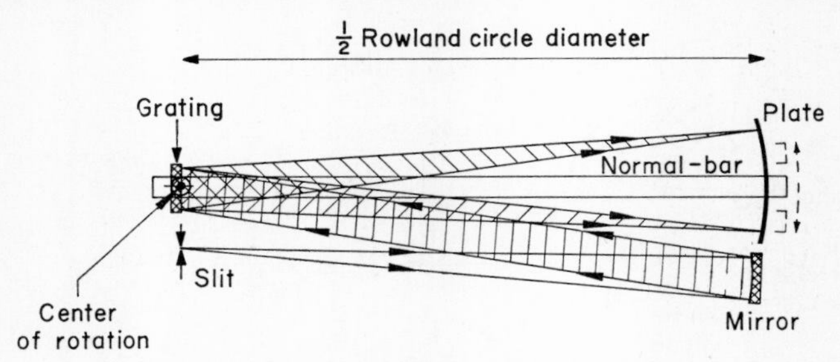

FIG. 35. Wadsworth stigmatic mounting.

spherical aberration and astigmatism with angle of diffraction. The resultant stigmatism makes the Wadsworth mounting especially favorable for use with Fabry–Perot etalons or in photometric work with a stepped slit or sector (Section 7.6.2). A disadvantage of this mounting is that the dispersion is only about half that obtained with the same grating in other mountings and is not as linear. In addition the adjustment necessary upon changing spectral regions is more complicated than in the Rowland-circle mountings.

7.3.3. Interference Spectrographs

Interference spectrographs provide high resolution by having a large optical path difference, of the order of 10^4 to 10^5 wavelengths, between interfering rays. The most common instruments are the Lummer–Gehrcke plate and Fabry–Perot etalon, both of which are multiple beam devices, and the reflection and transmission echelons, which operate on the principle of wavefront division. All are used in conjunction with ordinary spectrographs or filters.

[28] F. A. Jenkins and L. W. Alvarez, *J. Opt. Soc. Am.* **42**, 699 (1952).

The Lummer–Gehrcke plate makes use of fringes of equal inclination produced in a thick plane-parallel-sided plate (Fig. 36). The high reflectivity necessary to obtain a large number of interfering beams is obtained by having the light strike the surfaces internally at almost the critical angle; thus no metallic coating is necessary and the instrument gives good resolution and intensity throughout the spectral range in which the plate is transparent. It is most useful in the ultraviolet, where

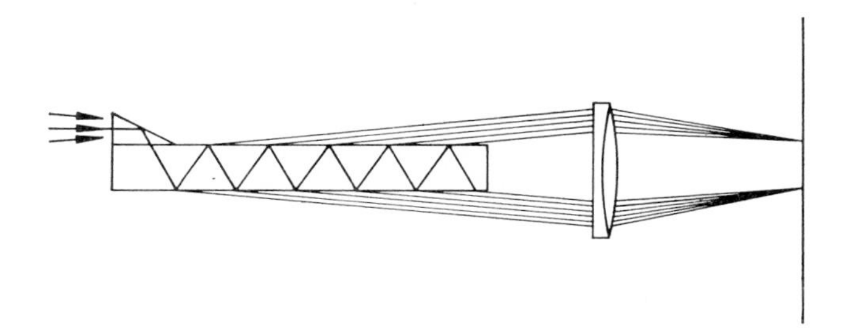

FIG. 36. Lummer–Gehrcke plate.

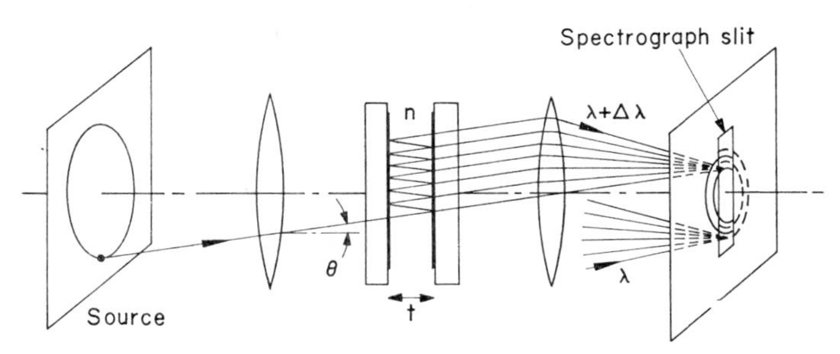

FIG. 37. Fabry–Perot interferometer, externally mounted, showing fringes of equal inclination.

metallic reflectivities are low. Because of its cost, lack of adaptability, exceptional care required, and small tolerance to temperature changes during use the Lummer–Gehrcke plate has now been almost completely superseded by the Fabry–Perot etalon.

The Fabry–Perot etalon consists of two optical flats, usually crystal quartz, separated by a parallel-sided ring spacer, either quartz or Invar, and with the two inner surfaces of the flats coated with a high reflectance film (Fig. 37). Light enters at or near normal incidence and is usually collimated so that Haidinger fringes are obtained in transmission. Each wavelength produces a series of concentric, sharp, bright rings at angles satisfying the relation

$$m\lambda = 2nd \cos \phi \tag{7.3.10}$$

where m is the integral order number, n the refractive index of the gas between the plates,* d the thickness of the spacer, and ϕ the angle at which the ring is formed. (Note that phase change at the reflecting surface has been neglected.) If a spectral line of several ordinarily unresolved components is examined each component will form a set of rings so that in each order the components will be separated according to wavelength, with the shorter wavelengths farther from the center. The spectral range without overlapping, i.e., the change in wavelength corresponding to a displacement from one order to the next is very small, and the instrument must be used in series with some apparatus to isolate spectral lines. Expressed in wave numbers, the spectral range equals the reciprocal of twice the optical separation of the plates and is thus a constant of the etalon depending only on the spacer. This illustrates the general rule that the free spectral range in cm^{-1} of any interferometer is the reciprocal of

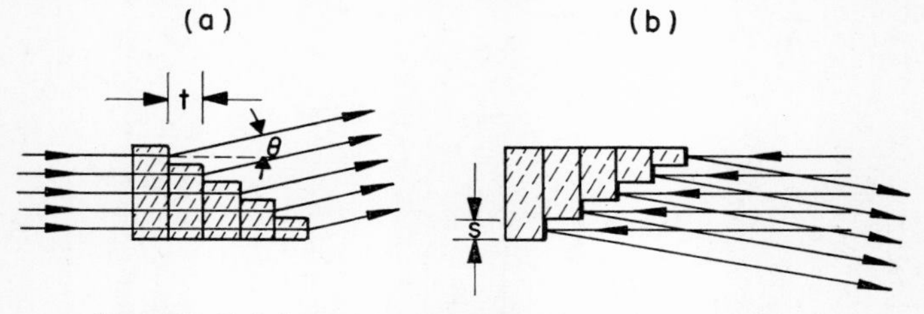

FIG. 38. Echelon; (a) transmission, and (b) reflection.

the path difference between successive beams. The resolving limit, in wave numbers, equals the free spectral range divided by the effective number of reflections in the etalon. Assuming no absorption in the films, the effective number of reflections equals $3.03R^{\frac{1}{2}}/(1 - R)$ for an etalon of unlimited area, where R is the reflectance of the film.

An echelon is a stack of plates of identical thickness, t, arranged in steps of width s in order to provide a constant optical path difference between light beams from successive steps (Fig. 38). For transmission, the plates of index n are transparent and a parallel beam of light is passed through them so that at normal incidence the optical path difference between light rays going through the successive steps is $(n - 1)t + s\theta$, where θ is the angle of observation, Fig. 38a. Thus the order of inter-

* In an ingenious application of Eq. (7.3.10) the n of the gas between the plates is varied linearly with time by changing the gas pressure so that the rings dilate from the center. Thus, if the center of the rings is focused on a pinhole behind which there is a detector connected to a recording galvanometer, the galvanometer trace obtained while the pressure is varied is a reproduction of the shape and structure of the spectral line. See C. Dufour and P. Jacquinot, *J. recherches centre natl. recherches sci.* **6**, 91 (1948).

ference, m, i.e., the optical path difference divided by the wavelength, may be very large. The resolving power, mN, of any such instrument can be considered as the path difference between the beams incident at the ends of the instrument, measured in wavelengths, and therefore can be very large, of the order of 500,000.

If the echelon is used in reflection (Fig. 38b), the steps must be coated with a metal of high reflectivity so that normally incident parallel light is reflected with an optical path difference $n(2t - s\theta)$ between successive steps, where n is the refractive index of the surrounding medium. For reinforcement this must equal an integral number, m, of wavelengths. Since in the reflection echelon the factor $(n - 1)$ of the transmission echelon is replaced by a factor 2, the resolving power for a given t should be about four times as great. The useful spectral range of an echelon is fixed when the thickness of the plate is chosen, and therefore it may be unsuited to a particular problem. In addition, the use of multilayer dielectric high-reflecting films in modern etalons has vitiated some of the comparative advantages the echelon possessed.

7.3.4. Light Sources*

Spectroscopic light sources may be divided into two main classes: (1) thermal sources, in which the radiation is due to high temperature; (2) sources depending on an electrical excitation of gases. The sun is an excellent example of the first class and is still used as an intense spectroscopic source, especially in tropical regions, but has been generally superseded by tungsten-filament lamps, various electric arcs at high pressure, and the flame. In the second class are high-voltage sparks, glow discharges at low pressure, and low-pressure arcs. Thermal sources usually produce a wavelength-continuous spectrum, while electrical discharges usually excite the emission of discrete lines or bands. The positive pole of the carbon arc, at about 4000°C, constitutes one of the brightest laboratory thermal sources.

Low-voltage metallic arcs are a useful source of line spectra of the metals. Two rods connected to a direct current source are touched together and drawn apart so that a brilliant arc forms between them. A resistance of high current capacity must be connected in series with the arc and, if necessary, a large self-inductance in series to stabilize it. Similarly, in the case of low current, low-pressure arcs, an air-core self-inductance seems best. In the vertically burning iron arc, commonly used as a source of standard wavelengths, the positive pole should be the lower electrode for maximum stability.

In the case of volatile and reactive metals the arc is enclosed in an

* See also Volume 4, Chapter 1.5.

evacuated glass envelope as in the common mercury and sodium arcs. A low-pressure mercury arc which gives very sharp, bright lines, uses a water cooled mercury pool as the cathode in conjunction with an iron anode and is run at a low current density. It is started by holding a small Tesla coil next to the glass and simultaneously short-circuiting the series inductance.

Gases and vapors are conveniently excited in high-voltage, low-pressure ac glow discharges. The usual arrangement is to seal the gas at a pressure of about 0.5 to 10 mm Hg in a glass or quartz tube provided with two electrodes. With a few thousand volts ac a bright glow discharge of the order of 10 ma develops. During operation most gases gradually disappear from the tube because of electrical clean-up, in which the gas is driven into the electrodes or walls and is thus removed from the discharge. Rare gases are subjected least to such clean-up.

Electrodeless glow discharges can be obtained by omitting the electrodes in the tube and applying a high-frequency voltage at 10–2000 Mc by coupling the tube to the field of an electrical oscillator, for example by placing it inside the oscillator plate inductance (tank coil) or inside the high-frequency cavity. Such a mode of excitation has many advantages including a drastically reduced rate of clean-up, improved purity of the gas because of absence of electrode contaminants, excitation at very low pressure and hence with very small amounts of gas, and sharpness of the emitted lines. In general, the higher the frequency the better, so that commercial microwave oscillators are commonly used for exciting spectroscopic sources.[29]

For the excitation of elements of low vapor pressure the hollow-cathode discharge tube used first by Schuler[30] is most successful. This is a low-voltage dc source in which the cathode is a hollow cylinder. With a few hundred volts applied, a glow discharge occurs which concentrates almost completely inside the hollow cathode if the gas pressure is adjusted properly. Noble gases at a pressure of a fraction of a mm Hg are usually used inside the tube because of the rapid electrical clean-up experienced with other gases. The light emitted from the cathode region shows the spectrum of the carrier gas plus that of the material on the inner wall of the cathode, even including some spark lines of the cathode material. The virtues of such a source include extreme line sharpness because of low pressures, weak electric field within the cathode, low-current density, and ease of cooling of the discharge by immersing the cathode

[29] A. T. Forrester, R. A. Gudmundsen, and P. O. Johnson, *J. Opt. Soc. Am.* **46,** 339 (1956).

[30] H. Schuler, *Z. Physik* **35,** 323 (1926); see also O. H. Arroe and J. E. Mack, *J. Opt. Soc. Am.* **40,** 386 (1950).

in a refrigerant. Very small amounts of an element, of the order of a fraction of a microgram, can be excited since the atoms can be sputtered into the glow repeatedly by the carrier gas.

For those particular problems where extreme monochromaticity of spectral lines is essential atomic and molecular beam sources may be used.[31] The arrangement is usually such that atoms, vaporized from a furnace and delineated into a beam by two slits, travel at right angles across the optic axis of a condenser lens. At the point of crossing the atoms are excited by electron impact, or otherwise, and their light can be focused on a spectrograph slit. Because of the absence of Doppler-, collision-, field-, and pressure-broadening the lines are very sharp, and extremely high resolving powers, of the order of several million, are required to show their width.

For exciting the spectra of ionized atoms, sparks are most frequently used. The voltage required to initiate a spark depends on the gas used and its pressure, the shape of the electrodes, and other factors. In air at atmospheric pressure the breakdown potential is about 12,000 volts/cm when sharply pointed electrodes are used. The power supply may be a transformer rated at 0.25 to 1.0 kva and 15,000 or 20,000 volts. It is customary to place a capacitor in parallel with the gap, in order to store more energy at the breakdown potential and thereby enhance the intensity of the spark. With the transformer described above and a gap of 4 to 5 mm, the optimum capacitance lies between 0.003 and 0.03 μf. Introducing a self-inductance of from 15 μh to 1 mh in series with the spark gap suppresses the spectral lines of the atmosphere in which the spark operates and gives rise to a hotter spark. Thus, in addition to spark lines from ionized electrode material its entire arc spectrum appears also. If the spark is operated in vacuum, high breakdown potentials are required and, by using large capacitors (0.1 to 1.0 μf) and transformers of considerable power ($\sim$2 kva), sparks are obtained of high emissivity in the far ultraviolet (Section 7.9.3).

The most common source of a wavelength continuum is an incandescent surface which obeys closely the laws of blackbody radiation. In this connection a blackbody is defined as one which absorbs all radiation falling upon it and may be approximated by small hole in a large cavity at a uniform temperature. For an incandescent source Kirchhoff's law (Section 7.1.1) gives the ratio of emittance to absorptance, while the Stefan–Boltzmann law relates the total energy emitted by a blackbody per unit surface and unit time, P_B, to the absolute temperature, T:

$$P_B = cT^4 \tag{7.3.11}$$

[31] D. A. Jackson and H. Kuhn, *Proc. Roy. Soc.* **A148,** 335 (1935).

where c is a constant, 5.6686×10^{-5} erg cm^{-2} sec^{-1} °K^{-4}. The spectral distribution is described in Planck's law:

$$P_\lambda \, d\lambda = \frac{c_1}{\lambda^5}\left(\frac{c_2}{e^{\lambda T}} - 1\right)^{-1} d\lambda \tag{7.3.12}$$

with $c_1 = 3.7413 \times 10^{-1}$ erg cm^2 sec^{-1} and $c_2 = 1.43884$ cm $\times$ °K if λ is in cm. Upon maximizing this yields Wien's displacement law,

$$\lambda_{\max} T = \text{const} = 0.28979 \text{ cm} \times \text{°K} \tag{7.3.13}$$

which states that the wavelength, $\lambda_{\max}$, of maximum emittance for a blackbody times the absolute temperature of the body is a constant.

From these laws it can be shown, that a black incandescent surface at 6000°K will produce its maximum emittance in the visible at 5000 Å, while at 4000°K, 3000°K, 2000°K, and 1000°K it will be at 7500 Å, 10,000 Å, 15,000 Å, and 30,000 Å respectively. Thus, while it is possible to use such sources in the visible and infrared (Section 7.9.1) regions, it would require much higher temperatures to concentrate radiant energy in the ultraviolet. A spark between metallic electrodes under water, however, yields a bright continuous spectrum extending to about 2000 Å.

One of the most convenient sources of a continuous spectrum in the ultraviolet is the hydrogen discharge tube. Such tubes are operated at pressures from 1 to 10 mm Hg, at voltages of about 1000 and at currents from milliamperes to several amperes. Such tubes usually contain large metallic surfaces to catalyze the recombination of hydrogen atoms, since the continuous spectrum (in reality a closely spaced line spectrum) is emitted by molecules.

An electron synchroton is also useful as source of a continuum in the far ultraviolet region.

7.4. Polarization*

7.4.1. Types of Polarization

Interference and diffraction indicate conclusively the wave nature of light but do not discriminate between transverse and longitudinal waves. The universally accepted electromagnetic theory of light specifically requires that the vibrations be transverse, entirely confined to the plane of the wavefront, and the existence of complete polarization (Section 7.4.3) corroborates this.

* See also Vol. 6, Chapter 9.4.

Unpolarized light possesses symmetry about its direction of propagation, and when it is forced to take on some asymmetrical character it is said to be polarized. If, for example, all directions of vibration except one are quenched the light is plane polarized, where in this and in all subsequent usage the plane of vibration is defined as one which contains the direction of propagation of light and the electric vector, E (Section 7.2.2). In addition to this there can be circular or elliptical polarization or mixtures of all these. If the light vector rotates at the frequency of the wave without changing its magnitude, it is said to be circularly polarized; if

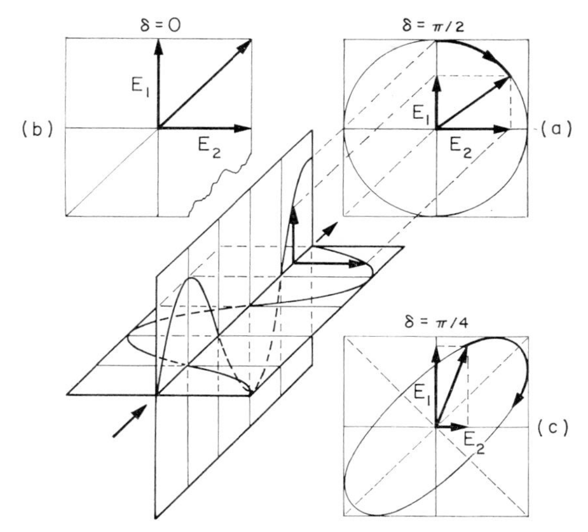

FIG. 39. Electric vector in polarized light in terms of two perpendicular components and their relative phase, δ; (a) left circular, (b) plane, and (c) left elliptical polarization.

it also changes its magnitude, it is elliptically polarized. When the direction of rotation is clockwise looking against the direction of propagation the light is called right polarized, while if the rotation is counterclockwise it is called left polarized.

It will be convenient for subsequent discussions to visualize plane, circularly, and elliptically polarized light in terms of their components, E_1 and E_2, which are mutually perpendicular (Fig. 39). If the phase difference, δ, between E_1 and E_2 is zero or π, plane polarized light is obtained; and if $\delta = \pi/2$ or $3\pi/2$ and $E_1 = E_2$, circularly polarized light results. At any other phase angle there is elliptical polarization with the major and minor axes of the ellipse changing direction and magnitude with different phase angles.

7.4.2. Polarization Phenomena

If unpolarized light falls on a dielectric surface at other than normal incidence the reflected beam is partially plane polarized, the fraction of polarization depending on the angle of incidence and the optical properties of the medium. This effect can be described by considering unpolarized light to consist of two incoherent components (without constant phase relation to each other), one parallel to the plane containing the incident

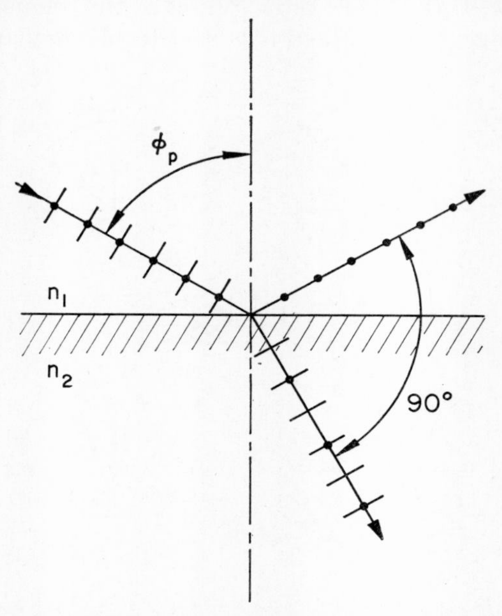

FIG. 40. Polarization by reflection at Brewster's angle, ϕ_P.

ray and the normal to the surface, the plane of incidence, and one perpendicular to this. The former is called the p (parallel) component and the latter the s (*senkrecht*) component. At an angle of incidence other than zero or 90° the reflectivity is different for the p and s components and therefore the reflected and refracted beams contain unequal mixtures of these components and are partially plane polarized. In case of a dielectric there is one angle of incidence, the polarizing angle, ϕ_P, for which the reflectivity of the p component is zero and thus the reflected beam is completely plane polarized containing only the s component. This angle is given by Brewster's law,

$$\tan \phi_P = n_{1,2} \tag{7.4.1}$$

where $n_{1,2}$ is the refractive index of the second medium relative to the first, and ϕ_P is sometimes called Brewster's angle (Fig. 40). Thus by measuring ϕ_P, the refractive index of, for example, an opaque material may be determined.

In case of reflection at the polarizing angle from transparent materials, the transmitted light is partially plane polarized, and for a stack of m plates the percentage polarization, PP, in the transmitted light is given by

$$PP = \frac{m}{m + [(2n)/(1 - n^2)^2]}. \tag{7.4.2}$$

This is a convenient way to produce polarization for wavelengths at which commercial Polaroid (see below) cannot be used. The black surface of selenium, melted onto a glass plate, is excellent for polarizing light by reflection, especially in the red and infrared. Because of its high refractive index as much as 49% of the incident radiation is reflected at $\phi_P \sim 70°$.

Polarization may also be observed in scattered radiation (Section 7.7.1), as seen in sky light, and is almost complete in a direction at right angles to the incident light, in which case the vibrations in the scattered light are perpendicular to the direction of the incident ray (Fig. 49).

7.4.3. Birefringence

Certain materials in which the molecular orientation is neither random nor spherically symmetric possess refractive indices which are functions of the polarization of the light and of the direction in the material. Thus when unpolarized light is incident on such a material there may be two refracted beams instead of one, an effect called double refraction or birefringence. Snell's law (Section 7.1.1) holds for one of the rays, called the ordinary or O ray, but not for the other, the extraordinary or E ray. The O ray is always in the plane of incidence, but only for special directions in the crystal is this true for the E ray. The most common examples of birefringent crystals are calcite and quartz, which have one, and only one, direction, the optic axis, in which the O and E rays behave alike in all respects, and such crystals are called uniaxial. (Materials with two optic axes are called biaxial crystals and will not be treated here.) In a uniaxial crystal the principal plane of the O ray is defined as a plane containing the optic axis and the O ray. The principal plane of the E ray is defined similarly. When unpolarized light enters a calcite crystal so that the principal planes are parallel it separates into an O ray vibrating perpendicular to the principal plane, and an E ray vibrating parallel to the principal plane. This provides a means of producing polarized light, and the Nicol prism is such a device in which the separation of the E and O rays is accomplished by total reflection (Fig. 41). In this arrangement the calcite crystal is cut obliquely into halves and then cemented together with Canada balsam, which has a refractive index midway between that of the O and E rays. The E ray is refracted into the balsam and on through the crystal, whereas the O ray is totally reflected. Thus, Nicol prisms are excellent polarizers throughout the entire visible spectrum. In the ultraviolet, however, they are useless because of the opacity of the balsam for wavelengths below about 3400 Å. If the Canada balsam is replaced by an air film, as in the Foucault prism, such a device may be used down to

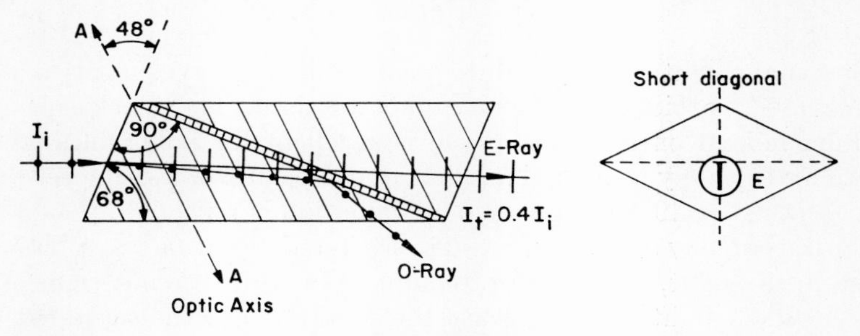

FIG. 41. Nicol prism.

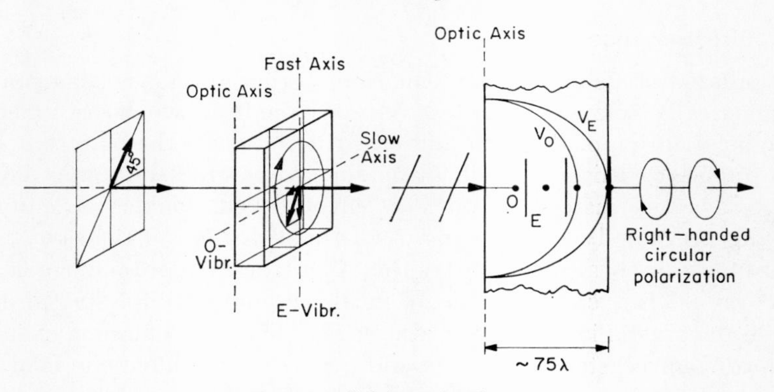

FIG. 42. Calcite λ/4 plate.

2300 Å. The angular aperture (acceptance) of a Nicol is about 24°, whereas for a Foucault prism it is about 8°.

If a parallel beam of plane polarized light is incident normally on a thin birefringent plate, such as calcite, it is found that there are two orientations of the plate (privileged directions) for which the emergent light is plane polarized (Fig. 42). In the case of calcite the E-vibration travels faster than the O-vibration ($n_E < n_O$). The direction of the E-vibration is called the fast axis, the other is the slow axis. Any plate for which the difference of optical thickness for the fast and slow axes is a quarter of a vacuum wavelength is called a quarter-wave plate for that specific wavelength. When the fast axis of a λ/4 plate is oriented at 45°

with the plane of polarization of plane polarized incident light the emerging light will be circularly polarized (Fig. 42).

The properties of quarter-wave plates are additive, that is, if two are superimposed with their fast axes parallel the combination will act as a half-wave plate, introducing a difference of $\lambda/2$ in the optical paths of the fast and slow components.

Quarter-wave plates may be made conveniently from a mica sheet by orienting an axis of the mica at 45° to the axis of a polarizer and laying

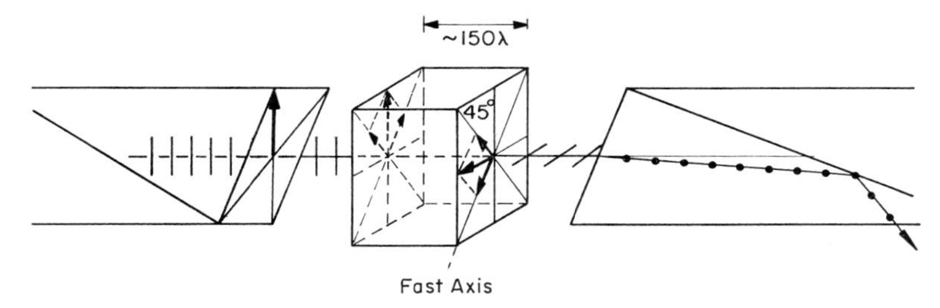

FIG. 43. Calcite $\lambda/2$-plate between parallel Nicols.

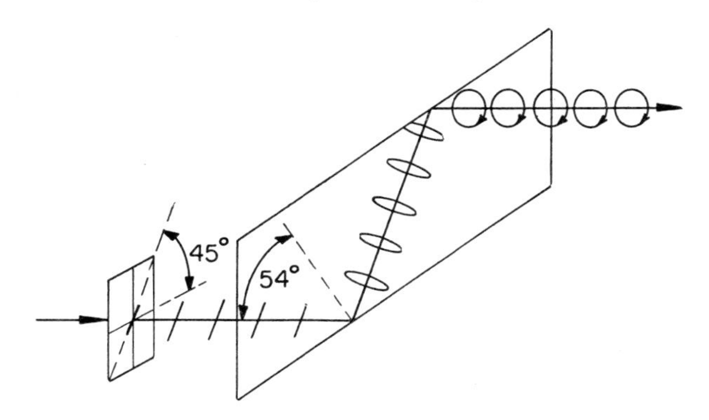

FIG. 44. Fresnel rhomb.

the combination on a mirror, mica next to the mirror. If the mica is a quarter-wave plate, or an odd multiple, the surface should appear dark. The proper thickness for yellow light is 0.032 mm and in practice the sheets should be micrometered to be sure they approximate this value. A half-wave plate may be similarly made and tested by inserting it between parallel polarizers (Nicols); at an orientation of 45° with them it should produce extinction (Fig. 43).

Although the quarter-wave plate is inexpensive its extreme chromatism is a disadvantage, and therefore other devices which act as quarter-wave plates over a wide wavelength range are used in more precise work.

The only requirement is that a phase difference of $\pi/2$ be introduced between the two components of plane polarized light. In the Fresnel rhomb (Fig. 44) this is done by using the difference in phase change on reflection of the two components at a glass–air surface.* Two internal reflections are required, each at an angle of about 54°, so that there is a phase difference of $\pi/4$ at each reflection. The phase difference introduced depends somewhat on the refractive index, but it varies so slowly with wavelength that a Fresnel rhomb designed for the middle of the visible spectrum will still function at the violet and red ends.

7.4.4. Compensators

In the Babinet compensator (Fig. 45) a phase difference is introduced between two components of the incident plane polarized light by utilizing the birefrigerence of quartz. The compensator consists of two wedge-shaped prisms of quartz of very small angle, and the optic axes in these

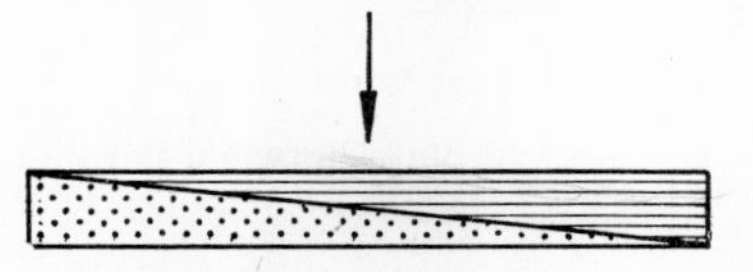

FIG. 45. Babinet compensator.

two wedges are parallel and perpendicular, respectively, to the two refracting edges. Thus, if plane polarized light is incident normally with its plane of vibration at an angle θ to the optic axis it will be broken up into two components. The E component, parallel to the optic axis in the first crystal, travels slower than the O component until it enters the second crystal. At this point, the E vibration becomes an O vibration, and the O vibration from the first crystal becomes the E vibration in the second. The result is that one prism tends to cancel the effect of the other along the center. On each side of the center one vibration will be behind or ahead of the other. Thus there are periodic regions where the compensator acts as a $\lambda/4$, $\lambda/2$, $3\lambda/4$, etc. plate. The chief disadvantage is that a specified retardation is confined to a narrow region along the plate.

The Soleil compensator is a modification which provides a controllable retardation over a large field. It consists of two quartz wedges with parallel optic axes mounted on top of a quartz flat with its axis perpendicular to the other two (Fig. 46). The thickness is varied by a calibrated screw which slides the top prism along the other. By making the prism

* The electric vector experiences a phase change, δ, upon reflection from a surface. The phase change depends on the angle of incidence, ϕ, and the index, n, and differs also for the p and s components. For dense to rare reflections in glass, $n = 1.51$, at $\phi = 54°$ the difference in phase amounts to $\delta_p - \delta_s = \pi/4$.

angles very small a careful adjustment to a thickness for a $\lambda/4$ or $\lambda/2$ plate is readily made for any wavelength.

There is a class of double refracting substances that absorb the O and E rays differentially, thereby being capable of producing plane polarized from unpolarized light. This phenomenon, dichroism, is exhibited, e.g., by tourmaline and by herapathite (iodosulfate of quinine). Polaroid consists of herapathite crystals embedded in a transparent matrix with

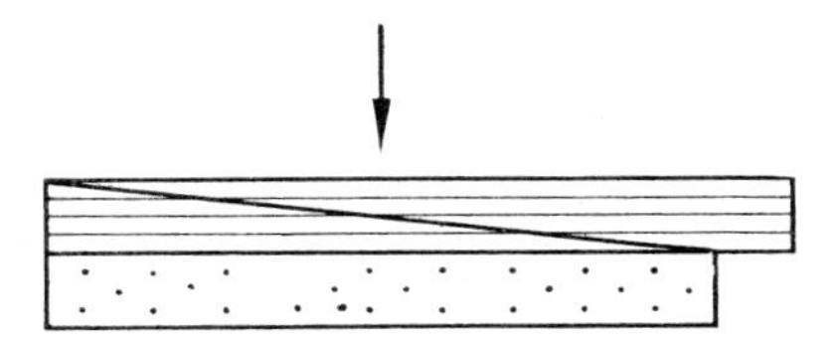

FIG. 46. Soleil compensator.

their optic axes all parallel, so that the result is a polarizing sheet of large area and low cost. Such sheets are effective polarizers throughout most of the visible spectrum but not at the extremes of the visible range.

7.4.5. Optical Activity

Some substances cause a rotation of the plane of polarization of light passing through them, an effect called optical activity (see also Section 7.5.5). If, for example, plane polarized light is directed along the optic axis of quartz it experiences a rotation of the plane of polarization which is proportional to the distance traveled in the quartz and depends strongly on the wavelength of the light. Such optical activity is also exhibited by cinnabar, sodium chlorate, turpentine, sugar crystals, sugar in solution, and sulfate of strychnine. Some quartz crystals and sugar solutions rotate to the right, dextrorotary or right-handed, and some to the left, levorotary or left-handed. By right-handed rotation is meant that, observed against the direction of propagation, the plane of vibration is rotated in a clockwise direction. Thus, with optical elements of crystal quartz it is necessary to make half the element of right-handed and half of left-handed quartz (Cornu prism, Section 7.3.1). The problem is accentuated by the strong dependence of the optical activity on wavelength.

An explanation of optical activity may be obtained by regarding plane polarized light as consisting of two components, one right and one left circularly polarized, traveling in phase to produce a single plane of vibration. Then the refractive index of the optically active material must be different for the two circular polarizations so that on passage through the medium a phase difference is introduced between them such that upon emergence they recombine to form a vibration plane at a different orientation from the one that entered.

7.4.6. Technique of Analysis of Polarization

In order to analyze a beam of light for polarization properties, some sort of polarizer and a quarter-wave plate or its equivalent are necessary. Then, the procedures shown in Table IV may be used to make a systematic qualitative investigation.

TABLE IV.[32] Analysis of Polarized Light

<table>
<tr><td colspan="4">A. No Intensity Variation with Analyzer Alone</td></tr>
<tr><td>If with $\lambda/4$ plate in front of analyzer</td><td colspan="3">If with $\lambda/4$ plate in front of analyzer one finds a maximum, then</td></tr>
<tr><td>one has no intensity variation, one has natural unpolarized light</td><td>if one position of analyzer gives zero intensity, one has circularly polarized light</td><td colspan="2">if no position of analyzer gives zero intensity, one has mixture of circularly polarized light and unpolarized light</td></tr>
<tr><td colspan="4">B. Intensity Variation with Analyzer Alone</td></tr>
<tr><td>If one position of analyzer gives zero intensity, one has</td><td colspan="3">If no position of analyzer gives zero intensity, insert a $\lambda/4$ plate in front of analyzer with its optic axis parallel to position of maximum intensity</td></tr>
<tr><td rowspan="2">plane-polarized light</td><td rowspan="2">if one gets zero intensity with analyzer, one has elliptically polarized light</td><td colspan="2">if one gets no zero intensity,</td></tr>
<tr><td>but the same analyzer setting as before gives the maximum intensity, one has mixture of plane-polarized light and unpolarized light</td><td>but some other analyzer setting than before gives a maximum intensity, one has mixture of elliptically polarized light and plane-polarized light</td></tr>
</table>

To determine whether a beam of circularly polarized light is right- or left-handed it should be passed through a quarter-wave plate and then a Nicol. The quarter-wave plate converts the light to a plane-polarized beam, and its direction of vibration is determined by the Nicol. If, observed against the direction of propagation, the plane polarized vibration makes an angle of plus 45° with the fast axis of the quarter-wave plate (where the plus direction is counterclockwise) the original beam is right-handed; if the angle is minus 45° it is left-handed.

[32] Prof. F. A. Jenkins, personal communication (1957).

With plane-polarized light normally incident upon a half-wave plate and making an angle ψ with one of the privileged directions, the emergent beam will also be plane-polarized at an angle $-\psi$ with the same direction, i.e., the E vector has been rotated through 2ψ (Fig. 43). A half-wave plate suitably oriented can change a right-handed elliptically polarized beam into a left-handed one, and the same holds for circularly polarized light except that in this case there is no need for any preferred orientation with the privileged directions.

7.5. Magneto- and Electrooptical Effects

7.5.1. Zeeman Effect

In the Zeeman effect a splitting of spectral lines with wavelength differences of the order of 1 Å for 20,000 gauss is observed when the light source is placed in a magnetic field. The emitted light is polarized and the polarization depends on the direction of observation with respect to that of the magnetic field. If viewed at right angles to the field (transverse Zeeman effect) the light is plane polarized, with the polarizations and intensities of the split components such that they would recombine to form unpolarized light. The components vibrating parallel to the field, usually the center lines of a pattern, are called p, and those vibrating perpendicular s (*senkrecht*) components. In the normal transverse Zeeman effect each spectral line is split into an undisplaced p component and two symmetrically displaced s components. The displacements, measured in wave number units ($\Delta\bar{\nu} = \Delta\nu/c$), are proportional to the applied field strength:

$$\Delta\bar{\nu} = \pm \frac{eH}{4\pi mc^2} = 4.66879 \times 10^{-5}\ \text{cm}^{-1}\ \text{gauss}^{-1} \tag{7.5.1}$$

and may be used to measure unknown fields, e.g., those of stars. In most spectral lines the Zeeman effect is anomalous in that an applied magnetic field splits the line into more than three components.

In the longitudinal Zeeman effect, observed parallel to the field, only frequencies corresponding to the s components are found and these are circularly polarized. Observed against the magnetic field the component of lower frequency is right-handed and the other left-handed.

If an absorbing material is placed in a magnetic field a corresponding splitting of the absorption lines is observed, the inverse Zeeman effect.

As far as the inverse longitudinal effect is concerned, unpolarized light passing through the material may be considered as consisting of right and left circularly polarized components with no constant phase relations. At a frequency at which the material would emit right circularly polarized light, if it were emitting, it now will absorb the corresponding component and therefore transmit only the left circular component. Thus the Zeeman components of a spectrum line observed in absorption along the field direction are not completely absorbed, and the residual light is circularly polarized in directions opposite to those in emission.

7.5.2. Stark Effect

When light-emitting hydrogen or ionized helium atoms are subjected to an electric field of the order of 100,000 volts/cm, a symmetrical splitting of each spectral line, called the Stark effect, is observed. Viewed perpendicular to the field the components are plane-polarized, some s and some p, components. Viewed parallel to the field only frequencies corresponding to the s components appear, but always as unpolarized light. The splitting is proportional to field strength and has been used to measure fields in electrical discharges, solar flares, etc. In the case of other atoms one usually observes only a slight shift of the lines to longer or shorter wavelengths, a phenomenon called the quadratic Stark effect since it is proportional to the square of the field strength.

7.5.3. Electrooptic Kerr Effect*

This effect is one in which a material placed in an electric field behaves optically like a double refracting uniaxial crystal (Section 7.4.2) with the optic axis parallel to the field. The magnitude of this effect can be measured by inserting two parallel metal plates in a glass cell containing the material (a Kerr cell) and placing the cell between a crossed polarizer and analyzer, orienting the plates at a 45° angle to a privileged direction of the polarizer. When a high potential is applied between the metal plates, such an arrangement constitutes a very fast electrooptic shutter: with the voltage off, no light is transmitted, and with it on the material becomes double refracting and light is partially restored. If the applied voltage is large enough the Kerr cell acts as a half-wave plate (Fig. 43) and thus the light is completely restored.

The magnitude of the effect is given by the Kerr constant, j, defined by the relation

$$\Delta = \frac{jLE^2}{d^2} \tag{7.5.2}$$

* See also Vol. 6, Section 10.5.2.

where E is the potential difference between the plates, in esu

$$(1 \text{ esu} = 300 \text{ volts})$$

Δ the optical path difference for the two components measured in wavelengths, L the length of the plates, and d the separation of the plates, both L and d being measured in cm. Typical values of the Kerr constant are 0.6×10^{-7} cm/esu^2 for benzene and 220×10^{-7} cm/esu^2 for nitrobenzene at wavelength 5.893×10^{-5} cm. In piezoelectric materials there is a similar but larger effect which is proportional to the first power of the electric field rather than the second as in the Kerr effect.

7.5.4. Voigt Effect, Magnetic Double Refraction

When a strong magnetic field is applied to a vapor through which light is passing perpendicular to the field, double refraction occurs. This phenomenon, the Voigt effect, is related to the inverse transverse Zeeman effect in precisely the same way that the Faraday effect is related to the longitudinal Zeeman effect. The magnitude of both the Voigt and Faraday effects is approximately proportional to the magnetic field strength.

7.5.5. Faraday Effect

When plane polarized light is sent through a material in a direction parallel to an applied magnetic field the plane of vibration is rotated through an angle, θ (in minutes), expressed by

$$\theta = VHL \tag{7.5.3}$$

where H is the magnetic field strength, L the path in cm, and V a constant (Verdet's constant) associated with each substance. Typical values of V are given in Table V. Verdet's constant is roughly inversely

TABLE V. Verdet's Constant, in Minutes of Arc per Gauss per Centimeter for $\lambda 5893$

Substance	t, °C	V
Water	20	0.0131
Glass (phosphate crown)	18	0.0161
Glass (light flint)	18	0.0317
Carbon disulfide, CS_2	20	0.0423
Phosphorus, P	33	0.1326
Quartz (perpendicular to optic axis)	20	0.0166
Iron		38 degrees*
Nickel		5.0 "

* The value varies with mode of preparation of iron films used. See H. König, *J. Optik* **3,** 101 (1948).

proportional to the square of the wavelength and may be either positive or negative. The rotation is positive if it is in the direction of the current producing the field; thus, since the rotation depends only on the field direction it may be multiplied by reflecting the light beam back and forth through the material. With large fields the effect can be used as a very fast light shutter. In ferromagnetic materials the constant is enormously large (Table V), positive, and greater at the red end of the spectrum. For most materials the Faraday effect is a necessary consequence of the existence of the Zeeman effect. In ferromagnetic substances, however, the connection is remote.

7.5.6. Cotton–Mouton Effect

This effect refers to the double refraction of light in a liquid in a transverse magnetic field. In some liquids, such as nitrobenzene, the effect is some thousand times greater than the Voigt effect and differs in being proportional to the square of the field strength. In addition, it decreases rapidly with a rise in temperature, whereas the Voigt effect is temperature independent. The Cotton–Mouton effect is the magnetic analogue to the electrooptic Kerr effect even though it is much smaller, and is not related to the Zeeman effect.

7.6. Photometry

7.6.1. Definitions

The lumen is the *luminous flux* emitted per unit solid angle in a direction θ by a uniform point source whose intensity is one candle in the direction θ. Thus the amount of light radiating from a point source, P, and contained within a small solid angle $d\omega$ is the *luminous flux*, dF, through an element of area dA a distance r from the source (Fig. 47),

$$dF \sim d\omega = \text{const}\,\frac{dA\,\cos\beta}{r^2}. \tag{7.6.1}$$

For a small element of source, dS, the flux will be due to all cones of light originating at the source and will, if the source obeys Lambert's law, be also proportional to the area of dS projected in the direction of observation θ. Thus

$$dF = B\,dS\,\cos\theta\cdot\frac{dA\,\cos\beta}{r^2} \tag{7.6.2}$$

where B, a constant characterizing the source, is called the luminance or brightness (see below).

The unit of luminous flux, the lumen, is formulated in such a manner that the relative visibility of the radiation is taken into account, i.e., the definition is designed to include the fact that the human eye does not give the same response to the same amount of energy at different wavelengths. Therefore, the lumen may be defined as the luminous flux in a pencil of light at a wavelength of 5550 Å and of amount equal to 0.00146 watt. At other wavelengths the Illuminating Engineering Society has adopted ratios of lumen/watt which take into account the spectral sensitivity of the human eye.

The *luminous intensity* of a point source P in a given direction θ is the amount of luminous flux radiated per unit solid angle in that direction, and its unit is the lumen per steradian, $I = dF/d\omega$, or the candle. Thus, for a point source radiating uniformly in all directions the total luminous

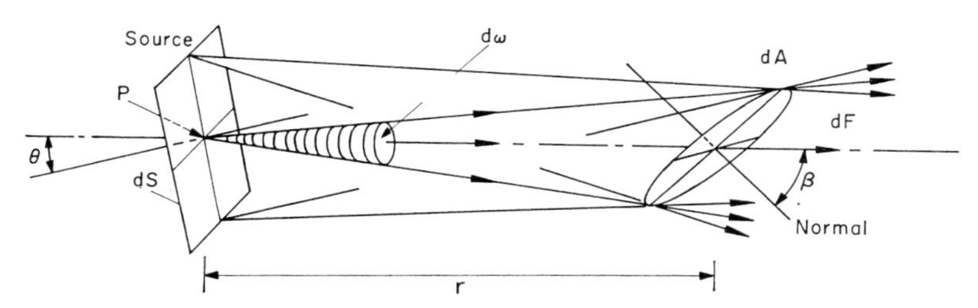

FIG. 47. Relations between luminous flux, dF, brightness, $(dF/d\omega)\ 1/dS \cos\theta$, and illuminance, dF/dA.

flux equals 4π times the luminous intensity, and consequently the mean spherical candle power for any source equals the total luminous flux radiated divided by 4π. Since measurements of intensity are simpler than those of flux it is more practical to maintain a laboratory standard of intensity rather than of flux. Therefore, the present standard, the new international candle, is defined as one-sixtieth of the luminous intensity of one square centimeter of a blackbody (Section 7.3.4) at the freezing temperature of platinum.

If the source has a finite area, the term *luminance* (or brightness, B in Eq. (7.6.2)) specifies the luminous intensity in a given direction from one square centimeter of projected area of the source surface. Thus, if dS (Fig. 47) is an element of area of the source, then the luminance B in a direction of θ with the normal is

$$B = \frac{dF}{d\omega}\frac{1}{dS\cos\theta} = \frac{I_\theta}{dS\cos\theta} \tag{7.6.3}$$

where I_θ is the luminous intensity of the elemental source in the direction θ. The units of B are candles per square centimeter. In the case of diffuse surfaces $I_\theta = I_n \cos\theta$, where I_n is the luminous intensity along the normal. Thus, $B = I_n/dS$, and the source appears equally bright from all directions (Lambert's radiation law).

The *illuminance* E refers to the amount of light flux that is incident upon a unit area of a surface (Fig. 47) and is defined by

$$E = dF/dA = B\,dS\cos\theta \cdot \frac{\cos\beta}{r^2} \tag{7.6.4}$$

measured in lumens per square centimeter.

7.6.2. Methods of Photometry

The photometer is used to measure the luminous intensity of light sources and may be based on the inverse-square law of the illuminance, Eq. (7.6.4), obtained with a point source. Thus, by measuring the source

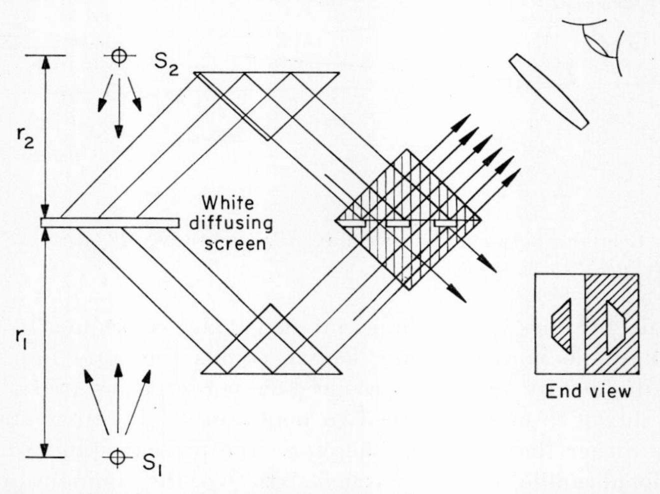

FIG. 48. Lummer–Brodhun photometer; eyepiece is focused on hypotenuse of cube.

distances at which the illuminances of two sources, an unknown and a standard, appear equal the ratio of their intensities can be calculated.

The Lummer–Brodhun photometer (Fig. 48) permits high precision in comparing illuminances. It contains a screen whose surfaces, illuminated by the two sources S_1 and S_2, are made of a white diffusing substance. Two totally reflecting prisms reflect the light onto the opposite faces of a photometric cube, which consists of two right-angled prisms in optical contact. The hypotenuse of the left-hand prism is ground away in part so that it forms the pattern shown in the end view of Fig. 48. Light from S_1 is transmitted through the cube by those regions of the hypotenuses that are in contact, while light from S_2 is totally reflected by the others. When the two sides of the screen have equal illuminance the pattern disappears. A wide variety of modifications of this basic idea have been used, some with lenses to parallelize the light falling on the screen and with variable collimator slits in front of the sources. Others employ rotating sector disks in the beams to change the effective aperture; the alternating flicker caused by this rotation also serves to minimize color differences between S_1 and S_2 if the correct frequency is chosen. In still other approaches a monochromator is used to map relative intensities for various colors.

If a photographic plate is used as detector, the relative intensity of spectral lines may be determined from the plate blackening as measured by a micro-densitometer or microphotometer. In this instrument the plate is mounted on a carriage and moves at a uniform rate past an extremely fine pencil of light which passes through it and falls on a photocell. The output of this cell may be connected to an electromechanical recorder thus forming a graphical record of the light transmitted at every point. Since the blackening of an emulsion is not linear with light intensity, a calibration curve must be prepared for each plate by measuring the densities produced by known light intensities. For the calibration, one must be careful to use light of the same wavelength and the same intensity range as the radiation under examination.

Relative light intensities are most frequently measured with photoemissive cells,[33]* either the simple phototube consisting of a light-sensitive cathode and an anode for collecting photoelectrons or the more complex photomultiplier tube containing 9 to 16 accelerating stages which serve to amplify the photoelectric current from the cathode by factors from 10^5 to 10^7 through the mechanism of secondary electron emission. Such devices are commercially available in a variety of spectral sensitivities. Some such tubes can be used for ultraviolet work if suitable fluorescent screens are placed immediately in front of them. A 1-mm thick layer of clear potassium- or ammonium-uranyl sulphate crystals, for example, will convert at constant quantum efficiency ultraviolet radiation shorter than 4600 Å into green fluorescent light (Section 7.8.1) of fixed wavelength distribution. In the vacuum ultraviolet sodium salicylate is used as a converter (Section 7.9.3).

In vacuum phototubes the primary emission current is of the order of

* See also Vol. 2, Part 11.

[33] V. K. Zworykin and E. G. Rambert, "Photoelectricity and Its Application," p. 269. Wiley, New York, 1949.

50 μa/lu or about 10^{-4} electrons/photon. For most cathodes this current is very closely proportional to the illumination and the "dark current" due to thermionic emission is essentially zero, except for the case of highly sensitive complex layer cathodes ($\sim 10^{-2}$ electrons/photon) which should be chilled to the temperature of solid carbon dioxide. This decreases the thermionic emission from 10^{-10} amp at room temperature to about 10^{-15} amp. These tubes should also be operated at low levels of illumination to avoid "fatigue" effects. Because of small time lag in vacuum phototubes, less than 10^{-10} sec, they are suitable for measurement of light intensities alternating at correspondingly high frequencies.

For less accurate work, simple "gas-filled" phototubes may be appropriate. In their case, the primary photoelectrons are accelerated by an anode potential of about 100 volt and liberate secondary electrons by ionizing collisions with gas atoms, thus increasing the initial current tenfold or more. Their output, however, is sensitive to the applied voltage and not linear with the light intensity. In addition, there is a time lag due to a transit time for the electrons to cross the gas-filled gap which is much longer than in a vacuum tube, and thus such tubes cannot be employed to measure intensities alternating at frequencies above 10^3 cycles/sec.

For certain specialized applications, image amplifier tubes (astronomical problems) and photoconductive devices (infrared, see Section 7.9.2) are also used.

If there is need to measure light intensities (in units of watt per square centimeters of receiver area), it is necessary to let radiation fall on an absorbing surface (preferably "black") and to measure the heating effect. (Because of the requirement of complete, or known, absorption of such radiation receivers the previously discussed devices are inappropriate.) The minute temperature changes so produced are most readily determined by the variation with temperature of the electrical resistance of fine wires (or more recently of certain semiconducting elements or thermistors) or by the thermoelectric effect where a potential difference is generated when a junction of two dissimilar wires is heated.[34] Such devices, bolometers and thermocouples, which are also available commercially, exhibit sensitivities of the order of 10 μv output/1 μw/cm^2 of incident radiation with a detection limit of 10^{-8} to 10^{-9} watt. Their time of response is better than 0.01 sec. In order to eliminate environmental temperature changes or instability of the external measuring circuit it is customary to employ two detectors as nearly identical as possible, one as the compensator and the other as the active radiation receiver. Modulating (chopping) the incident beam at a given frequency and having the

[34] D. F. Hornig and B. J. O'Keefe, *Rev. Sci. Instr.* **18**, 479 (1947); see also V. Z. Williams, *ibid.* **19**, 135 (1948).

detecting circuit tuned to respond to signals of only this frequency further enhances the signal to noise ratio.

If still higher sensitivities are required, a pneumatic detector[35] (Golay cell) or even a superconducting bolometer[36] may be employed (Section 7.9.1).

7.7. Scattering of Light

7.7.1. General Scattering Phenomena

When an electromagnetic wave is incident on a medium it induces in the atoms or molecules electric dipole moments oscillating with the frequency of the incident wave. The wave reradiated by the induced dipoles

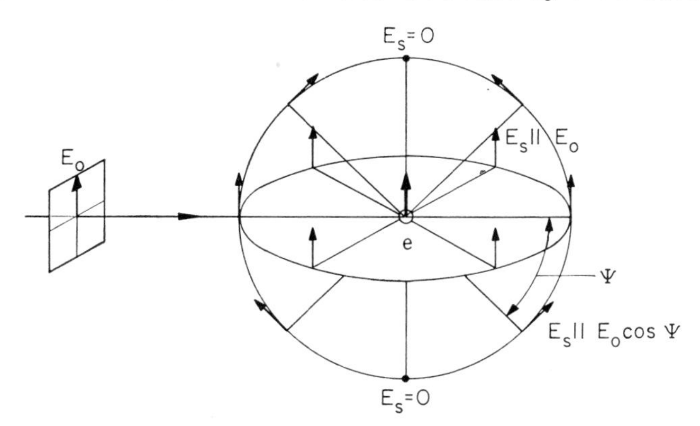

FIG. 49. Scattering of polarized light by a free electron, *e*.

is called scattered radiation and has usually the frequency of the incident light but, in contrast to it, is polarized (Fig. 49). If we call the scattering power, P_s, of the material the ratio of the total amount of light scattered per unit volume of material to the intensity of the incident wave, then the electromagnetic theory* gives

$$P_s = \frac{8\pi e^4}{3m^2c^4} \frac{N}{[(\nu_0^2/\nu^2) - 1]^2} \tag{7.7.1}$$

* The derivation of Eq. (7.7.1) neglects absorption, this is justified as long as ν is far from ν_0. Equation (7.7.1) is, in general, only approximate since it ascribes the same natural frequency to each electron in an atom.

[35] M. Golay, *Rev. Sci. Instr.* **18,** 357 (1947).

[36] D. H. Andrews, R. M. Milton, and W. DeSorbo, *J. Opt. Soc. Am.* **36,** 518 (1946).

where e and m are the electronic charge and mass, respectively, c is the velocity of light, N the number of dipoles per cm^3, ν_0 the natural frequency of the dipole, and ν the frequency of the incident wave. In the derivation, the N dipoles have been assumed to scatter independently (incoherent scattering) since the scattering centers are small compared with the wavelength and compared with their separation. Note that the scattering power has the dimension of inverse length and may be expressed as

$$P_s = \frac{32\pi^3}{3\lambda^4 N} (n - 1)^2 \tag{7.7.2}$$

where n is the refractive index.

For most atoms the natural frequencies lies in the ultraviolet, so that for visible light $(\nu_0/\nu)^2$ is very much greater than one, and Eq. (7.7.1) becomes

$$P_s = \frac{8\pi Ne^4}{3m^2\nu_0^4} \frac{1}{\lambda^4} = \frac{\text{constant}}{\lambda^4} \tag{7.7.3}$$

where λ is the wavelength of the incident radiation. This is Rayleigh's scattering formula. If $(\nu_0/\nu)^2$ is very much less than one, Eq. (7.7.1) becomes

$$P_s = \frac{8\pi Ne^4}{3m^2c^4} \tag{7.7.4}$$

which is Thomson's expression for the scattering of X-rays, and is independent of wavelength. In case ν and ν_0 are nearly equal the scattering becomes very large (resonance scattering) and Eq. (7.7.1) does not apply.

The scattering power is also called the attenuation coefficient for scattering, σ, and it enters as such into Eq. (7.1.4), with $\mu = \sigma + \tau$, so that

$$I = I_0 e^{-(\sigma+\tau)x}. \tag{7.7.5}$$

This, then shows that the total absorption coefficient μ takes account of the removal of light from a beam by two separate mechanisms which are additive: by scattering, σ, and by true absorption, τ, which converts a fraction of the incident radiation into heat. This differentiation between σ and τ is of particular significance for X-rays, where these two quantities change in a different way with wavelength of the radiation and with the atomic number of the absorbing material. In the visible, near, and far ultraviolet regions (see Section 7.9.3) the larger absorptions, $\mu > 1$, are caused by electronic transitions in the absorber and are therefore a measure of τ, since they involve degrading of the incident radiation. For very short wavelength X-rays σ becomes dominant.

Equation (7.7.1) is a good approximation only for gases because of the assumption of random positions of the scattering centers. In the case of solids the atoms are not arranged at random, and in the case of liquids the atoms are so closely packed that randomness cannot be complete. Thus, the scattering power of gases is generally greater than that of liquids, which in turn is greater than that of solids.

7.7.2. Raman Scattering

Raman scattering refers to the appearance in the scattered light of frequencies other than in the incident radiation. These frequencies are discrete rather than continuous and may be lower than the frequency of the incident light, Stokes lines, or higher, anti-Stokes lines. The quantum theory accounts for the Raman spectrum on the basis that an incident photon may give up part of its energy to a scattering molecule and thus be scattered as a photon of lower energy and lower frequency. On the other hand, a molecule may give up part of its rotational or vibrational energy to the photon it scatters, so that the photon has a higher energy and higher frequency. Consequently, there may be a rotational Raman spectrum consisting of a group of lines. By measuring the separations of the Raman lines from the much more intense scattered exciting line the positions of the rotational and vibrational molecular energy levels can be determined. Since the Raman lines are measured relative to the exciting line, the experimenter may work in the most convenient part of the spectrum instead of in the infrared region corresponding to these energy levels, for example, by using the green 4358 Å line of mercury to excite the Raman spectrum one gains the advantages of high transmission, high photographic activity, and good intensity for the Raman spectrum. The intensity of the Raman lines relative to the Rayleigh line may be as small as one in 10,000 for vapors, one in 1000 for liquids, and equal in clear solids because of the weakness of Rayleigh scattering here. Raman lines frequently show marked depolarization, whereas the Rayleigh line is highly polarized.

7.7.3. Atmospheric Optics

The blue color of the sky is attributable to Rayleigh scattering (Section 7.7.1) from molecules of the atmosphere. In fact, shorter wavelengths are so strongly favored in the scattered light that the total intensity of erythemal ultraviolet (between 2900 and 3100 Å) in the sky light is about equal to that of the direct sunlight. Strong polarization is also found in the sky light (Fig. 49).

If particles larger than λ are present they produce scattering which is independent of λ, Tyndall scattering, and the medium is turbid. The turbidity of photographic emulsions is an important limiting factor in the quality of photographic images. Similarly the Tyndall scattering in the atmosphere when there is a slight haze seriously detracts from the visibility either in daylight or by searchlight. Because of its longer wavelength the penetration of haze by infrared is much better than that by visible light and consequently infrared photographs of distant objects may be much superior to photographs by visible light. Particles present in a fog are much larger than near infrared wavelengths and hence the penetration of fog by infrared light is little better than by visible light.

In clear air the attenuation of visible radiation is very small. At 5000 Å nearly 90% of the light passes through the entire earth's atmosphere. Attenuation over distances of a few meters may be entirely neglected in the wavelength region 3000 to 10,000 Å. At less than 3000 Å attenuation becomes appreciable and increases rapidly toward shorter wavelengths. For example, the transmission of 1 meter of air is about 10% at 2050 Å and 99% at 2200 Å. It must be realized, however, that the solar spectrum at the earth's surface is cut off at 3000 Å because of the strong absorption of ozone formed in the upper atmosphere by a photochemical reaction induced by shorter wavelengths.

7.8. Fluorescence and Phosphorescence*

7.8.1. Definitions and Decay Time Measurements

Light emission excited by light absorption is known as photoluminescence and is found in certain solids, liquids, and gases. In the simplest case (resonance radiation) the emitted frequency is the same as the exciting frequency. If the intensity of the luminescence is proportional to the incident light intensity and if the decay time is independent of the temperature the effect is called fluorescence. In this case the time lag is usually short, ranging from about 10^{-3} sec in solids to about 10^{-8} sec in liquids. If the rate of decay of the re-emitted light increases with increasing temperature, the phenomenon is classed as phosphorescence and usually involves a longer decay time than in fluorescence. There is also a saturation effect in phosphorescence so that beyond a certain incident intensity the luminescence no longer increases. The temperature dependence and other characteristics of phosphorescence imply a more complicated mechanism involving molecular collisions prior to re-

* See also Vol. 6, Part 10.

emission. Both fluorescence and phosphorescence follow exponential laws of decay:

$$I = I_0 e^{-t/\tau} \tag{7.8.1}$$

where τ is the decay time.

Instruments designed for measuring the decay times of photoluminescence are called phosphoroscopes and fluorometers. In the Becquerel phosphoroscope the luminescent substance is placed between two disks which are mounted on a common axis and have sector shaped apertures shifted with respect to each other by an angle. The exciting light enters the phosphoroscope by the apertures in one of the disks and the luminescent light reaches the observer through the apertures in the other disk. The time interval between excitation and observation can be varied by altering the speed of rotation or the angle between the sectors. Average lifetimes as short as 10^{-4} sec can thus be determined.

In another apparatus a point image of the primary light source is thrown on a rotating disk uniformly coated with the luminescent material. If luminescence is persistent the point is drawn out into a bright streak, the length of which provides an approximate measure of decay times for a range from one to 10^{-5} sec. The fastest mechanical phosphoroscope seems to be one in which the fluorescence is excited by light from a spark and the luminous spot is drawn out into a long band by a high-speed rotating mirror; the spark is always initiated in the same position of the mirror by synchronization. The limit of resolution is claimed to be 10^{-6} sec. A similarly high resolving power is also obtained if an oscilloscope is used in conjunction with a photoelectric cell activated by the luminescence. The primary light may be a repetitive spark which triggers the sweep voltage of the oscilloscope.

For the measurement of short decay periods down to 10^{-9} sec fluorometers are used. In these instruments the rotating systems of the phosphoroscope are replaced by electrical or magnetic devices of practically no inertia. Common shutters are Kerr cells (Section 7.5.3) operated by high-frequency square-wave potentials and placed between crossed Nicols, or systems in which the incident light beam is diffracted periodically by supersonic waves. The time interval between excitation and observation is not produced by a phase difference between the first and the second shutter, but by varying the distance between the luminescent substance and the two shutters and thus the time in which the light traverses this distance.

7.8.2. Efficiencies of Photoluminescence

The quantum yield is defined as the ratio of the number of emitted to absorbed light quanta. In the special case of resonance fluorescence, i.e.

where the wavelength of the emitted light is the same as that of the incident light, the quantum yield is identical with the energy yield. In general, however, the energy yield is smaller.

Any photometer (Section 7.6.2) can be used for measurements of the intensity of fluorescence. For low intensities Geiger counters or photomultipliers can replace photocells. As in all photometric work it is of the greatest importance that any variations in the intensity of the primary light source are automatically compensated. It is, therefore, advisable to use as comparison light the radiation of a fluorescent substance which is excited by the same primary source.

The human eye, all photoelectric cells, and photographic plates have a more or less selective response to light of different wavelengths. The fact that the fluorescent yield of many compounds is independent of the wavelength of the exciting light over a wide range makes them useful for heterochromatic photometry. Thus, "integrating screens" of uranyl sulfate or cells filled with a luminescent solution covering the entrance window of a photoelectric cell stimulate a current in the cell which is proportional to the unknown intensity over a wide spectral range.

7.9. Nonvisible Light

7.9.1. Infrared

The radiation of longer wavelength than 7500 Å, ranging to about 1 mm, is called the infrared. It is a region of considerable importance in the study of molecular structure since absorption corresponding to rotational and vibrational excitation of molecules is likely to occur here.

For practical reasons, the infrared region may be loosely subdivided into the photographic infrared from 0.75 μ to 1.3 μ, the near infrared from 1.3 μ to 25 μ, and the far infrared above 25 μ. The near infrared is characterized by the ready availability of transparent materials and its long wavelength limit occurs roughly at a point where the absorption of suitable optical materials becomes prohibitively large.

Prism spectrometers are widely used in the near infrared, and regions of transparency of common prism materials are given in Table III (Section 7.3.1). Rocksalt is the most widely used as it is transparent to about 15 μ, but at less than 5 μ its dispersion is poor and other prism materials should be used for best performance. Calcium and lithium fluoride are excellent at less than 5 μ.

Crystals possess very high reflectances at characteristic wavelengths in the infrared. Thus, if radiation from a continuous source is reflected several times from a crystal, the reflected beam contains only the characteristic wavelengths or the "residual rays."

A conspicuous feature of the infrared is the low intensity of all available light sources. The most convenient source of an infrared continuum out to 2 μ is a tungsten arc in an evacuated bulb (pointolite lamp). At 2 μ the glass envelope begins to absorb and it is necessary to change to a rod of Carborundum (Globar), or to a filament coated with rare earth oxides which have a high infrared emissivity (Nernst filament). If the radiation from a Nernst filament is to be kept constant, a current of about 1 amp must be held to within narrow limits.

Because of the low intensity of infrared sources the design of the optical systems is dominated by the need for conserving energy and, therefore, the dispersing system must be of large aperture. Prisms are preferred because all the light appears in a single order spectrum. If gratings are employed they need to be blazed to concentrate the radiation in a desirable order (Section 7.3.2). Mirrors are used in preference to lenses for the reason that it is difficult, because of chromatic aberration, to locate the infrared image formed by a lens, whereas with a mirror the infrared image is coincident with the visible one.

7.9.1.1. Infrared Spectrographs. The two most widely used spectrometer arrangements for the infrared are the Wadsworth and the Littrow mountings (Sections 7.3.1 and 7.3.2). In the Wadsworth mounting the prism and the plane mirror (Fig. 50a) are rotated together about a vertical axis through the center of the prism base to pass various λ across a bolometer behind the exit slit. If the mirror is an extension of the prism base the angular deviation of a monochromatic beam traversing the prism at minimum deviation is exactly nullified by reflection from the mirror, although a lateral displacement results which is practically independent of wavelength. When arrangements have been made for radiation of one wavelength to traverse the prism at minimum deviation and focus at the exit slit, the whole spectral range of the prism may be scanned at minimum deviation merely by rotation of the prism–mirror combination.

In the Littrow mounting a plane mirror is placed nearly normal to the dispersed beam emerging from the prism (Fig. 50b). For one particular wavelength, which varies with the orientation of the plane mirror, the emergent beam is reflected exactly back upon itself and retraces its path through the prism. The spectrum may be scanned simply by rotation of the Littrow mirror about a vertical axis. If the prism remains fixed, however, only one wavelength passes through the prism at exactly minimum deviation. The chief advantages of the Littrow mounting are the doubled dispersion from a given prism train, and the compactness and economy resulting from the use of the same concave mirror as a collimator and a focusing mirror.

A 60° rock-salt prism, 6.25 cm high with a 7.5 cm base, in a Littrow mounting produces a resolving power of about 200 at 5 μ and about 400 at 14 μ, corresponding to a limit of resolution of 10 cm^{-1} and 2 cm^{-1}, respectively. At short wavelengths the quality of the optics governs the resolving power, while at longer wavelengths the small amount of energy

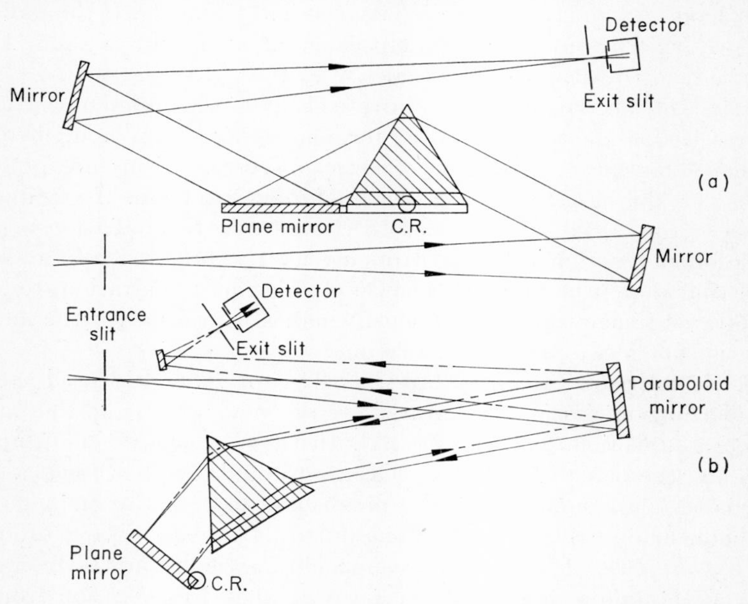

FIG. 50. Infrared prism spectrometer; (a) Wadsworth, and (b) Littrow mounting; C.R. indicates center of rotation.

available from the source forces the use of wide slits and a consequent reduction in resolving power.

Echelette gratings are used in the near infrared when high resolution is indispensable, and throughout the far infrared, because of the lack of prism materials. The optical arrangement can be similar to the Littrow mounting but the prism and Littrow mirror combination is replaced by a plane grating. Overlapping of spectral orders is especially troublesome in the infrared because of the high source intensity of the visible radiation, which in high orders overlaps the low intensity, first order infrared. The radiation must therefore be monochromatized partially in some way before it enters the spectrometer.

Monochromatization in the near infrared can be accomplished by a low dispersion prism in front of the grating monochromator. Beyond 25 μ, filters opaque to short waves must be used. Thin plates of fused quartz are suitable for the elimination of radiation between 10 and 50 μ, and hard paraffin, metallic blacks, and soot-blackened paper have also been employed. Multiple reflections from crystal surfaces, so that all wavelengths except the residual rays are absorbed, serve to isolate some spectral regions. Light choppers that are transparent to the visible have been used in combination with ac amplification of the detector signal.[34] Christiansen filters[37] are especially flexible with respect to wavelength.

A selective reflector of a different type uses the fact that a plane grating gives mirror reflection of wavelengths longer by a factor of roughly 1.5 than the grating spacing and disperses shorter wavelengths at angles different from the angle of reflection (Section 7.3.2).

Another method for selecting a narrow range of infrared radiation makes use of the rapid change of the refractive index as an absorption band is traversed. Thus, a quartz lens has such greatly differing focal lengths for the near and the far infrared that the long wavelengths can be made convergent while the short wavelengths are not. By placing a screen with a hole at the focal point of the long wavelengths they are separated from the short ones (focal isolation).

7.9.1.2. Infrared Detectors. Nonselective detectors of infrared radiation generally depend upon the heating effect, so they respond equally to a given amount of radiant energy regardless of wavelength. They are, therefore, particularly sensitive to ambient radiation and to change in temperature of the surroundings. Such detectors are exemplified by a well-blackened vane radiometer, Golay cell, thermopile, or bolometer. At wavelengths longer than 5 μ they are almost the only ones available.

The Golay cell[35] is a rapid and sensitive nonselective receiver. A potassium bromide window admits chopped radiation to a 3-mm diameter cell filled with a nonabsorbing gas, where it falls on a thin antimony film. The heating of the film warms the gas and causes a pressure change which produces a distortion of a curved flexible diaphragm with an outer mirror surface. The movements of this latter diaphragm transmit to a photoelectric cell corresponding changes in a light beam falling upon the mirror surface of the diaphragm. This cell, with a receiving area of about 0.03 cm^2 and a time constant between 0.001 and 0.003 sec has a sensitivity about that of the bolometer and may have a noise equivalent power of less than 6×10^{-11} watt.

Among the selective detectors are the photographic plate and the photoconductive, photoemissive, and photovoltaic cells.* The sensitivity of

* See also Vol. 6, Part 11.

[37] R. B. Barnes and L. G. Bonner, *Phys. Rev.* **49**, 732 (1936).

these instruments may vary enormously with wavelength and therefore must be calibrated at each wavelength by comparison with a nonselective detector. The photoemissive tube is unsatisfactory for most of the infrared since the sensitivity usually does not extend past 1.2 μ, and its high response to shorter wavelengths makes careful shielding against scattered light necessary.

Of the selective detectors in the near infrared the lead-sulfide photoconductive cell is the most important. Its sensitivity is usually expressed in terms of the noise equivalent power (nep), which is defined as the total power from a blackbody at 500°K required to give a signal equal to the noise of the cell. A typical nep value of 10^{-9} watt is obtained for a 0.5 by 0.5 mm PbS detector at room temperature, viewing a 500°K blackbody source chopped at 90 cps with a system bandwidth of 4 cycles. This sensitivity can be increased by an order of magnitude, at least, by cooling the cell to −40°C; thus, it is capable of several orders of magnitude greater sensitivity than the best thermocouples.

Over the wavelength region of 1.5–7 μ PbS, PbTe, and PbSe are by far the most sensitive and rapid infrared detectors now available.[38] Under usual operating conditions PbS covers the region 1–3 μ, PbTe covers 2–5.5 μ, and PbSe 3–7 μ. All three materials show an increase in the long wavelength limit on cooling, the shift being 4×10^{-4} ev/°C for all three. The total sensitivity varies considerably with temperature; for PbTe at room temperature it is immeasurably small. The rapid response time of all three cells, a few microseconds, makes them advantageous in the study of thermal transients. Germanium doped with small quantities of other elements shows promise as a photoconductive detector out to several microns if used at a low temperature.[39]

7.9.2. Near Ultraviolet (4000 Å–2000 Å)

Many substances which transmit visible light absorb in the ultraviolet; dense flint glass begins to transmit above 3950 Å, and a light crown glass at 3500 Å. Air absorbs wavelengths below 2000 Å, as does the gelatin of the photographic emulsion, so that below 2000 Å vacuum spectrographs with special detection methods (Section 7.9.3) must be used.

The ultraviolet transmission of glass is largely determined by the content of iron oxide, which absorbs strongly. Ordinary window glass in a thickness of 1 mm is practically opaque to wavelengths shorter than 3000 Å, whereas a 1-mm thickness of No. 774 chemical Pyrex transmits about 35% at 3000 Å. Corex D No. 9700 transmits strongly at 3000 Å

[38] T. S. Moss, *Proc. I. R. E.* **43,** 1869 (1955).

[39] E. Burstein, J. W. Davisson, E. E. Bell, W. J. Turner, and H. G. Lipson, *Phys. Rev.* **93,** 65 (1954).

but poorly at 2500 Å and is therefore used in the bulbs of sunlamps. Pyrex No. 9741 in a 1-mm thickness transmits as much as 70% at 2537 Å and hence is used in bactericidal lamps. Vycor glasses approach fused silica in their properties. In 1-mm thickness Vycor No. 791 has a transmission of 85% at 2537 Å and practically zero at 1849 Å. Both fused and crystalline quartz have a high transmission down to about 2000 Å. Below this the crystalline quartz has a considerably higher transmission than the fused material. Calcite, fluorite, and rock salt are used in addition to quartz as lenses and prisms for the near ultraviolet.

Metals which are good visible reflectors are frequently very poor ultraviolet reflectors. Front surface mirrors must be used in the ultraviolet because of the absorption in glass, and the best mirrors are formed by evaporating suitable metals onto very clean glass surfaces. Silver is unsatisfactory because of a pronounced minimum in its reflection curve at 3150 Å, and aluminum is most frequently used since it is easily shaped, light in weight, easy to evaporate, has a high reflectance, and is very resistant to corrosion. Between 1200 Å and 3600 Å its reflection is greater than that of any other metal, while above 3600 Å it is exceeded only by silver.

For absorption spectroscopy in the ultraviolet the low intensity of light sources creates a problem. A glow discharge at 1000 volts through hydrogen at 5 to 10 mm pressure in a 5-mm capillary provides a strong continuum (Section 7.3.4) extending to 900 Å (Section 7.9.3). The usual high-voltage glow discharge with cold electrodes exhibits drifts in brightness of the order of 5% and is therefore less suitable for photoelectric recording of absorption spectra. Low voltage hydrogen arcs in which the electron current is provided by an electronically controlled hot filament provide an ultraviolet continuum with a constancy within a few tenths of one per cent and are now commercially available.

Optical instruments for the near ultraviolet are generally constructed of fused or crystal quartz of the designs discussed in Section 7.3.1. For best definition fused quartz generally lacks sufficient homogeneity, and large components such as spectrograph prisms and lenses are made of crystal quartz. Care must then be taken that the light traverses the crystal along the optic axis and, in order to minimize the effect of optical activity (Section 7.4.2), that Cornu prisms and lenses are used. Only one wavelength traverses the optic axis but in practice this causes no difficulty. If any components are cemented together, e.g., a quartz-fluoride achromat, the cement must be transparent to the ultraviolet and also must not fluoresce. When it is necessary to polarize near ultraviolet light a modification of the Nicol prism is used in which the calcite halves are held together so that there is a film of air between them instead of balsam

(Foucault prism, Section 7.4.2). It has an angular aperture of only about 8°, and some difficulty is experienced with interference in the air film.

7.9.3. Vacuum Ultraviolet (2000 Å–200 Å)

The region from about 2000 Å towards shorter wavelengths has been termed the vacuum ultraviolet (vuv) because here the strong absorption of air makes it necessary that all optical components be placed in a vacuum. In Fig. 51 the absorption coefficients μ of oxygen[40] and nitrogen[41] are plotted versus the wavelength λ, where μ is in units of cm^{-1} and defined by Eq. (7.1.4). Since the absorption in strong resonance bands is much larger than in dissociation and ionization continua, it is apparent that

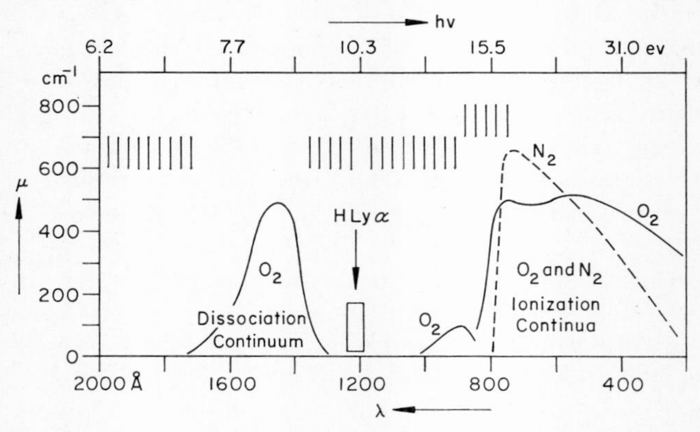

FIG. 51. Absorption of vacuum ultraviolet radiation by air, short vertical lines indicate resonance bands and open rectangle shows transparent region of about 3 Å width at 1216 Å.

the entire vuv region is essentially opaque to radiation, with the exception of a few narrow windows at or near H Lyman α (1216 Å). Other molecular gases show a trend similar to that depicted for O_2.

The choice of dispersing systems for the vuv is dominated by the optical properties of materials in this wavelength range, primarily transparency and reflectivity.[42] For refracting purposes, as in prisms and lenses,

[40] For absorption in the Schumann–Runge bands at and below 1900Å and in the adjacent strong dissociation continuum, see R. W. Ditchburn and D. W. O. Heddle, *Proc. Roy. Soc.* **A220,** 61 (1953); **A226,** 509 (1954); below 1300Å see Po Lee, *J. Opt. Soc. Am.* **45,** 703 (1955).

[41] G. L. Weissler, Po Lee, and E. I. Mohr, *J. Opt. Soc. Am.* **42,** 84 (1952); for other gases see G. L. Weissler, *in* "Handbuch der Physik—Encyclopedia of Physics" (S. Flügge, ed.), Vol. 21, p. 304. Springer, Berlin, 1956; R. W. Ditchburn, *Proc. Roy. Soc.* **A236,** 216 (1956).

[42] G. B. Sabine, *Phys. Rev.* **55,** 1064 (1939); M. Banning, *J. Opt. Soc. Am.* **32,** 98 (1942); S. Robin and B. Vodar, *J. phys. radium* **13,** 492 (1952).

only two are useful, calcium fluoride (fluorite) and lithium fluoride, with crystalline quartz a very limited third choice since its transmission limit ($\mu = 1$ cm^{-1}) is about 1800 Å. In clean and chemically pure specimens of fluorite this limit is about 1250 Å and for lithium fluoride about 1050 Å. At wavelengths shorter than 1000 Å all known solid or plastic materials exhibit absorption coefficients between 10^5 and 10^6 cm^{-1}, which excludes them except for very thin films which require a thickness of a few hundred angstrom units for 50% transmission.[42a]

7.9.3.1. Near Vacuum Ultraviolet. Because of these limitations in transparency it is convenient to divide the vuv into two regions, the near vacuum ultraviolet (nvuv) between 1900 Å and 1200 Å where calcium and lithium fluoride prisms and lenses may be employed, and the far vacuum ultraviolet (fvuv) between 1200 Å and 200 Å where the primary limitation is due to low reflectivities, R, from optical surfaces. Some characteristic values of R for evaporated aluminum surfaces,[43] commonly used for mirrors and gratings, are 74% at 1600 Å, 40% at 1200 Å, 17% at 900 Å, and about 2% at 584 Å, all values for $\phi = 20°$ and for freshly distilled surfaces. At grazing incidence, angles larger than 80°, the reflectivities are all above 80%. Most materials follow this trend, with glass, quartz, chromium, and platinum perhaps somewhat better than aluminum below 1000 Å.

The design of dispersing systems for the nvuv, using calcium or lithium fluoride prisms and lenses, follows along principles outlined previously (Section 7.3.1) and, in most instances, is characterized by the arrangement in Fig. 31. It must be kept in mind, however, that the dispersion,[44] $dn/d\lambda$, of CaF and LiF is rather large near their absorption limits, which makes it undesirable to construct achromats from these two materials. On the other hand, this high dispersion and the consequent large angle of tilt of the photographic plate with the optic axis of the camera lens result in a relatively large linear dispersion $dl/d\lambda$. For example, in the compact instrument of Cario and Schmidt-Ott,[45] 30 cm long and 7 cm in diameter, the place factor, $d\lambda/dl$, is 6 Å/mm at 1400 Å.

In the fvuv region focusing concave reflection grating spectrographs[42] afford the only practical means of dispersion, and in most instances the

[42a] Measurements on Al and other metals indicate their usefulness as windows since transmissions of the order of 30% below 800 Å were observed for unsupported films about 1000 Å thick. W. C. Walker, J. A. R. Samson, and O. P. Rustgi, *J. Opt. Soc. Am.* **48,** 71 (1958).

[43] G. Hass, W. R. Hunter, and R. Tousey, *J. Opt. Soc. Am.* **46,** 1009 (1956).

[44] See Handke, *in* "Physikalisch-chemische Tabellen" (H. Landolt and R. Börnstein, eds.), 5th ed., Vol. 2, p. 911. Springer, Berlin, 1923; E. G. Schneider, *Phys. Rev.* **49,** 341 (1936); H. Bomke, "Vacuumspektroskopie." Barth, Leipzig, 1937.

[45] G. Cario and H. D. Schmidt-Ott, *Z. Physik* **69,** 719 (1931).

optical arrangements are those shown in Fig. 34a with the primary slit, grating, and detector (e.g., photographic plate) on the Rowland circle. In the near normal incidence mounting the dispersion is closely linear, and also a wide spectral range is covered[46] in the first order, from $\lambda = 0$ to $\lambda = 2000$ Å. Thus, with a grating of 6,000 lines/cm and a radius of curvature of 1 meter (= diameter of Rowland circle), the plate factor is about 17 Å/mm. For an angle of incidence of 13.5° the grating astigmatism[47] results in a line image on the plate, which at 1000 Å is approximately 0.1 times the length of the grating groove for a point source. Because of the wide wavelength range of this mounting it is possible to focus it in air.

Normal incidence mountings also can be adapted for monochromator[48] use with fixed entrance and exit slits. The wavelength passing through the exit slit may either be changed by rotating the grating about the Rowland circle center or by rotating the grating about its own vertical axis.

Grazing incidence spectrographs are of significance because the reflectivity at such angles becomes much larger and thus shorter wavelengths may be recorded in spite of very strong astigmatism, of the order of 10 to 20 cm.

Edlén[49] and associates have described such instruments, and wavelengths down to about 10 Å have been recorded. Since the grating presents to the slit a projected width of $w \cos \phi$, the effective line spacing is decreased by the same factor and, therefore, the dispersion is correspondingly larger, though not uniform as in the case of normal incidence. Thus, a grating of 5,700 lines/cm and a 2-meter radius of curvature has a plate factor, $d\lambda/dl$, of 3.5 Å/mm at 1300 Å and of 2.5 Å/mm at 500 Å for $\phi = 83°$. Focusing in this case is, in principle, the same as that described previously, only it is much more critical. Thus it is advisable to use short-length photographic plates inclined at an angle of say 30° with the plane of the Rowland circle instead of their usual 90° position. The central image or spectral lines will then appear in the shape of hourglass figures with their narrowest portions indicating the true focus. A wavelength calibration of the plate may be obtained by measuring distances between

[46] R. A. Sawyer, *J. Opt. Soc. Am.* **15,** 303 (1927); K. T. Compton and J. C. Boyce, *Rev. Sci. Instr.* **5,** 218 (1934).

[47] H. G. Beutler, *J. Opt. Soc. Am.* **35,** 324 (1945).

[48] R. Tousey, F. S. Johnson, J. Richardson, and N. Toran, *J. Opt. Soc. Am.* **41,** 696 (1951); see also Y. Fujioka and R. Ito, *Sci. of Light* (*Tokyo*) **1,** 1 (1951).

[49] B. Edlén and A. Ericson, *Z. Physik* **59,** 656 (1930); B. Edlén, *Nova Acta Regial Soc. Sci. Upsaliensis* [4] **9,** No. 6 (1934); F. Tyren, *Z. Physik* **111,** 314 (1938); *Nova Acta Regial Soc. Sci. Upsaliensis* **12,** No. 1 (1940).

lines and identifying them on a basis of trial and error with standard wavelengths.[50]

A number of light sources for the vuv consist of electrical discharges through an open-ended tube with its axis precisely in line with the slit and the grating and containing gases at low pressures. A glow discharge through H_2 at a potential of about 1000 volts dc and a current of the order of 1 amp yields a very intense spectrum, which shows a peak at H Lyman α (1216 Å) and decreases rapidly to nearly zero intensity at 900 Å (Section 7.9.2). Because of ease of operation and stability it represents the obvious choice for the nvuv.

7.9.3.2. Far Vacuum Ultraviolet. In order to obtain an intense line spectrum in the fvuv, it is necessary to use repetitive sparks[47] (Section 7.3.4) through a capillary of about 5 cm in length and 3 mm diameter, with quartz or preferably a high melting point ceramic (aluminum oxide, etc.) as material for a capillary.[51] For wavelength continuous radiation, the Lyman source[52] provides the greatest intensity down to about 200 Å. It is operated by discharging a large capacitor, of about $\frac{1}{2}$ to 1 μf at 10,000 to 30,000 volts, through a quartz capillary of the above dimensions at the rate of one spark per second. Other, weaker wavelength continua may be obtained by making use of the molecular recombination spectra of the rare gases,[53] which for the case of helium lies between 600 Å and 1000 Å.

The vacuum spark, which was originally developed by Millikan, Sawyer, Edlén, and others, is a prolific source of emission lines, approximating a continuum, in the region from 2000 Å to 100 Å. In this arrangement, a high-voltage spark, energized by a power supply similar to that of the Lyman source, bridges the vacuum gap between two electrodes arranged vertically and therefore parallel to the spectrograph slit. Edlén and particularly Vodar and his group[54] have developed this idea further by permitting the spark to glide on such materials as carbon, alumina, and other refractories (*Halbleiterfunken*) and have obtained intense and closely spaced emission lines, down to 160 Å, which were

[50] J. C. Boyce and H. A. Robinson, *J. Opt. Soc. Am.* **26,** 133 (1936); Charlotte Moore, *Natl. Bur. Standards (U.S.), Circ.* **488,** Section 2 (1952).

[51] Po Lee and G. L. Weissler, *J. Opt. Soc. Am.* **42,** 80 (1952); R. W. Ditchburn, *Proc. Roy. Soc.* **A229,** 44 (1955); N. Wainfan, W. C. Walker, and G. L. Weissler, *J. Appl. Phys.* **24,** 1318 (1955).

[52] R. E. Worley, *Rev. Sci. Instr.* **13,** 67 (1942).

[53] Y. Tanaka, *Sci. Papers Inst. Phys. Chem. Research (Tokyo)* **39,** 465 (1942); *J. Opt. Soc. Am.* **45,** 710 (1955).

[54] J. Romand, G. Balloffet, and B. Vodar, *Compt. rend.* **240,** 412 (1955); B. Vodar and N. Astoin, *Nature* **166,** 1029 (1950); J. Romand and G. Balloffet, *J. phys. radium* **16,** 489 (1955).

primarily characteristic of the electrode materials, uranium or steel. The time constancy of this source is sufficiently good to permit photometric work. An additional advantage is gained because of the fact that it operates in a high vacuum in contrast to discharges requiring a carrier gas.

Before 1939, detection of vuv radiation relied exclusively on photographic plates, sensitized to this wavelength region by fluorescent oil and lacquer layers.[55] Because of its integrating characteristics this technique is still of use today; it must be kept in mind, however, that for photometric work each plate must be calibrated in terms of density versus intensity (Section 7.6.2). More rapid scanning of the spectrum, though not with the same resolution, may be achieved by mounting behind an exit slit in the focusing position, either: (a) a commercial photomultiplier tube coated with a layer of fluorescer, sodium salicylate;[56] or (b) an open phototube with a Pt-cathode[57] having a yield of 0.02 to 0.1 electron per photon below 1000 Å; or (c) an open and compensated thermocouple (Section 7.6.2) for absolute photon flux measurements.[58] Photon counters,[59] because they necessitate a window, are primarily of significance in the nvuv.

In general, it is desirable to have certain external controls to facilitate the operation of the spectrograph while under a vacuum: (a) motion of the plateholder for successive exposures; (b) exposure shutter; (c) vacuum valve to permit changing of the plate; (d) vacuum valve between the grating and the slit (or both slits in the case of a monochromator) in order to facilitate cleaning the slit, changing its width, and replacing the light source. In addition, differential pumping chambers should be used wherever pressure gradients occur, e.g., between the source and the main tank, in order to assure a good vacuum in the body of the instrument. The degree of vacuum required depends on the length of the actual light path and on the type of measurement to be undertaken. If the purpose is solely for the determination of wavelength, higher pressures may be tolerated than for intensity work. In the latter case pressures of about 10^{-4} mm Hg are adequate for most instruments.

[55] G. R. Harrison and P. A. Leighton, *J. Opt. Soc. Am.* **20,** 313 (1930); Po Lee and G. L. Weissler, *J. Opt. Soc. Am.* **43,** 512 (1953).

[56] F. S. Johnson, K. Watanabe, and R. Tousey, *J. Opt. Soc. Am.* **41,** 702 (1951); for open Cu-Be photomultipliers see also V. Schwetzoff, *Rev. gén. elec.* **63,** 71 (1954).

[57] W. C. Walker, N. Wainfan, and G. L. Weissler, *J. Appl. Phys.* **26,** 1366 (1955).

[58] D. M. Packer and C. Lock, *J. Opt. Soc. Am.* **41,** 699 (1951).

[59] T. A. Chubb and H. Friedman, *Rev. Sci. Instr.* **26,** 493 (1955).

7.10. X-Rays*

7.10.1. Physics of X-Rays

7.10.1.1. General Properties of X-Rays.† X-rays are transverse electromagnetic waves like visible light except that their wavelengths are much shorter. The long wavelength portion of the X-ray spectrum overlaps the extreme ultraviolet, whereas the short wavelengths are of the same order of magnitude as the γ-rays emitted by radioactive bodies. The range of wavelengths used for radiography extends from about 0.05 to 1.0 Å and that used for crystallography from about 0.5 to 2.3 Å. (One angstrom unit $= 10^{-8}$ cm.)

One may consider any kind of radiant energy, and therefore X-rays, from two complementary viewpoints. Sometimes it is convenient to think of X-rays in terms of their wave properties and sometimes in terms of a bundle of particles, called photons, moving with the velocity of light and having an energy $h\gamma = h(c/\lambda)$.

From the small value of wavelengths and the large value of the photon energy associated with X-rays stem significant differences between X-ray and light optics. Since for X-rays the index of refraction of all media is close to but less than 1, the rays cannot be bent significantly nor can they be reflected unless the incident and reflected rays graze the surface at an angle of only a fraction of a degree. Although X-rays, like light, can be polarized, the plane of polarization cannot be modified by the transversed medium. Consequently, optical polarimetry has no analog in the field of X-ray studies, and X-ray analysis is solely restricted to the physical phenomena of scattering and diffraction.

7.10.1.2. Production of X-Rays. X-rays are generated whenever high-speed electrons are suddenly stopped by a target material, which is usually a metal. Production of X-rays depends, therefore, on the generation and acceleration of electrons to a sufficiently high speed in the direction of the target by an electric potential difference applied to suitable electrodes in a vacuum tube. Elaborate high voltage equipment with rectifying and filtering circuits is required for radiography work, but is not essential for diffraction analysis.

A large variety of X-ray tubes is used for radiography and diffraction work, but basically one may distinguish two types: (a) gas tubes, and (b) electron tubes.

The gas tubes are operated with the aid of a vacuum pumping system,

† See also Vol. IV.

* Chapter 7.10 is by **Sigmund Weissmann.**

and since the production of electrons depends on the electrical discharge through low-pressure residual gas, some form of gas leak is incorporated. A steady gas pressure of about 0.001 mm Hg is required for proper operation. These tubes are very inexpensive and produce pure X-ray spectra, since the target cannot become contaminated with evaporation products of a hot filament.

In the electron tubes the electrons are emitted from a hot filament in a high vacuum. The essential features of this tube are schematically shown in Fig. 1. Two types of electron tubes are in use: the sealed-off, and the demountable tubes. The chief advantage of the sealed-off tubes lies in the freedom from maintenance difficulties but, on the other hand, only one radiation is obtainable from a single tube which has a limited life, and replacement of the tube is expensive. The demountable tubes offer greater maintenance problems than the sealed-off tubes, but their great merits lie in their interchangeable target arrangement, which

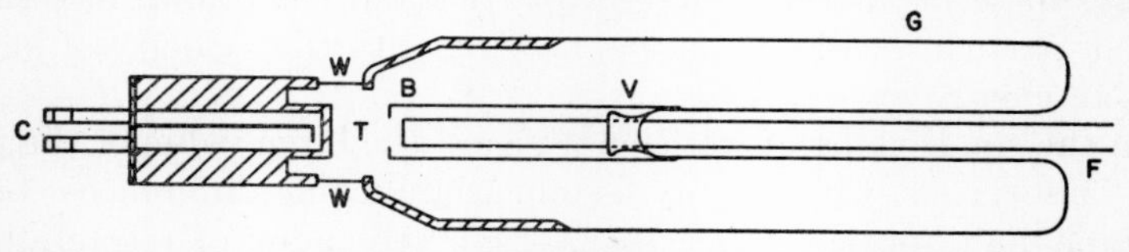

FIG. 1. Diagram showing principal features of an electron X-ray tube. *T*, target; *V*, glass vacuum seals; *G*, glass envelope; *B*, cathode-focusing tube; *F*, filament leads; *W*, windows; and *C*, cooling water. From H. S. Peiser, H. P. Rooksby, and A. J. C. Wilson, "X-Ray Diffraction by Polycrystalline Materials," Chapter 2. Institute of Physics, London, 1955.

permits a rapid change of wavelength, high continuous rating, production of clean spectra insured through repeated cleaning and resurfacing of the targets, and low replacement costs.

Important advantages are obtained in diffraction work by the use of fine focal spots and, consequently, increasing attention has been paid to the problem of electrostatic and electromagnetic focusing of the electron beam in demountable tubes.[1,2] Microfocusing tubes have found important applications in microanalysis by X-ray spectroscopy (Section 7.10.2.1), X-ray micrography (Section 7.10.2.2.2), studies of crystal imperfections (Section 7.10.2.4), and small-angle scattering (Section 7.10.2.5).

Recent developments in X-ray tube design have concentrated also on

[1] A. Guinier, "X-Ray Crystallographic Technology." Hilger & Watts, London, 1952; see also A. Guinier and J. Devaux, *Compt. rend.* **217**, 539 (1943); K. Drenk and R. Pepinsky, *Rev. Sci. Instr.* **22**, 539 (1951); W. Ehrenberg and W. E. Spear, *Proc. Phys. Soc.* (*London*) **B64**, 67 (1951); W. E. Spear, *ibid.*, p. 233.

[2] V. E. Cosslett, A. Engström, and H. H. Pattee, Jr., eds., "X-Ray Microscopy and Microradiography," pp. 49–195. Academic Press, New York, 1957.

the perfection of the rotating-target demountable tubes. The developments have been stimulated not only by the requirements of increased intensity output in diffraction work but also by new demands in medical therapy and radiography, in which fields the tubes have been widely used for a considerable time.[3]

It has been argued by various investigators that for diffraction work little advantage is gained if the time of exposure is reduced through the development of elaborate equipment which rotating anode tubes require, since the same effect may be obtained through microfocusing tubes, small cameras, and short focus-to-specimen distance.[4] It has been conceded, however, that further gains in beam brilliancy could be obtained by combining moving targets and microfocusing ideas.[5]

7.10.1.3. Detection of X-Rays. The intensity of an X-ray beam may be registered in a variety of ways. Probably the most common method used is the photographic method. When a film is blackened by an X-ray beam the density D of the exposed film is a criterion of the proportion of X-rays absorbed by the film. D is defined as $\log I_o/I$, where I_o and I are the incident and transmitted light beams, respectively. The *characteristic* curve of the film is obtained when the density is plotted as a function of $\log_{10} E$, where E is the intensity of the X-ray beam times the exposure time (Fig. 2). The density with zero exposure is known as the "fog" in the emulsion, and the most useful range over which the linear relationship between density and exposure is maintained lies between the fog level and the density of 1.0. Some screenless types of commercial X-ray films maintain linearity up to density values of 3.0 and more.

The inertia is often regarded as an indication of the film speed, but it is not the only criterion for it. The speed of the film is more properly defined as the ratio of D/E at a given density value above the fog. The slope, or γ, of the linear portion of the characteristic curve is a measure of the contrast of the film. The steeper the slope, the greater is the contrast.

X-ray densities recorded on a film can be analyzed photometrically or compared with a series of graded densities prepared on a calibration strip. The calibration strip of a film is prepared by varying the exposure time, since for X-rays the reciprocity law is obeyed, that is, exposure is

[3] R. E. Clay, *Proc. Phys. Soc.* (*London*) **B46,** 703 (1934); A. Müller and R. E. Clay, *J. Inst. Elec. Engrs.* (*London*) **84,** 261 (1939); D. P. Riley, *Brit. J. Appl. Phys.* **1,** 305 (1950); P. Gay, P. B. Hirsch, J. S. Thorp, and J. N. Kellar, *Proc. Phys. Soc.* (*London*), **B64,** 374 (1951); A. Taylor, *J. Sci. Instr.* **26,** 225 (1949); **27,** 757 (1956).

[4] C. S. Barrett, *in* "Modern Research Techniques in Physical Metallurgy," p. 78. American Society of Metals, Cleveland, Ohio, 1953.

[5] X-Ray Analysis Group, "X-Ray Diffraction by Polycrystalline Materials" (H. S. Peiser, H. P. Rooksby, and A. J. C. Wilson, eds.), p. 72. Institute of Physics, London, 1955.

proportional to the product of intensity and time of exposure. When a high degree of precision and maximum detail is not required, intensifying screens are used which are placed in close contact with the X-ray film. These screens reduce the exposure time by a factor of 10 or even more, particularly if short-wavelength radiation is used, and consequently they find frequent application in radiography.

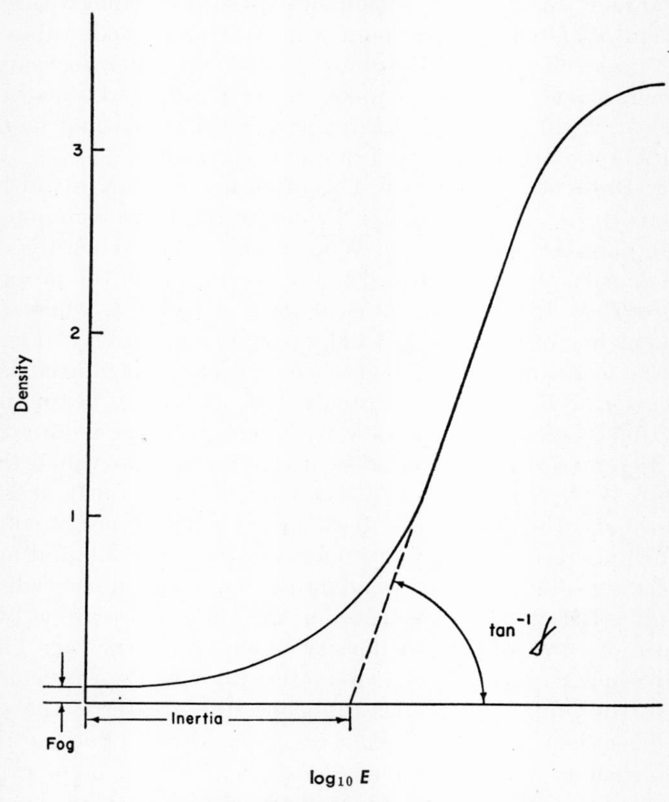

FIG. 2. Characteristic curve of X-ray film.

For precision measurements of intensities, electrical methods gain increasingly in importance. The most commonly used electrical detector is the Geiger counter. In its modern form, with halogen used as quenching gas, the sensitivity of this detector is great enough to cover a considerable range of the spectrum employed in diffraction and fluorescent analysis. One of the chief drawbacks of the Geiger counter is, however, its low sensitivity to the short-wavelength portion of the X-ray spectrum. Another severe limitation is its long dead time, which extends to about

270 μsec and which is mainly responsible for the considerable counting losses at high intensities.

In diffractometer work, Geiger counters are gradually being replaced by proportional counters which have similar spectral sensitivity characteristics. The dead time period of proportional counters is, however, much smaller, amounting to less than 1 μsec.

Scintillation counters are extremely sensitive for the detection of weak radiation. The most important advantages offered by these counters are: extremely low dead time and wide range of spectral sensitivity, especially for short wavelengths. When a photon is absorbed by a scintillating crystal such as thallium-activated NaI, a flash of light emanates from this crystal which is directed to the photocathode of a photomultiplier tube, giving rise to the ejection of photoelectrons. These are amplified by the cascaded secondary emission dynodes and emerge at the output as a current pulse analogous to the pulse that emerges from other counters.

For proportional and scintillation counters the pulse height is proportional to the energy of the absorbed photon, and consequently is inversely proportional to the wavelength of the impinging radiation. If the pulses are fed through a high-gain linear amplifier into a scaling circuit it is possible to increase the peak-to-background ratio and to discriminate between pulses deriving from different wavelengths.[6]

7.10.1.4. Spectral Distribution and Intensity of X-Rays. The spectral distribution of X-rays has been studied by absorption measurements, crystal diffraction, and ruled grating methods.[7,8] It has been shown that the spectrum is composed of a continuous (white) background radiation upon which the characteristic line spectrum is superimposed. The spectrum of the continuous radiation is cut off sharply at a minimum wavelength which is a function of the peak voltage applied. This relation is expressed by

$$\lambda_{\min} = \frac{12{,}395}{V_{\text{peak}}} \tag{7.10.1}$$

where $\lambda_{\min}$ is the minimum wavelength expressed in angstrom units and V_{peak} is the peak potential in volts.*

* See also this volume, Part 2.

[6] S. C. Curran, J. Angus, and A. L. Cockroft, *Phil. Mag.* [7] **40,** 36 (1949); A. R. Lang, *Nature* **168,** 907 (1952); *Proc. Phys. Soc.* (*London*) **A65,** 372 (1952); A. W. Arndt and D. P. Riley, *ibid.* **A65,** 74 (1952); G. T. Wright, *J. Sci. Instr.* **29,** 157 (1952); G. A. Morton, *Proc. Intern. Conf. on Peaceful Uses Atomic Energy, Geneva, 1955,* Vol. 14, p. 246, Paper 61 (1956).

[7] M. Siegbahn, "Spektroskopie der Röntgenstrahlen," 2d ed. Springer, Berlin, 1931.

[8] A. H. Compton and S. K. Allison, "X-Rays in Theory and Experiment," 2nd ed. Van Nostrand, New York, 1935.

The higher the atomic number Z of the target material, the greater the maximum intensity of the continuous radiation will be. Considering the same target material and constant tube current, increasing voltage gives rise to two effects: (1) the minimum wavelength decreases; (2) the maximum intensity of the radiation increases rapidly (Fig. 3).

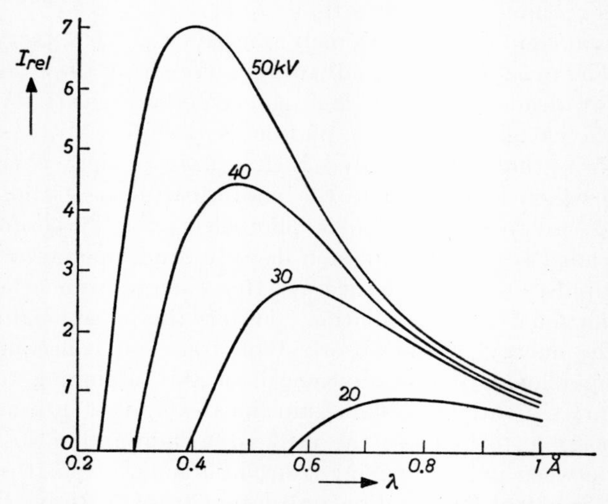

FIG. 3. Continuous spectrum of tungsten target X-ray tube (relative intensity I_{rel} versus wavelength λ) obtained with various voltages (peak values) and same tube current. Curves redrawn from separate rate meter recordings. Experimental conditions: full-wave rectification, silicon crystal analyzer (111-plane), scintillation counter. From W. Parrish, *Philips Tech. Rev.* **17,** 269 (1956).

The relation between the intensity of the continuous radiation I, the atomic number of the target material Z, and the applied voltage V, is expressed by the empirical equation:

$$I_{\text{continuous}} = KZV^2 \qquad (7.10.2)$$

where K is a constant. Consequently, in the field of radiography where continuous radiation is exclusively used, target materials of high atomic number, viz., tungsten, and high voltages are employed.

The line spectrum is characteristic of the target material. The wavelengths of the lines are independent of the applied voltage; the appearance of the lines is, however, dependent on the voltage. The production of the characteristic spectrum can be satisfactorily explained on the basis of the Bohr and the quantum mechanical model of the atom. According to this model the electron orbits of each atom are arranged in shells, viz., the K, L, M . . . shells. The characteristic lines originate in an ionization process of the inner electron shells followed by radiating transitions of electrons from outer shells to the vacant sites in the inner shells. Each jump of the electron from an outer shell to an inner shell is accompanied by the emission of a photon. The emission lines fall into wavelength groups designated as *K*, *L*, *M*, etc., in order of increasing wavelength, and correspond to the initial ionization of the respective shells concerned (*K*, *L*, *M*, etc.). The characteristic emission spectra consist of comparatively few lines owing to the small number of electron orbits which participate in the radiation process.

The intensity I of a spectral line is a function of the tube current i and of the potential V, and can be empirically expressed by the relation

$$I = ci(V - V_0)^n \tag{7.10.3}$$

where c is a proportionality constant and V_0 is the critical excitation voltage for the particular spectral line. The value of n is slightly below 2, provided the tube is operated at a constant potential and at voltages which are approximately two or three times that of the excitation potential. At higher voltages the value of n decreases towards unity.

7.10.1.4.1. UNITS OF INTENSITY. In X-ray crystallographic measurements the intensity of the X-radiation is expressed in cgs units—that is, in ergs per square centimeter per second. In X-ray therapy, however, it is customary to express the intensity in terms of roentgen units or *r* units. The roentgen unit has been defined by the 1956 International Commission on Radiological Units and Measurements in Geneva as "the exposure dose of X- or γ-radiation such that the associated corpuscular emission per 0.001293 gram of air produces, in air, ions carrying 1 esu of quantity of electricity of either sign."

7.10.1.5. Absorption and Scattering.[8] If high energy X-rays pass through a sheet of material of thickness t, a number of physical processes will occur:

(a) Some of the X-rays will get through without change in direction but reduced in intensity from I_0 to I.

(b) The photons which do not appear in the emerging beam have undergone various transformations. They may remain photons but may have been deviated from their original path. If the photons have been deviated without loss of energy, which corresponds to radiation without change of wavelengths, we speak of unmodified scattering. If the photons have been deviated with a loss of energy by colliding with free electrons, scattering associated with a change of wavelengths results. The latter type of scattering is termed modified Compton scattering.

The unmodified scattering is due to the forced oscillations generated in the electrons of the specimen by the incident X-ray beam. These forced oscillations are of the same frequency as that of the incident beam and, consequently, the electrons radiate in all directions X-rays of the same frequency as that of the incident beam. This scattered radiation will be polarized and may form Bragg reflections if the sheet of material is crystalline (see Section 7.10.2.3).

(c) The photons may have been absorbed by the atoms of the sheet material, giving rise to the photoelectric effect (see Section 7.10.1.4).

The combination of processes described under (b) and (c) is known as the absorption of X-rays and causes the reduction of the primary beam

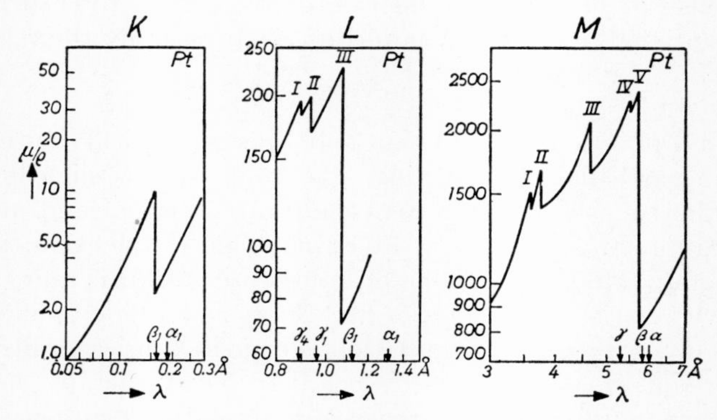

FIG. 4. Mass absorption coefficient (μ/ρ) of platinum as a function of wavelength λ in regions of K, L, and M absorption edges. Some characteristic emission lines of K, L, and M series of platinum shown with arrows. From W. Parrish, *Philips Tech. Rev.* **17,** 269 (1956).

from I_0 to I. For monochromatic radiation the reduction of intensity is governed by the following relation:

$$I = I_0 e^{-\mu t} \tag{7.10.4}$$

where μ is the linear absorption coefficient. Since μ is dependent on the state of the absorber, it is often convenient to use instead μ_m, the mass absorption coefficient; $\mu_m = \mu/\rho$ (ρ = density), which is independent of the physical or chemical state of the absorber and for a given absorber is only dependent on the wavelength of the incident primary beam.

If one plots the μ_m values for a given element versus the wavelength, the graph will not be smooth but will exhibit definite discontinuities. It will be seen that μ_m will fall abruptly whenever the wavelength exceeds certain critical values (Fig. 4). These discontinuities, known as absorption edges, are principally due to the photoelectric effect and, to a negligible extent, to scattering. The effect of scattering acquires importance only for rays of short wavelengths and for very light elements.

Aside from the significant role which the characteristic absorption edges play in the filtration of X-rays, the absorption process finds important practical application in spectrochemical analysis (Section 7.10.2.1), thickness-gaging and porosity measurements,[9] chemical analysis by relative absorption (X-ray photometry),[10] in diagnostic radiology, and in radiography and microradiography (Section 7.10.2.2).

7.10.2. Application of X-Rays

7.10.2.1. X-Ray Spectrochemical Analysis. The characteristic spectra of the various elements are very similar in structure. They contain emission lines which exhibit a frequency separation and an intensity distribution of remarkable regularity. The frequency, ν, of a line in a particular position varies with the atomic number according to the law discovered by Moseley:[11]

$$\nu^{\frac{1}{2}} = c(Z - \sigma) \tag{7.10.5}$$

where Z is the atomic number and c and σ are constants. This law forms the basis of the qualitative spectrochemical analysis by X-rays. A study of the intensities of the spectral lines of the elements present in a mixture or a compound will yield a quantitative analysis of their concentration.

One may distinguish two phases in the spectrochemical analysis by X-rays: (1) the excitation of the X-ray spectra, and (2) the analysis of the spectra.

The excitation process of the spectra can be achieved in two ways. The electrons can be dislodged from an inner shell either by the impact of a high-energy free electron or by the absorption of a high-energy photon. Consequently, the characteristic line spectrum of an element can be obtained either by bombardment of the specimen with high-speed electrons or by irradiation with a beam of X-rays. When an unknown substance is made the target of an X-ray tube, the K, L, M, etc., emission lines can be produced and analyzed. Historically, this was the standard method of spectral excitation applied for a long time by Siegbahn[7] and by von Hevesy.[12] Although this method is very sensitive, it has many drawbacks for practical chemical analysis. It is seldom that the requisite steadiness of emission can be achieved, and in the event of volatilization

[9] H. Friedman and L. S. Birks, *Rev. Sci. Instr.* **17,** 99 (1946).

[10] M. V. Sullivan and H. Friedman, *Ind. Eng. Chem., Anal. Ed.* **18,** 304 (1946); T. S. Rich and P. C. Michel, *Gen. Elec. Rev.* **50,** 45 (1947); H. A. Liebhafsky and E. H. Winslow, *ibid.* **48,** 36 (1945); H. A. Liebhafsky, J. M. Smith, H. E. Tanis, and E. H. Winslow, *Anal. Chem.* **19,** 861 (1947).

[11] H. G. J. Moseley, *Phil. Mag.* [6] **26,** 1024 (1913).

[12] G. von Hevesy, "Chemical Analysis by X-Rays and Its Applications." McGraw-Hill, New York, 1932.

of the constituents of the sample an erroneous analysis would inevitably result.

The second method of spectral line excitation by X-irradiation is the method widely employed today. The process of excitation is called fluorescence and the spectrochemical analysis is known as "X-ray fluorescence analysis." This method may be applied to materials in any form: bulk materials, powder specimens, liquids, or gases. Its major disadvantage lies in the great loss of intensity, which may be more than a hundredfold as compared to that obtained by direct excitation. The complete absence of continuous background radiation offsets, however, the disadvantage of intensity loss, since it is considerably easier to measure a weak line

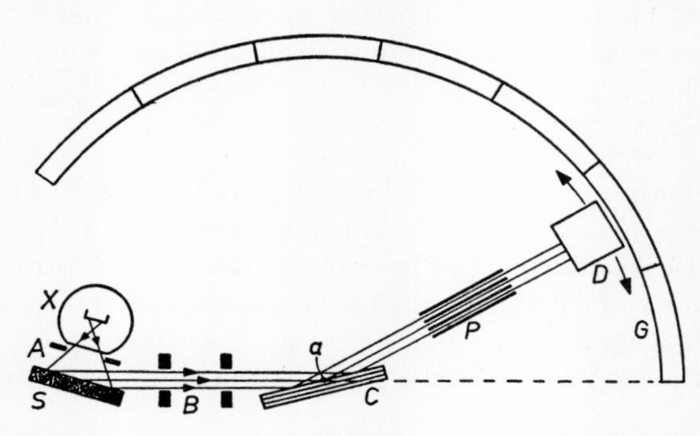

FIG. 5. Nonfocusing X-ray optics used for fluorescence analysis of large specimens. S, specimen excited by primary radiation from X-ray tube, X; C, analyzing crystal rotated around axis a in order to scan different wavelengths. Detector D rotated on goniometer G at twice angular speed of the crystal. Parallel slit system P insures required angular resolution. From W. Parrish, *Philips Tech. Rev.* **17**, 269 (1956).

against a zero background than a strong line against a high continuous radiation background.

The analysis of the excited spectra is usually carried out by applying the principle of X-ray diffraction. A single crystal with known interplanar spacing is selected to analyze the spectrum for all wavelengths present (Fig. 5). The crystal is rotated, and the different angles at which the diffracted energy is detected are related to the different wavelengths of the elements present in the test specimen. The most efficient detector for fluorescent analysis is the scintillation counter (Section 7.10.1.3).

A number of practical difficulties are encountered in the spectrochemical analysis by X-rays. Because of the finite thickness of the specimen a considerable portion of the fluorescent radiation may be absorbed within the specimen. Furthermore, the fluorescent radiation emanating from one element may in turn generate fluorescent radiation in another element of the specimen. For these reasons it is generally very difficult to insure a linear relationship between measured fluorescent intensity and concentration of an element in the specimen. Consequently, the quantitative analysis follows closely the method employed in the optical domain. The intensities of selected X-ray emission lines from a sample, viz., *K* or *L* lines, are compared to the corresponding lines of reference standards of known composition.

The spectrochemical analysis by X-rays is capable of fairly good precision. Minor constituents smaller than 0.001% have been determined by this method,[13] and specimens containing only a few milligrams have successfully been analyzed by special focusing techniques.[14] All elements of higher atomic number than titanium can be conveniently analyzed without surrounding the specimen by a helium atmosphere, whereas for lower elements, owing to the high absorption of the soft X-rays in air, special helium techniques are required. Castaing has combined the principle of electron microscopy with spectrochemical analysis by X-rays. By a system of electromagnetic reducing lenses the electron beam in an electron microscope housing is focused on the test specimen acting as the source of X-rays. The analysis of the spectrum emitted by the specimen is carried out by a curved crystal spectrograph. If the specimen is traversed across the electron beam the chemical composition of areas 1 μ^2 in size of metallographic and mineralogical specimens can be determined.[15]

7.10.2.2. Radiography.[16] 7.10.2.2.1. SENSITIVITY OF RADIOGRAPHY. Radiography depends on the differential absorption of X-rays by nonhomogeneous elements in a specimen. The experimental procedure consists of passing radiation through the test specimen and detecting and measuring by some experimental means the various intensities of the emergent beam.

A number of important factors govern the quality of a radiograph. Among these factors are (1) the photographic effect of X-rays, (2) the effects of X-ray tube voltage on the quality of the beam, (3) scattering of X-rays, and (4) focus-to-film distance.

If minute defects are to be detected in a radiograph, a high degree of film contrast is essential. The steeper the slope, γ, of the linear portion of the characteristic curve, the greater is the ability of the film to discriminate between slightly different X-ray intensities and, therefore, between slight variations in specimen thickness (Fig. 2). If the γ-value is high the range of specimen thickness which will keep the density values

[13] T. N. Rhodin, *Anal. Chem.* **27,** 1857 (1955).

[14] L. S. Birks and E. J. Brooks, *Anal. Chem.* **27,** 1147 (1955).

[15] R. Castaing, *Recherche aéronaut.* **23,** 41 (1951).

[16] G. L. Clark, "Applied X-Rays," 4th ed. McGraw-Hill, New York, 1955.

within the linear portion of the curve will be small. Conversely, low γ-values of the characteristic curve will provide the film with a wide latitude and will insure detection of big differences in thickness, but will exhibit a low contrast for the detection of small defects. In actual practice, a compromise between contrast and latitude has to be worked out, and it is often necessary to take a series of radiographs to cover the range in specimen thickness.

Metal foils, particularly lead, decrease the exposure period in certain voltage ranges and have a beneficial effect on the quality of the film. If placed on each side of the film, the lead foils not only reduce film fogging by absorbing a portion of the scattered radiation, but also produce secondary electrons under the action of the primary beam which have an activating effect on the film emulsion. Thus, a tube potential of 150 kv liberates sufficient secondary electrons to reduce the exposure time by a factor of 2. The scattered secondary radiation, particularly in heavy sections of metals, can also be reduced by placing a grid of the Potter–Bucky type between the object and the plate. The grid, which is slowly moved across the plate by a cam mechanism, permits the primary rays to pass straight through, but cuts off the obliquely traveling scattered rays.

Undesirable scattering may also be caused by the primary beam passing through holes of the specimen and then being scattered by the film itself. This type of scattering is eliminated by special shielding techniques with lead shots, lead sheets, or immersion of the object in suitable solutions.

The focus-to-object distance D as well as the object-to-film distance r have a marked influence on the sensitivity of radiography. It is evident that for the production of a sharp image the D/r ratio should be large. Since the intensity of the beam decreases in proportion to the square of the distance from the tube, the selection of the proper D/r ratio will depend largely on the details of the radiograph required.

7.10.2.2.2. Microradiography. Microradiography is a radiographic method by which microscopic discontinuities of an object can be studied. A relatively small volume of the object is radiographed on a fine-grained emulsion which is subsequently enlarged for inspection and interpretation. The fundamentals of the method have been described by various authors.[16–18]

The greatest advantages gained by this method over its optical counterpart are: (a) the removal of the two-dimensional limitation, and (b) the differentiation of various constituents of a radiographed substance by the skillful use of the X-ray absorption edges.

[17] A. Engström, *Acta Radiol.* **31**, 503 (1949).

[18] S. E. Maddigan, *J. Appl. Phys.* **15**, 43 (1944); *Ind. Radiography* **4**, 40 (1945).

The development of microfocusing X-ray tubes has given considerable impetus to new advances in microradiography.[2,19,20] By means of soft X-rays (200–2000 volts), Engström recently obtained contact micrographs of chromosomes and other cell components which exhibited a resolution of 0.3–0.5 μ.[2,21]

A method of high resolution has been developed by Cosslett and Nixon in which a point source is obtained through electromagnetic focusing, and this source is used for projecting shadow images at a primary magnification up to 100 times.[2,22] A resolving power approaching that of ultraviolet microscopy has been recently obtained with an exposure time of 1 min.[2]

The sensitivity and flexibility of microradiography have been extended through the skillful application of fluorescent radiation.[2,23]

7.10.2.3. Crystal Structure. 7.10.2.3.1. FUNDAMENTAL PRINCIPLES OF X-RAY DIFFRACTION.*[24] X-ray diffraction by crystals is essentially an interference phenomenon. This interference phenomenon stems from the fact that in every crystal the atoms or groups of atoms are more or less periodically arranged in three dimensions and that the wavelengths of X-rays are of the same order of magnitude as the interatomic spacings of the crystal. The interference of the X-rays scattered by the atoms will be destructive except for certain predictable directions where constructive interference will occur. W. L. Bragg has shown that the diffraction phenomenon may be considered as cooperative reflection from planes of atoms. He has shown that if an incident ray of wavelength λ irradiates a set of atomic planes with an interplanar distance d, at an angle θ, constructive interference can occur for the directions of specular reflection. The relation which governs the condition for specular reflection is expressed by

$$n\lambda = 2d \sin \theta$$

where n is an integer.

It may be seen that, unlike the case of diffraction of visible light by a ruled grating, X-ray diffraction is highly critical, and hence the name "selective reflection" has been frequently applied to this phenomenon.

The Bragg relationship is of fundamental importance. If the spacing of

* See also Vol. 6, Chapter 3.1.

[19] W. Ehrenberg and W. E. Spear, *Nature* **168,** 513 (1951).

[20] A. Engström, *Progr. in Biophys. and Biophys. Chem.* **1,** 164 (1950).

[21] A. Engström and R. C. Greulich, *J. Appl. Phys.* **27,** 758, 1956.

[22] V. E. Cosslett and W. C. Nixon, *J. Appl. Phys.* **24,** 616 (1953).

[23] T. H. Rodgers, *J. Appl. Phys.* **23,** 881 (1952); H. R. Splettstosser and H. E. Seeman, *ibid.* **23,** 1217 (1952).

[24] R. W. James, "The Optical Principles of the Diffraction of X-Rays" Macmillan, New York, 1948.

the crystal is known, the wavelength distribution of the X-ray spectrum can be studied (X-ray spectroscopy) and, conversely, if the wavelength of the probing beam is known the crystal spacings can be determined by measuring the diffraction angles (X-ray crystallography).

7.10.2.3.2. METHODS OF CRYSTAL ANALYSIS. If a randomly oriented crystal is analyzed by irradiation with a collimated monochromatic beam at an arbitrary angle, it is unlikely that the Bragg condition of reflection will be fulfilled by any lattice plane and, consequently, it is unlikely that a diffraction pattern will be observed.

To insure that the condition of reflection will be satisfied it is in general necessary to provide a range of values of either θ, the angle of incidence, or λ, the wavelength. Methods either varying θ and keeping λ constant, or varying λ and keeping θ constant form the bases of the standard diffraction methods of crystal analysis. This relationship between the variables offers, therefore, a convenient criterion for the classification of these methods.

7.10.2.3.3. ROTATING CRYSTAL METHODS.[25] For all rotating crystal methods, monochromatic X-rays are employed and a single crystal specimen is rotated or oscillated during the exposure to provide a variation in the angle of θ. The specimen is set with an important crystallographic axis parallel to the rotation axis of the instrument and usually perpendicular to the X-ray beam emerging from a pinhole system. The film is usually cylindrical with the axis coincident with the rotation axis. Frequently, however, a flat plate for transmission or back reflection photographs is employed. If a cylindrical film is used to record the diffracted beams the reflections are lined up on layer lines, perpendicular to the axis of rotation, whereas if a flat plate is employed, the reflections are arranged on a series of hyperbolae of different eccentricities. If the rotation axis is vertical, the spots on the equatorial layer line are reflections from planes the normals of which are horizontal. These planes comprise, therefore, a crystallographic zone having the rotation axis as the zone axis. The reflections on any subsequent layer line derive from planes that have the same intercept on the zone axis and, therefore, the distance between successive layers represents a measure of the identity period along the axis.

There exists a powerful concept in X-ray crystallography, the so-called reciprocal lattice concept,[26] with the aid of which it becomes easy to assign

[25] M. J. Buerger, "X-Ray Crystallography." Wiley, New York, 1942; N. F. M. Henry, H. Lipson, and W. A. Woster, "The Interpretation of X-Ray Diffraction Photographs." Van Nostrand, New York, 1951.

[26] P. P. Ewald, *Z. Krist.* **56,** 129 (1921); J. D. Bernal, *Proc. Roy. Soc.* **A113,** 117 (1926).

indices to the various spots on the rotation photographs, provided that confusion due to the overlapping of spots is avoided. The assigned indices (Miller indices) identify the relative orientation of the set of atomic planes which give rise to the reflections. Having indexed the reflections, it is possible to identify the space group of the crystal, or at least, to narrow down the selection of possible space groups to which the crystal may belong.* [25]

7.10.2.3.4. POWDER METHOD.[27] 7.10.2.3.4.1. *Film and Counter Diffractometer Methods.* In the powder method, monochromatic radiation is used and, therefore, λ is kept constant. Since the specimen consists of a myriad of tiny crystals, oriented in all directions, a variation in θ variable is provided for. Reflections will occur only at values of $\sin \theta = n\lambda/2d$. Consequently, for a given d-spacing of the specimen, reflection will occur when $\sin \theta$ is a constant. This condition translated into a physical picture implies that the loci of the reflected beams will lie on the surface of a cone subtending with the transmitted X-ray beam, which forms the central axis of the cone, an angle of 2θ. The coaxial diffraction cones emerging from the powder specimen are measured either on a film or by means of counters sensitive to the radiation (Section 7.10.1.3).

If a film is placed normal to the incident beam, the diffraction cones intersect the film in concentric circles. Each circle corresponds to one order of reflection from each set of planes having a definite d-spacing. If the film is cylindrical and coaxial with the specimen axis, the diffraction cones intersect the film in arcs. The dimensions of the diffraction rings or arcs—Debye–Scherrer lines—are directly related to the d-spacings of the reflecting planes. For flat films the diameter of the diffraction ring l is given by $l/s = 2 \tan 2\theta$, where s is the specimen-to-film distance. For cylindrical films, the distance between corresponding arcs l is given by $l/r = 4\theta$ (in radians), where r is the radius of the camera. Knowing the θ values, the d-spacings are computed from the Bragg relation. A great variety of powder cameras are in use, designed to fit a particular experimental need including high and low temperature work.

The most widely accepted diffractometer methods utilizing counters as detecting devices are based on the parafocusing principle first introduced by Brentano.[28] The entire assembly, consisting of the effective

* For a more detailed treatment of the rotating crystal methods including the oscillation and moving film methods, see "Methods of Experimental Physics," Vol. 6, Chapter 3.2.

[27] H. P. Klug and L. E. Alexander, "X-Ray Diffraction Procedures." Wiley, New York, 1954; X-Ray Analysis Group, "X-Ray Diffraction by Polycrystalline Materials" (H. S. Pieser, H. P. Rooksby, and A. J. C. Wilson, eds.). Institute of Physics, London, 1955.

[28] J. C. M. Brentano, *Proc. Phys. Soc. (London)* **37,** 184 (1925); **49,** 61 (1937).

X-ray source, represented by the entrance slit, the specimen and the detecting device, lies on the circumference of the focusing circle. The detecting device is synchronized with the specimen rotation and moves with twice the angular speed of the specimen.

Comparing the merits of the counter diffractometer method with those of the film method, it is apparent that the former will offer greater precision in intensity measurement. The counter diffractometer method is particularly suitable for continuous phase transformation studies and also for kinetic studies wherein a correlation is sought between the elapsing time and the variation of either the position, intensity, or profile characteristics of powder lines.

On the other hand, information concerning the texture of the specimen is usually suppressed by the diffractometer method and becomes more conspicuous on a film record. It is possible, however, to obtain accurate information on preferred orientation by special adaptation of diffractometer methods.[29]

7.10.2.3.4.2. *Identification of Polycrystalline Materials.* The powder method is particularly useful for the identification of unknown crystalline substances. Since every crystalline substance yields a diffraction pattern which is uniquely characterized by the d-spacings and relative intensity of the lines, one has merely to compare the pattern obtained with those of standard patterns. The X-ray diffraction cards of the ASTM constitute a card system of standard patterns in which the three strongest lines, as well as the interplanar spacings and their relative intensities, are listed. The procedure that is usually followed for the identification of a crystalline unknown entails: (1) the identification of a component of the mixture responsible for the strongest recorded line, (2) the subsequent elimination of all interplanar spacings pertaining to this component, and (3) the identification of the remaining strongest lines by an analogous procedure.

7.10.2.3.4.3. *Precision Determination of Lattice Parameters.* The powder method offers a tool *par excellence* for precision determination of lattice parameters. The precision measurements are greatly enhanced through the application of special focusing devices, internal standards, and extrapolation methods.

The differentiation of Bragg's law leads to the expression

$$\frac{\Delta d}{d} = -\cot\theta\, d\theta \tag{7.10.6}$$

from which it can be seen that the percentage error in spacing measure-

[29] B. F. Decker, E. T. Aps, and D. Harker, *J. Appl. Phys.* **19,** 388 (1948); L. G. Schulz, *J. Appl. Phys.* **20,** 1030 (1949); A. N. Holden, *Rev. Sci. Instr.* **24,** 10 (1953).

ment resulting from a given error in angle measurement will approach zero as cot θ approaches zero, that is when θ approaches 90°. A number of graphical and analytical extrapolation methods for precision lattice parameter determination are based on this fundamental concept.[30]

The precision determination of lattice constants finds application in a variety of important research problems. Among these are the determination of solubility limits in constitution diagrams, the determination of the coefficient of thermal expansion of solids, precipitation studies, and stress measurements.

7.10.2.3.5. THE LAUE METHOD.[31] In the Laue method continuous radiation is employed and the variable is, therefore, λ. Each set of crystal planes selects and diffracts rays of those wavelengths that satisfy the Bragg condition of reflection. If a film is placed normal to the incident beam, one obtains a series of spots. All the spots emanating from a family of planes, forming a zone, will lie on ellipses or hyperbolae. From the inspection of a Laue pattern, one obtains information concerning the symmetry of a crystal. This interpretation is enhanced if the incident beam coincides with one of the crystallographic axes of the crystal.

The Laue method is greatly used for the determination of the orientation of individual crystals, and also for studying the changes of orientation resulting from the deformation of crystals.

Gross lattice distortions can be revealed by the Laue method, since they give rise to a radial or circumferential spread of the spots known as asterism.[32]

A focusing Laue method in transmission has been developed by Guinier and Tennevin employing a narrow focal spot (effective width 40 μ) and a large crystal-to-film distance (up to 100 cm). This method is very sensitive to small orientation differences within the crystal and has been utilized for the study of the substructure of crystals ("polygonization").[33]

7.10.2.4. Crystal Imperfections.[24,34] The perfect three-dimensional periodicity of atomic arrangement applies only to ideal crystals, that is, to idealizations of the more complicated structure which actually exists

[30] A. J. Bradley and A. H. Jay, *Proc. Phys. Soc.* (*London*) **44,** 563 (1932); H. Lipson and A. J. C. Wilson, *J. Sci. Instr.* **18,** 144 (1941); A. Taylor and H. Sinclair, *Proc. Phys. Soc.* (*London*) **57,** 126 (1945); J. B. Nelson and D. P. Riley, *Proc. Phys. Soc.* (*London*) **57,** 160 (1945); A. J. C. Wilson, *J. Sci. Instr.* **27,** 321 (1950) (with special reference to counter diffractometer).

[31] C. S. Barrett, "Structure of Metals," 2d ed. McGraw-Hill, New York, 1952.

[32] P. Gay and R. W. K. Honeycombe, *Proc. Phys. Soc.* (*London*) **A64,** p. 844 (1951).

[33] A. Guinier and J. Tennevin, *Compt. rend.* **226,** 1530 (1948); *Acta Cryst.* **2,** 133 (1949).

[34] W. H. Zachariasen, "Theory of X-Ray Diffraction in Crystals," Chapter 4. Wiley, New York, 1945.

in real crystals. One of the principal deviations from this idealization is due to the fact that the atoms in a real crystal do not rest but vibrate continually about their mean position. The frequency of the vibrations is determined by the interatomic forces, and their amplitude by the temperature. Furthermore, impurities, i.e., foreign atoms, may be dispersed as individuals throughout the crystal, when they are said to be in solid solution, or they may cluster together to form particles of a second phase or inclusions.

Apart from vibrations and foreign atoms, the real crystals have irregularities which are generally known as crystal imperfections.

Many properties of crystals can be divided into either structure-insensitive properties or structure-sensitive properties. The structure-insensitive properties (specific heat, density, coefficient of thermal expansion, elastic constants) vary very little among different crystals of the same species, whereas the structure-sensitive properties (yield strength, breaking strength, magnetic permeability, electric conductivity) are intimately linked to the existence of crystal imperfections, such as cracks and dislocations, and exhibit, therefore, variations among different crystals of the same species.

7.10.2.4.1. LINE WIDTH AND LINE PROFILE MEASUREMENTS. Evidence for crystal imperfections has been gathered from measurements of the line width of reflected X-ray beams and from a comparison of measured and calculated values of the integrated reflections. Accurate studies of line width and integrated reflections have been made by means of the double-crystal spectrometer, in which the incident beam falling on the test crystal has been first reflected from another crystal, usually of the same species. Depending on the settings of the two crystals with respect to each other, the double-crystal spectrometer is a precision tool either for the analysis of the spectrum, or for the study of lattice imperfections. In the former setting (1, + 1) arrangement, the outward drawn normals of the two crystals subtend an angle of 2θ, whereas in the latter setting (1, − 1) arrangement they subtend an angle of 180°. For the study of lattice imperfections the (1, − 1), or parallel setting, is the preferred arrangement since it insures the absence of dispersion. The first crystal is set in peak reflecting position, and the test crystal is rotated through the Bragg angle over a small angular range. The reflected intensities registered by a radiation detector are plotted as a function of rotation angle. Parameters of the reflection curve such as the width at half maximum, the integrated intensity, or integral breadth (ratio of integrated intensity over peak height) are indicative of the angular misorientation in the crystal.

A modification of the double-crystal spectrometer method for the study of lattice imperfections has been recently introduced. This method combines a counter and supplementary film technique for the analysis of complex reflection curves which imperfect crystals exhibit. The film placed in front of the counter resolves small substructural details the intensity contributions of which are integrated by the counter. Furthermore, photographic tracing of the individual reflections to the specimen surface pins down the location of the reflecting lattice regions giving rise to a specific registered intensity value.[35]

Many X-ray investigations in the last three decades have attempted to unravel the nature of crystal distortions from the analyses of the broadening of the diffraction lines. Extreme caution is, however, necessary if one wishes to arrive at an unambiguous interpretation of the observed X-ray data. A number of factors either singly or in combination may contribute to the broadening of the lines, viz., orientation range, crystallite size, microstresses, stacking faults, precipitation, and thermal vibrations.

A very powerful method of line analysis is due to Warren and Averbach. By representing the peak shapes in terms of their Fourier coefficients, it is possible to separate the individual factors which contribute to the broadening effect.[36]

7.10.2.4.2. STUDY OF SURFACE TEXTURE. Lattice inhomogeneities of the surface layers of single crystals and polycrystalline specimens have been investigated by reflection methods.[37] Images of the surface texture of the specimen are produced by using a source of X-ray distant from the crystal and a fine-grained photographic plate close and parallel to the reflecting surface. The X-ray micrograph obtained may be considered as a map of the reflecting power varying from one lattice region to another. Although X-ray microscopy is a powerful tool for the study of surface texture, it lacks the element of diffraction analysis. A method which combines X-ray microscopy and diffraction analysis has, however, been recently developed. This method can be described by a two-stage process. During the first stage an X-ray micrograph of the specimen surface is obtained with a crystal-monochromatized beam and, by outward tracing of the reflections, the spots on the various Debye–Scherrer lines are correlated to the lattice regions on the specimen surface giving rise to

[35] J. Intrater and S. Weissmann, *Acta Cryst.* **7,** 729 (1954).

[36] B. E. Warren and B. L. Averbach, *J. Appl. Phys.* **21,** 595 (1950); **23,** 497 (1952); **23,** 1059 (1952). See also W. Shockley, "Imperfections in Nearly Perfect Crystals," Chapter 5. Wiley, New York, 1952; "Modern Research Techniques in Physical Metallurgy." American Society for Metals, Cleveland, Ohio, 1953.

[37] W. Berg, *Naturwissenschaften* **19,** 391 (1931); *Z. Krist.* **89,** 286 (1934); C. S. Barrett, *Trans. Am. Inst. Mining Met. Engrs.* **161,** 15 (1945); L. G. Schulz, *J. Met.* **6,** 1082 (1954).

them. During the second stage, an analysis of the individual reflections based on the double-crystal spectrometer principle is carried out.[38]

Utilization of the principle of X-ray reflection from a cylindrically bent crystal focusing the image of characteristic radiation of a given wavelength to a point has led to the development of a low-power X-ray microscopy.[39] The development of high-power X-ray microscopy is based on the principle of specular reflection of X-rays from curved mirrors.[40] Considerable advances have been made in this field through improved mirrors which reduce errors in astigmatism, spherical aberration, and which increase the resolving power of the instrument.[2]

7.10.2.5. Particle Size. Materials which consist of submicroscopic particles or clusters with voids between them exhibit small-angle scattering. This scattering depends only on the size and shape of the particles or clusters and is independent of their internal structure. The intensity of the small-angle scattering depends on the difference between the electron density of the particles and that of the surrounding medium. The scattering is continuous and may display scattering maxima. From the profile of the scattering curve or from the scattering maxima one may deduce the particle size.[41] If the particles form aggregates of small crystallites the average dimensions of the crystallites can be determined by line broadening measurements at large diffraction angles (Section 7.10.2.4.1). Consequently, both the individual crystallite size and the crystallite aggregates can be measured by diffraction methods.

The range of particle size which can be studied by small-angle scattering methods extends from about 20 to 1000 Å. Since this scattering is restricted to the region close to the undeviated primary beam, considerable experimental difficulties are encountered and great care must be given to the design and adjustment of the instrumentation so as to avoid any parasitic scattering. Vacuum techniques are often employed, and for the diffuse type of small-angle scattering monochromatic radiation is essential. The requirement of an intense, monochromatized beam for this type of study has logically led to the adaptation of the curved-crystal focusing principle.[42] Guinier and Fournet extended this principle to two curved crystals, mounted in a vacuum chamber in series and employing a rotating anode tube as a powerful X-ray source. Compared to the curved single-

[38] S. Weissmann, *J. Appl. Phys.* **27**, 389, 1335 (1956).

[39] Y. Cauchois, *Ann. phys.* [2] **1**, 215 (1934); *Rev. opt.* **29**, 151 (1950); L. V. Hamos, *Am. Mineralogist* **23**, 215 (1938); *J. Sci. Instr.* **15**, 87 (1938).

[40] P. Kirkpatrick and A. V. Baez, *J. Opt. Soc. Am.* **38**, 766 (1948).

[41] For a detailed treatment see A. Guinier and G. Fournet, "Small-Angle Scattering of X-Rays." Wiley, New York, 1955. This book also contains a comprehensive bibliography of the small-angle scattering literature by K. Yudowitch.

[42] A. Guinier, *Ann. phys.* [11] **12**, 161 (1939); *J. chim. phys.* **40**, 133 (1943).

crystal arrangement the double-crystal apparatus achieves a higher degree of monochromatization and focusing of the beam. The intensity loss which usually accompanies the introduction of a second crystal is partly compensated by the use of wide apertures which the double-crystal arrangement permits.[43]

Many investigators have successfully applied the double-crystal spectrometer method (Section 7.10.2.4) to the study of small-angle scattering.[44] The second crystal used in conjunction with the radiation detector functions as the analyzer of the radiation scattered by the specimen near zero angle. The crystals usually used are cleaved lamellae of calcite although, in this author's opinion, the (111) faces of silicon or germanium crystals are more effective. The search for increased resolution in this field of study has even led to the application of a triple-crystal diffractometer method.[45]

The property of total external reflection from curved surfaces has found an interesting and effective application to small-angle scattering by Ehrenberg and Franks.[46] This principle was applied by Henke and Du Mond to the long wavelength region (8 to 25 Å), from which a series of advantages could be gained. Since the scattering angles are proportional to the wavelength, the application of large wavelengths permits a higher resolution of the intensity distribution and better separation of the important central portion of the intensity curve from the direct beam. Furthermore, the parasitic background scattering can be controlled more effectively by stops, thus allowing the measurement of relatively weak patterns.[47]

[43] A. Guinier and G. Fournet, *Compt. rend.* **226,** 656 (1948).

[44] I. Fankuchen and M. H. Jellinek, *Phys. Rev.* **67,** 201 (1945); J. W. M. Du Mond, *ibid.* **72,** 83 (1947); W. W. Beeman and P. Kaesberg, *ibid.* **72,** 512 (1947); P. Kaesberg, H. N. Ritland, and W. W. Beeman, *ibid.* **74,** 71 (1948); H. Daams and J. J. Arlman, *Appl. Sci. Research* **B2,** 217 (1951); R. Wild, *Rev. Sci. Instr.* **22,** 537 (1951); L. Broussard, *ibid.* **21,** 399 (1950).

[45] H. N. Ritland, P. Kaesberg, and W. W. Beeman, *J. Appl. Phys.* **21,** 838 (1950).

[46] W. Ehrenberg, *Nature* **160,** 330 (1947); *J. Opt. Soc. Amer.* **39,** 741, 746 (1949); W. Ehrenberg and A. Franks, *Nature* **170,** 1076 (1952); *Proc. Phys. Soc. (London)* **B68,** 1054 (1955).

[47] B. L. Henke and J. W. M. Du Mond, *J. Appl. Phys.* **26,** 903 (1955).

8. ELECTRICITY*

8.1. Electrostatics

8.1.1. Introduction

Much of the experimental data on electric and magnetic fields and the interaction of electromagnetic fields with electric charges and electric and magnetic dipoles can be correlated, at least formally, by means of Maxwell's equations (8.1.1) to (8.1.4) and certain constitutive equations (8.1.5) to (8.1.8). These equations written in both the mks and Gaussian system of units for isotropic media at rest become:

mks Units	*Gaussian Units*	
$\nabla \cdot \mathbf{D} = \rho$	$\nabla \cdot \mathbf{D} = 4\pi\rho$	(8.1.1)
$\nabla \cdot \mathbf{B} = 0$	$\nabla \cdot \mathbf{B} = 0$	(8.1.2)
$\nabla \times \mathbf{E} = -\dfrac{\partial \mathbf{B}}{\partial t}$	$\nabla \times \mathbf{E} = -\dfrac{1}{c}\dfrac{\partial \mathbf{B}}{\partial t}$	(8.1.3)
$\nabla \times \mathbf{H} = \mathbf{J} + \dfrac{\partial \mathbf{D}}{\partial t}$	$\nabla \times \mathbf{H} = 4\pi/c\,\mathbf{J} + \dfrac{1}{c}\dfrac{\partial \mathbf{D}}{\partial t}$	(8.1.4)
$\mathbf{D} = \mathbf{P} + \epsilon_0\mathbf{E} = \epsilon\mathbf{E}$	$\mathbf{D} = 4\pi\mathbf{P} + \epsilon_0\mathbf{E}$	(8.1.5)
$\mathbf{B} = \mu_0\mathbf{H} + \mu_0\mathbf{M} = \mu\mathbf{H}$	$\mathbf{B} = 4\pi\mu_0\mathbf{M} + \mu_0\mathbf{H}$	(8.1.6)
$\mathbf{J} = \sigma\mathbf{E}$	$\mathbf{J} = \sigma\mathbf{E}$	(8.1.7)
$\mathbf{F} = q[\mathbf{E} + \mathbf{v} \times \mathbf{B}]$	$\mathbf{F} = q\left[\mathbf{E} + \dfrac{\mathbf{v} \times \mathbf{B}}{c}\right].$	(8.1.8)

Equations (8.1.1) and (8.1.2) represent Gauss' laws in differential form with **D** the electric displacement vector measured in mks units in coulombs per square meter, ρ the charge density in coulombs per cubic meter and **B** the magnetic flux density vector in webers per square meter. In Gaussian units, **D** is in statcoulombs per square centimeter, ρ in statcoulombs per cubic centimeter and **B** in gauss.

Equation (8.1.3) is Faraday's law of electromagnetic induction, with **E** the electric field in mks units measured in volts per meter. Equation (8.1.4) is Maxwell's modification of Ampere's law. The vector **H** is the magnetic field measured in amperes per meter. In Gaussian units **E** is in statvolts per centimeter, and **H** is in oersteds.

In Eq. (8.1.5) the vector **P** is the polarization of the media, i.e., the dipole moment per unit volume and is measured in coulombs per square meter in mks units and in statcoulombs per square centimeter in Gaussian units. For magnetic media, Eq. (8.1.6), the vector **M** is the magnetic moment per unit volume and is measured in webers per square meter in

---

* Part 8 is by Michael Ference, Jr. and L. J. Giacoletto.

mks units and in maxwells per square centimeter in Gaussian units. The simple relations $\mathbf{D} = \epsilon\mathbf{E}$ and $\mathbf{B} = \mu\mathbf{H}$ are valid for isotropic media.

Ohm's law for isotropic media is given by Eq. (8.1.7) with σ the conductivity expressed in mhos per meter in mks units and in mhos per centimeter in the Gaussian unit.

Equation (8.1.8) is the force equation giving the force (newtons) on a charge (coulombs) moving with a velocity (meters per second) in a combined electric and magnetic field. Measurements of the conductivity σ, permittivity or absolute dielectric constant ϵ, and permeability μ are treated in some detail at a later point in the text.

TABLE I. Conversion Table for Units

Quantity	mks Units	cgs Units
Current, i, I	amp	0.1 emu, 3×10^9 esu
Charge, q, Q	coulomb	0.1 emu, 3×10^9 esu
Potential, V	volt	10^8 emu, $\frac{1}{300}$ esu
Resistance, R	ohm	10^9 emu, $\frac{1}{9} \times 10^{-11}$ esu
Conductivity, σ	mho/meter	10^{-11} emu, 9×10^9 esu
Electric field, E	volt/meter	10^6 emu, $\frac{1}{3} \times 10^{-4}$ esu
Electric displacement, D	coulomb/sq meter	$4\pi \times 10^{-5}$ emu, $4\pi \times 3 \times 10^5$ esu
Capacitance, C	farad	10^{-9} emu, 9×10^{11} esu
Polarization, P	coulomb/sq meter	10^{-5} emu, 3×10^5 esu
Inductance, L	henry	10^9 emu, $1/(9 \times 10^{11})$ esu
Magnetic induction, B	weber/sq meter	10^4 emu, $1/(3 \times 10^6)$ esu
Magnetic field, H	amp turn/meter	$4\pi \times 10^{-3}$ emu, $12\pi \times 10^7$ esu
Magnetization, M	weber/sq meter	$10^4/4\pi$ emu, $1/(12\pi \times 10^6)$ esu

For the case of moving media Maxwell's equations are still valid provided a correction is applied to the current density equation (8.1.4) to include a convective term and a correction to the polarization current due to motion of the dielectric.[1]

Equations (8.1.1) to (8.1.8) form the theoretical basis for a good deal of the electrical and magnetic measurement techniques and equipment described in this book. Magnetic field determinations and measurements of the magnetic properties of materials are described in Part 9. Measurements of electric fields and the apparatus associated with the measurement of electrical phenomenon will be described in this and the following chapters. Most of the formulas that follow will be given in the mks system of units. Conversion factors of these formulas to esu or emu units are given in Table I.

[1] For a discussion of these modifications see W. K. H. Panofsky and M. Phillips, "Classical Electricity and Magnetism." Addison-Wesley, Reading, Massachusetts, 1955.

8.1.2. Fundamental Concepts in Electrostatics

The general problem of electrostatics is the evaluation of the field distribution in dielectrics and the determination of the surface charge distribution on conductors with known electric potential or field values. In practice electrostatic problems resolve themselves into measuring electric fields **E**, or differences in potential $(V_1 - V_2)$ between charged conductors, the dielectric constant of insulators, capacitances of conductors, and electric charges on materials. These quantities are interdependent and we shall describe both theoretical and experimental methods for their evaluation and measurement.

Formal electrostatics divides materials into ideal conductors and ideal insulators, a division that is somewhat arbitrary though very useful for the development of electrostatic theory. From a measurement point of view it is important to know whether a given material is to be considered as a conductor or an insulator. A useful criterion for making this determination is the comparison of the "relaxation time" of the material with the time over which an observation is made. The manner in which an initial density of charge ρ_0 placed in a conducting medium decays with time is given by

$$\rho = \rho_0 e^{-\frac{\sigma}{\epsilon} t} \tag{8.1.9}$$

where

$$T_r = \frac{\epsilon}{\sigma} \tag{8.1.10}$$

is the "relaxation time." For copper $T_r \sim 1.5 \times 10^{-19}$ sec; for salt water $T_r \sim 10^{-6}$ sec; for a typical petroleum oil[2] $T_r \sim 2$ sec and for fused quartz $T_r \sim 10^6$ sec.

Returning to Maxwell's equations in the mks units we observe that for electrostatic phenomena

$$\frac{\partial \mathbf{D}}{\partial t} = 0, \qquad \frac{\partial \mathbf{B}}{\partial t} = 0, \qquad \mathbf{J} = 0$$

so that for a homogeneous dielectric

$$\nabla \cdot \mathbf{E} = \rho/\epsilon. \tag{8.1.11}$$

Also, the electrostatic field can be represented as the negative gradient of the electrostatic potential V,

$$\mathbf{E} = -\operatorname{grad} V = -\nabla V. \tag{8.1.12}$$

Surfaces for which $V = \text{const}$ are equipotential surfaces and are orthogonal to the electrostatic flux lines. Since the field **E** within a conductor

[2] W. F. Cooper, *Brit. J. Appl. Phys.* **4**, 571 (1953).

is zero, the electrostatic potential throughout the conductor must be constant. It follows from Gauss' law that any surface charge density q_s (coulombs per square meter) on the conductor and the corresponding electric field $\mathbf{E}$ normal to the surface of the conductor are related by

$$\mathbf{E}_n = \frac{-\partial V}{\partial n}\mathbf{n} = \frac{q_s}{\epsilon}\mathbf{n} \tag{8.1.13}$$

where $\partial V/\partial n$ is the space derivative of V along the outward normal n to the surface of the conductor.

From Eqs. (8.1.11) and (8.1.12) we obtain Poisson's equation

$$\nabla^2 V = -\rho/\epsilon. \tag{8.1.14}$$

If the region is free of charge and the field is produced by charged surfaces or by charges that can be excluded from the region by closed surfaces drawn about them, then Poisson's equation reduces to

$$\nabla^2 V = 0 \tag{8.1.15}$$

the second-order differential equation of Laplace. If, therefore, we are given a set of conductors at potentials $V_1 \ldots V_i$, the determination of the potential field in the region surrounding the conductors resolves itself into finding a solution of Laplace's equation (8.1.15) which reduces to V_1 at the surface of conductor 1; V_2 on conductor 2 and so forth. Such a solution is unique.[3] The corresponding field intensity $\mathbf{E}$, may be obtained by an application of Eq. (8.1.12), and the surface charge density on the surface of a conductor by Eq. (8.1.13).

Mathematical techniques for solving Laplace's equation for certain simple geometrical configuration of conductors and dielectrics are given in such texts as Smythe,[3] Stratton,[3] Weber,[4] and Jeans.[5] Further discussion of experimental methods is given in Section 8.1.5.

8.1.3. Capacitance and Capacitors

In many problems in electrostatics it is useful to know the relation between charges carried by a group of conductors and the potentials of the conductors, for if this relation is known then a measurement of potential will yield information on charges and, conversely, a measurement of charge will yield potentials. The potential on any given conductor

[3] For a complete discussion of these points, the reader may consult: W. R. Smythe, "Static and Dynamic Electricity," 2nd ed. McGraw-Hill, New York, 1950; J. A. Stratton, "Electromagnetic Theory." McGraw-Hill, New York, 1941.

[4] E. Weber, "Electromagnetic Fields," Vol. 1. Wiley, New York, 1950.

[5] J. H. Jeans, "Mathematical Theory of Electricity and Magnetism," 4th ed. Cambridge Univ. Press, London and New York, 1948.

depends linearly on the magnitude and position of all charges present. In the case of a single conductor remote from all other conductors and the earth,

$$Q = CV. \tag{8.1.16}$$

Here, C is the capacitance of the conductor measured in farads if Q is in coulombs and V in volts. In the electrostatic system of units with Q in statcoulombs and V in statvolts, C is measured in centimeters. If the charge distribution on a conductor is given, then the potential V at point P on the surfaces can be computed from basic principles,

$$V = \int_s \frac{q_s\, ds}{\epsilon r} \tag{8.1.17}$$

where q_s is the surface density of charge and r the distance from P to each surface element ds and ϵ the dielectric constant. Such computations can be made for conductors of simple geometries. A few typical formulas are shown in Table II.[6]

TABLE II. Capacitance of Single Conductors Remote from the Earth (mks)

Sphere or radius a	$4\pi\epsilon a$
Oblate spheroid of semiaxes a and c, $a > c$	$4\pi\epsilon[a^2 - c^2]^{\frac{1}{2}}[\tan^{-1}(a^2c^{-2} - 1)^{\frac{1}{2}}]^{-1}$
Circular disk of radius a	$8\epsilon a$
Elliptic disk of semiaxes a and b, $a > b$	$4\pi\epsilon a\{K[(1 - b^2a^{-2})^{\frac{1}{2}}]\}^{-1}$ with K the complete elliptic integral

In the case of several conductors present in an electrostatic field there is no single or unique capacitance that can be assigned to each conductor since the charge on each depends on the potentials of the others.

For example, for a series of n conductors with charges $Q_1 \ldots Q_n$, the potential of the ith conductor is given by

$$V_i = \sum_{j=1}^{n} p_{ij} Q_j \qquad (i = 1 \cdots n) \tag{8.1.18}$$

and inversely the charge on the ith conductor is given by

$$Q_i = \sum_{j=1}^{n} c_{ij} V_j \qquad (i = 1 \cdots n). \tag{8.1.19}$$

The p_{ij} are known as the coefficients of potential; c_{ij} with $i = j$ are called the coefficients of capacitance and the constants c_{ij} with $i \neq j$ the coeffi-

[6] For a more complete compilation see D. E. Gray, ed., "American Institute of Physics Handbook," pp. 5–12. McGraw-Hill, New York, 1957.

cients of induction. The coefficient system p_{ij} and c_{ij} are related as coefficients of a system of linear equations.[7]

To have a precise meaning for a series of conductors, the dimensions, positions, and symmetry of the conductors must be exactly known. However, for the special and important case of two conductors with equal and opposite charges, a capacitance C can be ascribed to this system known as a capacitor. If V_1 and V_2 are the potential and q_1 and $-q_1$ the charge then

$$C = q_1/(V_1 - V_2). \tag{8.1.20}$$

Again for simple geometries of conductors appropriately arranged relative to each other, calculation can be made of the capacitance of the system; formulas for a few geometries are given in Table III.[8] Where

TABLE III. Formulas for Simple Capacitors (mks)

1. Concentric spheres radii a, b, $a > b$	$C = 4\pi\epsilon ab(a - b)^{-1}$
2. Concentric circular cylinders of radii a, b, with $b > a$ and length $L \gg (b - a)$	$C = 2\pi\epsilon \ln (a^{-1}b)L$
3. Confocal elliptic cylinders with semiaxes a, b and c, d with $b > a$, $d > c$, and $a > c$	$C = 2\pi\epsilon[\tanh^{-1}(b^{-1}a) - \tanh^{-1}(c^{-1}d)]^{-1}$
4. Two parallel plates of area A a distance d apart	$C = A\epsilon/d$
5. Parallel circular cylinders with D = separation of centers, radii a, b with $L \gg (b - a)$ and $L \gg D$	$C = 2\pi\epsilon L \left\{\cosh^{-1}\left[\dfrac{D^2 - a^2 - b^2}{2ab}\right]\right\}^{-1}$
6. Cylinder and infinite plane; with h the separation of center of cylinder and plane and $L \gg h$ and a radius of cylinder.	$C = 2\pi\epsilon L \left[\cosh^{-1}\dfrac{h}{a}\right]$

such absolute capacitors are required for precision experimental work (0.1% for 1 $\mu\mu$f air capacitor) the reader should consult the several papers of the National Bureau of Standards,[9] for construction details.

For modest voltage measurement in the laboratory (less than 1 kv) multiple plate, adjustable air capacitors are generally employed. Such capacitors require calibration and their accuracy is dependent on the closeness with which the angular position of the multiple plates with respect to the fixed plates can be maintained. Well-constructed com-

[7] J. C. Maxwell, "A Treatise on Electricity and Magnetism," 3rd ed., Vol. 1, Section 87. Clarendon Press, Oxford, 1892.

[8] For more complete compilation see C. Snow, "Formulas for Computing Capacitance and Inductance." *Natl. Bur. Standards (U.S.), Circ.* **544** (1954).

[9] C. Moon and C. M. Sparks, *J. Research Natl. Bur. Standards* **41,** 497 (1948); and C. Snow, *ibid.* **42,** 287 (1949).

mercial precision air-dielectric capacitors have accuracies of about ± 0.1 to $\pm 0.2\%$ over the ranges of 1 to 1000 $\mu\mu$f provided due care is given to the manner in which connections are made to the capacitor and the shielding that is used. Quartz insulation is used with these standard capacitors with insulating resistance of about 10^{12} ohms.

Capacitors used as precision circuit elements are usually of the solid-dielectric type and consist of thin mica sheets interleaved with metal foil. Such capacitors are available in sizes ranging from 10 $\mu\mu$f to 1 μf with accuracies of about $\pm 0.5\%$ and can withstand peak voltages of the order of 500 to 1000 volts.

If the capacitors described are to be used with low-frequency alternating voltages with angular frequency ω, it is necessary to take into account dielectric losses and phase angle changes (see Section 8.5.4). The equivalent lumped constant circuit of a variable air capacitor[10] is shown in Fig. 1, where R represents the resistance of the leads and plates, L the

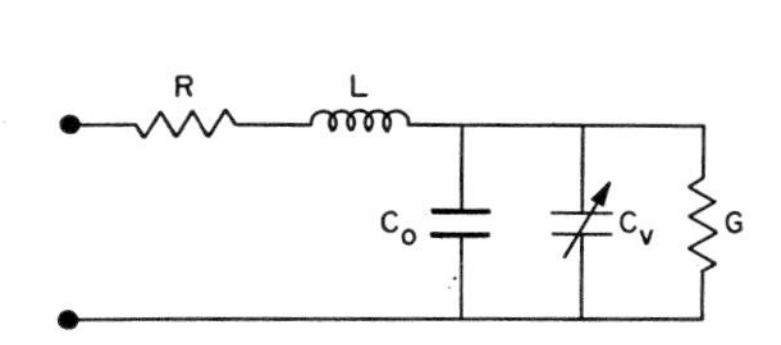

FIG. 1. The equivalent lumped constant circuit of the variable capacitor C_v.

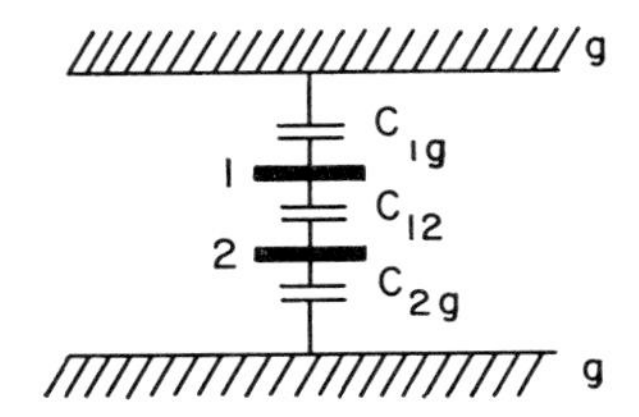

FIG. 2. The equivalent circuit of an air-capacitor showing interelectrode capacitance and stray capacitance to ground.

series inductance of the structure, C_v the variable low-frequency capacitance, C_0 the capacitance of the dielectric supporting structure, and G the total conductance including dc leakage conductance and dielectric losses in the supports. At low frequencies the capacitance at the terminals is $C_0 + C_v$. For dc uses the leakage conductance is the important factor. For audio-frequencies the dielectric losses in the insulating structure are important. As the frequencies increase to the radio range and beyond, the impedances of R and L become significant in comparison with the reactance C_v and must be taken into account.[11]

In practice a capacitor represents a system of conductors—the plates of the capacitor and the ground or shield that is provided. The total capacitance, therefore, is that of the plates plus the capacitance between the plates and ground as shown in Fig. 2. The total or working capacitance

[10] R. F. Field and D. B. Sinclair, *Proc. I.R.E.* **24,** 255 (1936).

[11] A. V. Astin, *J. Research Natl. Bur. Standards* **22,** 673 (1929); F. W. Grover, *Natl. Bur. Standards (U.S.) Bull.* **7,** 495 (1911).

C between conductors 1 and 2 then is given by

$$C = C_{12} + \frac{C_{1g}C_{2g}}{C_{1g} + C_{2g}} \qquad (8.1.21)$$

where C_{1g} and C_{2g} represent capacitances between the plates and ground. In this analysis it was assumed that the air capacitor was entirely shielded, an objective of a well-constructed capacitor. In practice this is not exactly true and therefore a certain indefiniteness is present in the value of C which may amount to several micro-microfarads.

It has been assumed in the discussion above that each conductor of a capacitor is an equipotential surface and that the earth and inter-capacitances are definitely localized quantities. There are many important cases in practice where these conditions are not fulfilled and the concept of a localized capacitance must be modified. For example, in the case of a coil there is a distributed capacitance from turn to turn and from turn to ground. The concept of distributed capacitance among conductors is especially pertinent when ac voltages are used (see Section 8.5.7).

Measurement of capacitance by bridge methods is given in Sections 8.3.2 and 8.5.6.

The construction of high-voltage capacitors where the voltage range is approximately 10 kv to 1000 kv presents the additional problems of dielectric breakdown and corona losses. Because of the large spacings required for the electrodes such capacitors are of small capacitance (10 to 100 $\mu\mu$f) and large bulk. Construction details for high voltage capacitors may be found in the papers by Schering and Vieweg[12] and Keller.[13]

8.1.4. Electrostatic Potential

The value of electrostatic potential may be measured in terms of the fundamental units of mass, length, and time by means of an absolute voltmeter or electrometer. This instrument measures charge or potential difference in terms of mechanical units or it can be used to measure capacitance and electrical currents.

The Kelvin absolute voltmeter illustrated schematically in Fig. 3 consists essentially of a parallel plate condenser in which the central portion of one plate is movable. With a given difference of potential V (volt) between the plates, the central disk of area A (square meter) and at a distance d (meter) from the lower plate is attracted to the lower plate

[12] H. Schering and R. Vieweg, *Z. Tech. Physik* **9**, 442 (1928).

[13] A. Keller, "Precision Electrical Measurements," Paper 19, Philosophical Library, New York, 1956.

with a force F (newton)

$$F = \frac{\epsilon_0}{2} \frac{V^2 A}{d^2} \tag{8.1.22}$$

where ϵ_0 is the absolute dielectric constant of the medium and equal to 8.85×10^{-12} farads/meter for vacuum. If the central disk is restrained from moving by gravitational force the value Mg then

$$Mg = \frac{\epsilon_0}{2} \frac{V^2 A}{d^2} \quad \text{or} \quad V = \sqrt{\frac{2Mg}{\epsilon_0 A}}\, d \tag{8.1.23}$$

giving the potential in terms of measurable mechanical quantities of mass, length, and time. With careful construction the attracted-disk voltmeter can measure voltages up to 250 kv or more with accuracies of the order of 0.02%. Such absolute measurements are tedious and not used for routine laboratory measurements,[14] though in some instances they have

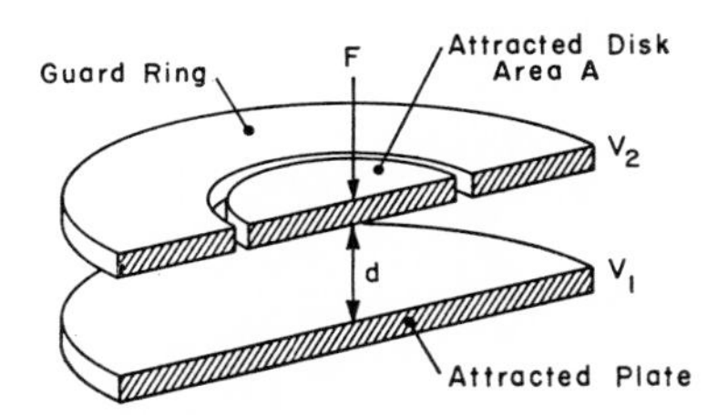

FIG. 3. Schematic drawing of the Kelvin absolute electrometer.

been employed for high potential measurements. In their stead, calibrated electrometers or voltmeters are used.

For the measurement of small differences of potential of less than 1 volt an electrometer is generally employed, particularly if it is desired to draw essentially no current. The basis for measurement is the displacement of a charged conductor in an electrostatic field. Of the many electrometers that have been described in the literature, perhaps the simplest practical electrometer is the string electrometer of which the Wülf[15] shown diagrammatically in Fig. 4 is typical. The instrument consists essentially of a fine fiber of sputtered quartz or a Wollaston wire stretched at right angles to an electric field produced by knife-edge electrodes. When an unknown potential V_x is applied to the wire it is displaced toward one or the other electrode depending on the polarity. The displacement is viewed with a microscope or is projected onto a

[14] For a detailed description of the absolute voltmeter, see H. B. Brooks, F. M. Defandorf, and F. B. Silsbee, *J. Research Natl. Bur. Standards* **20,** 253 (1938).

[15] T. Wülf, *Physik. Z.* **15,** 250, 611 (1914).

screen or photographed. In the Cambridge and Paul instrument the capacitance is about 2 $\mu\mu$f and voltage sensitivity is about 0.01 volts per division with 45 volts between the parallel plates. In the Lutz–Edelman type,[16] the wire is in the plane of two knife-edge electrodes and its lower end is fastened to a quartz filament bow. The capacitance is between 1.5 and 7.0 $\mu\mu$f and the sensitivity 0.001 to 100 volts per division. The insulation in these instruments is amber or fused quartz. The advantages of the string electrometer are compactness, small capacitance, wide sensitivity range, portability, and a short period.

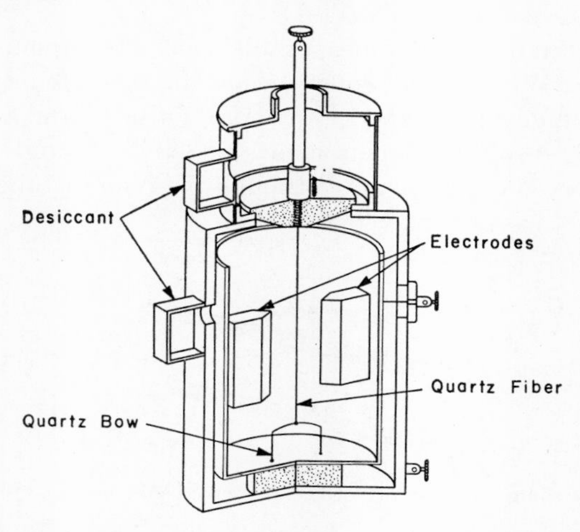

FIG. 4. Cross section of a Wülf-type string electrometer.

A widely used electrometer is the Dolezalek quadrant electrometer shown in Fig. 5. This instrument consists of a cylindrical box divided into insulated quadrants. A light metal sector-disk or needle is suspended in the hollow enclosure by means of a sputtered quartz fiber or Wollaston wire. Rotation of the fiber is measured by a lamp and scale arrangement similar to that used with a galvanometer. Diagonally opposite pairs are electrically connected. If V_1, V_2, and V_n are the potentials of a symmetrically displaced quadrant pairs and needle respectively, then the angle θ at which the needle rotates is given by

$$\theta = K\left[(V_1 - V_2)\left(V_n - \frac{V_1 + V_2}{2}\right)\right]. \tag{8.1.24}$$

The assumption is made that when the quadrants are grounded the position of the needle is the same whether charged or uncharged, that is,

[16] C. W. Lutz, *Physik. Z.* **17**, 619 (1916).

that the electrical and mechanical zeros are the same.[17] The Dolezalek electrometer has a capacitance of about 50 to 100 $\mu\mu$f and a sensitivity of about 0.5×10^{-3} volts per division on a scale 1 meter distant.

In the Compton electrometer[18] a certain amount of dissymmetry is deliberately introduced between the needle and quadrants by displacing one quadrant vertically with respect to the others and by also tilting the needle from a horizontal plane. By this modification Compton was able

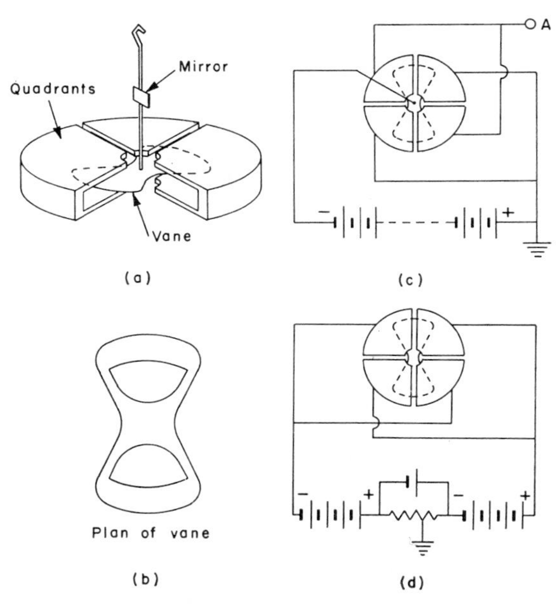

FIG. 5. (a) Cut-away of the Dolezalek electrometer showing position of quadrants and vane; (b) plan view of vane; (c) electrometer connected with the needle at high potential and charge to be measured to one set of quadrants at A; (d) fixed voltages are placed across the pairs of quadrants and the charge to be measured on the needle.

to obtain a sensitivity of about 0.5×10^{-5} volts per division with a capacitance of from 12 to 15 $\mu\mu$f.

Perhaps the greatest working sensitivity is that provided by the Hoffman electrometer.[19] In this instrument the quadrants are replaced by binants and the symmetrical needle with an unsymmetrical half-sector one as shown in Fig. 6. The voltage to be measured is applied to the needle and a fixed voltage placed across the binants. With a capacitance of about

[17] F. A. Laws, "Electrical Measurements," 2nd ed., p. 234. McGraw-Hill, New York, 1938.

[18] K. T. Compton and A. H. Compton, *Phys. Rev.* **14,** 85 (1919).

[19] G. Hoffman, *Physik. Z.* **13,** 480, 1020 (1912).

2 $\mu\mu$f a sensitivity in excess of the Compton electrometer has been obtained. If the capacitance is known small currents can be determined by measuring voltage changes. With the Hoffman electrometer currents of 10^{-19} amp have been measured with special attention to experimental details. Currents of 10^{-17} amp can be measured without any special precautions. An excellent review of experimental procedures to be followed with these electrometers is given by Neher.[20]

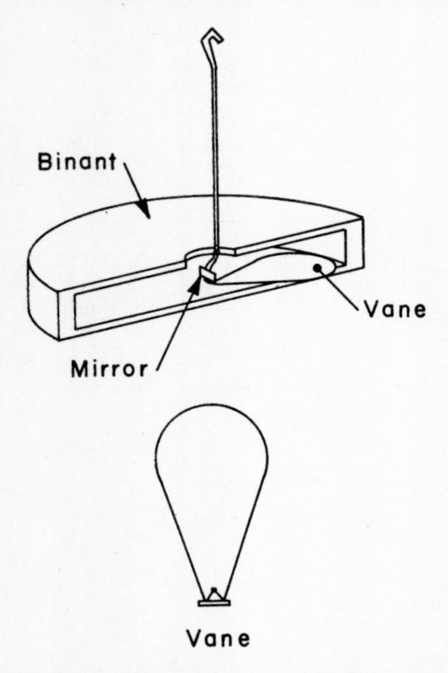

FIG. 6. Cut-away of the Hoffman electrometer showing a binant and half-sector vane.

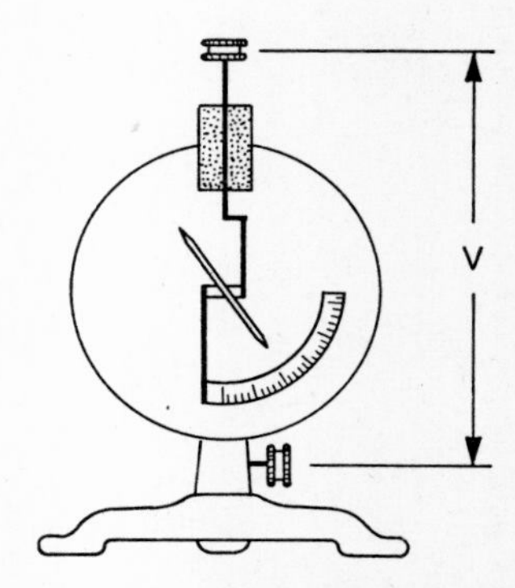

FIG. 7. Braun-type electrostatic voltmeter.

For a rough measurement of voltages in the range from 100 to about 10,000 the Braun-type electrostatic voltmeter is adequate.[21] This instrument is essentially a modification of the gold leaf electroscope with a leaf replaced with an aluminum metal (see Fig. 7). Although the instrument is not useful for precision work it is rugged and can be made with a very small capacitance of the order of a few micro-microfarads.

Techniques for the measurement of very high voltages, 100 kv to 1000 kv, can be divided into two classes—those that use low voltage instruments in conjunction with voltage dividers and those that measure

[20] J. Strong, H. V. Neher, A. E. Whitford, C. H. Cartwright, and R. Hayward, "Procedures in Experimental Physics," Chapter 6. Prentice-Hall, Englewood Cliffs, New Jersey, 1938.

[21] F. K. Harris, "Electrical Measurements." Wiley, New York, 1952.

the voltages directly. In the former class low voltage electrostatic voltmeters of about 100 volt range are used with carefully designed wire-wound resistors for precision work (accuracies of 1 in 10^4); for less precise work ceramic type resistors of good stability can be used. The use of direct voltage measuring techniques using a modification of the Kelvin absolute voltmeter is described by Browdler[22] and methods using sphere-gap instruments are described by Edwards.[23]

8.1.5. Electrostatic Fields

A direct determination of the magnitude and direction of the electrostatic field in terms of the defining equation, force per unit charge ($\mathbf{E} = \mathbf{F}/q$), is infrequently done. Rather, the electrostatic field is deduced from a knowledge of the potential field in accordance with Eq. (8.1.12)

$$\mathbf{E} = -\nabla V \tag{8.1.25}$$

or in terms of components

$$E_x = -\frac{\partial V}{\partial x}, \qquad E_y = -\frac{\partial V}{\partial y}, \qquad E_z = -\frac{\partial V}{\partial z}. \tag{8.1.26}$$

As we have indicated in Section 8.1.2, to determine the potential field produced by a set of conductors, it is necessary to solve the Laplace differential equation:

$$\nabla^2 V = \frac{\partial^2 V}{\partial x^2} + \frac{\partial^2 V}{\partial y^2} + \frac{\partial^2 V}{\partial z^2} = 0. \tag{8.1.27}$$

The potential must satisfy proper boundary conditions corresponding to the shape and potentials of the conductors. The scientific literature is rich in analytical methods and techniques for solving Laplace's equation. The reader should consult these texts for details.[3–5,24,25]* We shall, however, outline some of the experimental methods for measuring potential fields.

8.1.5.1. Direct Methods. To measure the potential V_p at some point in space we may place a sphere whose radius r is small compared with the

* See also Vol. 4.

[22] G. W. Browdler, "Precision Electrical Measurements," Paper 17. Philosophical Library, New York, 1956.

[23] F. S. Edwards, "Precision Electrical Measurements," Paper 20. Philosophical Library, New York, 1956.

[24] P. M. Morse and H. Feshbach, "Methods of Theoretical Physics," Vol. 2. McGraw-Hill, New York, 1953; L. M. Milne-Thomson, "Theoretical Hydrodynamics." Macmillan, London, 1938; O. D. Kellogg, "Foundations of Potential Theory." Springer, Berlin, 1939.

[25] R. V. Southwell, "Relaxation Methods in Theoretical Physics." Oxford Univ. Press; London and New York, 1956.

distance to the charged conductor at the point in question. By means of a fine wire the sphere is grounded, then insulated and carried to an electrometer completely screened and at ground potential. The potential of the sphere as measured by the electrometer is equal but of opposite sign to V_p.

In another method an isolated uncharged sphere is placed at the point at which the potential is to be measured as shown in Fig. 8. The sphere will take on the local potential V_p after a separation of charges on the sphere with, however, the resultant charge still zero. If the sphere is now connected by a fine wire to one pair of quadrants of an electrometer as shown in Fig. 8, the charge will be redistributed between the sphere and the electrometer for now the capacitance of the electrometer C_e is in parallel with the capacitance of the probe relative to the charged conductors. As a result the probe will have an excess of positive charge and will therefore distort the local field; the electrometer will not measure the

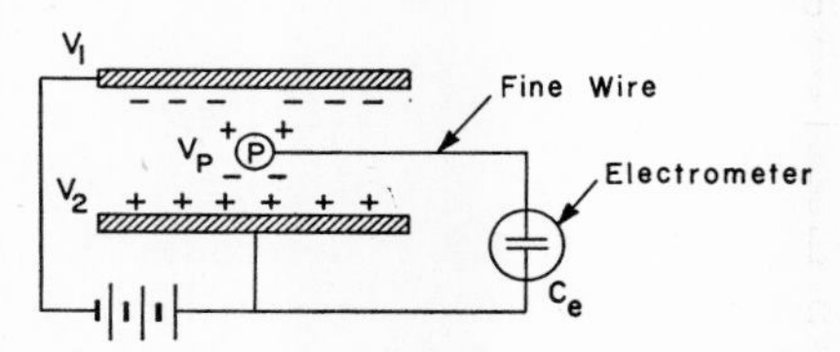

FIG. 8. Measurement of an electric field by means of small spherical probe.

correct potential. If the excess positive charge is dissipated by a suitable mechanism the electrometer and probe can then be made to read correct potentials. By moving the probe so as to maintain constant potential one can map out equipotential surfaces.

Various techniques have been evolved to dissipate excess charges on probes. Probes have been made in the form of a tiny Bunsen burner in order that the hot gases would provide an automatic charge dissipater. The amount of charge that leaves the burner depends on the difference in potential that exists between burner probe and surrounding space. When the probe is at the potential of the surrounding space no further charge is lost. Maxwell[26] suggested the use of small water sprays or a discharge from a fine point. Moulin[27] has summarized the effectiveness of some of these older techniques as potential equalizers. These techniques have been greatly improved and the reader should consult the work of Seminoff and Walther[28] who developed a technique of the incandescent probe,

[26] J. C. Maxwell, "A Treatise on Electricity and Magnetism," Vol. 1, p. 340. Dover Publications, New York, 1953.

[27] M. Moulin, *Ann. chim. et phys.* [8] **10,** 40 (1907).

[28] N. Seminoff and A. Walther, *Z. Physik* **17,** 67 (1923); **19,** 136 (1923).

Langmuir[29] who devised an emission probe for electrostatic fields in vacuum, and Street[30] who used the vacuum tube electrometer in conjunction with a probe for field mapping. The use of probes for potential measurement in plasma is described by von Engel and Steenbeck.[31] The obvious advantage of the direct approach to potential space measurement is that it permits measurement of fields about the apparatus under working conditions. The serious disadvantage is the disturbance of fields due to the probe and connecting leads.

The electrostatic field normal to the surface of a conductor can be determined by a measurement of surface charge density q_s. As shown by Eq. (8.1.13′)

$$E_n = q_s/\epsilon \tag{8.1.13′}$$

where q_s is the charge density in coulombs per square meter and ϵ is the absolute dielectric constant. The charge density can be approximated by placing a small insulated metallic disk in direct contact with the charged conductor so that the disk will assume the conductor potential and will have a charge density determined by the original conditions. The disk is then removed normal to the conductor so that the charge on the disk is equal to the charge on the area of the conductor covered by the disk. The charged density q_s of the conductor can then be approximated. For the case of a small circular disk probe of radius r and thickness $d \ll r$ Maxwell[26] (p. 344) deduced an approximate relation for the true density of charge q_s of a plane surface in terms of a measured surface charge density of the probe q_s',

$$q_s = q_s' \left[1 + \frac{8t}{r} \ln \frac{8\pi r}{t} \right]. \tag{8.1.28}$$

8.1.5.2. Use of the Electrolytic Tank. The mathematical problem of determining an analytical potential map for a system of charged conductors is very difficult. Although there are numerical techniques[32] that have been developed by which the potential distribution can be obtained for any boundary shape with the required accuracy, such methods are generally time consuming. Fortunately the use of an electrolytic tank has yielded useful potential maps of rather complicated systems of charged

[29] I. Langmuir, *J. Franklin Inst.* **196,** 751 (1923).

[30] R. Street, *J. Sci. Instr.* **23,** 203 (1946).

[31] A. von Engel and M. Steenbeck, "Elektrische Gasentladungen. ihre Physik und Technik," Vol. 2, §14. Springer, Berlin, 1934; A. von Engel, "Ionized Gases," p. 262. Oxford Univ. Press, London and New York, 1935.

[32] H. Leibmann and S. B. Bayer, *Sitzber. math.-physik. Kl. bayer. Akad. Wiss. München* **1918,** 385 (1918); G. H. Shortley and R. Weller, *J. Appl. Phys.* **9,** 334 (1938); **18,** 116 (1947); M. Kormes, *Rev. Sci. Instr.* **14,** 248 (1943).

conductors. This method has been especially useful in studying electric fields about accelerating electrodes of electron guns and electrostatic lenses.

The method consists of immersing models of the electrodes into a tank filled with a weak electrolyte of a proper conductivity and applying appropriate voltages to the electrodes. The potential distribution throughout the liquid is measured with a probe and bridge. A diagram of the basic system is shown in Fig. 9.

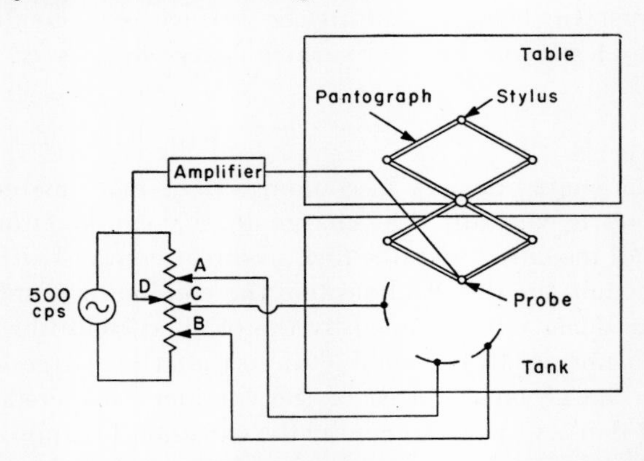

FIG. 9. Basic method of the electrolytic tank for obtaining potential maps of systems of conductors.

The theoretical justification for the method can be seen from the following. From Ohm's law, Eq. (8.1.7), with the conductivity σ as a constant,

$$\mathbf{J} = \sigma\mathbf{E}.$$

For a quasi-static field from Eq. (8.1.4)

$$\nabla \times \mathbf{H} = \mathbf{J}$$

and

$$\nabla \cdot \nabla \times \mathbf{H} = \nabla \cdot \mathbf{J} = 0$$

or

$$\nabla \cdot \sigma\mathbf{E} = 0. \tag{8.1.29}$$

The formal equivalence of Eq. (8.1.29) with Eq. (8.1.1), $\nabla \cdot \epsilon\mathbf{E} = 0$ (with $\rho = 0$), is clear. It follows that the potential distribution within the model satisfies Laplace's equation and, since solutions of Laplace's equation are unique and the boundary conditions are the same for those of the actual electrode system, the potential distribution is the proper one. Details of electrolytic tanks and tracing methods including automatic tracing may be obtained from the works of Hepp,[33] Zworykin,[34] Hutter,[35] and Green.[36]

8.2. Direct-Current Measurements*

8.2.1. General

The general measurement situation applicable to both dc and ac (see Chapter 8.3) is shown in Fig. 10 where the voltage and current of the load are to be determined. In order that these quantities can be determined without error, the ammeter must have zero internal impedance ($Z_A = 0$), and the voltmeter must have zero internal admittance ($Y_V = 0$). Electronic instruments utilizing electron-tube amplification

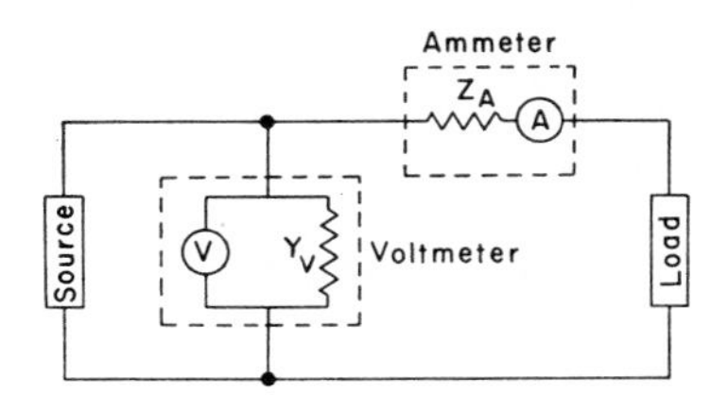

FIG. 10. A general measurement situation.

approach these ideal conditions rather closely. In the past, the accuracy of the electronic instruments has been somewhat poorer than electrodynamic instruments, but recent improvements have largely eliminated differences in accuracy.

Due to finite ammeter impedance, a correction must be made for the voltage drop thereof; likewise due to finite voltmeter admittance, a correction for the current thereof must be made. The necessary corrections can be made if the instrument parameters are known. The relative positions of the voltmeter and ammeter should be chosen so that the correction required is minimum.

Instruments require a small but significant energy to produce a readable change, and, generally, significant power to sustain the reading.

* See also Vol. 2, Section 9.5.1.

[33] G. Hepp, *Philips Tech. Rev.* **4,** 223 (1939).

[34] V. K. Zworykin, G. A. Morton, A. Ramberg, E. G. Hillier, and A. W. Vance, "Electron Optics and the Electron Microscope." Wiley, New York, 1945.

[35] R. G. E. Hutter, *J. Appl. Phys.* **18,** 797 (1947).

[36] P. E. Green, *Rev. Sci. Instr.* **19,** 646 (1948).

It is often desirable to maximize the change per unit energy factor of an instrument. It is for this reason that magnifiers as exemplified by galvanometer light beams, etc. are employed. Ultimately noise in one form or another determines the limits of measurement sensitivity,[37,38] and the signal energy to noise energy ratio of a given measurement task is of considerable importance. Instrument factors such as accuracy, reliability, ruggedness, etc. can be exchanged for sensitivity if the signal-to-noise ratio is more than adequate. If a measurement situation only requires the detection of a signal, the aforementioned ratio should be optimized by designing or choosing the measurement instrument so as to have an internal impedance which is the conjugate of the signal source internal impedance.

The quality of an instrument is often specified by its accuracy rating. Instrument accuracy, generally expressed in per cent of total-scale deflection, specifies the maximum scale error of the instrument when used under reference conditions.[39] Maximum error of a given measurement can be determined by:

(1) Multiplying the instrument accuracy by the scale length to determine the scale error.

(2) Adding the reading scale error to the instrument scale error to get the total scale error.

(3) Translating the total scale error to the reading point to determine the corresponding measurement error at that point.

The following factors may adversely affect the accuracy of an instrument: lead resistance, leakage, instrument position, instrument vibration, ambient temperature, temperature gradients, internal heating, external electric and magnetic fields, nearby magnetic materials, frequency, and waveform. Care must be exercised to eliminate in so far as possible these disturbing factors. A Julius suspension can be employed if vibration is a problem.[40]

Relevant material on Principles of Measurement will be found in Part 1. A publication containing a number of basic principles that are important in the operation of standardizing laboratories is available.[41]

[37] B. Barnes and S. Silverman, *Revs. Modern Phys.* **6**, 162 (1934).

[38] A. van der Ziel, "Noise." Prentice-Hall, Englewood Cliffs, New Jersey, 1954.

[39] American Standards Association, "Electrical Indicating Instruments: Panel, Switchboard and Portable Instruments." American Standard No. C 39.1-1955 (1955).

[40] F. A. Laws, "Electrical Measurements," pp. 41–42. McGraw-Hill, New York, 1938.

[41] F. B. Silsbee, "Suggested Practices for Electrical Standardizing Laboratories." *Natl. Bur. Standards (U.S.) Circ.* **578** (1956).

8.2.2. Current

A variety of principles have been used to measure direct current. These include: hot wire (force due to thermoexpansion); polarized and moving iron-vane (force on a ferromagnetic material in a magnetic field gradient); electrodynamic (force between two current conductors); and moving coil (force on a current conductor in a permanent magnetic field). The Kelvin balance used for primary calibration of current in terms of mechanical quantities employs the electrodynamic principle of force. The large majority of dc measurements are now made with moving coil (D'Arsonval) instruments. These instruments utilize the Faraday's law that the force F acting on a conductor of length L carrying a current I in a magnetic flux density B, is $F = BIL$. The flux density, produced by a permanent magnet, is made as large as possible—usually about 1 weber per square meter—consistent with a high order of stability with time. Since the resistance of the moving coil is proportional to L, the voltage across the moving coil in a given instrument can, in principle, be maintained constant independent of the current range being measured. In practice for a variety of reasons, the voltage of a dc instrument may vary considerably with the current range. A typical laboratory portable instrument has the following characteristics: $I = 10$ μa full scale; $R = 4000$ ohms; $V = 0.04$ volts; full-scale deflection energy $= 20 \times 10^{-7}$ joules; sustaining power $= 0.4$ μwatts; and accuracy $= \pm 0.5\%$ of full scale. These characteristics will change significantly as a function of the ruggedness of the indicating system suspension: i.e., galvanometers with fine wire supplying the restoring torque as compared with inexpensive panel instruments with relatively stiff spiral springs providing the restoring torque. Laboratory portable instruments are currently available which combine the high sensitivity (1 μa full scale with 0.005 μwatts sustaining power) of the galvanometer with the accuracy, portability, and ruggedness of conventional instruments.

The dynamic characteristics of an instrument are characterized roughly by a mass-spring-damping system; detailed solutions are available.[42] Ballistic galvanometers (Section 8.2.3) and vibration galvanometers (Section 8.3.1) are designed for special dynamic characteristics.

Shunts are used to change the range of dc instruments. If the instrument is to carry 1/10, 1/100, etc. of the total current, a shunt connected directly across the instrument terminals must have a resistance which is 1/9, 1/99, etc. of the instrument resistance. The temperature characteristics, temperature environment, lead resistance, and contact resistance

[42] F. A. Laws, "Electrical Measurements," pp. 14–21; 29–38, 95–114. McGraw-Hill, New York, 1938.

of the shunt must be considered. A simple shunt will change the damping of the associated instrument. An Ayrton-Mather universal shunt,[43] often used with multirange dc instruments, has the advantage that damping remains essentially constant if the resistance of the circuit external to the instrument is large compared to the shunt resistance.

A battery-resistance arrangement as shown in Fig. 11 can be employed to compensate for the voltage across the current instrument so as to approximate zero resistance. With this arrangement, the sensitivity of current measurement is determined by the null-detecting galvanometer, but the accuracy is determined by the current instrument.

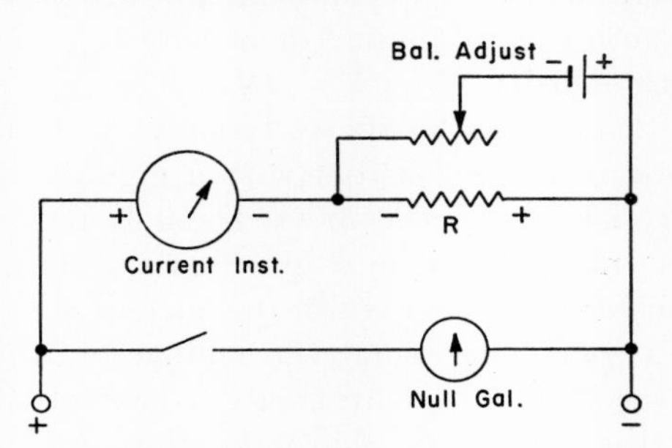

FIG. 11. Zero-resistance direct-current instrument.

Amplifiers have had a pronounced effect on measurements. The simplest arrangement is to use the amplifier to provide current amplification and thus increase the current sensitivity of the indicating instrument. As such, its role is similar to an optical (or similar) magnifier, but in contrast therewith an amplifier is capable of providing power amplification also. However, the basic limitations (see Section 8.2.1 above) of finite indicating energy, sustaining power, and noise remain. Noise, particularly in the form of drift, will be significant. A typical dc amplifier instrument has the following characteristics: $I = 3 \times 10^{-7}$ to 10^3 μa full scale in 20 ranges; $R = 10^{11}$ to 30 ohms; $C = 35$ $\mu\mu$f; $V = 0.03$ volts; input energy for full-scale deflection $= 1.6 \times 10^{-14}$ joules; sustaining power $= 9 \times 10^{-15}$ to 3×10^{-5} watts; zero drift $= \pm 0.003$ volts/hr after warmup; accuracy $= \pm 10$ to $\pm 3\%$ of full scale immediately following zero recheck. Direct current amplifier instruments generally have a multiplicity of ranges and can be grossly overloaded without damage; one of the measurement terminals must generally be at ground potential.

One method of circumventing the drift problem of dc amplifier instru-

[43] F. A. Laws, "Electrical Measurements," pp. 43–44. McGraw-Hill, New York, 1938.

ments is to use modulation[44] (chopper) to convert the dc into ac followed by an ac amplifier and synchronous demodulator.[45] Often a feedback system is employed to provide a null (zero resistance) current measurement and digital or recorded presentation of the reading. Typical characteristics are: $I = 10^{-3}$ to 2 μa full scale in 11 ranges; $R = 1000$ to 0.5 ohms; $C = 2$ μf; $V = 10^{-6}$ volts; input energy for full-scale deflection $= 10^{-18}$ joules; sustaining power $= 10^{-15}$ to 2×10^{-12} watts; zero error $= \pm 2\%$ of full scale after warmup; accuracy $= \pm 2\%$ of full scale immediately following zero recheck.

8.2.3. Charge*

Charge, Q, the time integral of current, is generally measured indirectly by measuring a voltage V across a capacitor C and using the relation: $Q = CV$. The instrument used for the voltage measurement must have a known or negligibly small input capacitance and input current. Electrostatic and amplifier instruments (see Section 8.2.4) generally satisfy these requirements.

In those cases where charge stored in a capacitor is to be determined, a ballistic galvanometer[46] may be employed provided that it is permissible to charge or discharge the capacitor for measurement purposes. The charge or discharge time must be small compared to the response time of the ballistic galvanometer, as the principle of operation is to have the discharge impulse (proportional to Q) produce a proportional angular momentum. If the galvanometer damping is negligible, the maximum deflection angle (point of zero kinetic energy and maximum potential energy) can be directly related to the charge transferred at zero time (point of maximum kinetic energy and zero potential energy). If damping is not negligible, a somewhat more complex relation exists between maximum deflection angle and charge.

In cases where charge measurement is to be made under variable current conditions and for specified time intervals, the integration, $Q = \int_{t_1}^{t_2} I\,dt$, must be performed. One method of doing this is to record the current (recording ammeter) and carry out the integration by well-known methods. A second method is to use an energy measuring (watt-hour) instrument (Section 8.2.6) operating with constant voltage. A third

* See also Vol. 2, Section 9.5.1.7.

[44] D. G. Tucker, "Modulators and Frequency Changers for Amplitude-Modulated Line and Radio Systems." MacDonald, London, 1953.

[45] A. J. Williams, Jr., R. E. Tarpley, and W. R. Clark, *Trans. Am. Inst. Elec. Engrs.* **67**, 47–57 (1948).

[46] F. A. Laws, "Electrical Measurements," pp. 88–116. McGraw-Hill, New York, 1938.

method is to use an electrolytic cell, determine the mass M (grams) deposited on an electrode by weighing and apply Faraday's law that 96,519.4 coulombs of charge corresponds to a gram-atomic weight of a univalent element or radical. A fourth and possibly most versatile method is the use of electronic integration circuits.[47] These circuits permit integration of fast transients that can be accurately timed by means of auxiliary pulse circuits.

8.2.4. Voltage*

By employing a suitable series resistor, any of the instruments described above for measuring current can be used for measuring voltage. Thus, quite frequently a moving-coil instrument with full-scale deflection for I amp is used with added series resistance forming a total resistance R. For 1 volt, $R = 1/I$ ohms, and this ohms per volt value is often used as a

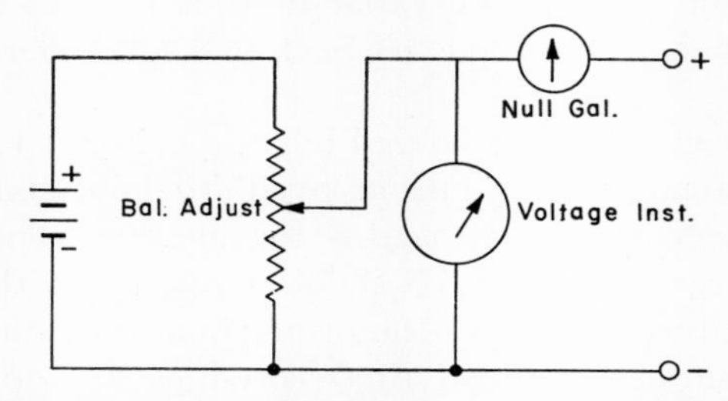

FIG. 12. Infinite-resistance direct-voltage instrument.

specification of this type of voltage instrument. The series resistance is changed to change the voltage corresponding to full-scale deflection. The minimum voltage for full-scale deflection is obtained when the added series resistance is zero; for this condition, the operating characteristics for a typical laboratory portable instrument is the same as for the current instrument indicated in Section 8.2.2. When the same instrument is used with series resistance to measure proportionally greater voltages, the total resistance, full-scale deflection energy, and sustaining power increase proportionally.

A battery-resistance arrangement as shown in Fig. 12 can be employed to compensate for the current taken by the voltage instrument so as to approximate infinite resistance. With this arrangement, the sensitivity of voltage measurement is determined by the null-detecting galvanometer, but the accuracy is determined by the voltage instrument.

* See also Vol. 2, Sections 9.5.1.2–9.5.1.6.

[47] G. A. Korn and T. M. Korn, "Electronic Analog Computers," 2nd ed. McGraw-Hill, New York, 1956.

Voltage instruments operating on the electrostatic principle approach rather closely the ideal infinite resistance. Typical characteristics are: $V = 120$ volts; $R = 10^{14}$ ohms; $C = 220\ \mu\mu f$; input energy for full-scale deflection $= 16 \times 10^{-7}$ joules; sustaining power $= 1.44 \times 10^{-10}$ watts; and accuracy $= \pm 1\%$ of full scale. If the instrument full-scale voltage is increased by a factor K the capacitance C is decreased in principle by a factor $1/K$ with the result that the input energy for full-scale deflection remains unchanged; at high voltages this relationship is altered by the presence of stray capacitance.

Amplifier instruments (Section 8.2.2) are readily applied to voltage measurement. A typical dc amplifier voltage instrument has the following characteristics: $V = 0.03$ to 10 volts full scale in six ranges; $R = 10^{14}$ ohms; $C = 35\ \mu\mu f$; input energy for full-scale deflection $= 1.6 \times 10^{-14}$

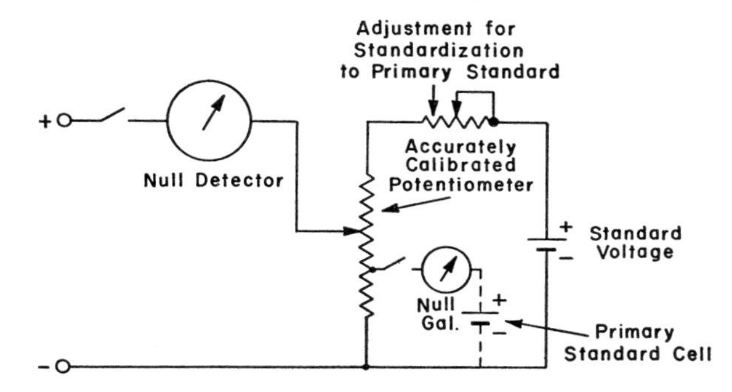

FIG. 13. Basic potentiometer arrangement for voltage measurements.

to 18×10^{-10} joules; sustaining power $= 9 \times 10^{-18}$ to 10^{-12} watts; zero drift $= \pm 0.003$ volts/hr after warmup; accuracy $= \pm 4$ to $\pm 2\%$ of full scale immediately following zero recheck. Typical characteristics of an amplifier voltage instrument employing the modulation-demodulation principle are: $V = 50$ to $2000\ \mu v$ full scale in six ranges; $R = 5 \times 10^{6}$ to 20×10^{6} (maximum) ohms; $C = 24\ \mu\mu f$; $I = 10^{-5}\ \mu a$; input energy for full-scale deflection $= 3 \times 10^{-14}$ to 48×10^{-12} joules; sustaining power $= 50 \times 10^{-11}$ to 2×10^{-8} watts; zero error $= \pm 0.5\ \mu v$ after warm up; accuracy $= \pm 2\%$ immediately following zero recheck.

Voltage measurements of high accuracy are generally made by a comparison technique employing an accurately known voltage ($\pm 0.01\%$), accurately calibrated potentiometer ($\pm 0.05\%$), and a high-sensitivity null detector (see Fig. 13). Generally, the standard voltage is a high-capacity battery which is periodically calibrated against a primary standard cell connected to the measurement terminals. In order to facilitate this periodic calibration without disturbing the measurement circuit, an auxiliary primary standard cell comparison circuit may be permanently built into the potentiometer as shown by the dotted circuit of Fig. 13. A switch arrangement can be provided so that a single null galvanometer serves both purposes.

For measurement of very small voltages, parasitic voltages (thermal, contact, etc.)—particularly those associated with the potentiometer switch or sliding contact—are serious sources of potential error. The basic circuit of Fig. 13 can be modified in the following ways to greatly reduce these potential errors.

(1) The measurement voltage and standard voltage can be interchanged so that the measurement voltage and null detector are connected directly to the ends of the potentiometer, and the standard voltage is

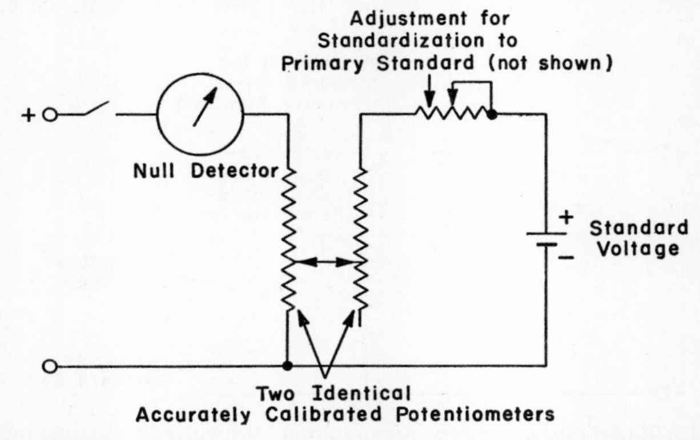

FIG. 14. Simplified potentiometer arrangement for reduction of parasitic voltage effects.

connected in at a variable point by means of the switch or sliding contact. The parasitic voltage associated with the variable point now is no longer directly in the measurement circuit, and errors therefrom are largely eliminated. A simplified version of this arrangement is shown in Fig. 14; two identical accurately calibrated potentiometers are employed so that as the variable point moves the total resistance and current in the standard battery circuit remains constant.

(2) By means of switches or sliding contacts, a high resistance can be shunted across a variable portion of a reference resistor directly connected into the measurement circuit. The parasitic voltages and contact resistance associated with the variable points are now in a high resistance circuit, and errors therefrom are greatly reduced. The adjustment range is generally rather small.

(3) The single null detector switch can be replaced by a reversing switch which permits periodic reversing of the polarity of any unbalanced

signal. In comparison any parasitic voltage in the null detector circuit is not reversed and does not affect the correct establishment of the null point. It is important, however, that the reversing switch be specially designed so that its parasitic voltages are negligible. This can be accomplished by (a) using metal parts whose thermal voltages are very small; (b) designing the switch to be thermally static by arranging adjacent junction pairs so connected that their thermal voltages cancel; and (c) providing a high-conductivity thermal shield in conjunction with thermal insulators so that heat entrance and thermal gradients arising therefrom are reduced. If fractional microvolts are being measured, it is generally advisable to reverse the polarities of both the measurement voltage and standard voltage and to average the resulting measurements. This procedure serves to average out certain residual parasitic voltages.

There are many special purpose potentiometers[48] that are commercially available. Their accuracies are largely determined by the voltage accuracy of the standard voltage. For highest accuracies close temperature regulation of both standard voltage and potentiometer is required.

Potentiometer methods are restricted to the measurement of voltages that are no larger than the standard voltage. Provided that the power handling capabilities of the potentiometer are adequate, the standard voltage can be moderately large; primary standard cells are generally about 1.5 volts. Commercial power supplies are available with moderate power capabilities and with variable output voltages electronically calibrated to $\pm 0.01\%$ against an unloaded internal primary standard cell.

A servo system can be used to automatically and continuously balance a potentiometer. An electronic circuit senses and amplifies the "null" voltage and drives a motor or stepping switch to operate the potentiometer variable point. Measurement read-out may be by: (a) calibrated scale; (b) continuous recording either analogue on chart paper or digital as printed data; or (c) digital display. Maximum resolution is about $\pm 0.1\%$.

High voltages can be measured by potential division techniques or by the use of electrostatic voltmeters which are readily available to about 100 kv. At much higher voltages measurements are frequently made with a variable spark gap. However, even with careful control the voltage at which a spark gap breaks down is somewhat variable. A better technique is to use an electrode arrangement of known field configuration and to operate below breakdown voltage. The voltage can then be computed by measurement of the electrostatic field using methods described in Chapter 8.1.

[48] F. A. Laws, "Electrical Measurements," p. 273. McGraw-Hill, New York, 1938.

8.2.5. Resistance

Resistance is generally measured by applying a voltage, measuring the resulting current, and computing using Ohm's law: $R = V/I$. This operation is performed in multipurpose instruments with the aid of an internal battery; variation in battery voltage is compensated in an approximate manner by adjustment for a standard condition, i.e., a full-scale deflection for zero resistance.

Bridge circuits permit accurate comparison measurement of resistance. The basic factors associated with bridge operation can be understood

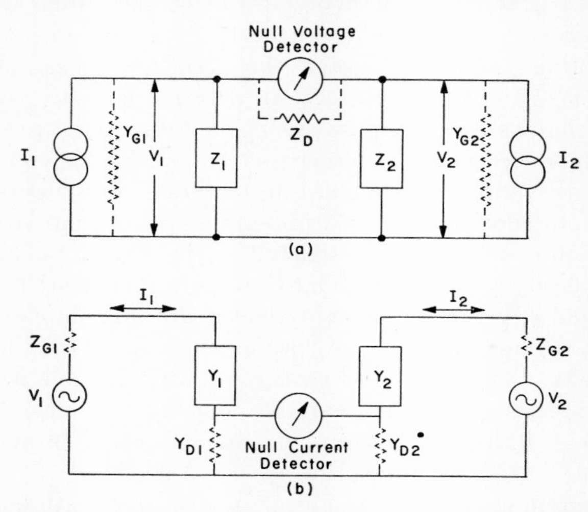

FIG. 15. Ideal bridge configurations: (a) constant-current test signals; (b) constant-voltage test signals.

in terms of ideal bridge configurations shown in Fig. 15. In Fig. 15(a), two constant-current test signals are connected to impedances. Clearly, if $I_1 = I_2$, then $Z_1 = Z_2$ if the null voltage detector indicates that the condition $V_1 = V_2$ exists. Scale factors such that $Z_1 = KZ_2$ can be introduced by choosing $KI_1 = I_2$, by having the voltage detector indicate a null condition when $V_1 = KV_2$, or by both means. For most sensitive bridge operation ($\Delta V = I\,\Delta Z$) it is necessary that the constant-current generators have zero internal admittance, $Y_{G1} = Y_{G2} = 0$, and that the voltage detector have infinite input impedance, $Z_D = \infty$. The ideal bridge configuration of Fig. 15(b) is the dual of that of Fig. 15(a) and employs constant-voltage test signals and a null current detector. Practical bridge circuits approximate more or less the ideal bridge configurations of Fig. 15, and in addition accommodate certain parasitic effects such as stray resistance, inductance, etc.

The well-known Wheatstone bridge, Fig. 16, is developed from the circuit of Fig. 15(a) by replacing the two constant-current sources by their approximate equivalent, a single large voltage source in series with two large resistors (R_3, R_4). The Wheatstone bridge consists of two potentiometers and, when balanced, the condition that $V_1 = V_2$ gives rise to the well-known equation, $R_1R_4 = R_2R_3$. With the aid of this equation, any resistor can be determined if the other three are known. To obtain maximum percentage sensitivity of bridge adjustment, i.e., maximum $\Delta V_1/V_1$ as a function of ΔR_1, R_3 and R_4 must preferably be large compared with R_1 and R_2 as otherwise $V_1 \rightarrow V$ and $V_2 \rightarrow V$ irrespective of bridge adjustment; also, the null voltage detector input resistance must be large, as otherwise $V_1 \rightarrow V_2$ irrespective of bridge adjustment. Increased sensitivity can be obtained by increasing V; the limit is set by

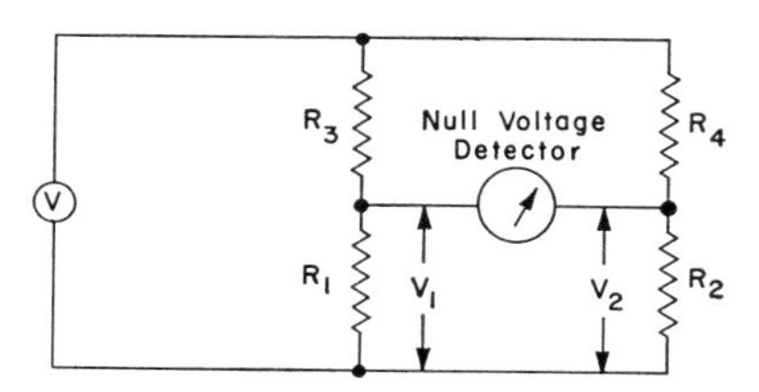

FIG. 16. Basic Wheatstone bridge.

the maximum power of bridge components, or by component nonlinearities. On the other hand if V is fixed, maximum ΔV_1 as a function of ΔR_1 is obtained when $R_3 = R_1$ and $R_4 = R_2$.

Since the Wheatstone bridge is closely related to potentiometers, the material in Section 8.2.4 relevant to parasitic effects is applicable here particularly when low resistances are being measured. A Kelvin or Thomson double bridge is particularly applicable to measurement of low resistance of a section of a conductor.

Very large resistances can be conveniently measured by connecting them in series or shunt with a capacitor (electrostatic voltage instrument) and evaluating the charge or discharge characteristics. The capacitor leakage must be negligibly small or known and accounted for.

Bridge accuracy is largely related to accuracy of standard resistors employed. Resistors invariably have temperature coefficients, aging characteristics, parasitic properties, and noise.

(a) *Temperature Coefficients.* The resistivity of pure metals generally increases with temperature. Resistivity and temperature coefficient thereof can be found in various handbooks. (See, for instance, "Handbook of Chemistry and Physics," 36th ed., pp. 2348–2357. Chemical Rubber, Cleveland, Ohio, 1954.) Special alloys such as manganin (Mn 13%–

balance Cu), Advance (Ni 43%–balance Cu), and Karma (Ni 73%, Cr 20% + Al + Fe) have very low temperature coefficients. (See booklet, "Nichrome," Driver-Harris Co., Harrison, New Jersey, for data). Semiconductors, either in polycrystalline form as in Thermistors[49] or in single crystal form as for germanium,[50] silicon,[51] etc., have large analytically definable, negative temperature coefficients as indicated by the general functional relationship: $\rho = \rho_0 \exp(B/T)$ where ρ_0 and B are constants and T is the Kelvin absolute temperature.

(b) *Aging Characteristics.* High-grade resistors are generally hermetically sealed or coated with a nonhygroscopic wax for protection against humidity and oxidation effects. Ambient temperature changes produce small changes attributable to annealing, and for highest accuracy standard resistors are aged over a long period of time before final calibration.

(c) *Parasitic Properties.* Resistors are never ideal but have associated with them inductance—particularly important for low resistance, and capacitance—particularly important for high resistance. Special construction techniques[52] can be employed for minimizing these effects.

(d) *Noise.* According to fundamental considerations of Nyquist,[53] any resistor of value, $R = 1/G$, in thermal equilibrium with its surroundings at Kelvin absolute temperature T has associated with it a mean-square fluctuation noise voltage, $\overline{V^2} = 4kTR\,\Delta f$, or alternatively a mean-square fluctuation noise current $\overline{I^2} = 4kTG\,\Delta f$; k is Boltzmann's constant, equal to 1.380×10^{-23} joules/°K, and Δf is the frequency band of interest in cycles/sec. Composition resistors generally have more noise, particularly at low frequencies, than indicated by these formulas.

8.2.6. Power, Energy

Power is the product of voltage and current VI and can therefore be determined by making simultaneous voltage and current measurements. Direct measurements are frequently made by means of an electrodynamometer mechanism; the fixed coil carries the current I and produces a magnetic flux proportional to I; the high-resistance moving coil is connected to V. The motive force diminishes as the coil turns, and this results in a nonlinear scale. If small power is being measured, the losses of the instrument must be taken into account. In some cases this com-

[49] J. A. Becker, C. B. Green, and G. L. Pearson, *Trans. Am. Inst. Elec. Engrs.* **65,** 711 (1946); *Bell System Tech. J.* **26,** 170 (1947).

[50] P. G. Herkart and J. Kurshan, *R C A Rev.* **14,** 427 (1953).

[51] E. M. Conwell, *Proc. I.R.E.* **40,** 1327 (1952).

[52] F. E. Terman, "Radio Engineers' Handbook," p. 27. McGraw-Hill, New York, 1943.

[53] H. Nyquist, *Phys. Rev.* **32,** 110 (1928).

pensation is provided by an auxiliary fixed coil in series with the moving coil. Full scale deflection corresponding to 0.05 watts represents a very sensitive instrument. Accuracies of $\frac{1}{4}\%$ are feasible at moderate powers.

A quadrant electrostatic instrument (Chapter 8.1) can be used to measure power by using a dropping resistor to sample the current. Measurements at high voltages are particularly feasible. Under carefully controlled conditions, the temperature of a load resistor can be used to measure its power input.

Energy (watt-hour) instruments most frequently employ a motor with associated arrangement to total the motor revolutions. The motor speed in turn is proportional to power.

8.3. Alternating-Current Measurements—Lumped Circuit*

8.3.1. Voltage, Current

The techniques and equipments used for measurement of either voltage or current are very similar and will be discussed simultaneously. The term signal will be used jointly for both voltage and current. Actually, power (Section 8.3.3) is basically the quantity being measured. If the power measurement is at a high impedance it would generally be considered a voltage measurement; if at a low impedance it would be considered a current measurement.

At low frequencies a galvanometer with a small time constant can be used to measure a signal. A vibration (string) galvanometer[54] has provision for varying the tension of the galvanometer string so that its resonant frequency is equal to the signal frequency.

The most common method of signal measurement uses a rectifier and dc instrument combination. For moderate signals the combination will measure the peak value of the signal if the dc instrument time constant is large; rms calibration is then applicable for a true sinusoidal voltage only, and significant errors will be encountered for complex wave shapes depending upon the magnitudes and phases of the component frequencies. Frequently, an ac electron-tube amplifier with stabilized voltage amplification is used with a rectifier at the output; voltages from about 100 μv to 1000 volts for frequencies from about 5 cycles/sec to 4 Mc/sec can then be measured. Since significant power is available to operate the

* See also Vol. 2, Section 9.5.2.

[54] F. A. Laws, "Electrical Measurements," pp. 35, 465. McGraw-Hill, New York, 1938.

rectifier, the rectifier can be used with a low time-constant load (dc milliammeter) so as to measure the average value of the rectified signal, or alternatively a multiplicity of biased rectifiers can be employed to approximate a parabolic characteristic so as to measure true rms values independent of wave shape. For measurements at higher frequencies it is advisable to rectify immediately by means of a probe arrangement and then use a dc amplifier. By this means, peak measurements of 1 volt full-scale are possible to about 1000 Mc/sec. The rectifier type instruments have accuracies of about $\pm 3\%$.

At low frequencies (up to a few hundred cycles/sec), moderate voltages (about 25 volts minimum), and significant currents (about 5 ma minimum), dynamometer type instruments with stator and moving coils in series can be used for rms measurements with good accuracies (0.25%); moving iron instruments have roughly the same performance. Electrostatic voltmeters (Section 8.2.4) will measure rms values, but the reactive charging current required will increase linearly with frequency. Electrometers (Section 8.1.4) specially constructed for small capacitance can be employed for measurements to about 1000 Mc/sec.[55]

Potentiometer measurements (Section 8.2.4), while feasible, are seldom used due, in part, to the fact that a simple accurate signal is not available. However, the basic comparison technique of the potentiometer is extensively employed particularly for small (microvolts or microamperes) signals. Comparison is usually carried out with the aid of a high-sensitivity, highly stable, and multiband receiver with a signal generator used to establish a reference. In some cases the receiver is calibrated by means of a built-in noise generator.[56] Accuracy, generally determined by the signal or noise generator, is about $\pm 12\%$.

Thermocouple instruments utilize the Seebeck effect (Section 8.7.1); ac power heats up a resistance element in contact with the "hot" thermocouple junction; the resulting thermoelectricity is measured by means of a dc moving-coil instrument. Thermocouple instruments measure true rms values irrespective of the wave shape. With care, accuracies of about $\pm 1\%$ can be obtained up to about 100 Mc/sec. Minimum power requirement for full-scale deflection is about 2 mw. Significant corrections for variations in ambient temperature may be required, particularly if the instrument does not use a temperature compensated junction.

Carefully designed transformers[57] with known ratios are often used to

[55] A. Hund, "High-Frequency Measurements," 2nd ed., p. 169. McGraw-Hill, New York, 1951.

[56] I.R.E. Standards Committee, 1952–1953, "Standards on Electron Devices: Methods of Measuring Noise." *Proc. I.R.E.* **41,** 890 (1953).

[57] F. K. Harris, "Electrical Measurements," p. 542. Wiley, New York, 1952.

extend the range of a measuring instrument. Transformer operation is somewhat frequency limited, as at low frequencies the inductive reactance becomes small compared to the resistance, and at high frequencies capacitance coupling effects predominate. Wide-band astatic transformers can be constructed with the aid of toroidal ferrite cores.[58]

8.3.2. Circuit Components*

Resistors, capacitors, and inductors, together with transformers, form basic circuit building blocks. These components are ideal over only a narrow frequency range at best. Their behavior over a broad range of frequencies can generally be represented by a complex equivalent circuit.[59–62] At a single frequency, this complex circuit will have an impedance, $Z = R + jX$, or an equivalent admittance, $Y = G + jB$, related by the equations:

$$R = \frac{G}{G^2 + B^2}; \qquad X = -\frac{B}{G^2 + B^2}; \qquad G = \frac{R}{R^2 + X^2}; \qquad \text{and}$$
$$B = -\frac{X}{R^2 + X^2}.$$

Formulas are available for computing values of resistance, capacitance, and inductance of a variety of simple configurations.[63–65] Calculations for more complex arrangements can often be carried out by graphical flux plotting techniques (Section 8.1.5.2).[66] At high frequencies, most of the current is carried by the surface of a conductor (skin depth). This has the effect of causing the high-frequency resistance and reactance of a conductor to be greater and smaller respectively than expected on the basis of low-frequency calculations.[67,68]

* See also Vol. 2, Chapter 2.5.

[58] M. M. Maddox and J. D. Storer, *Electronic Eng.* **29,** 524 (1957).

[59] Abd El-Samie Mostafa and M. K. Gohar, *Proc. I.R.E.* **41,** 537 (1953).

[60] U. Tiberio, *Proc. I.R.E.* **42,** 1812 (1954).

[61] C. T. Kohn, *Proc. I.R.E.* **43,** 951 (1955).

[62] H. M. Schlicke, *Proc. I.R.E.* **43,** 174 (1955).

[63] E. B. Rosa and F. W. Grover, "Formulas and Tables for the Calculation of Mutual and Self-Inductance." *Natl. Bur. Standards (U.S.) Sci. Paper* **S-169** (1916); **S-320** (1918).

[64] C. Snow, "Formulas for Computing Capacitance and Inductance." *Natl. Bur Standards (U.S.) Circ.* **C-544** (1954).

[65] F. W. Grover, "Inductance Calculations." Van Nostrand, New York, 1946.

[66] Massachusetts Institute of Technology, Electrical Engineering Staff, "Electric Circuits," p. 16. Wiley, New York, 1940.

[67] S. A. Schelkunoff, "Electromagnetic Waves," p. 260. Van Nostrand, New York, 1943.

[68] L. M. Vallese, *Proc. I.R.E.* **38,** 563 (1950).

Simple volt-ampere measurements can be used for determining the absolute magnitude of an impedance, $|Z|$. Bridge measurements (Section 8.2.5) are extensively used for determining the components of an impedance or admittance. The four arms of the generalized Wheatstone bridge (Fig. 16) are now impedances, and when the bridge is balanced so that $V_1 = V_2$, $Z_1Z_4 = Z_2Z_3$. A wide variety of bridges are possible depending upon the form of the bridge arms.[69,70] Some of the more common types of bridges together with associated balance equations are shown in Fig. 17.

The general-purpose bridge of Fig. 17(a) can be used for a direct comparison of an unknown impedance with a known impedance by choosing $R_3 = R_4$.

Small changes in impedance can be readily determined by balancing the bridge initially and noting the change required to restore balance when the unknown impedance is added. When the generator is a single frequency signal, only a resistive and reactive component of impedance can be determined. However, if the generator has a multiplicity of frequencies, as for instance with a square-wave, pulse, or swept-frequency generator signal, a complex configuration of Z_1 can be compared with an identical configuration of Z_2.[71,72] Thus, a broad-band representation of a complex circuit can be obtained directly without recourse to the usual point-by-point impedance measurement and subsequent analytical evaluation. For accurate bridge measurements careful shielding must be employed so that all current paths are known and confined; stray resistances, capacitances, and inductances must be considered and evaluated when they enter into the balance relations. Any point of the bridge may be grounded as convenient or as dictated for best accuracy. Often it is desirable to ground one end of the unknown impedance. In this event, either generator or detector must be accurately balanced with respect to ground, or their unbalance must be accommodated within the bridge balance equations. A highly astatic toroidal transformer is frequently used to provide a balanced to unbalanced coupling. Alternatively, a split-load electron tube amplifier[72] can be used to provide a balanced generator signal, and a differential amplifier[73] can be used as a balanced detector. If desired, the generator and detector can be interchanged; the balance equations remain the same.

The resonance bridge of Fig. 17(b) can be used to measure the series

[69] B. Hague, "Alternating Current Bridge Methods." Pitman, London, 1946.

[70] F. E. Terman, "Radio Engineers' Handbook," p. 902. McGraw-Hill, New York, 1943.

[71] T. Roddam, *Wireless World* **56,** 8 (1950).

[72] L. J. Giacoletto, *R C A Rev.* **14,** 269 (1953).

[73] W. E. Barnette and L. J. Giacoletto, *Electronics* **28,** 148 (1955).

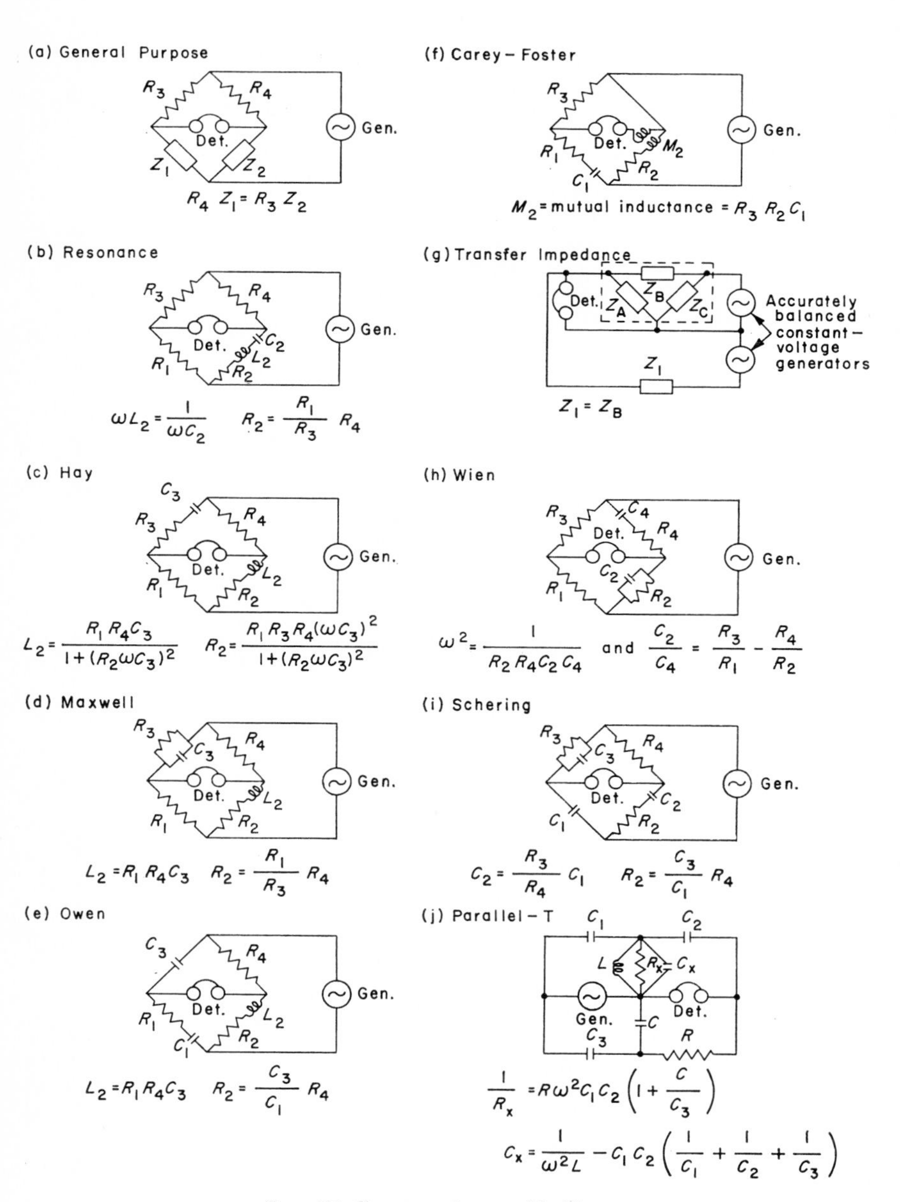

FIG. 17. Common types of bridges.

equivalent circuit of an inductor with the aid of a variable capacitor, or alternatively the series equivalent circuit of a capacitor with the aid of a variable inductor. Similar measurements are possible with the Hay, Fig. 17(c), Maxwell, Fig. 17(d), and Owen, Fig. 17(e), bridges. The latter two are particularly advantageous because the balance equations are independent of frequency.

The Carey–Foster bridge of Fig. 17(f) is particularly adapted to measuring the mutual inductance of a transformer. The transfer impedance bridge of Fig. 17(g) measures a mutual element of a π circuit, as, for instance, the direct resistance or capacitance between two terminals of a three terminal network. Various arrangements of the transfer impedance bridge can be used for measuring active and passive three terminal networks and devices.[74–76]

The Wien bridge, Fig. 17(h), is often used as a frequency measurement circuit. It can also be used to compare a series RC with a shunt RC. This can also be done with the Schering bridge, Fig. 17(i), which bridge has the added advantage that the balance is independent of frequency and can be used for measuring relatively low impedances.[77,78]

The Schering bridge can also be used for measuring direct capacitance between two terminals of a π array of loss-free capacitors. The two terminals are connected across C, with the third terminal connected to the R_3–R_4 junction point. The remaining two capacitors of the π are thus connected across the generator and across C_3.

The Parallel-T circuit[79,80] of Fig. 17(j) has the unique property that generator, detector, and element being measured have a common point that can conveniently be grounded. Null condition across the detector is established when the signal from the generator through the C_1–C_2 path gets shifted and attenuated just enough to be equal and opposite the signal transmitted through the C_3–R path. The unknown admittance to be measured is connected across L. C_x is adjusted to restore balance for measuring the unknown susceptance; R_x could be adjusted to restore balance for measuring the unknown conductance, but it is generally more convenient to adjust C for this purpose.

Often, circuit component measurements are required with current or

[74] W. N. Tuttle, *Proc. I.R.E.* **21,** 844 (1933).

[75] F. E. Terman and J. M. Pettit, "Electronic Measurements," p. 297. McGraw-Hill, New York, 1952.

[76] I.R.E. Standards Committee, 1949–1950, "Standards on Electron Tubes: Methods of Testing." *Proc. I.R.E.* **38,** 917 (1950).

[77] D. B. Sinclair, *Proc. I.R.E.* **28,** 497 (1940).

[78] J. H. Mennie, 1953 Convention Record of the I.R.E., Part 9, p. 79 (1953).

[79] D. B. Sinclair, *Proc. I.R.E.* **28,** 310 (1940).

[80] R. F. Proctor, *Proc. Inst. Elec. Engrs.* (*London*) **99,** *Pt. IV* 47 (1952).

voltage bias. A suitable bias can frequently be introduced in the bridges of Fig. 17 via transformers associated with either the generator or detector. Alternatively, the bias can be introduced directly by series or shunt circuit, and corrections determined for the bias circuit by making separate measurements thereof.

The quality factor Q can be defined in a variety of ways.[81]

(1) In an equivalent series circuit, Q is the ratio of the series reactance to series resistance,

$$Q = \frac{\omega L_s}{R_s} = \frac{1}{\omega C_s R_s}. \tag{8.3.1}$$

(2) In an equivalent parallel circuit, Q is the ratio of the parallel resistance to parallel reactance,

$$Q = \frac{R_p}{\omega L_p} = R_p \omega C_p. \tag{8.3.2}$$

(3) In either a series or parallel circuit,

$$Q = 2\pi \frac{\text{peak energy stored}}{\text{energy dissipated per cycle}}. \tag{8.3.3}$$

(4) In either a series or parallel circuit resonant at f_0, Q can be defined in terms of frequencies above and below resonance, f_2 and f_1 respectively, where the net circuit inductive or capacitative reactance equals the resistance of the circuit,

$$Q = \frac{f_0}{f_2 - f_1}. \tag{8.3.4}$$

The frequencies f_1 and f_2 also correspond to points at which the series circuit current or parallel circuit voltage is $1/\sqrt{2}$ times the value at resonance and to points where the power being dissipated is one-half that at resonance.

(5) The logarithmic decrement δ is the natural logarithm of two successive amplitudes (in the same direction) of free oscillation. Then, $Q = \pi/\delta$.

The Q relations given above approach equality as Q becomes larger.

A Q meter (Fig. 18) is widely used to measure circuit Q.[82] An accurately calibrated variable frequency voltage V' is introduced into a series circuit resonated by means of a low-loss calibrated capacitor. The voltage across the capacitor V measured with a vacuum-tube voltmeter is a direct measure of $Q = V/V'$. The same instrument can also be used to measure

[81] E. I. Green, *Am. Scientist* **43,** 584 (1955).

[82] V. V. L. Rao, *Proc. I.R.E.* **30,** 502 (1942).

inductance (by resonance), capacitance (by substitution to restore resonance), capacitor losses (by calculation), and resistance. The Q meter instrument of the type shown can be employed to about 50 Mc/sec.

The impedances (or admittances) associated with three terminals can be depicted in terms of a T or π network as shown in Fig. 19. The individual elements can be determined by making three independent two-

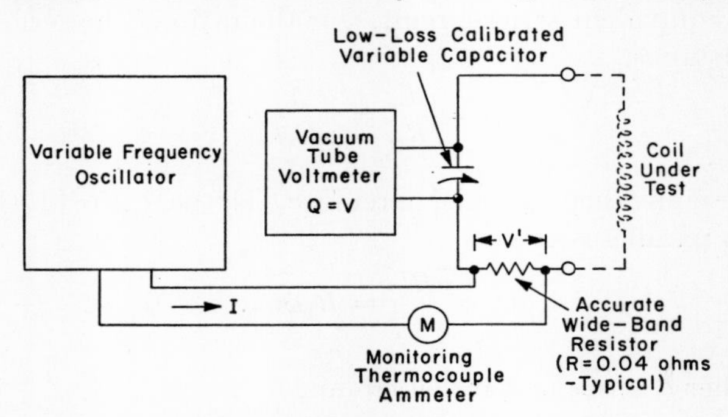

FIG. 18. Simplified Q meter circuit operation.

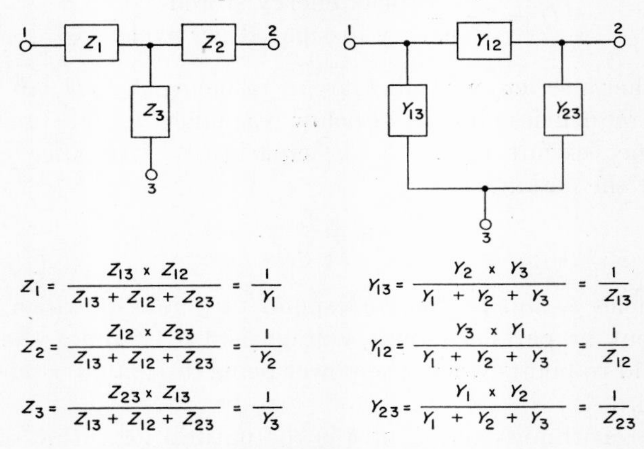

FIG. 19. Three-terminal T and π circuits.

terminal measurements. The third terminal for each independent measurement can be open-circuited, short-circuited, or connected through a known impedance. Special bridges of the general type shown in Fig. 17 can be used for the measurement of individual network elements. Active three-terminal devices such as electron tubes and transistors can be depicted for small-signal operation by a network as shown in Fig. 19 with one of the network elements having a negative resistance component. Similar measurement techniques can be applied in this case provided they will accommodate negative resistance measurements.[72,74–76] The special equipments for measuring three-terminal networks often employ a frequency conversion to facilitate measurement.[83–85]

8.3.3. Power, Energy

If

$$v = V_M \cos(\omega t + \psi) = Re V_M e^{j(\omega t + \psi)} = Re[(V_R + jV_X)e^{j\omega t}]$$

is the instantaneous voltage between two points and

$$i = I_M \cos(\omega t + \varphi) = Re I_M e^{j(\omega t + \varphi)} = Re[(I_R + jI_X)e^{j\omega t}]$$

is the instantaneous current through these points, then the instantaneous power

$$\begin{aligned} p = vi &= \frac{V_M I_M}{2} [\cos(\psi - \varphi) + \cos(2\omega t + \psi + \varphi)] \\ &= \tfrac{1}{2}(V_R I_R + V_X I_X) + \tfrac{1}{2}(V_R I_R - V_X I_X)\cos 2\omega t \\ &\qquad - \tfrac{1}{2}(V_R I_X + V_X I_R)\sin 2\omega t. \end{aligned} \tag{8.3.5}$$

In the relations above, V_M, ψ and I_M, φ are the peak value and phase angle of v and i, respectively, Re is an abbreviation for "real part of," $j = \sqrt{-1}$, and $V_R = V_M \cos \psi$, $V_X = V_M \sin \psi$ and $I_R = I_M \cos \varphi$, $I_X = I_M \sin \varphi$ are the resistive and reactive components of V_M and I_M.

Often, only the average (active) power P, which is the time average of the instantaneous power over one complete cycle of period T, is of interest.

$$\begin{aligned} P &= \frac{1}{T}\int_0^T p\, dt = \frac{V_M I_M}{2}\cos(\psi - \varphi) \\ &= \tfrac{1}{2}(V_R I_R + V_X I_X). \end{aligned} \tag{8.3.6}$$

In terms of a phasor peak voltage, $V = V_R + jV_X$, phasor peak current, $I = I_R + jI_X$, and impedance, $Z = R + jX$, or admittance, $Y = G + jB$,

$$P = \tfrac{1}{2}Re[VI^*] = \tfrac{1}{2}Re[V^*I] = \frac{I_M{}^2 R}{2} = \frac{V_M{}^2 G}{2}. \tag{8.3.7}$$

Electrodynamometer instruments (Section 8.2.6) can be used to measure power up to frequencies of a few thousand cycles per second with accuracies of about 1%. At higher frequencies conventional measurements of power being transmitted to a load can be made with the aid of electronic wattmeters[86,87] which perform mathematical multiplication[88]

[83] G. Thirup, *Philips Tech. Rev.* **14,** 102 (1952).
[84] D. A. Alsberg and D. Leed, *Bell System Tech. J.* **28,** 221 (1949).
[85] D. A. Alsberg, *Proc. I.R.E.* **40,** 1195 (1952).
[86] M. A. H. El-Said, *Proc. I.R.E.* **37,** 1003 (1949).
[87] M. Abdel-Halin Ahmed, *Trans. Am. Inst. Elec. Engrs.* **74,** Part I, 194 (1955).
[88] E. A. Goldberg, *R C A Rev.* **13,** 265 (1952).

and averaging. Two identical square-law devices such as thermocouples can be connected together to measure power being transmitted.[89,90]

Instead of measuring power being transmitted to a load, the measuring instrument itself can often be used as the load. Standard calorimeter techniques can then be employed.[91] In the bolometer method of power measurement the unknown power is absorbed in a temperature sensitive resistor (thermistor, instrument fuse, barretter, semiconductor, etc.). The change in resistance is a measure of the power absorbed. It is generally convenient and more accurate to use the bolometer element in a bridge circuit; bridge balance is restored by reducing the bridge excitation just sufficiently to compensate for the power from the source under measurement. Circuit feedback can be employed as shown in Fig. 20 to

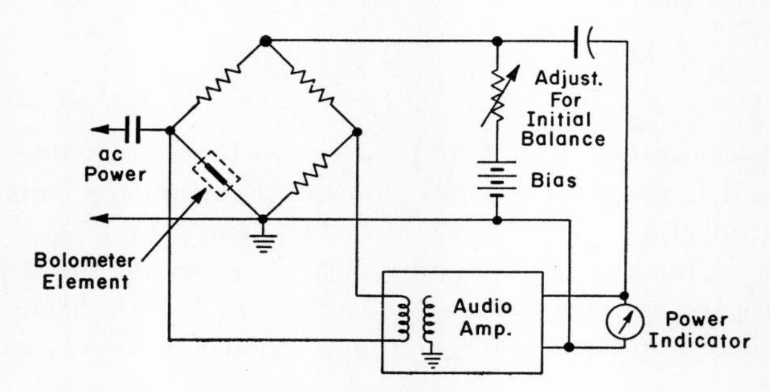

FIG. 20. Self-balancing bolometer bridge.

provide direct, automatic, and instantaneous readings.[92] Minimum power is about 10^{-4} watts full-scale, and accuracy is about $\pm 5\%$.

A matching network is an important auxiliary piece of equipment when making bolometer power measurements. The network should enable the bolometer resistance to be matched, as desired, to any source impedance. Allowance must be made for the network power losses. The matching network can be a transformer with additional reactances, a T, π, or L network,[93] or a transmission line with adjustable (stub) tuners. In some cases, wide-band matching networks may be useful.[94]

The measurement of energy requires the evaluation of the integral,

[89] G. H. Brown, J. Epstein, and D. W. Peterson, *Proc. I.R.E.* **31**, 403 (1943).

[90] F. K. Harris, "Electrical Measurements," p. 490. Wiley, New York, 1952.

[91] Radio Research Laboratory Staff, "Very High Frequency Techniques," Vol. 2, p. 581. McGraw-Hill, New York, 1947.

[92] L. A. Rosenthal and J. L. Potter, *Proc. I.R.E.* **39**, 927 (1951).

[93] F. E. Terman, "Radio Engineers' Handbook," p. 210. McGraw-Hill, New York, 1943.

[94] R. M. Fano, *J. Franklin Inst.* **249**, 57, 139 (1950).

$W = \int_{T_1}^{T_2} vi\, dt$. At low frequencies, particularly at 60 cycles/sec, this is done with a watt-hour meter,[95] an induction motor with rpm proportional to power and a totalizing arrangement. Various compensation techniques are employed to obtain a long-time accuracy of better than 1% for a wide variety of load and ambient conditions. Accuracy will vary significantly with frequency.

In those cases where the power varies slowly with time, the power, measured by methods discussed above, can be recorded, and graphical integration carried out. For the general case of complex wave shapes, small integration interval, etc., the integration must be carried out using an electronic multiplier[88] and integrator.[96]

8.3.4. Frequency, Waveform*

Frequency can be measured with a variety of resonance devices, tuned reeds, variable-tension vibrating string galvonometers, *LC* lumped circuits, *RC* twin-*T* circuits, sections of transmission lines, etc. The accuracy of measurement is determined by the "Q" of the device and typically might be 10^{-2} (1%).

Frequency is often measured by comparison with another signal. The accuracy of this method of measurement is largely determined by the "standard" of frequency employed. A piezoelectric oscillator[97–99] using a carefully mounted, temperature-controlled quartz bar generally operated at 100 kc/sec provides a frequency standard with an accuracy of $\pm 10^{-7}$, one-hour stability of about $\pm 10^{-9}$, and one-day stability of about $\pm 10^{-8}$. Molecular[100–102] and atomic[103,104] frequency standards are currently being developed and show promise of providing an accuracy of $\pm 10^{-9}$ and a long term stability of $\pm 10^{-12}$. At the present time astronomical observations[105] of the "mean solar day" are employed for long-time

* See also Vol. 2, Chapter 9.2.

[95] F. K. Harris, "Electrical Measurements," p. 512. Wiley, New York, 1952.

[96] G. A. Korn and T. M. Korn, "Electronic Analog Computers," 2nd ed. McGraw-Hill, New York, 1956.

[97] W. A. Marrison, *Bell System Tech. J.* **27,** 510 (1948).

[98] J. M. Shaull and J. H. Shoaf, *Proc. I.R.E.* **42,** 1300 (1954).

[99] F. D. Lewis, *Proc. I.R.E.* **43,** 1046 (1955).

[100] R. V. Pound, *Rev. Sci. Instr.* **17,** 490 (1946).

[101] C. H. Townes, *J. Appl. Phys.* **22,** 1365 (1951).

[102] J. P. Gordon, H. J. Zeiger, and C. H. Townes, *Phys. Rev.* **95,** 282 (1954).

[103] I. I. Rabi, S. Millman, P. Kusch, and J. R. Zacharias, *Phys. Rev.* **55,** 526 (1939).

[104] L. Davis, Jr., D. E. Nagle, and J. R. Zacharias, *Phys. Rev.* **76,** 1068 (1949); L. Essen and J. V. L. Parry, *Phil. Trans. Roy. Soc. London,* **A250,** 45 (1957).

[105] W. Markowitz, *Astron. J.* **59,** 69 (1954).

checks of frequency stability; due to a variety of variations the accuracy[106] is about $\pm 2 \times 10^{-8}$.

Standard-frequencies and time-signal radio broadcasts are carried out in several countries.[99] In the U.S., periodic standard signals are sent out by the National Bureau of Standards radio station, WWV, at Beltsville, Maryland on 2.5, 5, 10, 15, and 25 Mc/sec.

By means of frequency division, multiplication, and heterodyning a large number of reference frequencies can be generated with essentially

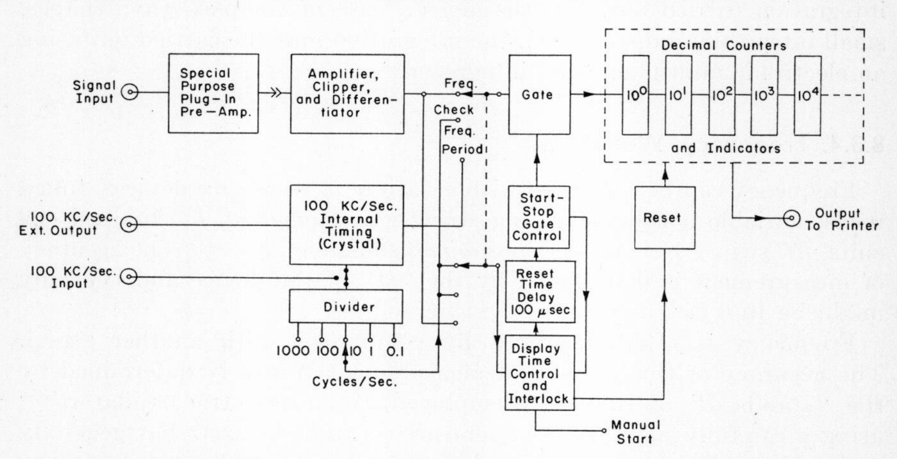

FIG. 21. Typical electronic counter. The special purpose plug-in pre-amp may include (1) wide-band (10 cycles/sec–10 Mc/sec), for increased sensitivity to 0.01 volts; (2) selective frequency converter (10 Mc/sec–100 Mc/sec) and selective frequency converter (100 Mc/sec–220 Mc/sec); (3) wide-band with adjustable clipping level; and (4) frequency divider, 1/10, 1/100, 1/1000.

the same accuracy as the original frequency standard. Continuous frequency interpolation[107] can be carried out with the aid of a mixer and calibrated low-frequency oscillator whose frequency error need only be small compared with the permissible frequency error of the high frequency being measured.

High-speed electronic counters are now frequently employed for frequency comparison.[108,109] A block diagram of a typical counter is shown in Fig. 21. The input signal is amplified, clipped, and differentiated so as to produce a pulse train which is gated to a decimal counter for an inter-

[106] H. M. Smith, *Proc. Inst. Elec. Engrs.* (*London*) *Pt. IV* **99,** Monograph 39, 273 (1952).

[107] J. K. Clapp, *Proc. I.R.E.* **36,** 1285 (1948).

[108] I. E. Grosdoff, *R C A Rev.* **7,** 438 (1946).

[109] R. L. Chase, *Electronics* **23,** 110 (1950).

val of time accurately determined by internal or external frequency standard. The decimal counters visually register the pulse count and either hold the count for manual reset or cyclically repeat the measurement after an adjustable hold period. Period measurements can be made by interchanging the signal input and frequency standard. Direct frequency measurements can be made from 10 cycles/sec to 10 Mc/sec with an indicating accuracy of ± 1 cycles/sec (10^{-8} accuracy with eight decade counters). Use of 10-sec gate instead of 1-sec decreases uncertainty to ± 0.1 cycles/sec. Indirect frequency measurements can be made to higher frequencies with the aid of plug-in frequency converters. A fast printer can be used to make a permanent record of the measurements.

Modern oscilloscopes[110,111] as shown in Fig. 22 can be used for quantitative as well as qualitative work. Their great versatility and ability to present and evaluate a large amount of waveform information in a very short time is one reason for their popularity. Some of the functions that can be performed with a typical oscilloscope of the type shown in Fig. 22 are:

(1) Display a complex signal with frequency components ranging from 0 to 10 Mc/sec. Vertical sensitivity is continuously variable and calibrated from 0.05 to 50 volts/cm deflection. Input impedance is 1 meg with about 10 $\mu\mu$f of shunt capacitance. The highly linear horizontal deflection has 24 fixed sweep speeds calibrated to about 3% accuracy. Sweep speed can be continuously varied from 0.02 μsec/cm to 12 sec/cm. The horizontal sweep can be operated free-running for viewing recurrent signals or individually or successively triggered either internally from the vertical signal or externally. Horizontal sweep can be initiated at any selected point of the signal under examination. An auxiliary delaying sweep in conjunction with an unblanking mixer and time delay (continuously variable and accurately calibrated from 2 μsec to 100 msec) permits variable points of the input signal to be timed and subsequently examined in detail. Alternatively horizontal sweep initiation can be delayed a calibrated period following a trigger signal.

(2) By means of an electronic switch, two input signals can be presented either on alternate sweeps or on a "chopped" time sharing basis with most of the other features as under (1).

(3) Balanced amplification can be provided to two input signals so that by employing the differential action inherent in the deflection of the oscilloscope beam only the difference between the two input signals

[110] I. A. Greenwood, Jr., J. V. Holdam, Jr., and D. MacRae, Jr., eds., "Electronic Instruments," M.I.T. Radiation Laboratory Series, Vol. 21. McGraw-Hill, New York, 1948.

[111] Y. P. Yu, H. E. Kallmann, and P. S. Christaldi, *Electronics* **24,** 106 (1951).

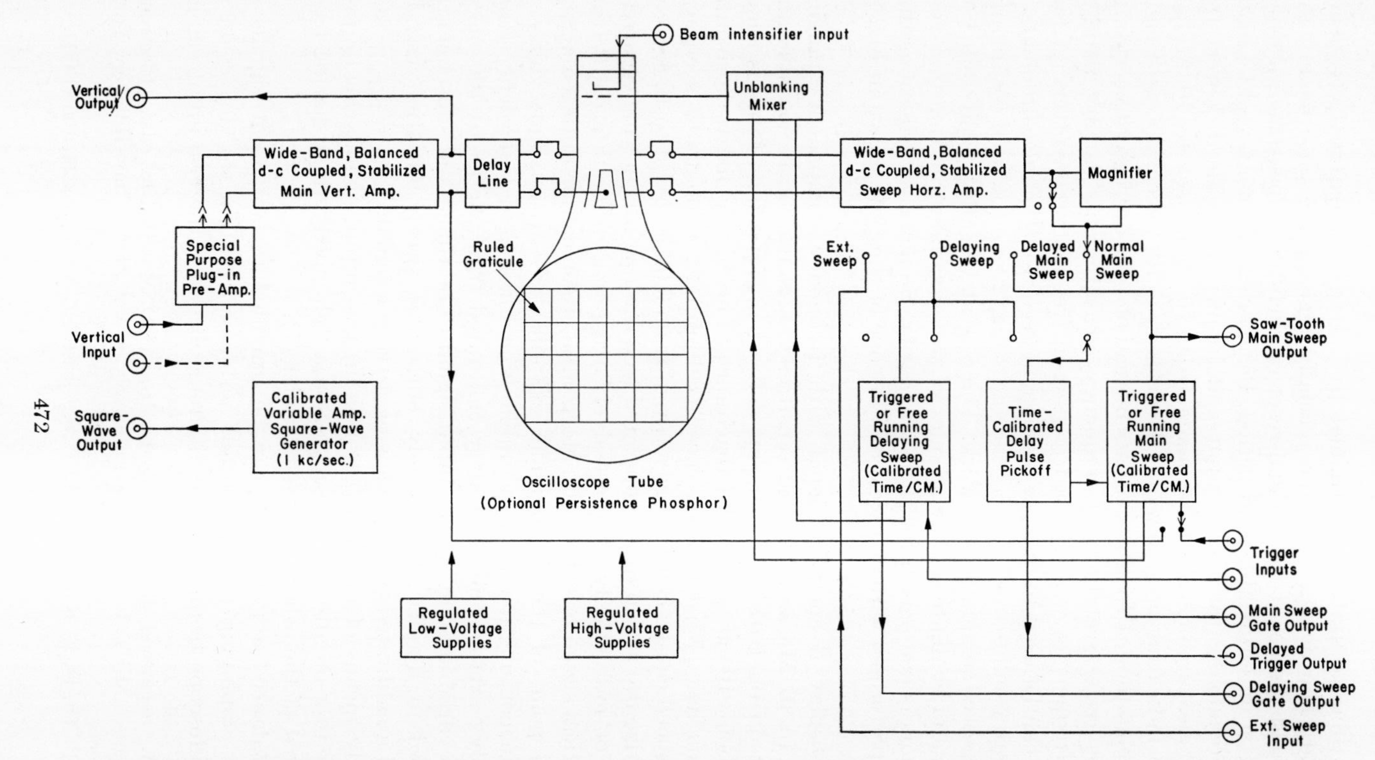

FIG. 22. Typical multipurpose precision wide-band oscilloscope. The special purpose plug-in pre-amp may include (1) wide-band dc; (2) wide-band high-gain; (3) dual-trace dc (triggered or free running electronic switch); and (4) differential dc high-gain.

is viewed. Rejection to in-phase signal is about 1:50,000. This feature is particularly advantageous for viewing and studying small differences between two signals as, for instance, in conjunction with bridge operation. Sensitivity of 50 μv/cm is provided with reduced frequency response.

(4) A multiplicity of output signals of different waveforms are available to facilitate synchronization, control, gating, and calibration. These include:

(a) Horizontal sweep saw-tooth voltage of calibrated duration.

(b) Gate pulse synchronized with and of same duration as horizontal sweep.

(c) Gate pulse synchronized with and of same duration as auxiliary delaying sweep.

(d) Delayed trigger pulse with calibrated delay time.

(e) Vertical deflection signal.

(f) Symmetrical square wave of 1 kc/sec repetition and amplitude adjustable in 18 calibrated steps from 0.002 to 100 volts peak-to-peak.

(5) A variety of wide-band probes can be used to accommodate larger voltages and increase input impedances.

A complex recurrent signal can be evaluated in terms of harmonic components by means of Fourier analysis. A variety of graphical-algebraic computations[112–114] can be carried out to determine the harmonic components. Mechanical,[115] optical,[116] or electronic[117] apparatuses can be used to carry out the computations. Through the audio spectrum, frequency-selective circuits can be used to determine component amplitudes without regard to phase.[118,119]

Phase-angle difference between two signals of the same frequency is directly related to the time difference of zero crossing.* Measurements can therefore be made by means similar to the electronic counter described above.[120,121] Measurement of phase-angle difference at higher frequencies

* See also Vol. 2, Chapter 9.4.

[112] F. W. Grover, "Analysis of Alternating-Current Waves by the Method of Fourier, With Special Reference to Methods of Facilitating Computations," *Natl. Bur. Standards (U.S.) Bull.* **9,** *Sci. Paper* 203 (1913).

[113] L. S. Cole, *Electronics* **18,** 142 (1945).

[114] J. B. Scarborough, "Numerical Mathematical Analysis," p. 477. Johns Hopkins Press, Baltimore, Maryland, 1955.

[115] J. H. Robertson, *J. Sci. Instr.* **27,** 276 (1950).

[116] H. C. Montgomery, *Bell System Tech. J.* **17,** 406 (1938).

[117] Jules Lehman, *Electronics* **22,** 106 (1949).

[118] C. R. Moore and A. S. Curtis, *Bell System Tech. J.* **6,** 217 (1927).

[119] F. E. Terman, R. R. Buss, W. R. Hewlett, and F. C. Cahill, *Proc. I.R.E.* **27,** 649 (1939).

[120] E. F. Florman and A. Tait, *Proc. I.R.E.* **37,** 207 (1949).

[121] E. R. Kretzmer, *Electronics* **22,** 114 (1949).

can most easily be made by heterodyning both signals simultaneously to a low frequency;[84] heterodyning, if properly carried out, changes frequency but not the phase relation. Thus, a single-frequency phasemeter can be used for measuring over a wide frequency range.

Pulse height and width measurements can conveniently be made with an oscilloscope described above.* Pulse height analyzers[122–124] are used extensively for energy studies of nuclear particles; they provide a count of different pulse amplitudes usually by means of a series of biased amplifiers with associated counters.

8.3.5. Oscillators†

Commercial oscillators are available that cover the frequency spectrum from about $1/100$ cycles/sec to 10^{10} cycles/sec.[125–127] Basically, an oscillator can be considered an amplifier with feedback to provide regeneration. Oscillations are obtained at the frequency at which the net phase shift around the amplifier-feedback loop is 360° provided the power amplification at that frequency is greater than unity.[128] The amplitude of oscillation is determined by nonlinearities in the loop and this basically implies some distortion of the signal. By suitable design the amount of distortion can be reduced to small values.[129]

With some care, variable frequency oscillators of good frequency stability can be built.[130,131] However, for frequency accuracies better than about 1% an independent measurement as described above should be carried out. Oscillator outputs are often calibrated in terms of voltage or power into a standard load resistance. Output is generally measured at a high level and then attenuated to low-levels usually used. Output

* See also Vol. 2, Chapter 9.6.

† See also Vol. 2, Chapter 6.3.

[122] P. W. Byington and C. W. Johnstone, *I.R.E. Convention Record* **Part 10,** 191 (1955).

[123] G. G. Kelley, *Proc. Intern. Conf. Peaceful Uses Atomic Energy, Geneva,* Paper 8/P/66 (1955).

[124] W. A. Higinbotham, *I.R.E. Trans. on Nuclear Science* **NS-3,** 3 (1956).

[125] F. E. Terman, "Radio Engineers' Handbook," p. 480. McGraw-Hill, New York (1943).

[126] F. E. Terman and J. M. Pettit, "Electronic Measurements," pp. 482–532. McGraw-Hill, New York, 1952.

[127] W. A. Edson, "Vacuum Tube Oscillators." Wiley, New York, 1953.

[128] A. A. Andronow and C. E. Chaikin, "Theory of Oscillations." Princeton Univ. Press, Princeton, New Jersey, 1949.

[129] L. A. Meacham, *Proc. I.R.E.* **26,** 1278 (1938).

[130] J. K. Clapp, *Proc. I.R.E.* **42,** 1295 (1954).

[131] F. B. Anderson, *Proc. I.R.E.* **39,** 881 (1951).

calibration accuracy is generally about ±10% when employed under the specified conditions.

Provisions are often made for the amplitude and/or frequency modulation of the output signal. Modulation accuracy is generally not better than ±10%. In some work residual modulation or unwanted modulation of one type accompanying the desired modulation may be troublesome.

Oscillators may have a significant amount of amplitude and/or frequency modulation noise.[132,133] That is, there is always a spectrum rather than a single frequency.

Pulse,[134,135] square-wave,[136] saw-tooth, stepped, and various other waveform oscillators[137,138] are now finding extensive use. For a versatile signal generator it is often necessary to have a variety of wave shapes available with variable and calibrated time relations. Thus a typical composite waveform generator can provide:

(1) Pulse variable in calibrated width from 0.1 to 1000 μsec (±2%) and in amplitude from 0 to ±50 volts with 50 ohms source resistance and synchronized with an external or internal trigger signal with a 3 μsec delay.

(2) Pulse as in (1) but with a calibrated delay variable from 0 to 10,000 μsec. (±2%) with respect to an external or internal trigger signal.

(3) Sum of (1) and (2) with either polarities, with indicated delay, and with independent pulse widths.

(4) Positive or negative gate signal starting at trigger signal and ending at termination of pulse (1).

(5) Same as (4) but ending at start of pulse (2).

(6) Linear negatively increasing voltage of same duration as (4).

(7) Same as (6) but of duration of (5).

(8) Linear negatively increasing voltage of same duration as (2).

(9) Signals indicated above can be obtained on a one-cycle operation by manual control.

[132] R. L. Beurle, *Proc. Inst. Elec. Engr. (London)* **Pt. B 103,** 182 (1956).

[133] W. W. Boelens and F. L. H. M. Stumpers, *Commun. News* **10,** 15 (1949).

[134] G. N. Glasoe and J. V. Lebacqz, eds., "Pulse Generators," M.I.T. Radiation Laboratory Series, Vol. 5. McGraw-Hill, New York, 1948.

[135] C. C. Cutler, *Proc. I.R.E.* **43,** 140 (1955).

[136] G. W. Gray, *Electronics* **25,** 101 (1952).

[137] F. E. Terman and J. M. Pettit, "Electronic Measurements," pp. 533–609. McGraw-Hill, New York, 1952.

[138] B. Chance, V. Hughes, E. F. MacNichol, D. Sayre, and F. C. Williams, eds., "Waveforms," M.I.T. Radiation Laboratory Series, Vol. 19. McGraw-Hill, New York, 1949.

8.4. Alternating-Current Measurements—Distributed Circuit

8.4.1. Transmission Systems[139–141] *

For frequencies above about 1000 Mc/sec, the lumped circuit concepts applicable at lower frequencies must be modified since every part of the circuit contributes significantly. Operation is then considered on a basis of a transmission system with discontinuities therein forming the circuit components. Actually, transmission theory may be applicable at much lower frequencies whenever the ratio, size (length) of the circuit to operating wavelength, approaches or exceeds unity.

Transmission systems are analyzed with the aid of Maxwell's equations (8.1.1) to (8.1.4). Because of spatial variations it is generally more convenient to consider the electric field, **E** (volts per meter) and magnetic field, **H** (amperes per meter) rather than voltage and current as is done at lower frequencies. The Poynting vector, $\mathbf{P} = \mathbf{E} \times \mathbf{H}$ (watts per square meter), gives the energy flow per unit area. When the energy flow takes place in a homogeneous isotropic medium defined by the ϵ, μ, and σ constants (Section 8.1.1), and both electric and magnetic fields vary sinusoidally at angular frequency ω with the factor $e^{j\omega t}$, Maxwell's equations in Cartesian coordinates are

$$\nabla \times \mathbf{E} = -j\omega\mu\mathbf{H} \quad \begin{cases} \dfrac{\partial E_z}{\partial y} - \dfrac{\partial E_y}{\partial z} = -j\omega\mu H_x & (8.4.1) \\[2ex] \dfrac{\partial E_x}{\partial z} - \dfrac{\partial E_z}{\partial x} = -j\omega\mu H_y & (8.4.2) \\[2ex] \dfrac{\partial E_y}{\partial x} - \dfrac{\partial E_x}{\partial y} = -j\omega\mu H_z & (8.4.3) \end{cases}$$

$$\nabla \times \mathbf{H} = (\sigma + j\omega\epsilon)\mathbf{E} \quad \begin{cases} \dfrac{\partial H_z}{\partial y} - \dfrac{\partial H_y}{\partial z} = (\sigma + j\omega\epsilon)E_x & (8.4.4) \\[2ex] \dfrac{\partial H_x}{\partial z} - \dfrac{\partial H_z}{\partial x} = (\sigma + j\omega\epsilon)E_y & (8.4.5) \\[2ex] \dfrac{\partial H_y}{\partial x} - \dfrac{\partial H_x}{\partial y} = (\sigma + j\omega\epsilon)E_z. & (8.4.6) \end{cases}$$

The transmission of energy is conveniently formulated in terms of combinations of modes of wave propagation characterized by particular field patterns. The individual modes arise from wave solutions of the

* See also Vol. 2, Chapter 10.2.

[139] J. A. Stratton, "Electromagnetic Theory." McGraw-Hill, New York, 1941.

[140] S. A. Schelkunoff, "Electromagnetic Waves." Van Nostrand, New York, 1943.

[141] G. C. Southworth, "Principles and Applications of Waveguide Transmission." Van Nostrand, New York, 1950.

above equations subject to suitable boundary conditions. For transmission in the z direction the wave propagation can be classified[142] in terms of

(a) *Transverse Magnetic Plane Waves* (TM waves) characterized by the fact that the magnetic field is always perpendicular to the direction of propagation; $H_z = 0$.

(b) *Transverse Electric Plane Waves* (TE waves) characterized by the fact that the electric field is always perpendicular to the direction of propagation; $E_z = 0$.

(c) *Transverse Electromagnetic Plane Waves* (TEM waves) characterized by the fact that both magnetic and electric fields are always perpendicular to the direction of propagation; $H_z = E_z = 0$.

A variety of structures can be used as transmission guides. Metal tubes (waveguides) are often employed. Waveguides act as high-pass filters and will propagate energy at frequencies above a certain cutoff frequency determined roughly as the wavelength corresponding to a guide cross-sectional dimension. Discrete higher frequency modes of transmission occur. Propagation is characterized by attenuation and phase shift per unit length in the direction of propagation.

In the case of TEM waves the propagation constant, Γ, is equal to the intrinsic propagation constant of the medium supporting the propagation, $\Gamma = \sqrt{j\omega\mu(\sigma + j\omega\epsilon)}$; also the wave impedance, η, (ratio of electric to magnetic fields) is equal to the intrinsic impedance of the medium, $\eta = \sqrt{\frac{j\omega\mu}{\sigma + j\omega\epsilon}}$. If the medium is free space, $\sigma = 0$, $\mu = \mu_0$, and $\epsilon = \epsilon_0$, $\Gamma = j\omega/c$ (c = velocity of light) and $\eta = \sqrt{\mu_0/\epsilon_0} = 120\pi$. TEM waves are uniform plane waves whose equiphase surfaces form a family of parallel planes normal to the direction of transmission z, and whose field intensities are independent of z. The integral of the electric field between two perfectly conducting boundaries yields the voltage between the conductors, and the integral of the magnetic field around the conductors yields the current through the conductor. Thus, TEM waves provide an easy transition to low-frequency phenomena and in fact TEM transmission can take place at any frequency down to dc. In transforming from electric and magnetic fields to voltage and current, one is led naturally to replacing the medium constants by conductor per unit length constants, C, L, and G, and the above Maxwell equations become the usual transmission line equations,

$$\frac{dV}{dz} = -(R + j\omega L)I \tag{8.4.7}$$

$$\frac{dI}{dz} = -(G + j\omega C)V. \tag{8.4.8}$$

[142] I.R.E. Standards Committee, 1952–1953, "I.R.E. Standards on Antennas and Waveguides; Definitions of Terms, 1953." *Proc. I.R.E.* **41,** 1721 (1953).

Here, as is common practice, a line resistance per unit length R has been introduced. For finite conductor resistance exact TEM waves are no longer possible and the above equations with R are therefore only approximately valid.

A convenient solution of Eqs. (8.4.7) and (8.4.8) is one which relates the voltage and current at the $z = l$ end, V_1 and I_1, to the voltage and current at the $z = 0$ end, V_2 and I_2,

$$V_1 = AV_2 + BI_2 \tag{8.4.9}$$

$$I_1 = CV_2 + DI_2 \tag{8.4.10}$$

where

$$A = \cosh \Gamma l \tag{8.4.11}$$

$$B = Z_0 \sinh \Gamma l \tag{8.4.12}$$

$$C = \sinh \Gamma l / Z_0 \tag{8.4.13}$$

$$D = \cosh \Gamma l \tag{8.4.14}$$

and Z_0 (characteristic impedance) and Γ (propagation constant) are

$$Z_0 = \sqrt{\frac{R + j\omega L}{G + j\omega C}} \tag{8.4.15}$$

$$\Gamma = \sqrt{(R + j\omega L)(G + j\omega C)}. \tag{8.4.16}$$

The matrix

$$\begin{Vmatrix} AB \\ CD \end{Vmatrix}$$

is the scatter matrix for the transmission line of length l, and conventional network analysis can be applied.[143] Calculations can be facilitated by graphical aids.[144–146] The ratio of Eq. (8.4.9) to Eq. (8.4.10) relates the impedance at one point in terms of the line properties and the impedance at another point. On this basis, a section of a transmission line can be used as a transformer. It is seen that the line characteristic impedance Z_0 can be defined as that impedance which when used across one place of the line will produce the same impedance at other places of the line. The termination of a line in its characteristic impedance gives rise to no reflections, and in fact the measurement of standing wave ratios (described below) can be used to evaluate an arbitrary impedance in terms

[143] N. Marcuvitz, ed., "Waveguide Handbook," M.I.T. Radiation Laboratory Series, Vol. 10, p. 101. McGraw-Hill, New York, 1951.

[144] A. E. Kennelly "Chart Atlas of Complex Hyperbolic and Circular Functions." Harvard Univ. Press, Cambridge, Massachusetts, 1924.

[145] P. H. Smith, *Electronics* **17,** 130 (1944).

[146] G. Deschamps, "A New Chart for the Solution of Transmission Line and Polarization Problems." *IRE Trans. Profess. Group on Microwave Theory and Tech.* **MTT 1,** 5 (1953).

of the characteristic impedance. Abrupt discontinuities in a line will produce reflections and potentially give rise to resonant sections (cavities). Preferably, the lines and resonant sections are made with minimum losses; if $R = G = 0$, the characteristic impedance is a resistance, whose value for a variety of lines has been calculated,[147] and propagation constant corresponds to phase shift without attenuation.

A TEM signal being transmitted along a uniform line has a ratio of voltage to current equal to the line characteristic impedance. If the line is not uniform but has a discontinuity in it, there will in general be a transmitted signal and a reflected signal in addition to the incident signal. Thus, the voltage and current of a transmission line with a discontinuity can, in general, be represented as sums of voltages and currents traveling in opposite directions which produce standing waves along the line. Measurement of the ratio of maximum to minimum of the standing wave together with the wave position can be used to determine the impedance at the discontinuity. Thus, the waves in opposite directions can be placed in evidence by expanding the hyperbolic functions of Eqs. (8.4.11) to (8.4.14) and substituting into Eqs. (8.4.9) and (8.4.10) to obtain (see Fig. 23(a) for polarity and direction convention)

$$V(z) = \frac{V_L + I_L Z_0}{2} [e^{\Gamma z} + \rho e^{-\Gamma z}] \tag{8.4.17}$$

$$I(z) = \frac{V_L + I_L Z_0}{2Z_0} [e^{\Gamma z} - \rho e^{-\Gamma z}]. \tag{8.4.18}$$

In these equations ρ is the reflection coefficient,

$$\rho = |\rho| e^{j\theta} = \frac{V_L - I_L Z_0}{V_L + I_L Z_0} = \frac{Z_L - Z_0}{Z_L + Z_0} \tag{8.4.19}$$

and is the ratio between the incident and reflected wave. The ratio of Eq. (8.4.17) to Eq. (8.4.18) gives an equation which relates the impedance, $Z(z)$ of a line of arbitrary length z, in terms of the line impedance Z_0, and load impedance Z_L,

$$Z(z) = \frac{e^{\Gamma z} + \rho e^{-\Gamma z}}{e^{\Gamma z} - \rho e^{-\Gamma z}} Z_0. \tag{8.4.20}$$

This equation can be used to compute the transformer properties of a length of line.

Equations (8.4.17) and (8.4.18) indicate that the variation of V and I are very similar, and either can be used in the development that follows.

[147] H. P. Westman, ed., "Reference Data for Radio Engineers," p. 588. International Telephone and Telegraph, New York, 1956.

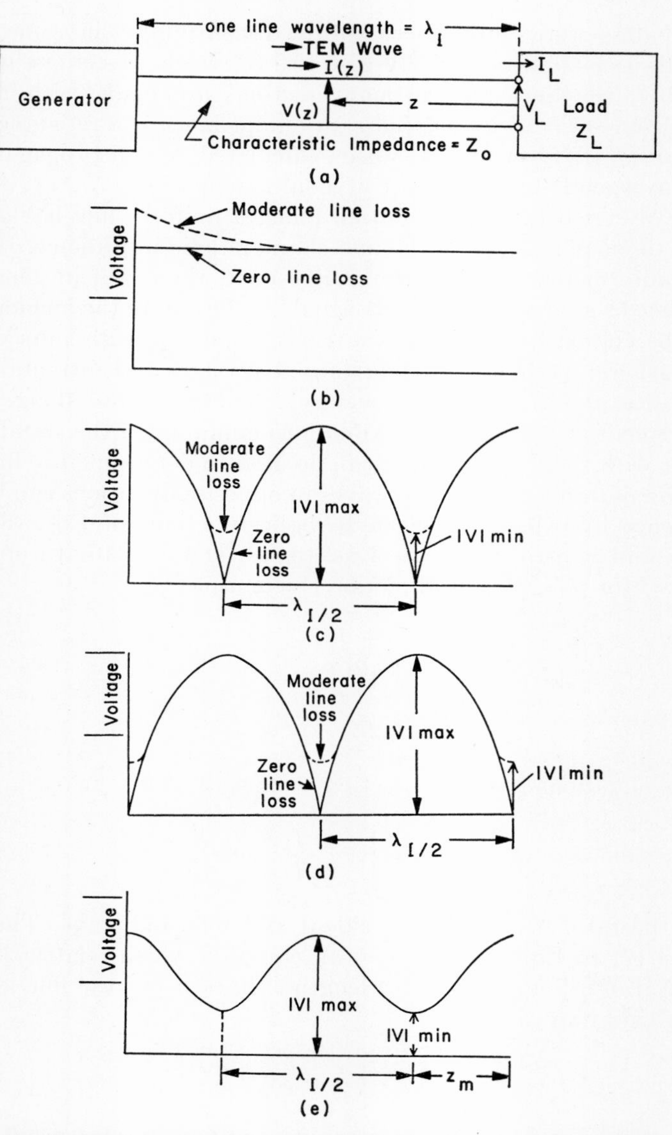

FIG. 23. Absolute voltage standing waves on lossless line for different line terminations: (a) line with generator and load; (b) line standing wave when $Z_L = Z_0$, $\rho = 0$; (c) line standing wave when $Z_L = \infty$ (open-circuited), $\rho = 1$; (d) line standing wave when $Z_L = 0$ (short-circuited), $\rho = -1$; and (e) line standing wave for arbitrary Z_L.

When $Z_L = Z_0$, $\rho = 0$, and $|V|$ is either exponentially decreasing or constant depending upon whether the line does or does not have losses; Fig. 23(b). When $Z_L = \infty$, $\rho = 1$, and $|V|$ varies as a rectified sinusoidal signal if line losses are zero; Fig. 23(c). When $Z_L = 0$, $\rho = -1$, and the variation of $|V|$, Fig. 23(d), is the same as for $Z_L = \infty$, but displaced $\lambda_l/4$. For an arbitrary Z_L, $|V|$ varies in a manner shown in Fig. 23(e); one can measure a voltage standing wave ratio (VSWR), $S = |V|_{\text{max}}/|V|_{\text{min}}$, and the location, z_m, of say the first $|V|_{\text{min}}$ from the load end of the line. There is a unique relation that exists for a given line between S, z_m, and the value of Z_L. This unique relation is particularly simple when line losses are zero. Thus, when

$$\Gamma = j\beta = j\frac{2\pi}{\lambda_l}$$

$$|\rho| = \frac{S-1}{S+1} \tag{8.4.21}$$

$$\theta = \left(4\frac{z_m}{\lambda_l} - 1\right)\pi. \tag{8.4.22}$$

These relations are often used to determine ρ and in turn Z_L from measurements of S and z_m. A similar development can be applied to transmission systems in general, but care must be exercised in defining the various quantities that enter into the formulation.[148]

8.4.2. Circuit Components

At higher frequencies circuits are often assembled with the aid of a variety of circuit components in much the same way as circuit elements are used at lower frequencies. There follows a brief description and performance characteristics of some of the circuit components.[149,150]

(a) *Transmission Lines* are transmission systems which propagate energy predominantly in a TEM mode. Lines have a variety of configurations, characteristic impedances, and operating characteristics.[151] Coaxial lines are most frequently employed.

(b) *Waveguides* are transmission systems of material boundaries capable of guiding waves. Although the boundary can be a variety of materials,

[148] S. A. Schelkunoff, "Electromagnetic Waves," pp. 319, 480. Van Nostrand, New York, 1943.

[149] C. G. Montgomery, R. H. Dicke, and E. M. Purcell, eds., "Principles of Microwave Circuits," M.I.T. Radiation Laboratory Series, Vol. 8. McGraw-Hill, New York, 1948.

[150] N. Marcuvitz, ed., "Waveguide Handbook," M.I.T. Radiation Laboratory Series, Vol. 10. McGraw-Hill, New York, 1951.

[151] H. P. Westman, ed., "Reference Data for Radio Engineers," pp. 549–644. International Telephone and Telegraph, New York, 1956.

a self-enclosed metal boundary of rectangular cross section is most frequently employed. Rectangular waveguides are generally operated in the $TE_{1,0}$ mode which has its electric field everywhere perpendicular to the z direction of propagation and parallel to the y dimension y_0 of the waveguide. The lower cutoff frequency f_c for this mode is

$$f_c = \frac{c}{2x_0} \tag{8.4.23}$$

where c is the velocity of light and x_0 is the x dimension of the waveguide. Tabular information of a variety of standard and nonstandard waveguides is available.[151] As a typical example, a brass waveguide with Army-Navy type number RG51U (R.E.T.M.A. number WR-112) has a $TE_{1,0}$ cutoff frequency of 5.26 kMc/sec, a useful operating range of 7.05 to 10.00 kMc/sec, internal cross-sectional dimensions of 2.850 cm $\times$ 1.262 cm, attenuation varying from 4.1 to 3.2 db/100 ft for lowest to highest frequency respectively, and a power rating of about $\frac{1}{3}$ Mw at standard pressure.

(c) *Delay Lines*[152] are transmission lines with increased values of per unit length inductance and capacitance so that the group velocity of wave propagation is reduced and essentially independent of frequency. Delay lines are often used at videofrequencies (0–4.5 Mc/sec) and typically: $Z_0 = 1000$ ohms, delay $= 4$ μsec/meter, and attenuation $= 20$ db/meter.

(d) *Line Stretchers* (phase shifters) are variable length transmission systems. Typically a coaxial line would have a $Z_0 = 50$ ohms, a 20-cm adjustable length, and a VSWR ranging from 1.03 to 1.25 over a frequency of 500 Mc/sec to 5000 Mc/sec. For convenience the variable length can be arranged as a "trombone" action. Reactive loading may also be employed for small phase shifts.

(e) *Rotary Joints* are transmission systems that permit rotary motion of one system with respect to another. For transmission systems that do not have circular symmetry, rotary joints are made by means of transformations to and from circularly symmetrical transmission systems.

(f) *Twisted Sections* are transmission systems that provide a rotation of the plane of polarization of a wave by mechanical means. VSWR is about 1.06 when a 90° twist takes place in a distance about 16 times the transmission system wavelength.

(g) *Elbows* are transmission systems that change the direction of wave propagation. The change in direction can be made by bending the waveguide in either the E or H planes. For either elbow the VSWR is

[152] H. E. Kallmann, *Proc. I.R.E.* **34,** 646 (1946).

about 1.1 for a 90° bend and a radius of curvature about equal to a waveguide wavelength.

(h) *Terminations* are sections of transmission systems that (1) totally absorb the incident wave without reflections (characteristic-impedance termination), (2) totally reflect the incident voltage wave (short-circuit termination), or (3) totally reflect the incident current wave (open-circuited termination). Characteristic-impedance terminations are preferably broad-band components; a 50-ohm coaxial cylindrical resistor termination has a VSWR = 1.05 from 0–5 kMc/sec. High-power characteristic-impedance terminations employ forced air or circulating coolant to remove heat from the termination. In some termination systems provision is made for moving the termination.

(i) *Radiators* are terminations that couple power to free space. Some radiators such as horns and lens propagate the free-space power in narrow beams.

(j) *Tees* provide a common junction of three transmission systems. Waveguide tees can be made either with E plane common to all three guides or with H plane common to all three guides. A wave propagated upward on the stem of the tee will divide equally at the top but the two outward traveling waves will be of opposite phases for an E-plane tee and of same phase for an H-plane tee.

(k) *Magic (Hybrid) Tees*[153,154] provide a common junction of four waveguide transmission systems. The junction consists of a combination of an E-plane tee and fourth side-arm forming an H-plane tee. The unique property of a symmetrical magic tee is that a wave propagated upward on the stem (1) of an E-plane tee is split equally with opposite phases at the top (2) and (3) of the tee, but no energy is propagated into the fourth (4) arm if there are no reflections from the outward traveling waves (2) and (3). Conversely a wave propagated towards the junction in (4) splits equally between arms (2) and (3) with same phase, but no energy is propagated down the stem (1) of the E-plane tee if there are no reflections from the outward traveling waves (2) and (3). Reciprocal conditions prevail, and two waves traveling towards the junction along (2) and (3) will propagate only along arm (4) if the original two waves have opposite phases, or only along arm (1) if the original two waves have the same phase provided in either case there are no reflections from energy propagated along arms (4) or (1) respectively. The unique property of a symmetrical magic tee is frequently employed when signals are to be mixed without interaction as in a receiver mixer or for phase dis-

[153] W. A. Tyrrell, *Proc. I.R.E.* **35**, 1294 (1947).

[154] L. D. Smullin and C. G. Montgomery, eds. "Microwave Duplexers," M.I.T. Radiation Laboratory Series, Vol. 14. McGraw-Hill, New York, 1948.

crimination. For a well-constructed magic tee the ratio of the power to the input arm to the power from the decoupled arm with the other two arms perfectly terminated will be about 60 db. The VSWR into any arm with the other three arms perfectly terminated will be about 1.7.

(l) *Directional Couplers*[155,156] provide coupling between two transmission systems in such a manner that a wave is propagated in the second system only if the wave is traveling in a particular direction in the first system. The power of the wave in the second system is a definite fraction of the power of the appropriately directed wave in the first system (coupling ratio) and is independent of oppositely traveling waves in the first system. A single-hole (Bethe coupler)[157] directional coupler makes use of the fact that both electric and magnetic field coupling results when a hole exists in the wall between two transmission systems. Typically a single-hole waveguide directional coupler would have a coupling ratio of 20 db and a directivity ratio (ratio of oppositely directed waves in second system) of 20 db. Multiple hole directional couplers may provide greater directivity but generally tend to be more frequency sensitive.

(m) *Attenuators* are transmission systems with significant attenuation which is often calibrated and variable. Waveguide variable attenuators usually employ a metallized vane (flap attenuator) which is progressively inserted into the waveguide to increase the attenuation. Typically, VSWR = 1.2, and attenuation is 0–20 db calibrated to an accuracy of ± 1 db, and maximum average power dissipation is 1 watt.

(n) *Stub Tuners* provide controlled discontinuities in a transmission system so that reflections in the system can be "tuned out" or cancelled. Generally, either two stub tuners are provided or provisions are made for varying both the position and length of stub. By this means a wide variety of impedances can be matched to provide maximum power transfer. Waveguide stub tuners can correct a VSWR of 20 to 1.02 with a 2 db insertion loss.

(o) *Phase Shifters* provide variable electrical lengths of transmission systems without changing physical length. Phase shifting is accomplished by progressively inserting a dielectric into the transmission system so as to alter the group velocity in the system. Maximum phase shift is generally about 180° over the operating range of frequencies with calibration accurate to $\pm 2°$. VSWR is about 1.1.

(p) *Transitions* are transmission systems used to transfer energy from one system to a second system of different dimensions with minimum reflections. In some cases transitions effect a change in the mode of

[155] W. W. Mumford, *Proc. I.R.E.* **35,** 160 (1947).
[156] B. M. Oliver, *Proc. I.R.E.* **42,** 1686 (1954).
[157] H. Bethe, *Phys. Rev.* **66,** 163 (1944).

propagation also. Minimum VSWR is obtained by providing a gradual tapering from the first to the second system.[158,159]

(q) *Cavity Resonators*[160] are enclosed structures that are frequency selective and therefore commonly used for approximate frequency measurements. Generally, the structures are sections of transmission systems with short-circuits at both ends. Cavity tuning can be affected by moving one of the short-circuits, by inserting a dielectric rod in regions of high electric field or magnetic rod in regions of high magnetic field, or by related methods for changing the volume of the cavity. Resonant frequency can be calibrated in terms of mechanical position. Absolute frequency accuracy of 0.05% is possible after suitable corrections are applied. Q values range from 1000–50,000 depending upon cavity size, construction, and use.

(r) *Transmission Filters* perform frequency selective operations on signal transmission. Generally the frequency selection is broad band in contrast to the selective action of a cavity resonator. Mode filters[161] are not frequency selective, but rather prevent the transmission of one or more unwanted modes of signal transmission.

(s) *Standing Wave Detectors*[162] are slotted transmission systems with associated pickup and detector used primarily for measurement as described above. Generally a probe pickup is employed and voltage standing wave ratios (VSWR) are measured. However, it is also possible to use a loop pickup in which case current standing wave ratios are measured. Probe travel must be a half wavelength minimum, but $\frac{3}{2}$ wavelength travel is generally desirable since the standing wave position will change with termination reactance. Precision mechanical construction is very important for elimination of sources of errors. The pickup slot should be perfectly parallel to the transmission system axis and of constant width. Probe depth must be uniformly maintained within $\pm 1\%$, and probe position must be indicated accurately to a few thousandths of a wavelength.

The listing of the above components is not intended to be exhaustive,

[158] J. C. Slater, "Microwave Transmission," pp. 42, 168. McGraw-Hill, New York, 1942.

[159] C. G. Montgomery, R. H. Dicke, and E. M. Purcell, eds., "Principles of Microwave Circuits," M.I.T. Radiation Laboratory Series, Vol. 8, pp. 334–364. McGraw-Hill, New York, 1948.

[160] C. G. Montgomery, ed., "Technique of Microwave Measurements," M.I.T. Radiation Laboratory Series, Vol. 11, p. 285. McGraw-Hill, New York, 1947.

[161] C. G. Montgomery, R. H. Dicke, and E. M. Purcell, eds., "Principles of Microwave Circuits," M.I.T. Radiation Laboratory Series, Vol. 8, pp. 347–364. McGraw-Hill, New York, 1948.

[162] C. G. Montgomery, ed., "Technique of Microwave Measurements," M.I.T. Radiation Laboratory Series, Vol. 11, pp. 473–514. McGraw-Hill, New York, 1947.

but rather includes the major items excepting generators that may be employed in building a microwave circuit.

8.4.3. Measurements*

Power or power density is the quantity of usual interest although measurements of electric and magnetic fields may be employed as evaluation thereof. It is important that measurements be related to a specific plane of a transmission system. Area integration over the plane, employing Poynting's vector, may be required when complex spatial variations are present.

Measurements of standing wave ratios described above are most frequently used. Details of practical techniques are available.[163,164] Standing wave ratios can be related to a variety of quantities including terminating impedance and transmitted and reflected power. Power delivered to a bolometer can be conveniently measured as described in Section 8.3.3.

Impedances can be measured by two techniques shown in Figs. 24(a) and (b). In the bridge equipment shown in Fig. 24(a), the sampling probe control and phase control are simultaneously varied until a null is obtained at the detector. The probe control setting indicates the magnitude of Z_x, and phase control setting gives the phase angle of Z_x. Typical operating characteristics are magnitude measurements of 2–2000 ohms $\pm 5\%$, phase-angle measurements of $\pm(90° \pm 3°)$ over a frequency range of 50–500 Mc/sec.

The bridge equipment of Fig. 24(b) measures the real and reactive components of the unknown impedance. The test oscillator excites the three branches with the same voltage. The respective branch currents as sampled by the coupling loops depend upon the respective input impedances. The three loops are connected in parallel and are rotated to produce zero current (null) at the detector. The short-circuited susceptance branch is adjusted so as to be one-eight wavelength in length at the measurement frequency so that its input impedance is a pure reactance. Likewise, the branch terminated in its characteristic impedance has a purely resistive input impedance. By rotating the two associated loops, appropriate calibrated amounts of resistive and reactive currents can be picked up to balance out corresponding currents at the input of the unknown impedance line. Knowing the length of this line, the unknown terminating impedance can be computed from the measured input impedance. Typical operating characteristics are resistance and reactance of 10–250 ohms $\pm 10\%$ over a frequency range of 70–1000 Mc/sec.

* See also Vol. 2, Chapter 10.5.

[163] C. G. Montgomery, ed., "Technique of Microwave Measurements," M.I.T. Radiation Laboratory Series, Vol. 11, pp. 473–514. McGraw-Hill, New York, 1947.

[164] F. E. Terman and J. M. Pettit, "Electronic Measurements," pp. 135–157. McGraw-Hill, New York, 1952.

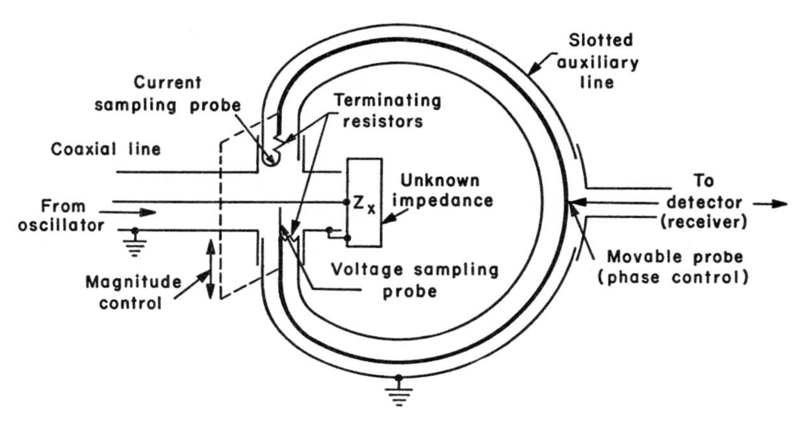

FIG. 24(a). Schematic diagram of transmission line bridge (from reference 164, p. 158).

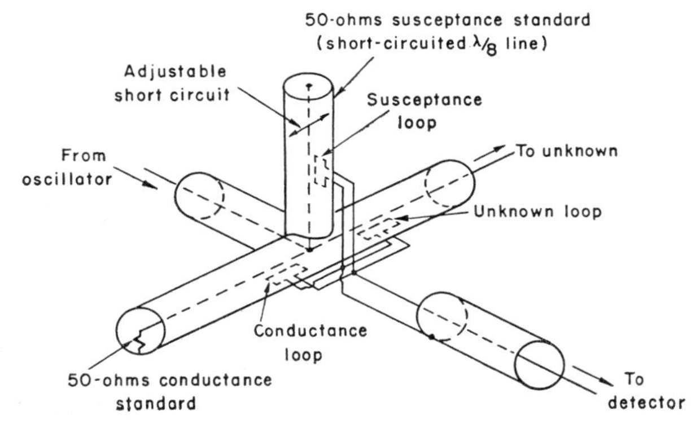

FIG. 24(b). Schematic diagram of the admittance comparator (from reference 164, p. 159).

Directional couplers with supplementary probes can be employed for impedance measurements.[165] Transmission systems such as the magic tee can be used for impedance comparison.

8.4.4. Oscillators*

Triodes,[166] klystrons,[166] and magnetrons[167] are extensively employed in oscillators at microwave frequencies. Backward wave traveling wave

* See also Vol. 2, Chapter 10.3.

[165] B. Parzen, *Proc. I.R.E.* **37,** 1208 (1949).

[166] D. R. Hamilton, J. K. Knipp, and G. B. H. Kuper, eds., "Klystrons and Microwave Triodes," M.I.T. Radiation Laboratory Series, Vol. 7. McGraw-Hill, New York, 1948.

[167] G. P. Collins, ed., "Microwave Magnetrons," M.I.T. Radiation Laboratory Series, Vol. 6. McGraw-Hill, New York, 1948.

tubes[168–170] are finding applications particularly where electronic tuning is employed. Induced emission between energy levels of solids is being employed to generate oscillations at millimeter wavelengths (far infrared).[171] Frequency stabilization is affected with an AFC circuit similar to lower frequencies with the aid of a microwave discriminator.[172] Power measurement, signal generator design, and calibration follow the principles described above for lower frequencies with adaptation to transmission systems.[173]

8.5. Dielectrics*

8.5.1. Permittivity

In Section 8.1.1 it was pointed out that one of the important constitutive equations is that which relates the polarization of a dielectric to an applied external field. A fundamental physical fact about a dielectric is that it does acquire a surface charge proportional to the applied field.

The polarization of an originally uncharged dielectric is due basically to a relative displacement of positive charges in the direction of **E** and negative charges in the opposite direction. These charges, equal and opposite in sign and displaced relative to each other, constitute an electric dipole. Polarization may also be due to the presence of permanent dipoles which become preferentially oriented by an external field. In all cases the polarization vector **P** is a measure of the dipole moment per unit volume and as such must be considered as a source of an electrostatic field.

The electrical property of a dielectric that is measured in the laboratory is the absolute dielectric constant or permittivity ϵ. The permittivity of a dielectric is always greater than the permittivity of vacuum, ϵ_0. It is convenient to use the relative permittivity or dielectric constant, ϵ_r, defined as

$$\epsilon_r = \epsilon/\epsilon_0 \tag{8.5.1}$$

* See also Vol. 6, Chapter 6.1.

[168] J. R. Pierce, "Traveling Wave Tubes." Van Nostrand, New York, 1950.

[169] R. Kompfner and N. T. Williams, *Proc. I.R.E.* **41,** 1602 (1953).

[170] Arthur Karp, *Proc. I.R.E.* **45,** 496 (1957).

[171] James P. Wittke, *Proc. I.R.E.* **45,** 291 (1957).

[172] C. G. Montgomery, ed., "Technique of Microwave Measurements," M.I.T. Radiation Laboratory Series, Vol. 11, pp. 58–78. McGraw-Hill, New York, 1948.

[173] *Ibid.*, pp. 79–281.

where $\epsilon_0 = 10^{-9}/36\pi \cong 8.85 \times 10^{-12}$ farads/meter in mks units. The relative dielectric constant, ϵ_r, is a dimensionless ratio. The word *dielectric constant* is used in the literature synonymously with relative *permittivity*. We shall also use the terms interchangeably and distinguish the relative and absolute values by the subscripts as shown in Eq. (8.5.1). The dielectric constant is not a constant but may depend on the temperature, pressure, humidity, and on the frequency of the applied fields. Typical values of the relative dielectric constant $\epsilon_r = \epsilon/\epsilon_0$ are given in Table IV.

8.5.2. Relation between P, E, and D

The electric field vector **E** is a measure of the local field due to all charges whether they be "free," "real" charges, or bound charges due to polarization of a dielectric. The electric displacement vector **D** is a measure of only the "free" charges, while **P** is a measure of polarization charges only. The relationship between these quantities can be illustrated by means of the simple parallel-plate capacitor of Fig. 25.

The capacitor plates with no dielectric present and neglecting fringe field effects have initially a surface "free" charge density $\pm\sigma$ coulombs/sq meter. Under these circumstances the electric field between the plates, $\mathbf{E}_0$, is given by

$$\mathbf{E}_0 = \frac{\sigma}{\epsilon_0}\mathbf{n} \text{ volts/meter} \qquad (8.5.2)$$

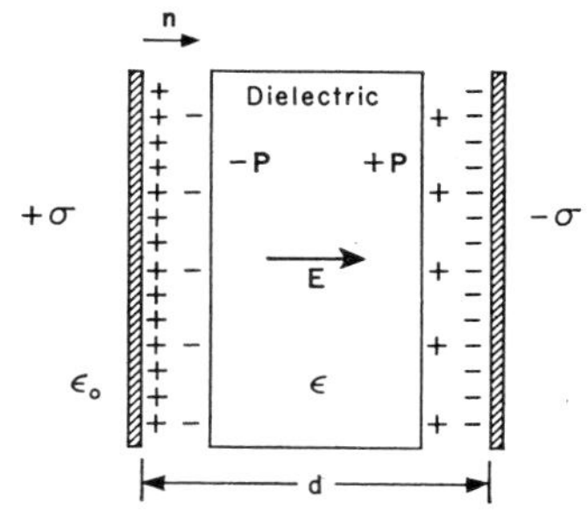

FIG. 25. Simple parallel-plate capacitor with dielectric of permittivity ϵ showing free surface charge density $\pm\sigma$ and surface polarization charge density $\pm P$.

where **n** is a unit vector directed normal to the plates. If a dielectric of permittivity ϵ is placed between the plates, a surface polarization charge appears at the plate dielectric interfaces due to the presence of the dielectric and this "bound" charge must be neutralized by equal "free" charge. The magnitude of the surface polarization charge density is $\mp P$ coulombs/sq meter. The macroscopic electric field **E** within the dielectric is now determined by the combined surface charges as shown in Fig. 25, i.e.

$$\mathbf{E} = \left(\frac{\sigma}{\epsilon_0} - \frac{P}{\epsilon_0}\right)\mathbf{n} \text{ volts/meter.} \qquad (8.5.3)$$

By definition above, the electric displacement **D** is determined only by σ so that

$$\mathbf{D} = \sigma\mathbf{n} \text{ coulombs/sq meter.}$$

Using this relation in Eq. (8.5.3)

$$\mathbf{D} = \mathbf{P} + \epsilon_0 \mathbf{E} \text{ coulombs/sq meter.} \tag{8.5.4}$$

For many gases, liquids, and solids the vectors, **E**, **P**, and **D**, have the same direction. This is interpreted to mean that the charge displacement in the dielectric has the direction of the applied field. Under these circumstances

$$\mathbf{D} = \epsilon \mathbf{E} \tag{8.5.5}$$

and ϵ is a scalar. For static (dc) fields the permittivity ϵ is dependent on molecular properties and is normally independent of the magnitude of the electric field. For high frequencies including optical frequencies a phase difference may exist between **P** and **E** and between **D** and **E**. This problem of dielectric dispersion is treated in Section 8.5.4.

The capacitance per unit area, C/A, of the capacitor of Fig. 25 is defined as the ratio of the magnitude of the "free" charge density ($\sigma = D$) to the magnitude of the voltage difference (Ed) between the plates, i.e.

$$\frac{C}{A} = \frac{D}{Ed}. \tag{8.5.6}$$

With no dielectric present, $C_0/A = \epsilon_0/d$. Therefore, with the aid of Eq. (8.5.5) it is seen that the capacitance of a capacitor with a dielectric is ϵ_r times that of the capacitance without a dielectric, in agreement with observations.

Experimental techniques for measuring permittivity of dielectric materials at various frequencies are described in the sections that follow.

From Eqs. (8.5.1), (8.5.4), and (8.5.5) it follows that

$$\mathbf{P} = \mathbf{E}\epsilon_0(\epsilon_r - 1)$$

or

$$\mathbf{P} = \mathbf{E}\epsilon_0 \chi. \tag{8.5.7}$$

The quantity $(\epsilon_r - 1)$ is called the dielectric *susceptibility* and is denoted by the letter χ.

For anisotropic media the relation given by Eq. (8.5.5) no longer holds, but can be replaced by

$$\mathbf{D} = \boldsymbol{\varepsilon} \cdot \mathbf{E} \tag{8.5.8}$$

where $\boldsymbol{\varepsilon}$ is now a tensor, or in component form

$$D_i = \sum_{j=1}^{3} \epsilon_{ij} E_j \qquad i = 1, 2, 3. \tag{8.5.9}$$

The development above and the discussion that follows is applicable over the linear range of the dielectric. For large electric fields the electric displacement and field are no longer linearly related, and higher order terms must be introduced.

8.5.3. Physical Theory of Dielectrics

We have described the formal relations that must exist between the various field vectors that appear in Maxwell's equations. Along with such a macroscopic formal theory there exists a physical theory of dielectrics that relates the measured values of permittivity ϵ, or the equivalent electric susceptibility χ, to atomic and molecular quantities and models. A description of such physical theories is beyond the scope of this article and the reader is referred to several comprehensive books that discuss dielectric theory. Van Vleck[174] and Fröhlich[175] treat of the quantum-statistical mechanical theory of dielectrics, especially as related to isolated molecules and their aggregates. Böttcher[176] has given a classical account of dielectric phenomena in terms of electrostatically interacting systems, while Smyth[177] summarizes the recent experimental work on dielectrics and dielectric losses. The present status of dielectric phenomena has been reviewed by Brown.[178] Experimental methods and applications of dielectric materials are contained in a survey prepared by von Hippel.[179]

8.5.4. Polarization at High Frequencies. Dielectric Loss*

Important information on the internal structure of dielectrics can be obtained from studies of the behavior of dielectrics under alternating electric fields. At sufficiently high frequencies there is an observable time lag in the attainment of equilibrium between the applied field **E**, and the corresponding electric displacement **D**, or polarization **P**. This means that there exists a measurable phase difference between **D** and **E** so that

* See also Vol. 6, Chapter 6.1.

[174] J. H. Van Vleck, "The Theory of Electric and Magnetic Susceptibility." Oxford Univ. Press, London and New York, 1932.

[175] H. Fröhlich, "Theory of Dielectrics: Dielectric Constant and Dielectric Loss." Oxford Univ. Press, London and New York, 1949.

[176] C. J. F. Böttcher, "Theory of Electric Polarization." Elsevier, Amsterdam, 1952.

[177] C. P. Smyth, "Dielectric Behavior and Structure." McGraw-Hill, New York, 1955.

[178] W. F. Brown, Jr., "Dielectrics," Handbuch der Physik (S. Flügge, ed.), Vol. 17. Springer, Berlin, 1956.

[179] A. R. von Hippel, "Dielectric Materials and Applications." Wiley, New York, 1954; "Dielectrics and Waves." Wiley, New York, 1954.

the simple relation, $\mathbf{D} = \epsilon\mathbf{E}$ is not valid. This phase difference may also be due to finite conductivity of the dielectric.

Polarization of a dielectric is a time-dependent phenomenon. For electronic polarization due to displacement of electrons within an atom, the order of magnitude of time required to attain equilibrium is about 10^{-15} sec, corresponding to optical frequencies in the ultraviolet. For atomic polarization due to the displacement of atoms within molecules the time is of the order of 10^{-13} to 10^{-14} sec. Orientation polarization of permanent dipoles requires time that depends on frictional resistances of the medium. The times vary from about 10^{-6} sec corresponding to radiofrequencies to 10^{-10} sec corresponding to microwave frequencies.

It is customary in describing phase dependent electrical phenomena to use complex number notations. We shall designate the time dependence of the electric field by

$$E = E_0 e^{j\omega t} \tag{8.5.10}$$

where the real part of E is the electric field intensity, $j = \sqrt{-1}$, and ω is the angular frequency equal to $2\pi f$ where f is the frequency measured in cycles/sec. If a phase difference exists between E and D of value δ, then

$$D = D_0 e^{j(\omega t - \delta)}. \tag{8.5.11}$$

The phase angle δ is independent of E, but is dependent on frequency, temperature, chemical composition, and structure of the dielectric. It is this dependence of δ on molecular parameters that makes its measurement and study of importance to physical theory.

On the assumption that δ is independent of E we may write

$$D = \epsilon E \tag{8.5.12}$$

and

$$\epsilon = \frac{D_0}{E_0} e^{-j\delta}. \tag{8.5.13}$$

The complex dielectric constant or permittivity ϵ can be written as

$$\epsilon = \epsilon_{Re} - j\epsilon_{Im} \tag{8.5.14}$$

with the subscripts denoting the real and imaginary parts. From Eqs. (8.5.12) and (8.5.13), we obtain

$$\epsilon_{Re} = \frac{D_0}{E_0} \cos \delta \tag{8.5.15}$$

$$\epsilon_{Im} = \frac{D_0}{E_0} \sin \delta \tag{8.5.16}$$

so that

$$\tan \delta = \frac{\epsilon_{Im}}{\epsilon_{Re}}. \tag{8.5.17}$$

The energy per cubic meter dissipated per second in a dielectric is

$$W = \frac{\omega}{2} D_0 E_0 \sin \delta \text{ (joules meter}^{-3}\text{ sec}^{-1}) \tag{8.5.18}$$

or from Eq. (8.5.16)

$$W = \frac{\omega}{2} E_0{}^2 \epsilon_{Im}. \tag{8.5.19}$$

The imaginary part of the complex dielectric constant is called the *loss factor*. The phase angle δ is sometimes called the *loss angle*, $\sin \delta$ the *power factor*, and $\tan \delta$ the *loss tangent* or dissipation factor.

As the frequency of the electric field approaches zero corresponding to the dc or static case, ϵ_{Im} approaches zero and ϵ_{Re} approaches the static dielectric constant ϵ_s. As the frequency approaches infinity, ϵ_{Re} approaches a value which we shall designate by ϵ_∞, the optical dielectric constant, related to the index of refraction n of nonferrous media by

$$\epsilon_\infty = n^2 \epsilon_0 \tag{8.5.20}$$

when absorption is negligible.

The polarizations thus far considered are those present in homogeneous materials. In a heterogeneous material an additional type of polarization, known as *interfacial polarization*, arises. Interfacial polarization must exist in any dielectric made up of two or more phases and will, therefore, contribute to the permittivity of the material. For a two-layer system, for example, the condition for the existence of interfacial polarization is that $\epsilon_1 \sigma_2 \neq \epsilon_2 \sigma_1$ where σ is the conductivity. The magnitude of the charge flow is determined by the difference in ϵ and σ of the two components. This charge accumulation is generally slow, requiring seconds or minutes, and observed at very low frequencies. However, if one phase has a high conductivity polarization may be observed at radiofrequencies.[180] The variation of permittivity with frequency for a material having interfacial, electronic, atomic, and dipole polarization is shown schematically by Fig. 26.

8.5.5. Relaxation Phenomena in Dielectrics

When the dielectric loss is due primarily to orientation of dipoles, it is convenient to introduce the concept of relaxation effect of the permanent dipoles and to associate a single *relaxation time* τ with the phenomenon. It is assumed that equilibrium is attained exponentially and that a function of the type,

$$f(t) \sim e^{-t/\tau} \tag{8.5.21}$$

[180] E. J. Murphy and S. O. Morgan, *Bell System Tech. J.* **16**, 493 (1937).

characterizes the relaxation mechanism. Here τ is the relaxation time independent of t but dependent on temperature and the structure of the dielectric.

For a simple dielectric model[175] in which it is assumed that the rate of polarization due to dipole orientation is proportional to the existing

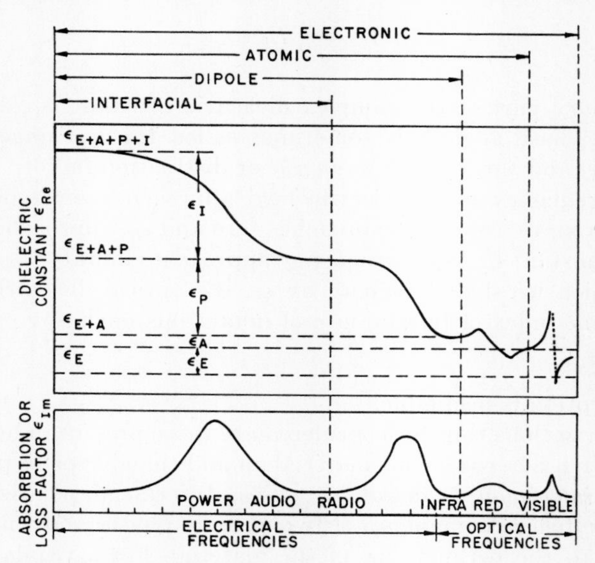

FIG. 26. Schematic diagram of variation of dielectric constant and dielectric absorption with frequency for a material having electronic, atomic, dipole, and interfacial polarizations (from reference 180, p. 497).

polarization, the relaxation time τ can be related to the complex permittivity by the equation

$$\epsilon = \epsilon_\infty + \frac{(\epsilon_s - \epsilon_\infty)}{1 + j\omega\tau}. \tag{8.5.22}$$

The real and imaginary parts can be separated to give

$$\epsilon_{Re} = \epsilon_\infty + \frac{(\epsilon_s - \epsilon_\infty)}{1 - \omega^2\tau^2} \tag{8.5.23}$$

$$\epsilon_{Im} = \frac{(\epsilon_s - \epsilon_\infty)\omega\tau}{1 - \omega^2\tau^2}. \tag{8.5.24}$$

It follows from these equations that ϵ_{Im} approaches zero for both small and large values of ω (anomalous dispersion), attaining a maximum value when

$$\omega\tau = 1. \tag{8.5.25}$$

By measuring the frequency at which the loss factor is a maximum, it is possible to obtain τ experimentally.

Debye[181] developed the classical interpretation of dielectric phenomena and suggested the relations that basically exist between permittivity, frequency, molecular size, and viscosity of the medium for the case of liquid dielectrics. According to this simple model of Debye an intrinsic relaxation time τ' can be deduced relating the spherical radius a, coefficient of viscosity η, and the absolute temperature T,

$$\tau' = \frac{4\pi\eta a^3}{kT}. \tag{8.5.26}$$

The relaxation time τ of Eq. (8.5.22) is related to τ' by

$$\tau = \tau' \left(\frac{\epsilon_s + 2}{\epsilon_\infty + 2} \right). \tag{8.5.27}$$

Modification, refinement, and extensions of the original Debye theory are described by Onsager,[182] Cole and Cole,[183] Fuoss and Kirkwood,[184] and Kauzmann.[185]

8.5.6. Lumped Circuit Measurements

The relative dielectric constant ϵ_r of a material is determined from measurements of the capacitance of a capacitor with and without the material under test, C and C_0 respectively. Thus,

$$\epsilon_r = \frac{C}{C_0}. \tag{8.5.28}$$

C_0 can be measured if the dielectric material can subsequently be introduced without altering the geometry. More frequently, the test capacitor is designed so that its air capacitance (except under unusual conditions air can be assumed to be equivalent to a perfect vacuum) can be accurately computed. For a parallel-plane capacitor of infinite extent (no fringe field) the per unit area capacitance (Section 8.1.3) is $C_0/A = \epsilon_0/d$ where d is the separation between planes. The per unit area capacitance of a parallel-plane capacitor of finite area is larger than that indicated by the simple formula due to capacitance arising from the fringe field. Corrections for the fringe field can be developed, but the usual technique

[181] P. Debye, "Polar Molecules." Chemical Catalogue, New York, 1929.

[182] L. Onsager, *J. Am. Chem. Soc.* **58,** 1486 (1936).

[183] R. H. Cole and K. S. Cole, *J. Chem. Phys.* **9,** 341 (1941).

[184] M. Fuoss and J. C. Kirkwood, *J. Am. Chem. Soc.* **63,** 385 (1941); *J. Chem. Phys.* **9,** 329 (1941).

[185] W. Kauzmann, *Revs. Modern Phys.* **14,** 12 (1942).

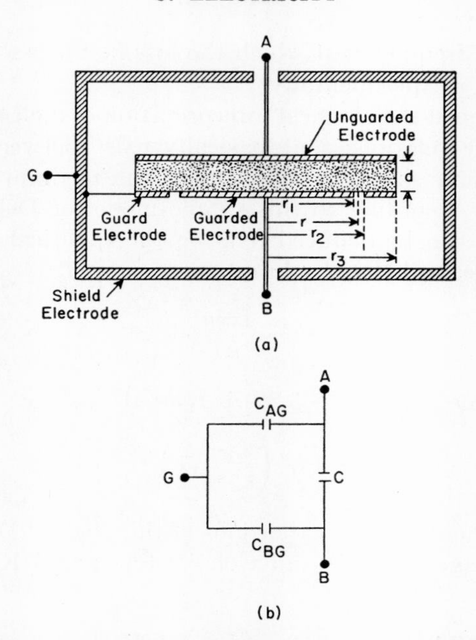

FIG. 27. Three-electrode test capacitor: (a) physical arrangement, (b) equivalent capacitances.

is to employ a guard ring as shown in Fig. 27(a) to greatly reduce the correction required.[186,187] For maximum effectiveness $(r_3 - r_2)$ should be at least $2d$, and $(r_2 - r_1)$ should be as small as possible. The effective radius of the guarded electrode r is used in computing the capacitance C_0 and is[188]

$$r = \frac{r_1 + r_2}{2} - \frac{2d}{\pi} \ln \left[\cosh \frac{\pi(r_2 - r_1)}{4d} \right]. \tag{8.5.29}$$

Evaluation of C of the test capacitor requires the use of three-terminal measurements as is seen from the equivalent circuit in Fig. 27(b). Reference 187 shows other test cell arrangements to accommodate gas, liquid, and solid dielectric materials.

Electrode attachment to the dielectric material is very important as even a small air gap between the material and electrodes may cause serious errors. The electrodes should preferably be very thin although not so

[186] Robert F. Field, *Am. Soc. Testing Materials* **54,** 456 (1954).

[187] *ASTM Standards* 1955 Part 6, D150-54T, p. 500 (1956).

[188] W. G. Amery and F. Hamburger, Jr., *Am. Soc. Testing Materials Proc.* **49,** 1079 (1949).

thin that lateral resistance is significant. Electrodes may be attached by the following techniques:

(1) Solid electrodes may be pressed against solid dielectric material. Even when material and electrodes are ground plane and parallel to 2.5 μ, the air gap may still be sufficient to introduce a measurement error of several per cent.

(2) Metal foil (7.5 to 75 μ) may be pressed on and made adherent with a minimum quantity of refined petrolatum, silicone grease, silicone oil, or other suitable low-loss adhesive. Foil can be trimmed at the edge with a finely ground blade, and guard electrode formed by cutting out a narrow strip by means of an ink compass ground to a narrow cutting edge.

(3) Conducting paint can be brushed or sprayed on. Commercial high-conductivity silver paints are easily applied and either air dried or baked at low temperatures. Paint solvent must not affect the dielectric material.

(4) Low-melting-point metal can be applied with a suitable spray gun.

(5) Metal may be evaporated onto the dielectric material with the aid of a vacuum system. The evaporated electrode is usually so thin that it must be backed with electroplated copper to provide adequate lateral conductivity.

(6) Silver electrodes can be fired onto dielectric materials that can withstand the firing temperature (about 250°C).

(7) Mercury can be used for the electrodes provided it does not react with the dielectric material. Confining rings with sharp edges can be used to retain the mercury. Usual precautions should be taken with regard to the toxicity of mercury vapor.

The losses of a capacitor are usually represented by a conductance in shunt with a capacitance. This representation will often be valid over a range of frequencies. If desired, a series resistance-capacitance combination which is equivalent at a given frequency can be determined with the aid of equations in Section 8.3.2. A circuit representation that is valid over a wide range of frequencies will generally be more complex, particularly if inductance, resistance, and capacitance of leads are to be introduced.

The dc capacitance of a test capacitor can be measured by applying a known dc voltage and measuring the resulting charge by methods discussed in Section 8.2.3. Three separate measurements must be made in order to determine and correct for fringe field and shield capacitance as shown in Fig. 27. The measured charge must also be corrected for leakage current if significant.

Interfacial polarization arises from charge carriers migrating through the dielectric; it is determined by applying a steady voltage and noting the decrease of current with time from an initial moderate value immediately following conventional charging current to a final small value corresponding to leakage current.[189] The interfacial polarization time variation may take place over minutes, days, and even months. Space charge effects play an important part in interfacial polarization, and tedious measurements are required for exact evaluation.

Generally, dc capacitance measurements are difficult to carry out and of limited accuracy as compared with ac measurements. Alternating current measurements evaluate $C = dQ/dV$ which may differ from $C = Q/V$ if nonlinear effects exist. Nonlinear effects can be investigated by either making dc measurements at different voltages or by making ac measurements with different dc biases. If significant nonlinearity exists, ac measurements must be made with a small enough signal so as to evaluate the slope at the point of interest.

Rapid approximate ac measurements of two-terminal capacitance can be made with a "Q" meter described in Section 8.3.2. A variety of bridges (Section 8.3.2) can be used for two-terminal capacitance measurements with somewhat improved measurement accuracy. Special techniques can be developed utilizing measurement of resonance frequency (Section 8.3.4) to determine two-terminal capacitance. Since frequency can be measured with high accuracy, this is an attractive method; however, although fractional capacitance changes of about 10^{-7} can be detected, absolute capacitance measurements have about the same accuracy as other measurement methods. Two-terminal capacitance measurements should be corrected for fringe-field capacitance (including stray capacitance).

As mentioned above, corrections to two-terminal capacitance measurements can be largely eliminated by making three-terminal measurements. Measurement of the direct capacitance C of Fig. 27 can be made with the use of a transfer impedance bridge of Fig. 17(g) or a Schering bridge of Fig. 17(i) both of which are discussed in Section 8.3.2. In the case of the Schering bridge a double balancing is required to complete the measurement.

With moderate care test capacitors can be made which require negligible corrections up to 1 Mc/sec for residual impedances such as lead resistance, capacitance, and inductance. Capacitance can be measured with an accuracy of $\pm(0.2\% + 0.04\ \mu\mu f)$, and loss tangent of $\pm(2\% + 0.00005)$. A special micrometer-electrode test capacitor[190] can be employed for measurements up to 100 Mc/sec without introducing significant errors. Capacitance accuracy is about $\pm(0.5\% + 0.1\ \mu\mu f)$, and loss tangent accuracy is about $\pm(2\% + 0.0002)$. The micrometer-elec-

[189] A. R. von Hippel, "Dielectric Materials and Applications," p. 52. Wiley, New York, 1954.

[190] L. Hartshorn and W. H. Ward, *J. Inst. Elec. Engrs. (London)* **79**, 597 (1936).

trode system can be incorporated in the axis of a re-entrant cylindrical cavity, and resonance-frequency measurements of capacitance can be extended to about 1000 Mc/sec.[191,192] Higher frequency measurements are generally made using distributed circuit techniques considered in the next section. Details concerning accuracy of different measurement methods for various frequencies are given in reference 186.

The dielectric strength of a material is the voltage gradient which causes electrical breakdown. The voltage corresponding to breakdown is not clearly defined but is generally considered to be the voltage marking the transition from essentially no current to short-circuit current. Thus, rather arbitrarily breakdown voltage could be specified as the voltage corresponding to a specified current. More reproducible data can generally be obtained by defining breakdown voltage as the point at which the current has a specified arbitrary rms noise (fluctuation) value. Breakdown voltage depends upon many factors such as electrode size, shape, and spacing, voltage wave shape and rate of increase, temperature, humidity, material conditioning, etc., and these factors must be carefully controlled for standard testing.[193]

8.5.7. Distributed Circuit Measurements[194] *

As the measurement frequency is increased, material size, electrode, and lead effects become more significant. In the vicinity of 1000 Mc/sec these effects become so dominant that suitable measurements can only be made by introducing the dielectric material as a discontinuity in a transmission system (see Section 8.4.1). Measurements at frequencies higher than about 20,000 Mc/sec are conveniently made using optical principles (Part 7).

Slotted transmission systems can be adapted in several ways to dielectric constant measurements. The most direct technique is to use a suitable standing wave detector to measure the input (characteristic) impedance,

$$Z_0 = \sqrt{\frac{R + j\omega L}{G + j\omega C}}$$

of a similar infinite-length transmission system. If the per unit length

* See also Vol. 2, Section 10.6.3.

[191] C. N. Works, T. W. Dakin, and F. W. Boggs, *Trans. Am. Inst. Elec. Engrs.* **63,** 1092, 1452 (1944); or *Proc. I.R.E.* **33,** 245 (1945).

[192] J. H. Beardsley, *Rev. Sci. Instr.* **24,** 180 (1953).

[193] *ASTM Standards 1955*, Part 6, D149–55T, p. 493 (1956).

[194] A. R. von Hippel, "Dielectric Materials and Applications," p. 63. Wiley, New York, 1954; see also P. A. Miles, W. B. Westphal, and A. von Hippel, *Revs. Modern Phys.* **29,** 279 (1957).

resistance R and inductance L of the transmission system are known, the per unit length conductance G and capacitance C of the dielectric material can be evaluated from the Z_0 measurement. A similar measurement can be carried out with a dielectric material transmission system of finite length. Equations (8.4.17) and (8.4.18) can be used to determine the input impedance of the dielectric system in terms of the characteristic impedance and the known terminating impedance.

Dielectric constant measurements can also be made by a comparison arrangement, where the attenuation and phase shift of a signal transmitted through the dielectric material is compared on a null basis with

TABLE IV. Values of ϵ_{Re}/ϵ_0 and tan δ for Some Dielectric Materials

Material*	Temperature (°C)	Frequency (cycles/sec)								
		10^2	10^3	10^4	10^5	10^6	10^7	10^8	10^9	10^{10}
Ice, from conductivity water	−12	—	—	—	4.8	4.15	3.7	—	—	3.17
		—	—	—	8000	1200	180	—	—	7
Sodium chloride, fresh crystals	25	5.90	5.90	5.90	5.90	5.90	5.90	—	—	—
		<1	<1	<1	<2	<2	<2	—	—	—
Glass, Corning 0010	24	6.68	6.63	6.57	6.50	6.43	6.39	6.33	—	5.96
		77.5	53.5	35	23	16.5	15	23	—	90
Mica, clear ruby muscovite	26	5.4	5.4	5.4	5.4	5.4	5.4	5.4	—	—
		25	6	3.5	3	3	3	2	—	—
Polyethylene	24	2.25	2.25	2.25	2.25	2.25	2.25	—	—	2.25
		5	<3	<3	<5	<4	<3	—	—	4
Teflon	22	2.1	2.1	2.1	2.1	2.1	2.1	2.1	2.1	2.08
		<5	<3	<3	<3	<2	<2	<2	1.5	3.7
Araldite, Ciba E-134	25	7.3	6.1	5.3	4.7	4.4	4.1	3.7	—	3.1
		1200	1050	920	760	770	1000	1300	—	390
Water, conductivity	25	—	—	—	78.2	78.2	78.2	78	—	55
		—	—	—	4000	400	46	50	—	5400
Carbon tetrachloride, purified	25	2.17	2.17	2.17	2.17	2.17	2.17	2.17	2.17	2.17
		60	8	0.4	<0.4	<0.4	<2	<2	<1	16
Silicone, Dow DC550	25	2.91	2.90	2.90	2.90	2.90	2.88	—	—	2.60
		1630	170	18	3.7	3.8	12	—	—	220

* For each material the top line contains values for ϵ_{Re}/ϵ_0; the bottom line, for tan δ.

the same signal transmitted through a calibrated attenuator and phase shifter.[195] The arrangement is essentially a bridge which can conveniently be constructed using magic tee components.[196]

Resonance methods have been adapted to cavities operating to about 20,000 Mc/sec.[197,198]

8.5.8. Data of Dielectric Materials

Table IV gives data of a few representative materials taken from an extensive tabulation of more than 600 important materials.[199] Additional data will be found throughout various publications and are reviewed annually.[200]

8.6. Atmospheric Electricity

Problems in atmospheric electricity of concern here consist primarily in measurements of electric field or potential gradient E, the conduction current density J, and electrical conductivity of the air λ. These measurements are made during fair weather and during thunderstorms, near the ground and in the upper atmosphere. During fair weather potential gradient near the ground averages about 126 volts/meter corresponding to a surface charge density of negative electricity of about 11×10^{-10} coulombs/sq meter. The corresponding conduction current density is about 3.2×10^{-12} amp/sq meter or about 1630 amp for an area equal to the whole surface of the earth.[201–203] The ionization of the lower atmospheric layer is caused by radiation from radioactive materials and by cosmic rays.[204,205]

[195] J. G. Powles and W. Jackson, *Proc. Inst. Elec. Engrs.* **96, Pt. III,** 383 (1949).

[196] H. G. Beljers and W. J. van de Landt, *Philips Research Repts.* **6,** 96 (1951).

[197] G. Birnbaum and J. Franeau, *J. Appl. Phys.* **20,** 817 (1949).

[198] S. Saito and K. Kurokawa, *Proc. I.R.E.* **44,** 35 (1956).

[199] A. R. von Hippel, "Dielectric Materials and Applications," p. 291. Wiley, New York, 1954; see also "Tables of Dielectric Materials," Technical Report 119, Vol. 5. Laboratory for Insulation Research, Massachusetts Institute of Technology, Cambridge, Massachusetts, 1957.

[200] "Digest of Literature on Dielectrics," National Academy of Science, National Research Council, Division of Engineering and Industrail Research, Washington, D. C.

[201] S. J. Mauchly, *Terrestrial Magnetism and Atmospheric Elec.* **28,** 61 (1923).

[202] J. Frenkel, *Zhur. Fiz. Khim.* **8,** 285 (1944).

[203] J. A. Chalmers, "Atmospheric Electricity." Oxford Univ. Press, London and New York, 1949.

[204] J. Elster and H. Gietel, *Physik Z.* **2,** 590 (1901).

[205] J. R. Wight and O. F. Smith, *Phys. Rev.* **2,** 459 (1915).

During thunderstorm activity the field near the ground and under a cumulus-nimbus cloud can vary within a few minutes from values of 100 volts/meter to upwards of 10,000 volts/meter accompanied by reversals in sign of the gradient.[206]

8.6.1. Measurement of Electric Field in the Atmosphere

The present techniques used to measure potential gradients in the atmosphere are but slight modifications of those described in Section 8.1.5 in electrostatics. If a sphere of radius r is exposed to the atmosphere with its center at the point at which the potential V is to be measured, then the potential of the sphere is given by

$$v = V + \frac{Q}{\epsilon_0 r} \tag{8.6.1}$$

if the radius of the sphere is small compared to the height above the ground. Q is the charge on the sphere, r the radius, and ϵ_0 is the universal constant equal to 8.854×10^{-12} farads/meter. It follows that if Q and v can be measured then V can be deduced. If the sphere is connected to the ground whose potential is zero, then

$$V = -\frac{Q}{\epsilon_0 r}. \tag{8.6.2}$$

In practice the sphere is momentarily grounded while exposed and then shielded from the earth's field and measurements made of the induced charge by an electrometer. If the capacity C of the entire conducting system—sphere plus electrometer—is known, then the induced charge is obtained from a measure of the potential V_1 to which the system is raised on shielding.

The test plate method of Wilson[207] for measuring the field at the ground is a simple modification of the above method as shown in Fig. 28. A horizontal conducting plate T of a few square centimeters area is connected to an electrometer by the rod W. The plate is surrounded by a larger guard ring G in the same plane as the plate. A conducting cover A is placed over the guard ring completely shielding the test plate. The electrometer and test plate are then grounded and the cover removed whereupon the test plate and electrometer will be raised in potential corresponding to existing conditions in the atmosphere. The potential can be brought back to zero as indicated by the electrometer by means

[206] J. M. Meek and F. R. Perry, *Repts. Prog. in Phys.* **10**, 314 (1944).

[207] C. T. R. Wilson, *Proc. Cambridge Phil. Soc.* **13**, 184 (1905); *Proc. Roy. Soc.* **A80**, 555 (1908).

of the variable capacitor P, electrically connected to the system by the rod R. The testing plate is now at zero potential, and the charge on its exposed surface is the same as if it were connected to the earth. By proper calibration the charge is given by reading the calibrated scale of the variable capacitor. The earth's field at the ground is then computed from the measured charge density of the test plate.

Several variations of the test plate techniques have been devised. If the test plate is rapidly exposed to and shielded from the field, an ac potential is generated and can be easily amplified and subsequently rectified for measurement. Although the potential is proportional to the field, the sign of the field is not known unless an auxiliary phase sensitive detector

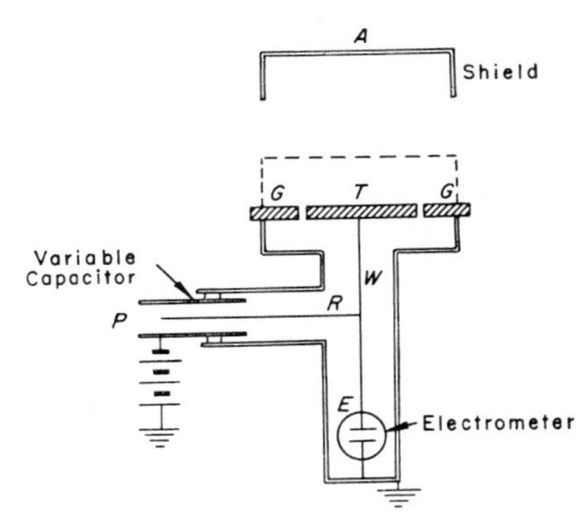

FIG. 28. Principle of the Wilson test plate technique for measuring electric fields near the ground.

is used. Waddel[208] used mechanical commutation by means of a switch on the rotating sector wheel. Cross[209] used a phase sensitive rectifier to determine the sign of the field. An electrical polarity indicating device used for field measurement from rockets is described by Clark.[210] Workman and Holzer[211] and Chalmers[212] measured field intensities by rotating the test plate from an open to a covered position and then measured the induced intermittent charge. Some of the problems associated with the interpretation of recorded field data are discussed by Israël and Lahmeyer.[213]

[208] R. C. Waddel, *Rev. Sci. Instr.* **19,** 31 (1948).

[209] A. S. Cross, *Brit. J. Appl. Phys.* **4,** Supplement 2, 547 (1953).

[210] J. F. Clark, Jr., *Instruments* **22,** 1007 (1949).

[211] E. J. Workman and R. E. Holzer, *Rev. Sci. Instr.* **10,** 160 (1939).

[212] J. A. Chalmers, *J. Atmospheric Terrest. Phys.* **4,** 124 (1953).

[213] H. Israël and G. Lahmeyer, *Terrest. Magnetism and Atmospheric Elec.* **53,** 373 (1948).

If the electric field varies with height, there must be a space charge density ρ that satisfies Poisson's equation

$$\frac{dE}{dh} = \frac{\rho}{\epsilon_0}.$$

8.6.2. Measurement of Atmospheric Currents

Measurements of the vertical electric current densities are difficult due principally to their minuteness, 10^{-12} amp/sq meter, and the disturbing effects of electrostatic induction due to changing electric fields. The current density J may be measured directly or may be computed from a knowledge of a vertical field E and the conductivity of the air $\lambda (E\lambda = J)$.

The most widely used direct method is basically the test plate method of Wilson[207] and consists of measuring the charge that accumulates on the test plate per unit time. In applying this method it is necessary to compensate for the bound charge on the plate due to changes in the field. Wilson[207] and Scrase[214] overcome this difficulty by the use of a compensating potential (Section 8.6.1) to obtain an effect equal and opposite to the change in bound charge. The use of a direct recording dc amplifier to measure atmospheric electric current is described by Kasemir.[215]

8.6.3. Atmospheric Conductivity

The conductivity of the air is due to the presence of positive and negative ions. The total conductivity λ is due to the contribution λ_+ and λ_- of each ion present so that

$$\lambda = \lambda_+ + \lambda_- \tag{8.6.3}$$

with

$$\lambda_+ = \Sigma q_+ n_1 \mu_1 \quad \text{and} \quad \lambda_- = \Sigma q_- n_2 \mu_2 \tag{8.6.4}$$

where n_1 and n_2 represent the number of positive and negative ions per unit volume, μ_1 and μ_2 their mobilities, and q_+ and q_- the charges on each ion species.

The method that is employed to determine the conductivity due to either ion species present in the atmosphere is to allow a stream of air to pass over a charged conductor and to measure the rate of loss of charge on the conductor. The apparatus used is the Gerdien aspirator[216] or a suitable modification.[217,218] In this apparatus the charged conductor is cylindrical and is surrounded by a coaxial cylinder kept at a zero poten-

[214] F. J. Scrase, *Geophys. Mem. Profess. Notes Meteorol. Office* **58** (1933).
[215] H. W. Kasemir, *J. Atmospheric Terrest. Phys.* **2,** 1 (1952).
[216] H. Gerdien, *Physik. Z.* **6,** 800 (1905).
[217] W. F. G. Swann, *Terrest. Magnetism and Atmospheric Elec.* **19,** 23 (1914).
[218] J. J. Nolan and P. J. Nolan, *Proc. Roy. Irish Acad.* **A43,** 79 (1937).

tial.[216] A current of air is drawn between the cylinder, and the potential charge of the inner cylinder is measured by an electrometer. The potential difference between the cylinders is kept small and the amount of air drawn large so that only a small number of ions present in the air stream is removed during its passage by the cylinder. If C_e is the capacitance of the exposed conductor measured in farads, C the capacitance of the entire system including the electrometer, V the potential of the exposed conductor measured in volts, and t the time in seconds, then

$$\lambda = -\frac{C}{C_e}\frac{\epsilon_0}{V}\frac{dV}{dt} \tag{8.6.5}$$

with $\epsilon_0 = 8.854 \times 10^{-12}$ farads/meter. Typical values for conductivity near the ground are $\lambda_+ = 1.26 \times 10^{-14}$ mho/meter and $\lambda_- = 1.1 \times 10^{-14}$ mho/meter.

8.6.4. Measurements in the Free Atmosphere

The techniques used to measure electric fields and air conductivities in the free atmosphere are basically similar to those described for ground measurements, suitably modified, however, to adapt them for use in aircraft, balloons, or rockets. For a summary of these techniques see Israël[219] for aerological applications and Newell[220] for applications to rockets.

8.6.5. Precipitation Electricity

Associated with the phenomena of atmospheric electricity previously discussed are important electrical effects that arise during the production of precipitation. Among these effects are the mechanisms responsible for the production of lightning and the mechanisms responsible for the maintenance of the observed free charge on the surface of the earth. The experimental techniques for studying basic mechanisms of charge production in clouds are described in papers by Findeisen,[221] Gunn,[222] Workman,[223] Simpson,[224] and Nakaya.[225] Methods for measuring fields within clouds are described by Gunn.[226]

[219] H. Israël, *Wiss. Arb. deut. meteorol. Dienstes französischen Besatzungsgebiet* **2,** 20 (1949).

[220] H. E. Newell, "High Altitude Rocket Research." Academic Press, New York, 1953.

[221] W. Findeisen, *Meteorol. Z.* **57,** 201 (1940).

[222] R. Gunn, *Geophys. Research Papers U.S.A.F.* **54,** 57 (1949); **55,** 171 (1950).

[223] E. J. Workman and S. E. Reynolds, *Phys. Rev.* **75,** 347 (1949).

[224] C. G. Simpson, *Terrest. Magnetism and Atmospheric Elec.* **53,** 33 (1948).

[225] U. Nakaya and T. Terada, *J. Fac. Sci. Hokkaido Imp. Univ., Ser.* **1,** 181 (1934).

[226] R. Gunn, *J. Appl. Phys.* **19,** 481 (1948).

8.6.6. Lightning Discharge

There exists a vast literature on the problem of electrical discharges from clouds and of the experimental techniques that have been developed for photographing and studying lightning discharges. The reader is referred to the detailed summaries on this subject prepared by the American Institute of Electrical Engineers[227] and to the papers by Bruce and Golde,[228] Hagenguth,[229] Wagner and McCann,[230] and McEachron.[231]

8.7. Thermoelectricity*

8.7.1. Thermoelectric Phenomena

A thermoelectric electromotive force (emf) is set up in a circuit composed of two dissimilar homogeneous metallic conductors A and B, whose junctions are kept at different temperatures T and T', as shown in Fig. 29. This emf called the Seebeck[232] effect can be empirically represented by the following equation

$$E = (at + 10^{-2}\, \tfrac{1}{2}bt^2 + 10^5\, \tfrac{1}{3}ct^3)10^{-6} \text{ volts} \tag{8.7.1}$$

where E is the emf, t the temperature in °C and a, b, and c constants for a given pair of metals; one junction is assumed to be kept at 0°C. In the case of an iron–constantan thermocouple with one junction at 0°C and another at 227°C the Seebeck emf is about 12.5 mv. By convention E is regarded as positive if the current produced flows from A to B at the junction which is at 0°C. A representative set of data is given in Table V. Only the order of magnitude of the constants is significant, since thermoelectric effects are sensitive to purity, mechanical, and thermal treatment.

Associated with the Seebeck effect is the Peltier[233] effect which can be described as follows. If the two conductors shown in Fig. 29 are kept at a constant temperature, and a dc current I passes through the circuit,

* See also Vol. 6, Chapter 6.6, and this volume, Section 6.1.4.2.

[227] Am. Inst. Elec. Engrs., Committee on Power Transmission and Distribution, "Lightning Reference Book." New York, 1937; *ibid.*, 1950.

[228] C. E. R. Bruce and R. H. Golde, *J. Inst. Elec. Engrs. (London)* **88**, 487 (1941).

[229] J. H. Hagenguth, *Gen. Elec. Rev.* **43**, 195 (1950).

[230] C. F. Wagner and G. D. McCann, *Trans. Am. Inst. Elec. Engrs.* **59**, 1061 (1940).

[231] K. B. McEachron, Lightning and Lightning Protection, *in* "Encyclopaedia Britannica," Vol. 14, p. 114. Chicago, 1948.

[232] T. J. Seebeck, *Abhandl. Königl. Akad. Wiss. Berlin*, 265 (1822).

[233] M. Peltier, *Ann. chim. et phys.* **56**, 371 (1834).

TABLE V. Thermoelectric emf in μv[a]

A	B	t_1 (°C)	t_2 (°C)	a	b	c
Ag	Pt	0	900	3.04	2.01	
Al	Pt	0	800	−0.80	−0.91	−0.67
Bi	Pt	0	268	−61.95	4.50	26.82
Cd	Pt	0	320	0.39	0.38	
Cu	Pt	0	900	3.13	2.46	
Ni	Pt	0	313	−2.89	0.62	
Rh	Pt	0	1,300	6.27	1.61	0.18
Sb	Pt	0	630	46.24	6.36	−14.33
W	Pt	0	1,200	9.4	3.68	
Zn	Pt	0	450	5.74	3.30	

[a] International Critical Tables, Vol. VI, p. 214. McGraw-Hill, New York, 1929.

then heat in addition to Joule heat is absorbed or generated at the junctions. The amount of heat, Q_{12}, generated or absorbed per second is proportional to the current,

$$Q_{12} = \pi_{12} I \tag{8.7.2}$$

where π_{12} is a Peltier coefficient whose value depends on the temperature of the junction and on the nature of the conductors. The Peltier heat is reversible heat and the coefficients

$$\pi_{12} = -\pi_{21}. \tag{8.7.3}$$

If the current passes in the same direction as the current produced by the Seebeck effect heat is absorbed at the hot junction while at the cold junction heat is generated. The fact that Peltier heat is reversible permits the

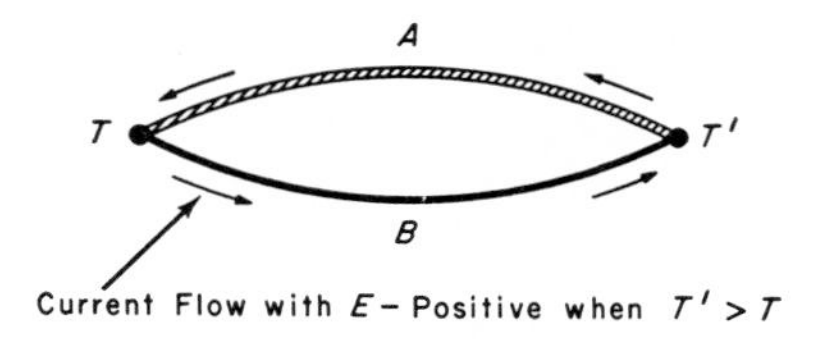

FIG. 29. The thermoelectric circuit of two dissimilar metals.

experimental possibility of separating the effect from the irreversible Joule heat I^2R. For the iron–constantan junction kept at 227°C (500° absolute) the Peltier effect is about 0.0275 joules/coulomb.

The third effect associated with thermoelectricity is the generation of Thomson[234] (Lord Kelvin) heat. In this case if a current I passes through an element of a homogeneous conductor whose temperature difference is

[234] W. Thomson, *Trans. Edinburgh Soc.* **21,** 153 (1847).

ΔT an amount of heat is absorbed or generated per second equal to

$$\Delta Q = \sigma \, \Delta T I \tag{8.7.4}$$

where σ is the Thomson coefficient, positive when the current is in the direction of increasing temperature. The Thomson heat like the Peltier heat is reversible in the sense that ΔQ changes sign when the current is reversed.

The thermodynamic relations that exist between these various thermoelectric effects are discussed in detail by Bridgman[235] and Herring.[236] We shall summarize some of the more important results. In the circuit of Fig. 29 let a unit charge pass around the circuit. Then by the first law of thermodynamics the total work done plus the heat given out is zero. Since the work done per unit charge is the emf E_{12} we have

$$E_{12} = \pi_{12} - \pi'_{21} + \int_T^{T'} (\sigma_2 - \sigma_1) \, dT \tag{8.7.5}$$

and for an infinitesimal temperature difference between junctions,

$$\frac{dE_{12}}{dT} = \frac{d\pi_{12}}{dT} + \sigma_1 - \sigma_2. \tag{8.7.6}$$

By the second law the total entropy change must be 0 since the heat processes are assumed to be reversible; therefore

$$\frac{\pi_{12}}{T} - \frac{\pi'_{21}}{T'} - \int_T^{T'} \frac{\sigma_2 - \sigma_1}{T} \, dT = 0 \tag{8.7.7}$$

or for infinitesimal temperature differences $dT = T - T'$,

$$\frac{d(\pi_{12}/T)}{dT} = \frac{\sigma_2 - \sigma_1}{T}. \tag{8.7.8}$$

By eliminating $\sigma_1 - \sigma_2$ from Eqs. (8.7.8) and (8.7.6) we have

$$\pi_{12} = T \frac{dE_{12}}{dT}. \tag{8.7.9}$$

Also by substituting π_{12} from Eq. (8.7.9) into Eq. (8.7.8) we obtain

$$T \frac{d^2 E_{12}}{dT^2} = \sigma_2 - \sigma_1. \tag{8.7.10}$$

The quantity dE_{12}/dT sometimes designated by the letter Q is called the

[235] P. W. Bridgman, "The Thermodynamics of Electrical Phenomena in Metals." Macmillan, New York, 1934.

[236] C. Herring, *Phys. Rev.* **59**, 889 (1941).

thermoelectric power. From the third law of thermodynamics,

$$\lim_{T\to 0}\left(\frac{\pi}{T}\right) = 0 \tag{8.7.11}$$

so that we may rewrite Eq. (8.7.10) as

$$Q_{12} \equiv \frac{dE_{12}}{dT} = \int_0^T \frac{\sigma_2 - \sigma_1}{T}\,dT. \tag{8.7.12}$$

This result must hold for any pair of metals. It suggests, therefore, that for each metal separately

$$Q \equiv \frac{dE}{dT} = \int_0^T \frac{\sigma}{T}\,dT \tag{8.7.13}$$

where dE/dT is the absolute *thermoelectric power* of the material.

Since we do not know precisely how π_{12} or σ vary with temperature we cannot integrate Eqs. (8.7.9) and (8.7.10) to obtain a general relation between E and T. However, Eqs. (8.7.9), (8.7.10), and (8.7.12) do represent the thermodynamic equations for a thermocouple. The experimental work of Borelius[237] indicates that these relations hold with an experimental accuracy of a few per cent. Moreover, there is no experimental evidence to indicate that these relations are not correct.

From measurements of the thermal emf as a function of temperature the relative Peltier and Thomson coefficients may be deduced. Although these same equations can be qualitatively deduced from the electron theory of metals[238] and correct order of magnitudes deduced for the Peltier and Thomson coefficients for certain metals, thermal emf's to be expected from a given pair of elements for a given temperature difference cannot yet be predicted with any degree of confidence from theory. It is necessary to resort to experimentally determined temperature–emf curves. It should be noted that the quantities E, dE/dT, π_{12}, and σ are greatly affected by impurities, heat treatment, stress, and magnetization, in fact by anything that affects the structure of the metals.

For methods for measuring the thermoelectric power of a single metal (the absolute thermoelectric power) the reader is referred to the papers of Borelius.[239] For a detailed discussion of thermoelectric phenomena in

[237] G. Borelius, *Ann. Physik* **52**, 398 (1917); **56**, 388 (1918).

[238] A. H. Wilson, "The Theory of Metals," 2nd ed. Cambridge Univ. Press, London and New York, 1953.

[239] G. Borelius, *in* "Handbuch der Metallphysik" (G. Masing, ed.), Vol. 1. Akademische Verlagsges. Leipzig, 1935.

terms of solid-state band theory the reader should consult the works of Wilson[238] and Jones.[240]

The dependence of the Thomson coefficient on temperature can be represented by the empirical formula

$$\sigma = (a + bt\,10^{-2} + ct^2\,10^{-5})10^{-6} \text{ volts deg}^{-1}. \qquad (8.7.14)$$

A few representative values are given in Table VI.

TABLE VI. Thomson Coefficient μv deg^{-1} [a]

Metal	Temperature range (°C)	a	b	c
	t_1 to t_2			
Ag	−123 to 127	−1.17	−0.50	
Al	71 to 322	+0.27	+0.08	
Au	−100 to 103	−1.49	−0.44	
Bi	25 to 32.5	6.76	2.8	
Cd	48 to 343	−9.0	−1.55	−1.5
Cu	−60 to 127	−1.42	−0.74	
Fe	32 to 182	7.66	4.1	17.0
Pb	−153 to 117	0.61	0.22	−0.38
Pt	−72 to 128	9.10	−0.48	+4.75
Zn	40 to 343	−3.11	−0.24	

[a] International Critical Tables, Vol. VI, p. 228. McGraw-Hill, New York, 1929.

In our discussion of thermoelectric effects we postulated that our conductors are homogeneous and isotropic. The problems associated with the anisotropy of thermoelectric effects are treated by Domenicali[241] and in some detail by Meissner.[242] These authors also discuss the Bridgman "internal Peltier effect" present in crystals of low symmetry.

8.7.2. Thermoelectric Laws

The use of thermoelectric effects in experimental work, especially thermometry,* is based on the validity of several thermoelectric principles or laws of thermoelectric circuits. One of the most important from a practical point of view is the law of Magnus[243] which states that for a given pair of homogeneous isotropic conductors the thermal emf developed in the closed circuit is dependent only on the temperatures of the

*See Chapter 6.1.

[240] H. Jones, *in* "Handbuch der Physik" (S. Flügge, ed.), Vol. 19. Springer, Berlin, 1956.

[241] C. A. Domenicali, *Revs. Modern Phys.* **26,** 237 (1954).

[242] W. Meissner, *in* "Handbuch der Experimentalphysik" (W. Wien and F. Harms, ds.), Vol. 11. Akademische Verlagsges., Liepzig, 1935.

[243] G. Magnus, *Ann. Physik* [2] **83,** 469 (1851).

junctions T and T' and is independent of the temperature gradient and distribution along the conductors. There have been attempts to show that a nonsymmetrical distribution of temperature and a homogeneous conductor will give rise to an emf, in this connection the work of Benedicks[244] being the most serious. However, the experimental evidence[245] is in favor of the validity of the law of Magnus and any observed emf's are to be attributed to local inhomogeneities.

TABLE VII. Thermal emf of Metals Relative to Platinum[a, 246]

Temperature (°C)	mv						
	Copper	Silver	Iron	Rhodium	Chromel P	Alumel	Constantan
−200	−0.19	−0.21	−3.10	−0.20	−3.36	+2.39	+5.35
−100	−0.37	−0.39	−1.94	−0.34	−2.20	+1.29	+2.98
0	0	0	0	0	0	0	0
+100	+0.76	+0.74	+1.98	+0.70	+2.81	−1.29	−3.51
+200	1.83	1.77	3.69	1.61	5.96	−2.17	−7.45
600	8.34	8.41	8.02	6.77	19.62	−5.28	−25.47
800	12.84	13.36	11.09	10.16	26.23	−7.08	−34.86
1000	18.20	—	14.64	14.05	32.52	−8.79	−43.92
1200				18.42	38.51	−10.74	

[a] A positive sign means that in a simple thermoelectric circuit the resultant emf given is in such a direction as to produce a current from the element to platinum at the reference junction (at 0°C).

Two additional theorems, the law of intermediate metals and the law of intermediate temperatures, can be deduced from our earlier thermodynamic analysis. The law of intermediate metals can be stated as follows: any number of intermediate metals may be inserted in a thermoelectric circuit without affecting the resultant emf, provided each inserted conductor has the same temperature at both ends. The validity of this statement permits the introduction of copper leads and measuring devices into a circuit without affecting the emf of the measuring junction. As another consequence of this law it is seen that if the thermal emf of each of several metals A, B, C, etc., is known with respect to a reference metal, say platinum, then the emf of any combination of the metals can be obtained by taking the algebraic difference of the emf's of each metal with respect to the reference metal. Some representative emf's of common metals with respect to platinum are shown in Table VII.

[244] C. Benedicks, *Ann. Physik* **55**, 1 (1918).

[245] N. Fuschillo, *Proc. Phys. Soc.* (*London*) **65B**, 896 (1952).

[246] American Institute of Physics, "Temperature, Its Measurement and Control in Science and Industry," p. 1308. Reinhold, New York, 1941.

The third law states that if two homogeneous conductors produce an emf E_1 when the junctions are at temperatures T_1 and T_2 and the emf of E_2 when the junctions are T_2 and T_3, then the emf produced when the junctions are T_1 and T_3 is $E_1 + E_2$.

8.7.3. Measurement of Thermal emf

Perhaps the most widely known application of thermoelectric phenomena is for precision thermometry. A comprehensive treatment of the problems and techniques of thermoelectric thermometry can be found in reference 246.

In practice, thermal emf's are measured either by a millivoltmeter taking advantage of Ohm's law ($E = RI$) or by a potentiometer as shown

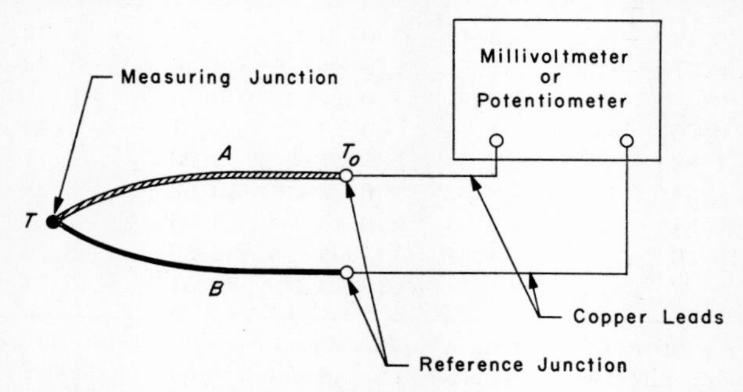

FIG. 30. Circuit diagram for the measurement of thermal emf.

in Fig. 30. The advantages of the millivoltmeter are simplicity of measurement, portability, and cost. The disadvantage is the lack of precision due to errors in the millivoltmeter and errors due to variation of resistance with temperature of the junction. For precise measurement of small thermal emf's recourse is made to a null method using the potentiometer.

For most measurements of thermal emf's to obtain accuracies of about 5 μv, slidewire potentiometers of the laboratory type will suffice. For refined measurements involving very small temperature differences or for the calibration of secondary thermometric standards requiring the highest accuracy (0.1 μv) it is advisable to use special potentiometers such as those designed by White[247] or by Wenner[248] (see also Section 8.2.4). The Wenner potentiometer is considered the most precise instrument now available for the measurement of thermal emf's.

Since thermal emf developed by a thermocouple depends on the temperature of a reference junction (usually kept at 0°C) as well as upon the temperature of the measuring junction, provision must be made to cor-

[247] W. P. White, *Z. Instrumentenk.* **34,** 147 (1914); *Rev. Sci. Instr.* **4,** 142 (1933).
[248] L. Behr, *Rev. Sci. Instr.* **3,** 109 (1932).

rect readings if the reference junction temperature varies. For general laboratory and industrial use portable potentiometers (accuracy 40 to 100 μv) are available with an automatic reference-junction compensator, usually in the form of a nickel coil whose resistance varies as the temperature changes and thus compensates for changes in the reference junction temperature.[249]

8.7.4. Some Useful Thermocouple Systems

Although many combinations of metal elements and alloys have been used for temperature measurements, only a few combinations of metals have survived the test of time.

The platinum–platinum + 10% rhodium[250] is used for precise temperature measurements in the range of 0° to 1500°C. It is also used to define the international temperature scale from 630.5° to 1063°C. It is not useful below 0°C since the thermoelectric power becomes too low (5 μv/°C at 0°C) for accurate measurement reaching 0 at about −138°C.

The Chromel P ($Ni_{90}Cr_{10}$)–Alumel($Ni_{94}Mn_3Al_2Si_1$)[250] thermocouple has excellent oxidation resistant properties at high temperatures. Its thermoelectric power is relatively high, about 40 μv/°C in the range of 250° to 1100°C. These thermocouples are best employed for temperature measurements from 500° to 1200°C.

The iron–constantan ($Cu_{55}Ni_{45}$) thermocouples[251] are extensively used for temperature measurements below 800°C. They have found wide use in industry because of the high thermoelectric power, good life in both oxidizing and reducing atmospheres and comparatively low cost. The reproducibility of the thermocouple is somewhat poor with errors of about ±1% in the temperature range from 200° to 800°C.

The copper–constantan thermocouple[250] is used for accurate temperature measurements from 350°C to liquid air temperatures (−188°C). Because of the rapid oxidation of copper this thermocouple is not used at temperatures higher than about 350°C. The thermoelectric power of the copper–constantan thermocouple varies from about 15 μv/°C at −200°C to about 60 μv/°C at 350°C. With suitable precautions a copper–constantan junction can be used to measure temperatures as low as 10°K (−262°C).[252]

[249] P. H. Dike, "Thermoelectric Thermometry," p. 56. Leeds and Northrup Co., Philadelphia, Pennsylvania, 1954.

[250] H. Shenker, J. I. Lauritzen, Jr., and R. J. Corruccini, *Natl. Bur. Standards* (*U.S.*) *Circ.* **508** (1951).

[251] R. J. Corruccini and H. Shenker, *J. Research Natl. Bur. Standards* **50,** 229 (1953).

[252] J. G. Aston, *in* "Temperature, Its Measurement and Control in Science and Industry," by American Institute of Physics, p. 219. Reinhold Publishers, New York, 1941.

For use in molten metals (up to 1800°C) a thermocouple using rod silicon carbide and graphite has been used.[253] For temperatures above −200°C and up to 1000°C Chromel–constantan[250] gives larger thermoelectric powers than those described above and has been used in radiation pyrometers.

8.7.5. Thermoelectric Properties of Semiconductors

Some of the largest thermoelectric effects are to be found in the semiconductors. Interpretation of these effects in terms of recent theoretical

TABLE VIII. Thermoelectric Powers of Various Semiconductors

Material	dE/dT in μv deg^{-1}	Remarks
n-type Ge	−800	300°K ($n \sim 2 \times 10^{15}/cm^3$)
p-type Ge	+625	300°K ($n \sim 2 \times 10^{15}/cm^3$)
Bi_2Te_3		
p-type	200–265	Room temperature
InSb		
n-type	−350	Room temperature
InSb		
p-type	+800	$T \sim 100$°K
GaAs	−195	$n \sim 4.3 \times 10^{17}/cm^3$
CdS	550	$n \sim 4 \times 10^{16}/cm^3$
PbSe		
n-type	−290	$T = 300$°K
p-type	+260	$T = 300$°K
PbTe		
n-type	−420	$T = 300$°K
p-type	+445	$T = 300$°K
Se doped with		
Cu	1240	$n \sim 0.15 \times 10^{15}/cm^3$
Ag	1040	$1.17 \times 10^{15}/cm^3$
Cd	870	$8.71 \times 10^{15}/cm^3$
Sb	1000	$1.91 \times 10^{15}/cm^3$
Bi	1190	$0.2 \times 10^{15}/cm^3$

models are given in papers by Frederikse,[254] Herring,[255] and Johnson.[256] Application of semiconductor and metallic materials to thermoelectric generators are described by Telkes.[257] In Table VIII are given thermoelectric powers of various semiconductors. When known, the concentra-

[253] G. R. Fitterer, *Am. Inst. Mining Engrs. Iron and Steel Division* **105,** 290 (1933).

[254] H. P. R. Frederikse, *Phys. Rev.* **92,** 248 (1953).

[255] C. Herring, *Phys. Rev.* **96,** 1163 (1954).

[256] V. A. Johnson, *in* "Progress in Semi-Conductors" (A. F. Gibson, ed.), Vol. I, p. 65. Wiley, New York, 1956.

[257] M. Telkes, *J. Appl. Phys.* **25,** 765 (1954).

tion of carriers and the temperature at which the measurements were made are listed under remarks.

Thermoelectric powers (dE/dt) of semiconductors are measured by maintaining a temperature gradient along a bar of the material and measuring the potential difference with a potentiometer.[258] For most semiconductors the thermoelectric power is nearly independent of the metal electrodes, usually copper, used for the potential measurement.

The thermoelectric power (dE/dt) is:

$$\left(\frac{dE}{dt}\right) = \frac{V}{T_1 - T_2}\ \text{mv/°C} \tag{8.7.15}$$

where T_1 and T_2 are the temperatures at the two ends of the sample, and V the potential difference.

[258] A. E. Middleton and W. W. Scanton, *Phys. Rev.* **92,** 219 (1953).

9. MAGNETISM*

This part starts with a discussion of the measurement and production of magnetic fields (Chapters 9.1 and 9.2), then considers the magnetic properties of matter by introducing the concept of magnetization and the means of measuring it (Chapter 9.3), and ends with an examination of the means of determining the relationship between magnetization and field for different classes of materials (Chapters 9.4 and 9.5). The magnetic field may be measured by means of any of its many effects on particles, currents, and material substances, the principal effects being forces on charges and conductors, voltages induced in coils by changing fields, and alignment of fundamental particles, nuclei, and atoms as in resonance experiments. Magnetization is measured principally by means of forces or torques on magnetized bodies and by determining the changes in the field upon introduction of magnetizable materials. The pertinent parameters used in describing the relationships of magnetization and field span many orders of magnitude. Thus a wide variety of techniques have been perfected to determine these relationships in the different ranges of possible academic or technological interest. For example, initial susceptibility (ratio of magnetization to field) of magnetizable bodies may vary from $\sim|10^{-7}|$ in common diamagnetic organic materials to $\sim 10^{+5}$ for supermalloy. This chapter is intended to serve as an introduction to the concepts useful in the field of magnetism and as a guide to the possible methods of measurement and the available literature rather than as a complete description of the multitude of techniques with which measurement of fields and magnetization is possible.

9.1. The Electromagnetic Field

The electromagnetic field is a tensor whose components are defined in a fixed coordinate system in terms of the two field vectors, one polar and one axial, which appear in the Lorentz equations for the force on a charged particle

$$\mathbf{F} = q\mathbf{E} + q\frac{\mathbf{v} \times \mathbf{B}}{c}. \tag{9.1.1}$$

---

* Part 9 is by **Anthony Arrott** and **J. E. Goldman.**

The magnitude* of these vectors depends on the choice of the unit of charge q, but is independent of the units of v/c, the velocity of the charge with respect to the velocity of transmission of electromagnetic fields. The polar vector $\mathbf{E}$, called the electric field intensity, exerts a force independent of the motion of the particle. The axial vector $\mathbf{B}$, called the magnetic induction, exerts a force only if the charge is in motion relative to the coordinate system in which $\mathbf{E}$ and $\mathbf{B}$ are evaluated. $\mathbf{E}$ and $\mathbf{B}$ represent integrations of the contributions arising from all other charges and are directly obtainable from observations of accelerations once the charge to mass ratio is specified for a test charge or set of test charges. A beam of electrons such as in a cathode ray tube[1] can be made visible by means of either the phosphorescent screen or ionizing gas within the tube. A study of the beam allows determination of the fields acting on the moving electrons. The magnetron when used as a field measuring device involves the accelerations in crossed electric and magnetic fields.[2]

9.1.1. Forces on Conductors

In conductors, e.g., a metallic wire, electrons are set in motion by the application of an applied voltage. If the conduction takes place in a magnetic field, the force on the electrons in motion will be imparted to the wire as a whole, through collisions with the lattice, giving a force per unit length of wire carrying a current, i

$$\frac{d\mathbf{F}}{dl} = \frac{\mathbf{i} \times \mathbf{B}}{c}. \tag{9.1.2}$$

The net force on two identical current-carrying wires running close together, either parallel or twisted together, is zero if the currents are in opposite directions.

Consider a plane-rectangular coil of wire of length a and width b with n turns connected to a source of current i through twisted leads. The loop may experience net forces or torques depending on its relationship to the magnetic field. If the loop is in a uniform field, it can only experience a torque. The maximum torque occurs when the plane of the loop is parallel

* The units used in this chapter are Gaussian: current and voltage are in electrostatic units (i (esu)$/c = i$ (amps)$/10$) and magnetic induction $\mathbf{B}$ is in electromagnetic units. Though the magnetic field intensity $\mathbf{H}$, and the magnetization $\mathbf{M}$, have the same units as $\mathbf{B}$ it is customary to use the term Gauss for $\mathbf{B}$, Oersted for $\mathbf{H}$, and simply emu's for $\mathbf{M}$.

[1] E. Bruche, Concerning the determination of magnetic fields by means of electron beams. *Z. tech. Physik* **12,** 94 (1931).

[2] P. M. Weinzierl, *Rev. Sci. Instr.* **21,** 492 (1950).

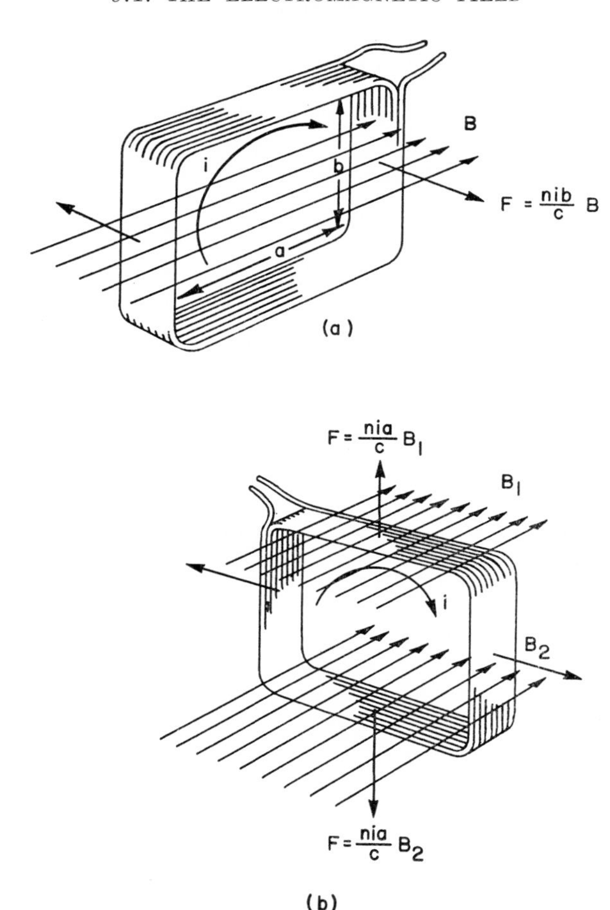

FIG. 1. (a) If a coil of n turns of wire carrying a current i is placed parallel to a uniform field of strength B, a torque results from the forces on the wires which carry current perpendicular to the field. (b) If a coil of n turns of wire carrying a current i is placed perpendicular to a field with a gradient from top to bottom, the forces on the sides cancel but the difference from top to bottom results in a net force on the coil as a whole.

to the magnetic field and is given by; see Fig. 1(a)

$$\tau_{\max} = \frac{Niab}{c} B. \tag{9.1.3}$$

Torque exerted on a current carrier is the basis for the moving coil galvanometer, the field being supplied by a permanent magnet. The calibration of a galvanometer in principle measures the field of the permanent magnet, after which it can be used to measure current. The torque on an ordinary galvanometer movement placed in an unknown field is well suited to the measurement of fields.[3,4] Another means of measuring field with the galvanometer movement is to pass an alternating current through the coil of the proper frequency, ω, to cause resonance. In this case

$$\omega = NabB/(LI)^{\frac{1}{2}} \tag{9.1.4}$$

where L is the inductance of the coil and I is its moment of inertia. This result is independent of the amplitude of the current.[5] The reduction of the measurement of magnetic field to a measurement of frequency permits convenience and accuracy.[6]*

If there is a variation of **B** across the coil, then the coil can also experience a net force. Let the coil be placed in the position of zero torque, that is, with its plane perpendicular to the field. The force is then given by (see Fig. 1(b))

$$F_z = \frac{Nia}{c}[B_{y_2} - B_{y_1}]. \tag{9.1.5}$$

In the limit where the change of B from top to bottom of the coil is small

$$F_z = \frac{Niab}{c}\frac{dB_y}{dz}. \tag{9.1.6}$$

The quantity $Nabi/c$ which appears in both Eqs. (9.1.3) and (9.1.6) is termed the magnetic moment of a coil, and because of its importance when generalized to describe the properties of matter, it is given a symbol $\boldsymbol{\mu}$. Equation (9.1.5) forms the basis for one of the most accurate means of determining the absolute value of a magnetic field. Laboratory fields are arranged so that the magnetic field is uniform over some region of space and negligible in another. The upper end of the coil is placed in the negligible field and the lower end in the uniform unknown field. The force is measured with several values of the current and with the current in both directions to correct for possible errors such as contamination of the wire by ferromagnetic impurities, magnetic properties of the materials used, and bowing out of the wires at the sides. This method has been used[7] to determine the value of the field from which the magnetic moment of the proton (see Section 9.1.3) has been most accurately calculated.

* See also Vol. 2, Part 9.

[3] C. C. Lauritsen and T. Lauritsen, *Rev. Sci. Instr.* **17,** 41 (1946); **19,** 916 (1948).

[4] H. S. Jones, *Rev. Sci. Instr.* **5,** 211 (1934).

[5] S. Butterworth, *Proc. Phys. Soc.* (*London*) **24,** 88 (1912).

[6] G. W. Green, R. C. Hanna, and S. Waring, *Rev. Sci. Instr.* **28,** 4 (1957).

[7] H. A. Thomas, R. L. Driscoll, and J. A. Hipple, Measurements of the proton moment in absolute units. *J. Research Natl. Bur. Standards* **44,** 569 (1950).

The sensitivity of such a measurement depends on the sensitivity of balances and current measurement and has been used to a relative accuracy of one part in 10^6. The absolute accuracy is limited by the uncertainty in the measurement of the effective width of the coil.

The rectangular loop of wire is not the only geometry which provides a convenient force measurement of field; for example, the force exerted on a solenoid with one end in an unknown and the other in a known field provides a convenient geometry.[8] A wire carrying a current will tend to bend in the field in order to conform to the circular orbits the electrons would follow if unimpeded. A measurement of this bending moment has been used for field measurements in small spaces.[9] The bending of a flexible current-carrying wire under tension is a powerful tool for determining particle trajectories in accelerators.[10]

9.1.2. Magnetic Induction

While **E** and **B** are independent, their space and time derivatives are connected by Maxwell's equations (see Part 8). In particular, one of these equations summarizes the facts concerning the voltage induced as the result of a changing field, namely,

$$V = \underset{\substack{\text{around}\\\text{the contour}\\\text{of the surface } S}}{\int} \mathbf{E} \cdot d\mathbf{l} = \frac{d}{dt} \underset{\substack{\text{over the}\\\text{surface } S}}{\int} \mathbf{B} \cdot d\mathbf{S}. \tag{9.1.7}$$

If the contour is formed by a coil with n turns of wire, this voltage is induced in each turn giving the total voltage

$$V = n \frac{d}{dt} \int \mathbf{B} \cdot d\mathbf{S}. \tag{9.1.8}$$

The integral on the right is called the flux ϕ through the surface S. When the coil is small enough for **B** to be considered uniform

$$\phi \equiv \int \mathbf{B} \cdot d\mathbf{S} = B_n S \tag{9.1.9}$$

where B_n is the field normal to the surface S. If the coil is rotated through 180° about an axis of symmetry in the plane of the coil, the change in flux will be given as a time integral of the induced voltage

$$\int V\, dt = 2n\phi = 2nB_n S. \tag{9.1.10}$$

[8] A. R. Kaufmann, *Rev. Sci. Instr.* **9**, 369 (1938).
[9] J. E. Parton, *Philips Tech. Rev.* **15**, 49–62 (1953).
[10] L. Cranbrook, U. S. Atomic Energy Comm. Rept. AECU 1670 (1951).

A coil used this way is often referred to as a flip coil. Through it, the measurement of flux becomes a problem of determining the $\int V\,dt$, a quantity which plays an important role in many magnetic measurements. In principle, the integral of an induced voltage can be obtained graphically using any method that permits the recording of the time–voltage pulse, e.g. oscilloscope or high speed recorder. Also electrical networks which generate a voltage proportional to $\int V\,dt$ can be used. Usually, however, such measurements are carried out with the aid of a long time-constant galvanometer, sometimes used ballistically but more often heavily damped in which case it is called a fluxmeter. The fluxmeter is treated in many standard texts. Particular reference is made to Harnwell[11] and to an article by Dunn.[12] The fluxmeter has been made into a recording instrument by means of photocells and also by adding an additional coil to sense the position of the measuring coil.[13–15] Automatic systems also permit feedback to the galvanometer to eliminate errors from the finite restoring torque of the suspension.[16] A fluxmeter can be calibrated by changing known currents in the primary of a mutual inductance whose secondary is in series with the fluxmeter and the measuring coil.

The flip coil method is made more convenient and also continuously indicating if the coil is rotated at constant angular velocity. This results in an ac voltage which can be measured with an electronic voltmeter. A useful commercial instrument uses this principle and measures the voltage by means of chopping the output and measuring the current in a meter with sufficient inertia to average out the pulsating dc. Much greater accuracy is obtained if the ac is compared with the ac generated by a second coil rotating in a standard field. Both coils are usually on the same shaft, although synchronous motors can be used. The alternating voltages are compared either by means of a potentiometer and a null detector or by the null detector alone when the reference field is variable.* [17,18]

* Here, as in several other places in this chapter, reference is made to only the recent papers as these contain adequate reference to the important earlier works.

[11] G. P. Harnwell, "Principles of Electricity and Electromagnetism," 2nd ed. McGraw-Hill, New York, 1949.

[12] H. K. Dunn, *Rev. Sci. Instr.* **10,** 368 (1939).

[13] R. F. Edgar, A new photoelectric hysteresisgraph. *Trans. Am. Inst. Elec. Engrs.* **56,** 805 (1937).

[14] R. S. Tebble, A recording fluxmeter. *J. Sci. Instr.* **30,** 369 (1953).

[15] R. H. Dicke, An electronic fluxmeter. *Rev. Sci. Instr.* **19,** 533 (1948).

[16] F. E. Haworth, A fluxmeter with counterbalanced restoring torque. *Rev. Sci. Instr.* **2,** 125 (1931).

[17] H. R. Fletcher and S. Rubin, Null reading flip-coil fluxmeters. *Rev. Sci. Instr.* **26,** 1108 (1955).

[18] S. M. Langer and F. R. Scott, The measurement of the magnetic field in a nuclear spectrometer. *Rev. Sci. Instr.* **21,** 522 (1950).

One of the difficulties with the flip coil method is that the result gives an average value of **B** over the coil. It often becomes necessary, however, to know field distributions and, therefore, to obtain the value of the field at a point. If the coil is made too small, sensitivity is sacrificed. If, however, the proper geometry is chosen for the coil, a finite coil gives the field at a point. A winding on the surface of a sphere such that there is a constant number of turns per unit length along an axis of the sphere is one such geometry. A sphere built up of such surface windings one on top of the other is known as a flux ball.[19,20] A more convenient geometry is the cylindrical coil with outside diameter at least three times the inside diameter and equal to 1.49 times the length of the coil.[21,22]

9.1.3. Nuclear Resonance Method*

The potential energy of a current carrying coil of magnetic moment, $\boldsymbol{\mu}$, as previously defined, is $-\boldsymbol{\mu} \cdot \mathbf{B}$. If the direction of the moment is reversed the energy is increased by $2\boldsymbol{\mu} \cdot \mathbf{B}$. The magnetic moment of the coil can be represented as a vector perpendicular to the plane of the coil in the direction of the advance of a right-handed screw when rotated in the manner of the current flow. It is a property of most fundamental particles that they possess energy differences depending upon their orientation in a magnetic field. Hence particles are said to possess magnetic moments. When a particle such as a proton is placed in a magnetic field B, it is possible to induce energy transitions by subjecting the proton to a beam of radiant energy of frequency ν, such that

$$h\nu = 2\boldsymbol{\mu}_H B. \tag{9.1.11}$$

The resulting absorption of energy from the beam occurs over a very small range of frequency so that once $\boldsymbol{\mu}_H$ is determined using a known B, the measurement of unknown fields can be reduced to a measurement of the resonance frequency of the proton in the unknown field. The above relation is written as $\gamma_H \equiv \dfrac{2\boldsymbol{\mu}_H}{h} = \nu/B$. The accepted value is

$$\gamma_H = (2.67523 \pm 0.00006) \times 10^4 \text{ sec}^{-1} \text{ gauss}^{-1}. \tag{9.1.12}$$

Nuclear magnetic resonance devices are commercially available and generally are used as laboratory standards. A rotating coil fluxmeter, stand-

* See also Vol. 2, Chapter 9.7 and Vol. 4, Section 4.3.2.

[19] W. F. Brown, Jr. and J. H. Sweer, *Rev. Sci. Instr.* **16,** 276 (1945).

[20] K. I. Williamson, *J. Sci. Instr.* **24,** 242 (1947).

[21] R. F. Herzog and O. Tischler, *Rev. Sci. Instr.* **24,** 1000 (1953).

[22] P. Gautier, *Compt. rend.* **242,** 1707 (1956).

ardized by a nuclear magnetic resonance measurement, is a convenient and accurate means of making most magnetic field measurements. For a comprehensive review of "Methods of Measuring Strong Magnetic Fields," see J. L. Symond's article in *Reports on Progress in Physics* (**18,** 83, 1955). Most of the methods listed above are discussed, and numerous references are supplied. In particular, the discussion and references on magnetic resonance methods are complete. The free precession of the proton in the earth's magnetic field has also been used for the precise measurement of that weak field.[23,24]

9.1.4. Measurements in Terms of Material Properties

The oldest useful field detector is the compass needle. This device still finds use as a null detector of field. Pairs of needles can be mounted so as to be insensitive to uniform fields yet quite sensitive to inhomogeneous fields.[25] In addition, use has been made of various effects of magnetic fields on the properties of material bodies. Such devices are, in principle, limitless in number, but in practice the problems in reproducing materials with nonvariable characteristics have limited such devices to applications where accuracy is not important or where induction, force, or resonance techniques are difficult to use. In particular, the magnetoresistance of bismuth and the Hall effect in bismuth and in some semiconductors have received attention.* Also high permeability magnetic materials are very sensitive to small fields. While not suited to absolute measurement of a field, they are ideal as null indicators to detect zero field. The condition of zero field is produced by a variable auxiliary field about the null indicator.[26] Such devices are greatly used in geomagnetic surveys and have sensitivities in the Γ range ($\Gamma = 10^{-5}$ oersteds).[27] The principle and use of such a device is discussed in Section 9.5.2.

* See also Vol. 6, Part 7.

[23] G. S. Waters, *Nature* **176,** 691 (1955).

[24] M. Packard and R. Varian, *Phys. Rev.* **93,** 941 (1954).

[25] R. M. Bozorth, "Ferromagnetism," p. 957. Van Nostrand, New York, 1951.

[26] J. M. Kelly, Magnetic field measurements with peaking strips. *Rev. Sci. Instr.* **22,** 256 (1951).

[27] V. B. Gerard, A simple, sensitive, saturated-core recording magnetometer. *J. Sci. Instr.* **32,** 164 (1955).

9.2. Production of Magnetic Fields

Magnetic fields used in the laboratory are generally produced by means of current carriers in the form of coils, by current carriers in conjunction with ferromagnetic materials (electromagnets), or by ferromagnetic materials with special properties (permanent magnets).

9.2.1. Coils

In the case of vacua except for wires carrying steady currents, Maxwell's equation connecting magnetic fields and currents is

$$\nabla \times \mathbf{B} = \frac{4\pi}{c} \mathbf{J} \tag{9.2.1}$$

or in the integral form

$$\mathbf{B} = \frac{1}{c} \int \frac{\mathbf{J} \times \mathbf{r}_0}{r^2} \, dV \tag{9.2.2}$$

where $\mathbf{r}_0$ is a unit vector from the current element $\mathbf{J}\, dV$ toward the point at which $\mathbf{B}$ is to be determined. For linear circuits this becomes

$$\mathbf{B} = \frac{i}{c} \int_{\substack{\text{around} \\ \text{the circuit}}} \frac{\mathbf{s}_0 \times \mathbf{r}_0}{r^2} \, dl \tag{9.2.3}$$

where $\mathbf{s}_0$ is a unit vector in the direction of the current element $i\, dl$.

The basic current pattern used in producing magnetic fields is the solenoid, a circular coil of insulated wire, usually wound uniformly with the turns of wire advancing along a given layer like the threads of a screw. Unfortunately, the fact that alternate layers of a multilayer coil must have opposite handed threads leads to some difficulty in winding uniform coils from round wires. Solutions to this problem range from using paper between layers or string wound in the grooves to the use of commercially available square wire. The field at the center of a coil of wire is at a maximum and, in general, the field can be written as a symmetric expansion in terms of even powers of the coordinates about the center of symmetry.[28] The uniformity of the field around the center of symmetry can be increased indefinitely by increasing the length of the coil. The field and its spatial variation have been calculated to consider-

[28] C. L. Bartberger, The magnetic field of a plane circular loop. *J. Appl. Phys.* **21,** 1108 (1950).

able accuracy for the long coil or solenoid as it is called.[29,30] If one specifies that a field be uniform to a certain degree over a given volume, and if one defines the efficiency of the field producing coils as the product of the field and the uniform volume per unit power, then the uniformly wound solenoid is not a very efficient configuration. It is possible to make one or a number of the derivatives of the field vanish exactly at the center

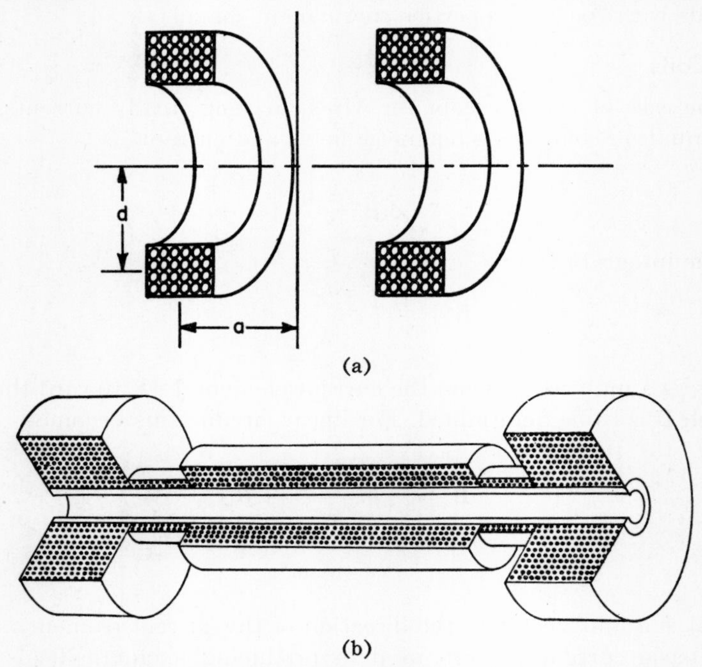

FIG. 2. Some means of producing fields from current configurations: (a) the Helmholtz coil arrangement giving maximum uniformity at the center of symmetry for $a = d$, but for uniformity over a greater volume $a \neq d$;[36] (b) Heddle's compound solenoid; (c) Clark's spherical magnet;[32] (d) Bitter's high current solenoid.

of symmetry by using multicoil configurations and by using nonuniformly wound solenoids. Helmholtz coils, for instance, make the second derivative of the field vanish by using two identical coils of diameter d placed a distance d apart (Fig. 2(a)). Analysis of various configurations has been carried out including sets of up to five coils and solenoids with end coils (Fig. 2(b)). A complete discussion of the problem is given by Garrett.[31]

[29] C. Snow, Magnetic fields of cylindrical coils and annular coils. *Natl. Bur. Standards (U.S.), Appl. Math. Ser.* 38 (1953).

[30] J. R. Barker, *Brit. J. Appl. Phys.* **1,** 65 (1950).

[31] M. W. Garrett, Axially symmetric systems for generating and measuring magnetic fields. *J. Appl. Phys.* **22,** 1091 (1951).

His analysis includes configurations for search coils which can be used to find the field at a point inasmuch as this is just the field generation problem in reverse. For instance, as the flux ball measures the field at a point exactly, so does a spherical winding produce a uniform field over the entire enclosed region. The field inside an ellipsoid is exactly uniform if the current density is such as to give a uniform current per unit length of

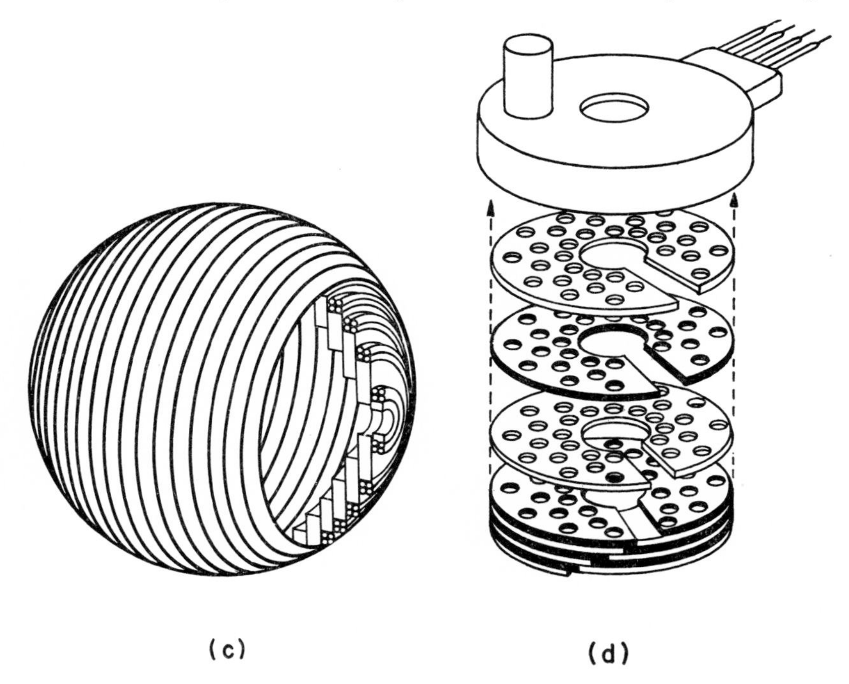

FIG. 2. (*Continued*).

axis of the ellipsoid (Fig. 2(c)). Though such fields are not readily accessible for many experiments, they have found uses.[32]

Any symmetric configuration which is used to produce a uniform field can be used to produce a uniform gradient if the current is reversed in one half of the configuration making it antisymmetric. References 33–43 are given as a bibliography.

[32] J. W. Clark, *Rev. Sci. Instr.* **9,** 320 (1938).
[33] J. R. Barker, *J. Sci. Instr.* **26,** 273 (1949).
[34] J. R. Barker, *J. Sci. Instr.* **27,** 197 (1950).
[35] M. Ference, Jr., A. E. Shaw, and R. J. Stephenson, *Rev. Sci. Instr.* **11,** 57 (1940).
[36] R. H. Lyddane and A. E. Ruark, *Rev. Sci. Instr.* **10,** 253 (1939).
[37] A. E. Ruark, M. S. Peters, *J. Opt. Soc. Am.* **13,** 205 (1926).
[38] L. W. McKeehan, *Rev. Sci. Instr.* **7,** 150 (1936).

The solenoid can be used in continuous operation to produce fields of up to approximately 1000 gauss without encountering cooling problems at room temperature. Solenoids immersed in liquid nitrogen or better still liquid hydrogen can be used for continuous operation to up to 20,000 gauss and higher for intermittent work.[44] The Bitter solenoid operates at 100,000 gauss using pancake shaped coils with high-speed water cooling directly through holes in the coils (Fig. 2(d)).[45] Pulsed field techniques are now producing fields greater than 1,000,000 gauss.[46]

Combinations of Helmholtz coils are used to cancel the earth's field where it is necessary to have field-free regions of space. The fluctuations of the earth's field with time may require the use of continuous feedback devices to control the current in instances where extremely low fields are desired.[47]

9.2.2. Electromagnets

Electromagnets are used to produce magnetic fields primarily in the range from 100 gauss to 40,000 gauss. The power dissipation and current control problems become important in solenoids when approaching the kilogauss range but with electromagnets relatively small currents are sufficient to produce fields with values approaching the saturation magnetization of iron or iron–cobalt alloys, i.e., 22,000 to 25,000 gauss.[48–50] The use of special shapes for pole faces in electromagnets allows concentrations of flux giving fields up to 40,000 gauss in limited volumes.[51]

[39] L. W. McKeehan, *Rev. Sci. Instr.* **7**, 178 (1936).

[40] L. W. McKeehan, *Rev. Sci. Instr.* **10**, 371 (1939).

[41] S. M. Rubens, *Rev. Sci. Instr.* **16**, 243 (1945).

[42] R. H. Bacon, *Rev. Sci. Instr.* **7**, 423 (1936).

[43] T. A. Heddle, *Brit. J. Appl. Phys.* **3**, 95 (1952).

[44] J. L. Olsen, Liquid helium cooled coils for making intense transient magnetic fields. *Helv. Phys. Acta* **26**, 798 (1953).

[45] F. Bitter, The design of powerful electromagnets. *Rev. Sci. Instr.* **7**, 479 (1936); **8**, 318 (1937); **10**, 373 (1939).

[46] H. P. Furth and R. W. Waniek, Production and use of high transient magnetic fields, I. *Rev. Sci. Instr.* **27**, 195 (1956); S. Foner and H. H. Kolm, Proceedings Magnetism Conference, Am. Inst. Elec. Engrs., Boston, 1956, p. 586.

[47] G. G. Scott, Compensation of the earth's magnetic field. *Rev. Sci. Instr.* **28**, 278 (1957).

[48] F. Bitter and F. E. Reed, A new type of electromagnet. *Rev. Sci. Instr.* **22**, 171 (1951).

[49] R. P. Hudson, A simply constructed small electromagnet of high performance. *J. Sci. Instr.* **26**, 401 (1949).

[50] O. Snellman and E. J. Burge, The Uppsala electromagnet. *J. Sci. Instr.* **26**, 331 (1949).

[51] H. B. Dwight and C. F. Apt, The shape of cores for laboratory electromagnets. *Rev. Sci. Instr.* **7**, 144 (1936).

Above 40,000 gauss, fields are obtained by means of solenoids either by using pulse techniques or by coping with the problem of large heat dissipation. Electromagnets can be made to give uniform fields by means of auxiliary coils about the field region or by means of special distributions of magnetic materials. The latter method is known as "shimming" the field. In resonance work where inhomogeneities of more than a few parts per 10^8 lead to line broadening and in large particle accelerators such as cyclotrons where inhomogeneity of field leads to defocusing, shimming is indispensable.[52] Electromagnets can also be designed to give uniform gradients.[53,54] Some interesting combinations of current sheets and electromagnets for producing special fields are discussed by Richardson.[55] Electromagnets play an essential part in the operation of particle accelerators and thus considerable contributions to the design of magnets have come from the building of accelerators.* [56]

In all electromagnet and solenoid operations there is the problem of current control. The approaches to this problem are:

1. Batteries which may drift with time but which have little if any short time fluxations.

2. Electronic regulators which give excellent stability against drift and in some cases negligible short time fluctuations. Generally such regulators are best suited for high voltage–low current operation.† [57]

3. Motor-generator sets which drift and give large amounts of short time fluctuations, yet are well suited for turning available ac power into low voltage high current dc power. The drift can be eliminated by feedback to the excitation windings of the generator. The noise can be slightly reduced by this means, but in general should be dealt with on the magnet side of the generator. The usual method is to filter. However, Surgan has pointed out that if only half of the turns in an electromagnet are used to produce a field, the other half can be used to eliminate fluctu-

* See also Vol. 5, Section 3.2.3.

† See also Vol. 2, Chapter 8.2.

[52] E. R. Andrew and F. A. Rudeworth, *Proc. Phys. Soc.* (*London*) **B65,** 801 (1952).

[53] A. Pacault, J. Hoarau, and J. Joussot-Dubien, The form of pole pieces for the measurement of magnetic susceptibilities. *Compt. rend.* **232,** 1932 (1951).

[54] W. Sucksmith and J. E. Thompson, *Proc. Phys. Soc.* (*London*) **A225,** 362 (1954). Discusses gradient from a current carrier in an electromagnet.

[55] H. O. W. Richardson, The use of current sheets in the design of magnets to give bounded fields of required form, free from edge distortion. *Proc. Phys. Soc.* (*London*) **B65,** 5 (1952).

[56] The essential part magnets play in accelerators is well illustrated in a series of articles in the September issue of *Rev. Sci. Instr.* **24,** 723–870 (1953); also see M. Foss, D. Sc. Thesis, Carnegie Institute of Technology, Pittsburgh, Pennsylvania, 1948.

[57] H. S. Somers, Jr., P. R. Weiss, and W. Halpern, Magnet current stabilizer. *Rev. Sci. Instr.* **22,** 612 (1951).

ations.[58] If the current is passed through the two halves in parallel opposition there will be no field produced and no fluctuations. If in one half of the parallel circuit a battery is placed so as to reduce the current flowing, a field is produced, but the fluctuations will be zero as the effective field is due only to the battery voltage. The battery is always being charged in this circuit.

9.2.3. Permanent Magnets

Permanent magnets supply a steady magnetic field and may be formed to give various shapes of magnetic fields of strengths up to 8000 gauss in reasonable volumes. They have many important applications in magnetic measurements besides the role played in various current-measuring devices. Both in magnetochemistry where stability to one part in 10^4 is required and in high resolution resonance work requiring stability of one part in 10^8, permanent magnets are most useful research tools. The main problems are temperature variations and accidental demagnetization but these are not insurmountable.[59] Some discussion of permanent magnet design is given at the end of Section 9.5.2.3 in connection with the concept of flux leakage from a magnetic circuit.

9.3. Magnetization and Its Measurement

Matter, in general, is made up of fundamental particles in dynamic equilibrium. The nuclear particles and the electrons which make up matter possess an intrinsic magnetic moment. In addition, the electrons are in motion in such a manner as to constitute current loops with which can be associated magnetic moments. It is not surprising, therefore, that the energy W of a substance depends in general on the magnetic induction $\mathbf{B}$ and that we can define a magnetic moment for the substance $\boldsymbol{\mu} = \text{grad}_{\mathbf{B}} W$. The effects of the magnetic moment may equally well be described in terms of currents, but it has been found useful, partly for historical reasons, to introduce the intensive variable *magnetization* $\mathbf{M}$, defined as the magnetic moment per unit volume. In the presence of magnetizable matter, then Eq. (9.2.1) may now be rewritten in the form

$$\begin{aligned} \nabla \times \mathbf{B} &= \frac{4\pi}{c}(\mathbf{J} + \mathbf{J}') \\ &= \frac{4\pi}{c}\mathbf{J} + 4\pi\nabla \times \mathbf{M} \end{aligned} \tag{9.3.1}$$

[58] G. Surgan, *J. Sci. Instr.* **29,** 335 (1952).
[59] J. T. Arnold, *Phys. Rev.* **102,** 136 (1956).

where $\mathbf{J}' \equiv c\nabla \times \mathbf{M}$ may be thought of as a pseudo current since magnetic moments are equally describable in terms of currents. A uniformly magnetized material, one where **M** is constant within the material, will contribute a pseudo current $\mathbf{J}'$ only at its surface. For instance a uniformly magnetized bar gives a field **B** which is identical to that which would be produced by a solenoid consisting of a current sheet conforming to the geometry of the sides of the bar. The current per unit length in the equivalent solenoid would be cM. This can be directly confirmed in the region external to the bar. The confirmation of this with a measurement of **B** within the bar is not really possible, as long as the bar is considered as a continuum. Measurements of **B** and **M** as they appear in Maxwell's and Lorentz's equations are made external to the bar. It should be noted that for reasons to be discussed below the relationship between **B** and **M** in materials is not generally quoted directly, but rather the relationships are given between **M** and a quantity $\mathbf{B} - 4\pi\mathbf{M} \equiv \mathbf{H}$.

9.3.1. The Magnetic Field Intensity, H

The use of the quantity $\mathbf{B} - 4\pi\mathbf{M} \equiv \mathbf{H}$ comes primarily from the large magnetizations found in the ferromagnetic materials. The simple experiment of the toroid of iron wound with two coils, one, the exciting coil, connected to a current source, and the other, the search coil, to a fluxmeter, is familiar to every student of physics. The large changes in **B** found for small changes in the current i can only be explained on the basis of a large change in the quantity $c\nabla \times \mathbf{M}$. By integration of Eq. (9.3.1) it follows that

$$\frac{4\pi ni}{c} = \oint_c (\mathbf{B} - 4\pi\mathbf{M}) \cdot d\mathbf{l} \equiv \oint_c \mathbf{H} \cdot d\mathbf{l} \tag{9.3.2}$$

where the integral is around any closed contour passing through the coil carrying the current i. Experiment shows that the placement of the search coil does not matter, hence **B** is uniform around the toroid. If **M** depends upon **B** then **M** is uniform around the toroid. Hence there exists in this case a quantity $\mathbf{H} \equiv \mathbf{B} - 4\pi\mathbf{M}$ which is directly proportional to the current in the exciting coil. Substituting $\mathbf{H} \equiv \mathbf{B} - 4\pi\mathbf{M}$ into Maxwell's equation gives

$$\nabla \times \mathbf{H} = \frac{4\pi}{c} \mathbf{J}. \tag{9.3.3}$$

Thus **H** is a quantity whose curl depends only on the real currents. There are no sources of **B** in the sense of magnetic poles since

$$\nabla \cdot \mathbf{B} = 0 \tag{9.3.4}$$

as follows from the induction equation. On the other hand, this means that

$$\nabla \cdot \mathbf{H} = 4\pi \nabla \cdot \mathbf{M} \tag{9.3.5}$$

so that **H** is a vector which has sources in the sense of magnetic poles. **H** is given the name magnetic field intensity, whereas **B** takes its name from the induction equation, namely, the magnetic induction. It should be made clear that, except in the case of the classical toroid experiment or in other simple geometries to be discussed below where the current is a measure of magnetic intensity, it is always **B** or **M** rather than **H** which in the strict sense is measured in an experiment.

9.3.2. Ellipsoid in a Uniform Field

The general problem of measurement of **M** follows most naturally from consideration of the very specific problem of an ellipsoidal specimen in a uniform field.[60] The field should be considered as one coming from a current configuration rather than electromagnets or permanent magnets, as the magnetization of these will be affected by the presence of the specimen in a way which depends on the specimen itself.* The current configuration can always be made not to depend upon the specimen. It is required that the volume of the field which is uniform in the absence of the sample be larger than the sample. In this discussion the solenoid is chosen as the current configuration.

If the ellipsoid is placed in such a field it will become uniformly magnetized if **M** is a single valued function of **B** for at least a certain range of values of **B**. The magnetic induction within the specimen will also be uniform. In the case where one of the axes of the ellipsoid coincides with the axis of the solenoid the magnetic induction within the ellipsoid is

$$\mathbf{B}_i = \mathbf{B}_0 + 4\pi(1 - D)\mathbf{M} \tag{9.3.6}$$

where $\mathbf{B}_0$ is the field in the absence of the sample and D is a constant depending on the ratios of the ellipsoidal axes. The intensity of magnetic field within the specimen will also be uniform and given by

$$\mathbf{H}_i = \mathbf{B}_0 - 4\pi D\mathbf{M}. \tag{9.3.7}$$

The effect of the sample is to decrease the intensity of the magnetic field at a point which lies inside the sample by an amount which depends on the magnetization and the geometrical factor D. For this reason D is called

* This is generally known as the "image effect."

[60] For the mathematical details see J. A. Stratton, "Electromagnetic Theory," Chapters 3 and 4, particularly pp. 211–217. McGraw-Hill, New York, 1941.

the demagnetizing factor.* D has the value $\frac{1}{3}$ for a sphere, approaches 0 in the limit of an infinitely long needle-shaped sample, and approaches 1 as the specimen approaches an infinitesimally thin plane perpendicular to the field. D is, in general, an elliptical integral. Some values are given in Bozorth's book along with references to original calculations.

Outside the specimen, **B** and **H** are equal though the sources must be considered as different. **B** will be a superposition of the fields from the solenoid and the pseudo or surface currents of the uniformly magnetized specimen whereas the **H** will be a superposition of the fields from the solenoid and the "magnetic poles" arising from $\nabla \cdot \mathbf{M}$ at the surface of the specimen.

9.3.3. Mechanical Measurements with Ellipsoids

If the uniformly magnetized ellipsoid is thought of in terms of its pseudo surface currents it is possible to see how uniform fields can exert torques on an ellipsoid whose axes do not coincide with the field direction and how field gradients can result in forces on an ellipsoid. If an ellipsoid is placed in a solenoid, **B** and **M** will be uniform in the ellipsoid regardless of the orientation of the axes with respect to the field but they coincide in direction only when one of the axes is aligned with the axis of the solenoid. If there is an angle between **B** and **M**, there is a torque given by

$$\tau = \int_v \mathbf{M} \times \mathbf{B} \, dV. \tag{9.3.8}$$

The torque on ellipsoids can be used as a measurement of **M** if the demagnetizing factors are known and the sample is isotropic. The frequency of oscillation of an ellipsoid suspended on a torsion member can also be used to determine **M**.[61]

If the magnetic field is in the x direction but varies with y, then the forces on the pseudo currents on the surface will not cancel each other and will result in a net force in the y direction according to

$$F_y = MV \frac{dB_x}{dy}. \tag{9.3.9}$$

The value of **B** which appears in this expression, like that for the torque above, is neither **B** nor **H** inside the sample but $\mathbf{B}_0 = \mathbf{H}_0$, the field in the absence of the sample. It is here assumed that the gradient of $\mathbf{H}_0$ is insufficient to cause the nonuniformity of the field to make inapplicable

* In the literature, $4\pi D$ is sometimes called the demagnetizing factor. We prefer, however, the terminology of D for demagnetizing factor and to refer to $4\pi D$ as the demagnetization coefficient, and $4\pi DM$ as the demagnetizing field.

[61] J. H. E. Griffiths and J. R. MacDonald, *J. Sci. Instr.* **28**, 56 (1951).

the solution of the problem of the ellipsoid in the uniform field. Whether or not this is so depends on a knowledge of dM/dB or dM/dH for the sample under consideration. If the magnetization is not constant over the whole sample, the pseudo currents appear also in the interior of the sample and the force must be given by an integral

$$F_y = \int M \frac{dH_x}{dy} dv. \tag{9.3.10}$$

Some methods of applying Eq. (9.3.10) will be discussed in Section 9.4.1. (An equivalent discussion of forces and torques on magnetized bodies can be carried out in terms of poles arising from div $\mathbf{M}$.)[62]

9.3.4. Moments Deduced from Measurement of Fields External to Ellipsoids

The fields outside ellipsoids in uniform fields have been calculated and as one might expect the solutions involve elliptical integrals. In particular the problems of flux linkage with search coils have been investigated.[63] At the equator of an ellipsoid of revolution the magnetic field just at the surface is tangential and equal to $\mathbf{H}$ inside the ellipsoid. Any method of measuring the field at such a point along with a measurement of $\mathbf{H}_0$, the field in the absence of the sample, is sufficient to determine $\mathbf{M}$ from Eq. (9.3.7) if D is known. On the other hand, a turn of wire placed tightly around an ellipsoid of revolution with the axis of the turn coinciding with the axis of the ellipsoid is linked by a flux equal to its enclosed area times the induction $\mathbf{B}$ inside the ellipsoid. This, with the knowledge of the field in the absence of the sample, is sufficient to determine $\mathbf{M}$ from Eq. (9.3.6)—again if D is known. In this latter fact lies the inherent advantage of induction measurements since they allow a direct determination of the fields within a sample. However, once the coil method is applied without conforming to the surface of the sample this advantage is lost, and the errors applicable to all other methods of measuring fields external to the ellipsoidal sample and deducing the moment from them are equally pertinent. Any method of field measurement can be used for absolute determination of magnetic moment of ellipsoids provided it allows determination of the field at a point in space. For relative measurements even

[62] H. Diesselhorst, Magnetic field and torques in the cases of a magnetized ellipsoid in a (magnetically) permeable medium and an applied field. *Ann. Phys. (Leipzig)* [6] **9,** 316 (1951); W. Doring, On the force and torque on magnetized bodies in a magnetic field. *Ann. Phys. (Leipzig)* **9,** 363 (1951).

[63] G. H. Hunt, The flux linkage with a search coil produced by a coaxial uniformly magnetized prolate spheroid. *Brit. J. Appl. Phys.* **5,** 260 (1954).

this restriction to a point in space is not necessary. The sensing element need but be calibrated in terms of a standard ellipsoid.

A particular advantage of a spherically shaped sample is that its external field is identical with that of a dipole of magnetic moment MV placed at the position of the center of the sphere. Hence, the field is readily calculable. In addition, the field depends only on MV and not on the radius of the sphere. This means that an apparatus calibrated for a sphere of one radius is calibrated for all spheres if a means for locating the center of the sphere is provided.

One approach to measurement of magnetic moments by external field measurements is to place the sample in one of two identical fields. The fields are arranged so that there is a region between them where the two fields completely cancel each other. When the field measuring device is then placed in that region it responds only to the field of the sample. A solenoid placed about the field measuring device can be used to cancel out the field of the sample thus requiring only that the device be null indicating.

Induction techniques are applied by causing a change in flux linkage with a coil. One method is to hold the sample and coil stationary and change the magnetization by changing the applied field. For instance, with a turn around an ellipsoid the reversal of applied field can cause a reversal of magnetization and the flux induced will be $2B_iS$, where S is the area of the turn and B_i is the induction within the sample. Another approach is to use coils which move with respect to the sample or vice versa, such that the flux due to the applied uniform field linked by the coil is not changed when the sample or coil is moved and therefore only the field due to the sample is detected. This can be carried out by moving a single coil with respect to the sample, moving the sample with respect to a single coil or by employing two or more coils with the possibility of either sample motion or coil motion. Double coils give discrimination against either space variations or time variations of the applied field. A pair of coils arranged so as to measure B_i and B_0 respectively can always be placed in series opposition to give a measure of $4\pi D\mathbf{M}$ directly. If the pair of coils measure B_i and H_i then their difference is just $4\pi\mathbf{M}$. Such methods can be carried out using a fluxmeter or any other device for integrating and measuring the induced voltage. As in the case of the rotating flip coil, it is possible to measure $\int V\,dt$ by producing an alternating voltage if the flux due to sample or coil motion can be made to change periodically. With the use of narrow band amplifiers the sensitivity of induction techniques can be increased by as much as 10^6 over that available with galvanometers. Some ways of producing alternating flux are vibrating the sample or the coils, revolving the sample or coils, or modulating the field or the magnetization.[64–66]

9.3.5. Null Determination of M[67]

The magnetization of a body can be measured directly in terms of its equivalent surface current $\mathbf{J}' = c\nabla \times \mathbf{M}$. If an infinite slab of material is placed perpendicular to a uniform magnetic field, **M**, **B**, and **H** will all be uniform, **B** will be the same inside and outside the slab, and inside the slab $\mathbf{H} = \mathbf{B} - 4\pi\mathbf{M}$. Any piece of isotropic material can be made to consider itself part of such a slab by putting the proper current distribution on its surface by means of, for example, the proper distribution of fine

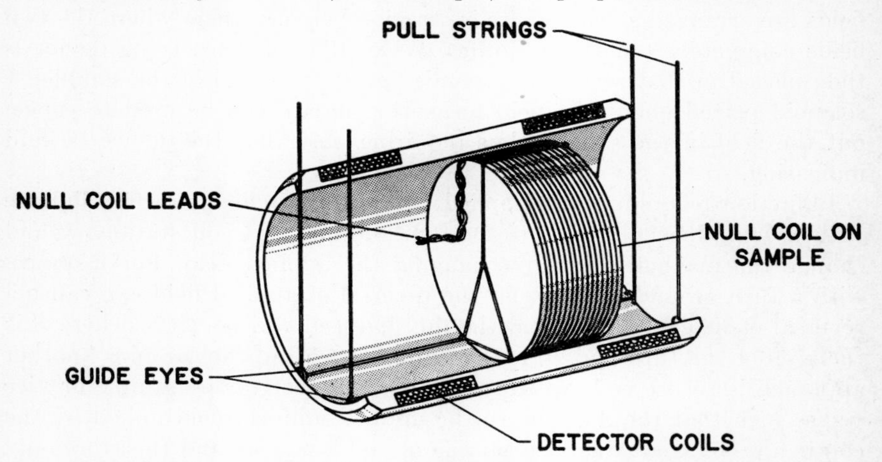

FIG. 3. Cylindrical sample with null coil attached is shown moving between detector coils which are connected in series opposition to cancel fluctuations from applied field.

wires. If any piece of isotropic material were uniformly magnetized, it would have a surface current on its boundaries given by $\mathbf{J}' = c\nabla \times \mathbf{M}$. If this current is cancelled by means of the proper current distribution the sample considers itself to have no boundaries in the sense of an infinite slab. The current is perpendicular to the direction of magnetization and such that the current per unit length of the sample in the direction of the magnetization is $c\mathbf{M}$. A useful application of this principle is to measurements on cylinders. Wound on the cylindrical surface is a fine

[64] D. O. Smith, Development of a vibrating coil magnetometer. *Rev. Sci. Instr.* **27** 261 (1956).

[65] J. F. Frazer, J. A. Hofmann, M. S. Livingston, and A. M. Vash, *Rev. Sci. Instr.* **26**, 475 (1955).

[66] S. Foner, *Rev. Sci. Instr.* **27**, 548 (1956).

[67] A. Arrott and J. E. Goldman, *Rev. Sci. Instr.* **28,** 99 (1957).

pitch solenoid simulating a current sheet (see Fig. 3). If such a cylinder were placed in the field of a large solenoid such as discussed in the case of the ellipsoids, the cylinder would not be uniformly magnetized. If, however, a current is passed through the fine pitch solenoid on the surface of the sample, a current will be reached for which the sample will be uniformly magnetized. The magnetization will then be given by I/c where I is the current per length of the specimen given by $i \cdot n/l$ where i is the current per turn and n/l the number of turns per unit length. The magnetic moment of the sample will be given by inS/c where S is the cross-sectional area of the sample. Under these conditions **B** will be uniform in all space. Thus, any of the methods of measurement of magnetization discussed above may be applied to determine when the proper current is passing through the coil, as the problem reduces to determining that the apparent magnetization of the sample is zero and that the uniformity of the original field $\mathbf{B}_0$ has been restored. The current required to apply this to the measurement of strongly magnetic materials presents a severe heating problem. A current of 1 amp with 100 turns/cm corresponds to a magnetization of 10 emu whereas some materials possess magnetizations larger than 1000 emu (iron = 1700 emu).

Sensitive methods of detection of voltage pulses which do not necessarily integrate well can be used, e.g., a dc chopper amplifier, when the null method is employed. Also all methods of field detection need only be methods which give no signal in uniform field without regard to linearity of signal with field. An oscillating sample need not have controlled amplitude.

9.4. Measurements of Susceptibility*

The vast majority of materials of interest to the physicist and chemist exhibit a constant susceptibility i.e., a linear relation between **M** and **H**. The many special techniques which allow a determination of the susceptibility of diamagnetic and weak paramagnetic materials make a very important contribution to the field of magnetic analysis. These methods fall into two categories: force methods of which there are two principal types; and induction techniques a few of which have been discussed above.

9.4.1. Force Measurements of Susceptibility

For a detailed discussion of the many methods of measuring susceptibility based on force measurements the reader is referred to the excellent

* See also Vol. 6, Chapter 8.1.

discussion in Selwood's "Magnetochemistry" and Bates' "Modern Magnetism,"[68] in which references are made to all the early work. To quote Prof. Bates, "So many different experimental arrangements have been used for the measurement of susceptibilities of liquids and solids that one might think that variety had been the main object in the minds of many workers in the field." In what follows the discussion is limited to general remarks on force measurements and to add references to advances made since 1950 with regard to the means of measuring small forces. Force measurements of susceptibility can be characterized according to whether the sample is subjected to an approximately constant $\mathbf{H} \cdot (d\mathbf{H}/dx)$, the Curie method, or the sample is of sufficient length that its two ends are in substantially different fields, the Gouy method. The basis of these is given below.

The force on a current loop of magnetic moment $\boldsymbol{\mu}$ was shown in Section 9.1.1 to be given by $F_y = \boldsymbol{\mu}(dH_x)/(dy)$. When the current loop is replaced by a sample of material with magnetic moment $\boldsymbol{\mu}$, this equation still is obeyed if H_x is the field which was present before the introduction of the sample. If the material has a volume susceptibility χ, the magnetic moment is given by $\chi H_i V$, where H_i is the field inside the sample, and is given by $H_i = H_x - 4\pi D\chi H_i$ assuming the shape of the sample is such as can be described by a demagnetizing factor. Thus

$$F_y = \frac{\chi V}{1 + 4\pi D\chi}\left(H_x \frac{dH_x}{dy}\right). \tag{9.4.1}$$

If $4\pi\chi \ll 1$ the uncertainty in D introduces only a second-order correction, hence the limitation to ellipsoidal shape is unnecessary. This expression is correct for a sample of infinitesimal size, the expression for a finite sample is an integral over the sample

$$F_y = \chi \iiint H_x \frac{dH_x}{dy}\, dV \tag{9.4.2}$$

where the term $4\pi D\chi$ has been dropped. If

$$\frac{dH_x}{dy} \cdot y \cdot \frac{1}{H_x} \ll 1$$

where y is the extent of the specimen in the y direction, $H_x(dH_x)/dy$ can be taken outside the integral. In most cases, however, large gradients and large specimens are desirable for sensitivity, and then the integration

[68] P. W. Selwood, "Magnetochemistry." Interscience, New York, 1943; L. F. Bates, "Modern Magnetism," pp. 113–143. Cambridge Univ. Press, London and New York, 1948.

must be carried out. There are two general means of experimentally simplifying this integration which are illustrated in Fig. 4. The first method (Fig. 4(a)), due to Faraday and Curie, is to make dH_x^2/dy constant over the specimen, giving

$$F_y = \frac{V\chi}{2}\frac{dH_x^2}{dy}. \tag{9.4.3}$$

The production of fields with constant dH_x^2/dy is generally accomplished by means of shaped pole faces in an electromagnet. Note that the Curie method involves an accurate knowledge of the value and the variation of a nonuniform field and of the position of the sample. Where relative measurements with respect to a known sample are made, sample placement is the major problem.

The second method (Fig. 4(b)), due to Gouy, is to use a cylindrical specimen with its axis in the y direction in which case dV is replaced by $S\,dy$ so that the integrand is an exact differential of H^2 and the force is

$$F_y = \frac{\chi S}{2}(H_x'^2 - H_x''^2) \tag{9.4.4}$$

where S is the cross-sectional area of the specimen and H_x' and H_x'' are the fields at the ends of the rod. By using a long rod which hangs with one end in a uniform field and the other end in a low field where $(H''/H')^2 \ll 1$, it is necessary only to measure a single uniform field. The advantage of the Gouy method is the insensitivity to sample placement and its principal disadvantage is in the need for long homogeneous samples of uniform cross section, and the fact that variations in magnetic properties that take place with small field increments cannot be measured.

The use of force techniques requires (1) a means of suspending the sample in the magnetic field (preferably with one degree of freedom), (2) a means of applying a restoring force, and (3) a means of detecting motion of the sample. The measurements are carried out both as a null method, by applying a known compensating force to return the sample to its original position, and as a deflection method, i.e., measuring the motion and calculating the force from a knowledge of the force constant of the system.

The suspensions most commonly used are: (a) the simple pendulum in which case the force can be exerted along the suspension or result in an angular deflection from the vertical (Fig. 4(b)); (b) the translational pendulum[69] which gives a horizontal displacement; and (c) the torque

[69] C. A. Domenicali, *Rev. Sci. Instr.* **21**, 327 (1950).

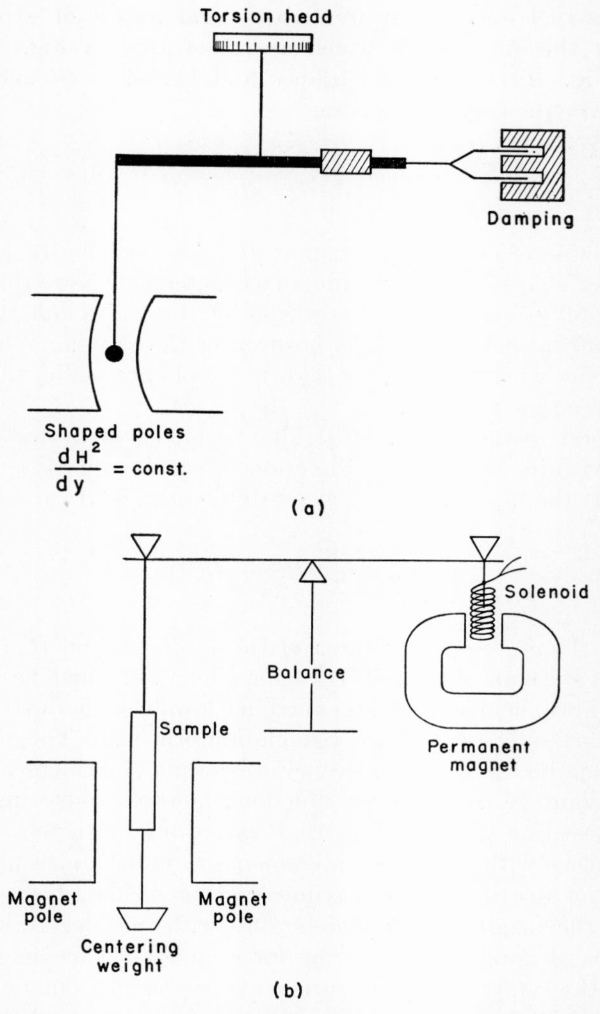

FIG. 4. Common methods of measuring the force on a magnetic body arising from a field gradient: (a) the Faraday or Curie method using a small sample suspended from a torsion balance with electromagnetic damping; (b) the Gouy method using a long sample suspended from a chemical balance with electromagnetic restoring force.

suspension in which the sample is on one end of a horizontal beam supported in balance by a torsion fiber (Fig. 4(a)).

The restoring force of gravity is used most often with the ordinary chemical balance and a vertical suspension of the sample. Where necessary the entire balance and region around the sample can be vacuum jacketed. The balance measurements are then carried out by mechanical or electrical connections through the vacuum case.[70,71] When the restoring force of gravity in a balance is replaced by the force exerted by a permanent magnet on a solenoid a convenient means of measuring force in terms of current is obtained (Fig. 4(b)).[72,73] An elastic restoring force is used in the Sucksmith ring balance where the distortion of a ring from which the sample is suspended is amplified by means of a light beam reflected successively from two mirrors mounted on the ring.[74,75] An elastic beam has been used by measuring the change in capacity of the beam with respect to a fixed member on flexure of the beam[76] and also by measuring the change of resistance of a strain gage on flexure of the beam. The greatest sensitivity with elastic members, however, is found by using the torsion balance where diamagnetic materials are measured to one part in 10^4.[77–80] If the upper end of the torsion fiber is rotated to return the sample to its original position, the torque may correspond to many rotations of the fiber. As an alternative, the force from a solenoid on a permanent magnet attached to the torsion beam on the arm opposite to the sample may be used to bring the sample back to its original position. The magnetic restoring force can also be used with flexure devices. In some cases current for null measurements can be taken as a fraction of the current producing the field thus making the susceptibility measurement insensitive to changes in field excitation current. Susceptibility measurements then reduce to the measurement of the ratio of two currents once the apparatus is calibrated. In balance measurements eddy-current

[70] S. T. Lin and A. R. Kaufmann, *Phys. Rev.* **102** 640 (1956).

[71] M. I. Pope, *J. Sci. Instr.* **34,** 229 (1957).

[72] T. R. McGuire and C. T. Love, *Rev. Sci. Instr.* **20,** 489 (1949).

[73] O. M. Hital and G. E. Fredericks, *J. Chem. Soc.* p. 785 (1954).

[74] W. P. Van Ort, A modification of Sucksmith's method for the measurement of susceptibilities of para- and ferromagnetic materials at temperatures between −100°C and 1100°C. *J. Sci. Instr.* **28,** 279 (1951).

[75] B. S. Chandrasekar, Improvement to the Sucksmith Balance. *Rev. Sci. Instr.* **27,** 967 (1956).

[76] Y. L. Yousef, R. K. Girgis, and M. Mikhail, Measurement of the magnetic susceptibility of manganese nitride Mn_5N_2 by a microvibration method. *J. Chem. Phys.* **23,** 959 (1955).

[77] S. K. Dutta Roy, An improved method of measuring the absolute susceptibilities of single crystals over wide ranges of temperatures. *Indian J. Phys.* **29,** 429 (1955). A review of the whole subject of force measurements is given in this article.

[78] R. Cini and L. Sacconi, An apparatus for accurate measurement of magnetic susceptibility. *J. Sci. Instr.* **31,** 237, 56 (1954).

[79] S. P. Yu and A. H. Morrish, Torsion balance for a single microscopic magnetic particle. *Rev. Sci. Instr.* **27,** 9 (1956).

[80] G. T. Croft, F. J. Donahoe, and W. F. Love, Automatic recording torsional magnetic susceptibility balance. *Rev. Sci. Instr.* **26,** 360 (1955).

damping is usually introduced by attaching a conductor to a moving part of the apparatus. Where measurements are made on conducting materials the damping provided by the sample can be so great as to make the actual measurement very difficult.[81] With most of these balance methods a photoelectric cell can be used as a null detector and can be made to drive a rebalancing mechanism giving automatic operation.

9.4.2. Mutual Inductance Measurements of Susceptibility

The electromagnetic coupling of two coils depends on the medium between them. In particular a changing current in one coil will produce a changing intensity of the magnetic field **H** which will be seen by the other coil as a changing magnetic induction **B**. The voltage induced in the second coil depends on the geometry of the setup, the current change in the first coil, and the relationship between the change of **H** and the change of **B** in the medium. A measurement of mutual inductance gives the permeability $\mu = dB/dH$ which is related to the susceptibility by $\mu = 1 + 4\pi\chi$. When χ is small, measurements of μ must be carried out with extreme accuracy to obtain χ. Fortunately, mutual inductance bridges give sufficient sensitivity for such measurements. For paramagnetic materials where $\chi \sim 10^{-4}$ such measurements have the great advantage of speed of measurement. In adiabatic demagnetization experiments where this is important, such methods are widely used. For use with diamagnetic materials the method can be applied successfully but only with many precautions as the requirement is a measurement of μ to 1 part in 10^9. The details are found in Broersma's book[82] and the listed recent references.[83–89]

9.4.3. Inductive Measurements

Induction techniques have been applied to measurements of susceptibility using moving samples. Such techniques are becoming more popular because of the speed and convenience of electrical measurements in contrast to mechanical. Moreover, it is preferable in many cases to make

[81] E. W. Pugh, Ph.D. Thesis, Carnegie Institute of Technology, Pittsburgh, Pennsylvania, 1956; R. Bowers, *Phys. Rev.* **100,** 1141 (1955).

[82] S. Broersma, "Magnetic Measurements on Organic Compounds." Nijhoff, The Hague, 1947.

[83] J. A. Beun, N. J. Steenland, E. D. de Klerk, and C. J. Gorter, *Physica* **21,** 651 (1955).

[84] E. D. de Klerk and R. P. Hudson, *J. Research Natl. Bur. Standards* **53,** 173 (1954).

[85] S. Broersma, *J. Chem. Phys.* **17,** 873 (1949).

[86] S. Broersma, *Rev. Sci. Instr.* **20,** 660 (1949).

[87] F. R. McKim and W. P. Wolf, *J. Sci. Instr.* **34,** 64 (1957).

[88] R. A. Erickson, L. D. Robert, and J. W. T. Dobbs, *Rev. Sci. Instr.* **25,** 1178 (1954).

[89] H. G. Effemey, D. F. Parsons, and J. O'M. Bockris, *J. Sci. Instr.* **32,** 99 (1955).

measurements in a uniform field giving directly the magnetization at each specific field (e.g., the de Haas–Van Alphen effect) and to avoid certain disadvantages of the use of a balance such as sensitivity to vibration, instability of the sample in the magnetic field because constraints might interfere with the accuracy of the balance, and the problem of attaching a temperature sensing device to the sample without disturbing the balance action. Techniques using a vibrating sample show a sensitivity suitable for measurements of χ as small as 10^{-9}.[66]

One necessary precaution in the use of moving sample induction techniques is consideration of an eddy current effect. If the field that the sample moves in is not perfectly uniform, moving through the gradient will induce currents in the sample if it is a good conductor. These currents tend to keep **B** constant in the sample hence the sample appears to carry an additional magnetization, which for a good conductor can be of the order of 1% or more of $\Delta\mathbf{B}$, the variation of **B**. These eddy currents integrate out to zero only if the detector used is a perfect integrator. Where $\Delta\mathbf{B} \gg \chi\mathbf{B}$ imperfections in integration can become a serious problem.[81]

9.4.4. Some Considerations in the Measurement of Susceptibility

In generally available laboratory fields, the magnetizations induced in most para- and diamagnetic materials are far smaller than those of ferromagnetic materials. This makes it necessary to consider effects of contamination. In magnetochemistry, particularly in the study of organic materials, where results are obtained to a tenth of a per cent and better, the accuracy is obtained by great precaution in preparing specimens. Measurements are usually made at a single fixed field without an attempt to use magnetization versus field data to correct for impurities. The apparatus is generally arranged so that measurements are relative to a standard sample. The choice of a standard differs with various experimenters. Two particular favorites are $NiCl_2$ in solution and ferrous ammonium sulfate.[90,91] The dangers of ferromagnetic contamination can be made clear by a typical example. Consider the contribution to the magnetization of a one gram specimen of an organic liquid made by a single dust particle (volume 10^{-8} cm^3) of iron oxide (Fe_3O_4) which has a saturation magnetization of 1000 emu/cm^3. A typical organic liquid would have to be measured in a field of 10,000 gauss to bring the accuracy of the determination to 0.1%. In preparation of "pure" metals contamination of one part in 10^5 is not uncommon, yet for a metal such as

[90] P. W. Selwood, "Magnetochemistry," pp. 28–30. Interscience, New York, 1943.

[91] S. Broersma, "Magnetic Measurements on Organic Compounds." Nijhoff, The Hague, 1947.

lithium the error in a field of 10,000 gauss in the presence of this much iron in the form of iron crystallites could cause a 100% error in the determination.

As the field dependence of ferromagnetic impurities depends on the unknown shape of the impurities and their alignment with respect to the field it is impossible in general to predict the behavior of impurities unless the field is sufficient to saturate the magnetization of the impurities. The applied field necessary for saturation of a crystallite of ferromagnetic material can be as high as 22,000 gauss in the case of platelets of iron aligned perpendicular to the field. As the vast majority of all magnetic measurements are made below 22,000 gauss, these results have an inherent error if such platelets exist. In liquids the problem is not so severe as the worst case would be that of a sphere which for iron saturates near 7000 gauss. Any other shape will rotate to minimize the demagnetizing field. If the ferromagnetic impurity is an iron compound, the fields required for saturation are reduced at least by a factor of 2. Workers in the field generally operate on the assumption that a large fraction of the impurities are saturated by 8,000 gauss. Hence, susceptibility results on contaminated specimens are generally obtained by treating the data above 8,000 gauss in the Curie method as being given by

$$F_y = \frac{\chi V}{2} \frac{dH_x^2}{dy} + |\mathfrak{u}| \frac{dH_x}{dy} \tag{9.4.5}$$

where $\mathfrak{u}$ is the magnetic moment of the impurities. The analysis is carried out in terms of an apparent susceptibility, χ', defined as

$$\chi' \equiv F_y/H_x \frac{dH_x}{dy} \cdot V. \tag{9.4.6}$$

Thus the true χ is given by

$$\chi = \chi' - \frac{|\mathfrak{u}|}{VH_x} \tag{9.4.7}$$

so that a plot of χ' versus $1/H_x$ gives in the limit of $1/H_x = 0$ the correct value of χ. The slope $|\mathfrak{u}|/V$ measures the magnetic moment of the impurities per unit volume of the sample. The problem is similarly treated for induction measurements.

The Gouy method is usually applied with one end of the sample in zero field, thus the impurities are never saturated over the entire sample. An exact treatment of impurities is possible only if the sample is placed so that the field at the upper end of the sample is still relatively high. In this case the force is given by

$$F_y = S \int_{H_0}^{H} \chi H \frac{dH}{dy}\, dy + \frac{|\mathfrak{u}| S}{V} \int_{H_0}^{H} \frac{dH}{dy}\, dy \tag{9.4.8}$$

where S is the cross-sectional area of the sample, $\boldsymbol{\mu}/V$ is the magnetic moment of the impurities per unit volume of the sample, and $H > H_0 > H_s$ where H_s is the field necessary for saturation of impurities. The apparent susceptibility is defined as

$$\chi' = 2F_y/S(H^2 - H_0^2). \tag{9.4.9}$$

The actual susceptibility is given by

$$\chi = \chi' - 2|\boldsymbol{\mu}|/V(H + H_0). \tag{9.4.10}$$

Here the plot of χ' versus $(H + H_0)^{-1}$ gives χ as the intercept for

$$(H + H_0)^{-1} = 0.$$

If the ferromagnetic impurities all have the same saturation magnetization and the same demagnetizing factor, a correction can be made to conventional Gouy measurements with one end of the sample in zero field.[92] In this case there is a well defined saturation field H_s and all impurities have a field dependence which is linear up to H_s and are field independent above H_s. The value of H_s depends on the saturation magnetization and the demagnetizing factor. In this analysis the apparent susceptibility is simply

$$\chi' = 2F_y/SH^2 \tag{9.4.11}$$

but the actual susceptibility is

$$\chi = \chi' - \frac{|\boldsymbol{\mu}|}{V}\frac{2H - H_s}{H^2}. \tag{9.4.12}$$

For spheres of iron $H_s \approx 7200$ oersted. For unknown impurities of random demagnetizing factors, Knappwost[92] has proposed standardization on $H_s = 3000$ oersted as a suitable approximation. This above equation then becomes an empirical law justified only on past experience and not necessarily applicable in any given case. The problem of ferromagnetic impurities is also important in field measurements by forces and torques on current carriers, though there they can be corrected for by measuring at several currents.

An additional consideration in all measurements of weakly magnetic materials is the effect of oxygen if the measurements are carried out in air. The effect at room temperature is that the susceptibility measured is actually the difference between the specimen and the air it displaces. The susceptibility of air at room temperature is of the order of 3×10^{-8} emu/cc. For measurements below the critical point of oxygen, condensation of oxygen can give rise to serious errors as the susceptibility of

[92] A. Knappwost, *Z. Elektrochem.* **59**, 561 (1955).

oxygen in the liquid state is of the order of 10^{-4} emu/cc. Absorption of oxygen can also give errors up to 1% in diamagnetic organic liquids. Magnetic investigations with a sample directly in liquid nitrogen are extremely hazardous as liquid nitrogen is easily contaminated by (5% or more of) oxygen.

9.5. Measurements on Ferromagnetic Materials*

The complex behavior of ferromagnetic materials requires many different measurements to specify completely the magnetic characteristics. For these materials there exist a range of fields within which the properties depend in part on the past history of the materials, with a result that for each such field there is a range, generally continuous, of possible values for the magnetization.

A ferromagnetic material can possess regions within which the magnetization is uniform, called domains, but yet possess many domains differing in orientation with respect to each other. Thus a sample may have no net moment, yet possess many domains each with a magnetic moment. The number of ways in which the state of zero net moment can be obtained is very large. At sufficiently high fields the ferromagnetic material approaches uniform magnetization, that is, becomes a single domain. The properties associated with the behavior of individual domains are often termed intrinsic magnetization properties while those associated with the multidomain nature of materials are termed technical magnetization properties.

9.5.1. Determination of "Spontaneous Magnetization"

The magnetization within a single domain is termed the "intrinsic magnetization." The value of the intrinsic magnetization when the intensity of the magnetic field inside the domain $\mathbf{H}_i$ is zero is called the "spontaneous" magnetization. Important as the concept of "spontaneous" magnetization is, it is not possible, except in a few cases, to have an entire sample which is uniformly magnetized with $\mathbf{H}_i = 0$. Hence the measurement of spontaneous magnetization takes some consideration of the properties of ferromagnetic materials. Consider a single crystal in the shape of a sphere placed in a solenoid. Let its magnetization be measured as a function of the field and its direction with respect to

* See also Vol. 6, Chapters 8.2 and 8.3.

the axes of the single crystal. It will be found that the net magnetization of the sample will increase linearly with applied field according to $\mathbf{M} = \mathbf{B}_0/(4\pi/3)$ where $\frac{1}{3}$ in the denominator is the demagnetizing factor for the sphere. As the field is increased the moment eventually will no longer increase linearly with field. The field at which this happens is different for different directions of the field relative to the crystal directions. There exists a set of directions for which the field at which the curve departs from linearity is a maximum. These are the directions of easy magnetization. The magnetization obtained for $\mathbf{H}_i$ sufficient to overcome the small impedance to domain boundary motions due to imperfections in the direction of easy magnetization is the "spontaneous magnetization." The intrinsic magnetization $\mathbf{M}_I$ is itself a function of internal field $\mathbf{H}_i$. In many substances at not too high temperatures the rate of change of $\mathbf{M}_I$ with $\mathbf{H}_i$ corresponds to a very small susceptibility. In these substances $\mathbf{M}_I$ may be considered independent of $\mathbf{H}_i$ and equal to $\mathbf{M}_s$. A sphere of such a substance should show a linear increase in magnetic moment with field in the easy direction until $\mathbf{B}_0 = (4\pi/3)\mathbf{M}_s$, after which $\mathbf{M} = \mathbf{M}_s$ at all higher fields. In any other direction it takes an infinitely intense field to align completely the magnetic moment in the direction of the field. It can be said that the crystal exerts a torque on the magnetization tending to rotate the magnetization toward an easy direction. The counter-torque exerted by the applied field can never completely overcome this. Actual torque measurements on single crystals can be used to investigate the nature of this magnetic anisotropy. The approach to complete alignment with the magnetic field is, in the case of anisotropy, given by

$$\mathbf{M} = \mathbf{M}_I\left(1 - \frac{b}{H^2}\right). \tag{9.5.1}$$

Curves (a) of Fig. 5 show the results of Honda and Kaya for iron in three principal directions. When the external field has been corrected to take into account the demagnetizing field, curves (b) are obtained. Note that there is a rounding off of the curve just before saturation in the case of the easy direction. This is always observed and is presumably due to lack of perfect single crystals or perfect orientation. The curvature is then the approach of the magnetization of the misaligned region or regions to the intrinsic magnetization. The rapid approach of $\mathbf{M}$ to $\mathbf{M}_I$ and the small variation of $\mathbf{M}_I$ with field make possible an accurate determination of $\mathbf{M}_s$ for iron by using a single crystal of iron aligned in the easy direction. Note that for iron the determination of $\mathbf{M}_s$ even in a hard direction is possible. Thus, measurements on polycrystalline iron could be used to determine $\mathbf{M}_s$. This, however, is not in general true as the combination of a large dM_I/dH_i and a large value of b in the expression

$$\mathbf{M} = \mathbf{M}_I\left[1 - \frac{b}{H_i^2}\right]$$

can lead to difficulties in determining M_s from polycrystalline samples. If as a first approximation $\mathbf{M}_I = \mathbf{M}_s + \chi_p\mathbf{H}_i$ where χ_p is termed the "parasitic" paramagnetism, the general expression for the approach of the measured magnetization to the intrinsic magnetization becomes

$$\mathbf{M} = \mathbf{M}_s\left(1 - \frac{\chi_p b}{M_s H_i} - \frac{b}{H_i^2}\right) + \chi_p\mathbf{H}_i. \tag{9.5.2}$$

From this expression the term in $1/H$ should be negligible. Actually a sizeable $1/H$ term is often noted and is thought to be a consequence of imperfections in the crystal.[93,94]

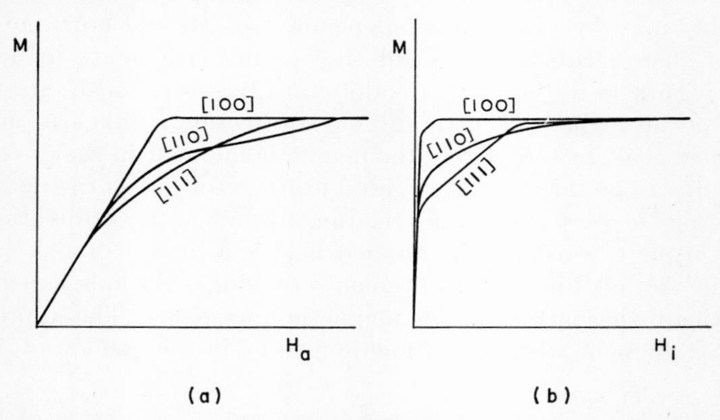

FIG. 5. Magnetization curves for iron single crystals: (a) as observed; (b) after correcting for the demagnetizing field (K. Honda and S. Kaya, *Sci. Repts. Tohoku Imp. Univ.* **15**, 721 (1926)).

The actual measurements of intrinsic magnetization can be carried out by any of the induction techniques discussed in Section 9.1.2 as those use uniform applied fields and ellipsoidal samples. The constant gradient force method (Curie method) can also be used if care is taken that

$$\frac{dM_I}{dH_i} \cdot \frac{dH_0}{dy} \cdot y \ll M_I$$

where y is the linear dimension of the sample in the direction of the gradient.

The magnetization of a ferromagnet decreases with increasing tem-

[93] L. Neel, *J. phys. radium* **9**, 184 (1948)

[94] W. F. Brown, Jr., *Phys. Rev.* **82**, 94 (1951).

perature. The Curie temperature, by definition the temperature at which the spontaneous magnetization goes to zero, is an important quantity but its determination is difficult to carry out with great accuracy. The work of Smith on Fe_3O_4 gives a conceptually clean-cut method of determining the Curie temperature (Fig. 6(a)). The break away of the moment from the linear dependence on applied field occurs at lower and lower fields as the Curie temperature is approached. Curie temperatures have

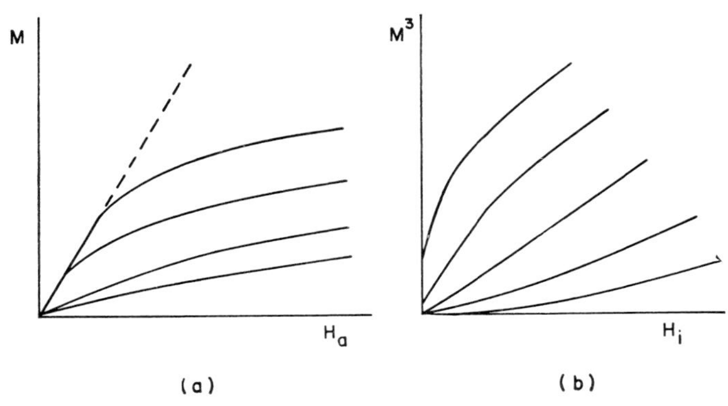

FIG. 6. Determinations of Curie temperatures: (a) from the disappearance of spontaneous magnetization; (b) from the curvature of M^3 versus **H** isotherms.

generally been determined by reliance on theoretical models of ferromagnetics. One method is to use Eq. (9.5.2) to obtain M_s at temperature T, below the Curie temperature θ, and the expression

$$T = \theta - aM_s^2 \tag{9.5.3}$$

to obtain θ. Another is to use the form of the magnetic isotherm for applied fields greater than $4\pi DM$ in comparison to the expression

$$\begin{aligned} H_i &= \epsilon M + bM^3 + cM^5 \\ \text{where} \qquad \epsilon &= a(T - \theta). \end{aligned} \tag{9.5.4}$$

The field is proportional to the cube of the magnetization at the Curie temperature. A plot of M^3 versus **H** gives positive intercepts for $H = 0$ below the Curie point (Fig. 6(b)). A plot of M^2 versus H/M gives the spontaneous magnetization as the intercept on the M^2 axis and $1/\chi$ as the intercept on the H/M axis for temperatures just below and just above the Curie temperature.

9.5.2. Technical Magnetization

A number of terms are introduced to describe the technologically significant properties of ferromagnetic materials. These terms can be understood by considering a typical magnetization curve of a sample as shown in Fig. 7. The sample starts in the demagnetized condition, namely a state of zero net moment, which can be achieved by cooling from above the Curie temperature in the absence of external fields or by subjecting the sample to an alternating field of ever decreasing amplitude. The application of a field will then give a magnetization as determined by the initial or virgin magnetization curve. Note that the field is the internal field $\mathbf{H}_i$. The initial permeability, given by $\mu_0 = dB/dH$ for the virgin curve at the origin, can depend on the method of demagnetization though the technologically significant slope is that which is observed after ac

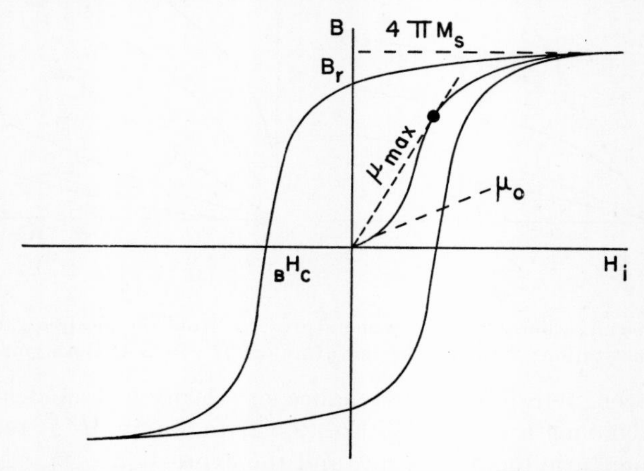

FIG. 7. A hysteresis curve for a ferromagnetic material showing the initial susceptibility μ_0, the virgin magnetization curve, the major magnetization curve (that is, the curve after reaching saturation magnetization $\mathbf{M}_s$), the remanence $\mathbf{B}_r$ and the actual coercive force ${}_B\mathbf{H}_c$.

demagnetization. The effective permeability, defined as $\mu_{\text{eff}} = B/H$, reaches a maximum at a finite field intensity. If the field is increased sufficiently, the magnetization saturates, and it is customary to refer to an intrinsic saturation induction $4\pi\mathbf{M}_s$ which if $\mathbf{H} \ll 4\pi\mathbf{M}_s$ is often called $\mathbf{B}_s$, the saturation induction. If $\mathbf{H}_i$ is returned to zero the residual induction is called the remanence, $\mathbf{B}_r$. The field necessary to reduce $\mathbf{B}$ to zero is called the coercive force ${}_B\mathbf{H}_c$. This coercive force is in general smaller than the intrinsic coercive force ${}_M\mathbf{H}_c$ which is the field necessary to reduce $\mathbf{M}$ to zero. The area of the loop represents the energy loss per cycle per unit volume. The shape of the loop varies greatly with materials but one may readily classify nearly all materials in three groups: (a) *Soft magnetic materials* are characterized by small values of $\mathbf{H}_c$ and a narrowness of the hysteresis loop; (b) *Permanent magnetic materials* are characterized by large values of $\mathbf{H}_c$ and $\mathbf{B}_r/\mathbf{B}_s$ and a large area for the loop; (c) Speciality materials are where the shape of the loop makes possible particular applications such as the square loop materials used in computing devices. In measurements of the static (dc) properties of all such materials the problem is that of deducing $\mathbf{H}_i$. As this is most troublesome in high permeability materials the discussion given below is aimed largely at the methods of measurement for soft magnetic materials.

9.5.2.1. Static Measurements on Soft Magnetic Materials.[95] As has been noted, in soft magnetic materials the principal uncertainty is that in determining precisely the field $\mathbf{H}_i$ in the sample. There are three different ways in which this can be accomplished and the choice is usually governed by such factors as the form in which the materials are available, the magnitude of the permeability, sensitivity to strain, etc.:

(a) *Using a geometry in which* $\mathbf{H}_i$ *can be precisely calculated from circuit constants that are obtainable with accuracy.* The simplest such geometry is a toroid where $\mathbf{H}_i$ is given directly by $4\pi NI/l$ where I is the current in the primary winding and N/l is the number of turns per unit length. In practice, materials are not usually available in a form that permits convenient mounting as a toroid. An alternative practice is to attempt to synthesize a toroid by the stacking of thin strips in the form of a square and mounting them inside a square array of four coils connected in series as in the Epstein ac method (Fig. 8(a)). The strips which are usually 30×3 cm $\times$ material thickness are known as Epstein strips and have become quite standard in the measurement of properties of electrical sheet steel. A principal disadvantage of this method is that there will always be gaps in the corners and in a very high permeability material the error introduced by their presence could be very serious. Since $\mathbf{H}$ in the gap is equal to $\mathbf{B}$ in the material adjacent to the gap, there is a substantial contribution to $\int \mathbf{H} \cdot d\mathbf{l}$ (taken around the closed path through all the strip) from the gap. If the permeability of the magnetic material is 10^6, a combined length of gap of the order of 10^{-3} cm will cause considerable error in the measurement of the initial permeability. There are other disadvantages to this method: errors in materials with directional anisotropy due to the flux direction having to turn out of the preferable direction near the corners; where the permeability of materials is strain sensitive, the geometry of stacking does not lend itself to strain-

[95] Several reference works are devoted primarily to a discussion of methods of measurement of technical magnetization. Of these particular mention should be made of the following: T. Spooner, "Properties and Testing of Magnetic Materials." McGraw-Hill, New York, 1927; P. Vigoureux and C. E. Webb, "Principles of Electric and Magnetic Measurements." Prentice-Hall, New York, 1937; N. F. Astbury, "Industrial Magnetic Testing." Institute of Physics, London, 1952.

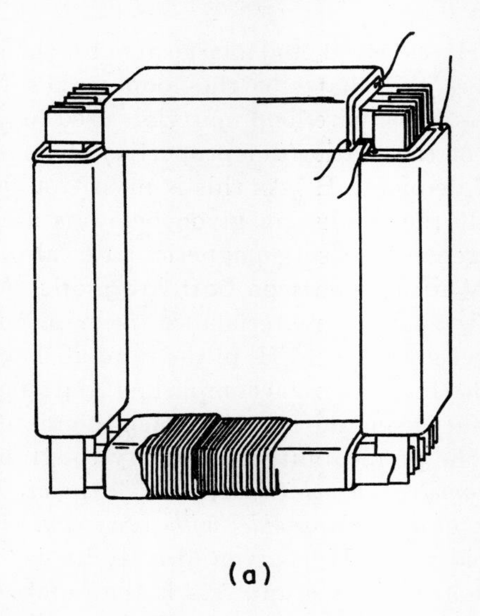

(a)

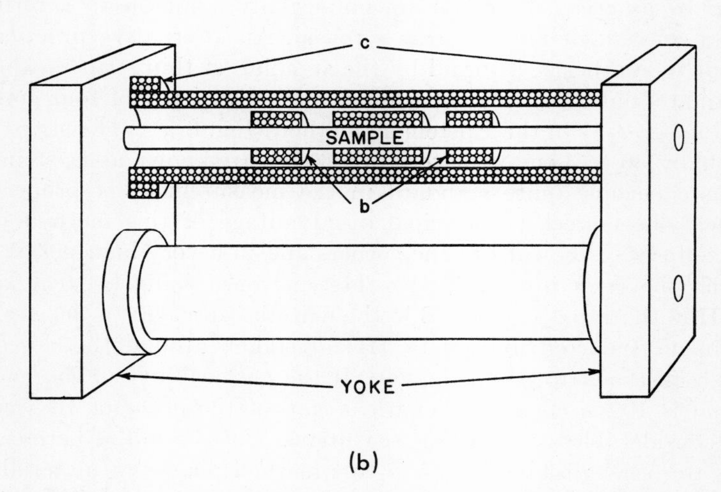

(b)

FIG. 8. Some means of determining properties of ferromagnetic materials are (a) the conventional Epstein strip method with interlocking strips forming a square; (b) the Burrows permeammeter showing symmetrical double yoke double sample arrangement with check coils *b*, the compensating field coils *c*; (c) the Brown and Rubens double yoke method where the field $\mathbf{B}$ in the region between plate-like samples corresponds to the cavity definition of $\mathbf{H}_i$; and (d) the Chattock magnetic potentiometer.

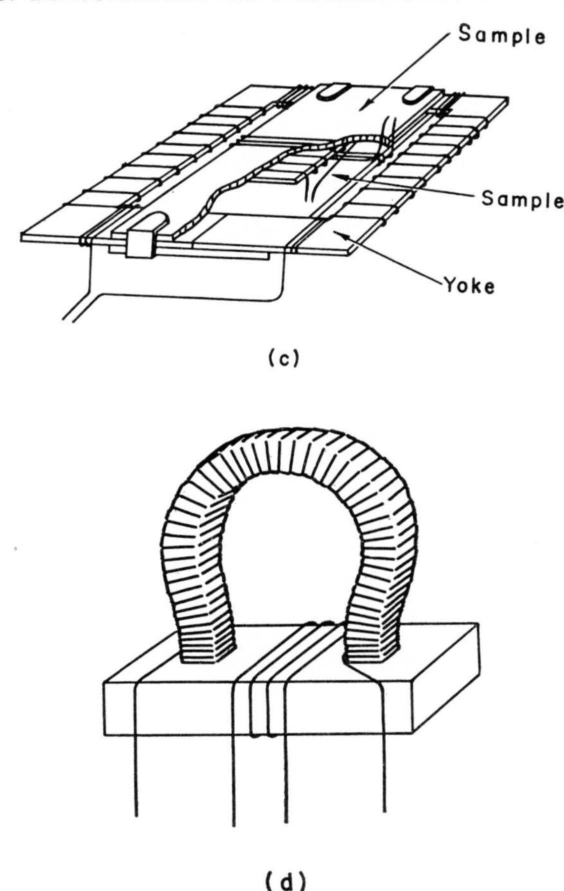

FIG. 8. (*Continued*).

free handling; and, finally, the method only gives an average of the properties for an entire stack and gives no indication of the possible variations within a batch of Epstein samples or within a single sample.

An alternative means of obtaining a geometry in which $\mathbf{H}_i$ can be calculated is to surround the sample with a coil and place the sample against a yoke of relatively wide cross section of high permeability material. Under this circumstance, the contribution to $\int \mathbf{H} \cdot d\mathbf{l}$ through the yoke is negligibly small so that H_i can be calculated from the current according to $H = 4\pi Ni/l$. In this method, as in the case of the Epstein method, care must be taken to obtain good magnetic contact at the yoke faces. In spite of that, there is an inevitable source of error in the measurement of high permeability materials with this method. In the Burrows[96] and Iliovici[97] permeammeters which are based on the above principle, a method of compensating for the errors is introduced (Fig. 8(b)).

Finally, the method of using ellipsoids and measuring the field either in the absence of the sample or utilizing a solenoid to obtain the field makes it possible to calculate $\mathbf{H}_i$ since tables of demagnetizing factors for ellipsoids are available. This method is useful where the material is available in the form of wires or thin films since a dimensions ratio of 10^3 would be required for the ellipsoid if acceptable accuracy is to be obtained in a material with a permeability of 10^6.

(b) *Measuring* **H** *near the surface of the sample at a point of uniform* **B** *and assuming constancy of the tangential component of* **H** *near the surface.* Any of the methods used for the measurement of fields can be used here. Since the field falls off very rapidly near the surface the field probe should be extremely small to assure maximum accuracy. For this reason a peaking strip is a particularly useful method as in the Carr permeammeter.[98] In this method, a wire of very high permeability with a fine coil wound on it is placed very close to the surface of the sample. If a high frequency alternating field is applied to the coil, the presence of a field, however small, will cause a second harmonic to be induced. The field is then nulled out by superimposing a dc in the coil. The dc in the coil when the detectable second harmonic goes to zero is proportional to **H**.

In the Brown and Rubens[99] permeammeter (Fig. 8(c)), the fall-off in **H** near the sample surface is made small enough to allow the use of a search coil. The **H** coil is sandwiched between two samples which are in the form of large plates. The plates are in turn in contact with a yoke to complete a magnetic circuit. The measurement of the field in the region between the plates in this manner corresponds to the cavity definition of **H**.

The field **H**, very close to the surface of a sample, can also be measured inductively by winding two coils on the sample, one on top of the other and connecting them in series opposition. If the coils have the same number of turns, the integrated voltage induced on changing the applied

[96] C. W. Burrows, The determination of the magnetic induction in straight bars. *Bull. Natl. Bur. Standards (U.S.)* **6,** 31 (1909).

[97] M. A. Iliovici, *Bull. soc. intern. electriciens* **3,** 581 (1913).

[98] W. J. Carr, Jr., "Symposium on Magnetic Testing," pp. 63–80. American Society for Testing Materials, Philadelphia, Pennsylvania, 1948.

[99] W. F. Brown, Jr. and S. M. Rubens, A double yoke plate permeammeter, *J. Appl. Phys.* **16,** 713 (1945).

field is given by $n\ \Delta\mathbf{H}(S_2 - S_1)$ where n is the number of turns per coil, S_2 and S_1 are the areas of the inside and outside coils and $\Delta\mathbf{H}$ is the change in $\mathbf{H}$ in the region between the two coils. Thus it is possible to determine $\mathbf{H}$ within the distance of the diameter of the wire used in the coil. This compares to a distance of one half the wire diameter for the magnetic potentiometer (discussed below) and three halves the wire diameter for the Carr device. However, in order to effect such accuracy, the areas of both coils must be fixed and known to a very high degree of accuracy and herein usually lies the principal source of error in this method.

(c) *The Magnetic Potentiometer.* It is possible to obtain H at the surface with considerable precision with the magnetic potentiometer[100] which is shown in Fig. 8(d). The principle is that, between any two points a and b, $\int_a^b \mathbf{H} \cdot d\mathbf{l}$ is independent of the path as long as any two paths being compared are not threaded by a current. If a and b are points right at the surface of a specimen in which $\mathbf{H}$ is uniform, between a and b the line integral is given by H_il where l is the distance between a and b. A path outside the specimen joining a and b must also give a line integral $\int_a^b \mathbf{H} \cdot d\mathbf{l} = H_il$. This integral is determined by measuring the flux induced in a coil of the shape shown in Fig. 8(d) when the coil is removed from the sample to a position of zero field (or a change of $\mathbf{H}$ when the coil is stationary and the field changed). The accuracy depends on using very fine wire and having the end turns lie flush with the parallel ends of the coil form. In the Fahy[101] permeammeter an $\mathbf{H}$ coil connected across the poles of the yoke parallel to the sample is used essentially as a potentiometer but with the questionable assumption that $\mathbf{H}$ is uniform along the entire length of the sample. In the Iliovici permeammeter previously mentioned, a potentiometer is used not to measure $\mathbf{H}$ but rather to detect the uniformity of $\mathbf{H}$ between the two points where $\mathbf{H}$ is calculated from the current. In that case the potentiometer is a null detector and its sensitivity may be increased appreciably by threading the coil with a core of high permeability material.

Many of the principles outlined above have been adapted to automatic techniques for producing static hysteresis loops. Because of the labor involved in making point by point determinations of $\mathbf{B}$ and $\mathbf{H}$, automatic

[100] A. P. Chattock, On a magnetic potentiometer, *Phil. Mag.* [5] **24,** 94 (1887); N. F. Astbury, "Industrial Magnetic Testing," p. 39. Institute of Physics, London, 1952; T. A. Margerison and W. Sucksmith, *J. Sci. Instr.* **23,** 182 (1946); L. F. Bates, "Modern Magnetism," p. 64. Cambridge Univ. Press, London and New York, 1948.

[101] F. P. Fahy, Permeammeter for general magnetic analysis. *Chem. & Met. Eng.* **19,** 339 (1918); R. L. Sanford, *J. Research Natl. Bur. Standards* **4,** 703 (1930).

devices are becoming more and more popular, particularly in view of the general availability of commercial x-y self-balancing potentiometers.[102]

9.5.2.2. Dynamic Measurements of Soft Magnetic Materials. Most soft magnetic materials are designed for applications where they will be subjected to alternating magnetic fields. A useful figure of merit for such materials is the magnitude of the ac energy loss. The sources of energy absorption by a magnetic circuit exposed to alternating fields are: (a) the hysteresis loss, defined per cycle as $\oint \mathbf{M}\, d\mathbf{H}$ where the integral is taken over a complete magnetization cycle and the resultant loss is in ergs/4π cm^3 if M is in emu and $\mathbf{H}$ in oersteds; (b) the eddy current loss which is present only in materials of high electrical conductivity, i.e., metals, and originates in the ohmic loss due to the current loops induced in the material by the varying magnetic field; (c) the "anomalous loss" apparently present in most materials. The "anomalous loss" covers the present inability to account for the observed total loss in materials even at power frequencies by the calculation of the eddy losses and computation of the hysteresis losses in the manner indicated. Separation of the total loss into its component parts which is often a useful way of investigating the source of losses in technologically interesting materials can be effected by plotting the frequency dependence. The hysteresis loss by the above definition depends linearly on frequency while the eddy loss depends quadratically on frequency. These losses are often measured directly as ac power rather than by determining the area of the dynamic loop. For a material in the form of a closed magnetic circuit and in which $\mathbf{B}$ is always parallel to $\mathbf{H}$ it can be shown[103] that the power measured externally by measuring the product of i, the current in the magnetizing winding, and E, the voltage across a secondary winding threading the sample is proportional to within the geometrical constants of the coils to the *total* loss in the material. This is most readily accomplished by using a dynamometer type wattmeter or its electronic variant.

Measurement of most ac properties, at least for frequencies below the radio-frequency range, is somewhat simpler than dc measurements. This is because integrating and measuring networks are more readily devised and rendered operative. For example, presentation of a BH loop on an oscilloscope with ac excitation is readily accomplished with two coils wound on the sample. A resistor is placed in series with each of the coils

[102] See references in R. M. Bozorth, "Ferromagnetism," p. 861. Van Nostrand, New York, 1951; others of pertinent interest are: D. Gall and J. D. Watson, An instrument for the automatic recording of hysteresis loops. *J. Sci. Instr.* **33,** 265 (1956); H. C. Lehde, A ballistic hysteresis graph. *Rev. Sci. Instr.* **2,** 16 (1931).

[103] N. F. Astbury, "Industrial Magnetic Testing," p. 16. Institute of Physics, London, 1952.

and a voltage is tapped off from the resistance in series with the magnetizing coil and connected to the horizontal plates of the oscilloscope. The voltage across the secondary coil circuit is passed through an integrating network and put on the vertical terminals. The simplest such circuit consists of a resistance and capacitor in series, with the vertical terminals connected across the condenser. If the secondary voltage, $V_s \propto (dB/dt)$, is applied across the series network, the voltage that appears across the condenser is $\frac{1}{RC} \int V_s \, dt \propto B$ provided that the effective impedance of the capacity at the frequency of measurement is small compared to that of the resistor so that the current is determined primarily by the latter. Because of eddy current effects, the samples either must be thin or in laminated form. At 60 cycles/sec the skin depth (penetration range of magnetic flux into a solid body) is less than 1 mm at permeabilities found in soft materials.

The ac peak induction can also be measured directly by means of an *average* voltmeter (as distinct from a rms voltmeter). If only odd harmonics are present in the flux wave, it can be shown[103] that the average voltage per half cycle in the secondary winding is proportional to the peak induction. Commercial instruments are available which can give this quantity accurately. The requirement that the magnetizing voltage have only odd harmonics present is normally met if the source of excitation is a rotating machine. For electronic excitation, corrections are necessary in order to obtain the flux from a measurement of average voltage.

9.5.2.3. Determination of Permanent Magnetic Properties. To illustrate the principal quantities used to describe the properties of permanent magnet materials we return to the toroidal example. Let the toroid first be saturated by means of a current configuration which gives a large field following the contour of the toroid. A typical "charging" operation consists of using a single conductor threaded through the hole in the toroid and passing a large current in a single pulse as from the discharge of capacitors. The removal of the applied field leaves $\mathbf{H} = 0$ inside the toroid and the induction is the remanent induction $\mathbf{B}_r$. As $\mathbf{B}_r$ is always tangential to the surface of the toroid it is necessary to make a gap in the toroid in order that the remanence manifest itself in an external field. The introduction of the gap will produce a negative value of H inside the material. For simplicity consider the toroid to have a cross-sectional area S, a circumference l, and the gap length d. From $\oint \mathbf{H} \cdot d\mathbf{l} = 0$ in the absence of currents it follows that

$$\mathbf{H}_m(l - d) = -\mathbf{H}_{\text{gap}} d. \tag{9.5.5}$$

From the continuity of B expressed by $\oint \mathbf{B} \cdot d\mathbf{S} = 0$ it follows that

$$\mathbf{B}_m A_m = \mathbf{B}_{\text{gap}} A_{\text{gap}} \tag{9.5.6}$$

where A_m and A_{gap} are the cross-sectional areas of the magnet and the gap, respectively. The product of these two equations gives

$$(BH)_m V_m = -(BH)_{\text{gap}} V_{\text{gap}} = -H_{\text{gap}}^2 V_{\text{gap}}. \tag{9.5.7}$$

Since H_{gap}^2 is proportional to the energy density of the magnetostatic field in the gap, the quantity $(BH)_{\max}$ is often called the energy product for a permanent magnetic material since when it is a maximum, optimum energy is realizable in the gap. The second quadrant of a plot of $\mathbf{B}$ versus $\mathbf{H}$ on decreasing from saturation is called the demagnetization curve. Its determination characterizes a permanent magnet material.

A single observation of the field in the gap and the use of Eqs. (9.5.5) and (9.5.6) permits a determination of a single point on the demagnetization curve. A series of toroids with different ratios of gap length to circumference would allow in principle a determination of the demagnetization curve. If the gap is held constant and the volume of the magnetic material is varied from sample to sample, the maximum field in the gap per unit volume of magnetic material occurs when HB in the magnet is a maximum. As $(HB)_{\max}$ occurs for different values of $\mathbf{B}$ in different materials it is clear that the best material for a given application depends on the field desired in the gap. It is usual then to specify both $(HB)_{\max}$ and $\mathbf{B}_d$ where the latter is the induction at $(HB)_{\max}$. These quantities can be determined by any of the methods outlined above for the determination of hysteresis loops of soft magnetic material. The only difficulties arise from the much larger exciting currents necessary for producing fields equal to the coercive force or sufficient for saturation.

A determination of $\mathbf{B}_r$ by measuring $\mathbf{B}$ in zero applied field necessitates a geometry with small demagnetizing factor, such as a toroid with a small gap, an ellipsoid with large length to diameter ratio, or a sample making good contact to a high permeability yoke. As the differential permeability of permanent magnets near remanence is not very large, the requirements are far less severe than with soft magnetic materials. For small gaps the field in the toroid gap is very close to the remanence. Ellipsoid axis ratios of 10 to 1 are generally sufficient. A suitable yoke for such measurements is an ordinary electromagnet except when the remanence is quite high. In that case the regions of the pole faces of the electromagnet next to the sample will be near saturation and hence have low differential permeability and effectively act as an air gap. The electromagnet with a sample bridging the gap is a most convenient means of determining saturation hysteresis loops. The induction is measured with a coil about the samples. $\mathbf{H}_i$ in general differs only slightly from the field in the electromagnet gap provided the area of the pole faces is large compared to the sample area. This can be checked by bringing any suitable field measuring probe close to the sample. The measurements can be carried out either as a point by point operation or can be adapted to x-y recorders. For the older permanent magnet materials, where the fields required to saturate the material are under 5000 gauss, the Sanford–Bennett permeammeter[104] is useful. This utilizes the "isthmus" method in which the sample, usually in the form of a bar, is wedged into grooves cut into the pole pieces of a double yoke magnetizing magnet.

In connection with permanent magnets some note is here taken of the concept of a leakage factor. In the discussion of the toroid made of permanent magnetic material use was made of the two equations $\oint \mathbf{B} \cdot d\mathbf{S} = 0$ and $\oint \mathbf{H} \cdot d\mathbf{l} = 0$ and the assumption of uniformity. For a general magnetic circuit the assumption of uniformity should be replaced by boundary conditions more suitably describing the situation. This is a difficult problem involving the knowledge of the demagnetizing curve of all materials in the circuit and numerical integration. A reasonable approximation can be made by introducing geometry dependent leakage factors into the Eqs. (9.5.5) and (9.5.6), describing the uniform case; thus

$$\mathbf{B}_m S_m = K \mathbf{B}_g S_g \tag{9.5.8}$$

and

$$\mathbf{H}_m L_m = -K' \mathbf{H}_g L_g. \tag{9.5.9}$$

Leakage factors for different configurations and geometries have been tabulated in the literature and their use has made possible effective design at least for simple configurations.[105] Where greater accuracies in design are required, relaxation methods[106] and iterative procedures may be used. In this connection electronic computers accelerate the process considerably.*

* See also Vol. 2, Chapter 9.8.

[104] R. L. Sanford and E. G. Bennett, *J. Research Natl. Bur. Standards* **23,** 415 (1939).

[105] E. M. Underhill, *Electronics* **16,** 126, 316 (1943); C. A. Maynard, *Elec. Mfg.* **34,** 93 (1944); S. Evershed, *J. Inst. Elec. Engrs.* **58,** 780 (1920); **63,** 786 (1925); D. J. Desmond, *ibid.* **92,** 229 (1945); see also K. Hoselitz, "Ferromagnetic Properties of Metals and Alloys." Oxford Univ. Press, London and New York, 1952.

[106] R. V. Southwell, "Relaxation Methods in Engineering Science." Oxford Univ. Press, London and New York, 1940; S. S. Attwood, "Electric and Magnetic Fields," 3rd ed. Wiley, New York, 1949.

AUTHOR INDEX

Numbers in parentheses are footnote numbers. They are inserted to indicate the reference when an author's work is cited but his name is not mentioned on the page.

A

B

C

D

E

F

G

H

I

J

M

R

T

U

V

W

Y

Z

SUBJECT INDEX

E

F

H

N

O

P

T

U

V

W

X

Y

Z